PLANT-ANIMAL INTERACTIONS

Plant–Animal Interactions

An Evolutionary Approach

Edited by

Carlos M. Herrera

Estación Biológica de Doñana
Consejo Superior de Investigaciones Científicas
E-41013 Sevilla
Spain

Olle Pellmyr

Department of Biology
Vanderbilt University
Nashville, TN 37235
USA

Blackwell
Science

© 2002 by Blackwell Science Ltd
a Blackwell Publishing company

Editorial Offices:
Osney Mead, Oxford OX2 0EL, UK
    Tel: +44 (0)1865 206206
Blackwell Science, Inc., 350 Main Street, Malden, MA 02148-5018, USA
    Tel: +1 781 388 8250
Blackwell Science Asia Pty, 54 University Street, Carlton, Victoria 3053, Australia
    Tel: +61 (0)3 9347 0300
Blackwell Wissenschafts Verlag, Kurfürstendamm 57, 10707 Berlin, Germany
    Tel: +49 (0)30 32 79 060

First published 2002 by Blackwell Science Ltd

Library of Congress Cataloging-in-Publication Data has been applied for.

ISBN 0-632-05267-8

A catalogue record for this title is available from the British Library.

Set in 9.5/12 pt Garamond
by SNP Best-set Typesetter Ltd., Hong Kong
Printed and bound in Great Britain by
TJ International Ltd., Padstow, Cornwall.

For further information on
Blackwell Science, visit our website:
www.blackwell-science.com

Contents

Contributors, vii
Preface, viii
Acknowledgements, ix
Geochronological perspective, xii

Part 1   Introduction

1   Species interactions and the evolution of biodiversity, 3
    Peter W. Price
2   The history of associations between plants and animals, 26
    Conrad C. Labandeira

Part 2   Mostly Antagonisms

3   Plant–insect interactions in terrestrial ecosystems, 77
    Sharon Y. Strauss and Arthur R. Zangerl
4   Mammalian herbivory in terrestrial environments, 107
    Kjell Danell and Roger Bergström
5   Granivory, 132
    Philip E. Hulme and Craig W. Benkman

Part 3   Mostly Mutualisms

6   Pollination by animals, 157
    Olle Pellmyr
7   Seed dispersal by vertebrates, 185
    Carlos M. Herrera

Part 4   Synthesis

8   Ant–plant interactions, 211
    Andrew J. Beattie and Lesley Hughes
9   Plant–animal interactions: future directions, 236
    John N. Thompson

Appendix: Supplementary information for Chapter 2, 248
References, 263
Index, 294

Colour plate falls between p. 84 and p. 85

Contributors

Andrew J. Beattie
Key Centre for Biodiversity and Bioresources, Department of Biological Sciences, Macquarie University, NSW 2109, Australia
abeattie@rna.bio.mq.edu.au

Craig W. Benkman
Department of Biology, New Mexico State University, Las Cruces, NM 88003, USA
cbenkman@nmsu.edu

Roger Bergström
Research Unit, Swedish Association for Hunting and Wildlife Management, Bäcklösavägen 18B, SE-75242 Uppsala, and Department of Animal Ecology, Swedish University of Agricultural Sciences, SE-90183 Umeå, Sweden
roger.bergstrom@sjf.slu.se

Kjell Danell
Department of Animal Ecology, Swedish University of Agricultural Sciences, SE-90183 Umeå, Sweden
kjell.danell@szooek.slu.se

Carlos M. Herrera
Estación Biológica de Doñana, Consejo Superior de Investigaciones Científicas, Avenida de María Luisa s/n, E-41013 Sevilla, Spain
herrera@cica.es

Lesley Hughes
Key Centre for Biodiversity and Bioresources, Department of Biological Sciences, Macquarie University, NSW 2109, Australia
lhughes@rna.bio.mq.edu.au

Philip E. Hulme
NERC Centre for Ecology and Hydrology, Hill of Brathens, Banchory, Aberdeenshire, AB31 4BW, Scotland, UK
pehu@ceh.ac.uk

Conrad C. Labandeira
Department of Paleobiology, National Museum of Natural History, Smithsonian Institution, Washington, DC 20560-0121, and Department of Entomology, University of Maryland, College Park, MD 20742, USA
labandeira.conrad@nmnh.si.edu

Olle Pellmyr
Department of Biology, Vanderbilt University, VU Station B Box 351812, Nashville, TN 37235, USA
olle.pellmyr@vanderbilt.edu

Peter W. Price
Department of Biological Sciences, Northern Arizona University, Flagstaff, AZ 86011-5640, USA
peter.price@nau.edu

Sharon Y. Strauss
Section of Evolution and Ecology, University of California, Davis, CA 95616, USA
systrauss@ucdavis.edu

John N. Thompson
Department of Ecology and Evolutionary Biology, Earth and Marine Sciences Building, University of California, Santa Cruz, CA 95064, USA
thompson@biology.ucsc.edu

Arthur R. Zangerl
Department of Entomology, University of Illinois at Urbana-Champaign, Urbana, IL 61801, USA
zangerl@uiuc.edu

Preface

More than 2000 years ago, the philosopher Aristotle wrote on many aspects of the life of animals. He produced no less than five major works dealing in considerable detail with varied aspects of their anatomy, behaviour, movement and reproduction. His dedication to the study of animals stood in sharp contrast to his apparent lack of interest in plants, to the study of which he devoted only one minor work. Advances in plant science were left to his pupil Theophrastus. As a counterpart to Aristotle's zoological work, Theophrastus provided exquisitely detailed accounts of the life history and reproduction of plants, and their responses to variations in the physical environment. The pioneering contributions of these two philosophers laid the earliest foundations of the two disciplines that would eventually come to be known as zoology and botany.

But this ancient story also illustrates one perpetual characteristic of the study of plants and animals until quite recently: the knowledge of these two groups of organisms has traditionally progressed along separate lanes, under the leadership of different researchers and independently of each other. This historical separation crystallized in Europe by the late eighteenth century, when botanical gardens and museums of natural history became the designated locations where studies of plants and animals respectively were separately pursued. To some extent, this separation still persists in the way departments of our modern universities are organized.

As with so many other aspects of biology, there is also before and after Darwin in respect to prevailing attitudes towards the study of animals and plants. By providing a brand-new conceptual scenario in which organismal features can be explained as adaptive responses to selective pressures exerted by other organisms, Darwin laid the first building blocks for bridging the long-standing gap between animal and plant studies. In *The Origin of Species*, (1859), he emphasized that 'plants and animals, most remote in the scale of nature, are bound together by a web of complex relations' (p. 73), and his famous 'tangled bank' analogy in the closing paragraph of the book is all about the importance of species interactions. More than a century of subsequent research in ecology and evolution has confirmed Darwin's foresight, and proved beyond any doubt that we will never fully understand the evolution of the morphology, behaviour and life history of plants and animals unless we understand in sufficient detail their reciprocal influences in ecological and evolutionary time.

The importance granted to plant–animal interactions, as both powerful evolutionary forces and influential factors in ecological communities, has increased considerably in recent years, as revealed by the prominence reached in ecological literature by plant–animal studies in the last few decades. This book, aimed at upper-division undergraduate students and those starting graduate studies, attempts to provide a manageable synthesis of recent developments in the field of terrestrial plant–animal interactions. The chapter topics were chosen to reflect the variety and richness of the implications of such interactions from an evolutionary perspective. The amount of information available on plant–animal interactions is already vast, and it is increasing at a very fast rate, in such a way that any of the subjects of the chapters of this book would justify a whole textbook on its own. To keep the book within a reasonable size and to facilitate reading, we asked the authors of the chapters to avoid thorough reviews of the literature, to keep the number of references within what seemed to us a reasonable limit, and to use citations in a representative rather than a comprehensive way. We hope that the emphasis on recent synthetic papers will help in tracking the next level of papers on any given topic of particular interest to the reader. Our apologies if these rather stringent rules have sometimes precluded an explicit mention of some relevant work.

Carlos M. Herrera
Olle Pellmyr

Acknowledgements

From Blackwell Science Ltd we thank Ian Sherman, former commissioning editor, for the initial invitation that sparked this project, enthusiastic support at all stages and wise counselling when the need arose. Sarah Shannon, his successor, provided expert advice and support during the latter stages of the book's production. The authors of the various chapters would like to thank the following people for their help. Credits for photographs are given in their captions.

Chapter 2   Finnegan Marsh, David Weishampel and the Walcott Fund of the US National Museum of Natural History.

Chapter 3   May Berenbaum and the US National Science Foundation.

Chapter 4   Göran Björnhag, John Pastor, Swedish University of Agricultural Sciences, Swedish Association for Hunting and Wildlife Management, Swedish Environmental Protection Agency, Swedish Council for Forestry and Agricultural Research, and Swedish Natural Science Research Council.

Chapter 5   Clare Brough, Chris Smith, the Royal Society, the British Ecological Society and the US National Science Foundation.

Chapter 6   Jeff Conner, James Thomson, Paul Wilson, US National Science Foundation and the National Geographic Society.

Chapter 7   Conchita Alonso, Carlos M. Duarte, Juan F. Haeger, José Guitián, Carlos A. Herrera, Mónica Medrano, José R. Obeso, Pedro Rey, Miguel Salvande, Rodrigo Tavera, Dan Wenny, Regino Zamora and the Estación Biológica de Doñana.

Chapter 9   US National Science Foundation.

The authors and publisher gratefully acknowledge permission to reproduce copyright material in this book.

Fig. 2.2: Reprinted with permission from Labandeira, C.C. and Sepkoski, J.J. Jr (1993) Insect diversity in the fossil record. *Science*, **261**, 310–315, copyright 1993 American Association for the Advancement of Science; Fig. 2.3: Reprinted with permission from: Wilf, P. and Labandeira, C.C. (1999) Response of plant–insect associations to Paleocene–Eocene warming. *Science*, **284**, 2153–2156, copyright 1999 American Association for the Advancement of Science; Fig. 2.7c: From Amerom, H.W.J. and Boersma, M. (1971) A new find of the ichnofossil *Phagophytichnus ekowskii* van Amerom. *Geologie en Mijnbouw*, **50**, 667–669, with kind permission from Kluwer Academic Publishers; Fig. 2.7f,g: Reprinted from Beck, A.L. and Labandeira, C.C. (1998) Early Permian insect folivory on a gigantopterid-dominated riparian flora from north-central Texas. *Palaeogeography, Palaeoclimatology, Palaeoecology*, **142**, 139–173, copyright 1998, with permission from Elsevier Science; Fig. 2.7l: From Ash, S. (1999) *International Journal of Plant Science*, **160**, 211, with permission from the University of Chicago; Fig. 2.9e: Reprinted from Zhou, Z.-Y. and Zhang, B. (1989) A sideritic *Protocupressinoxylon* with insect borings and frass from the Middle Jurassic, Henan, China. *Review of Palaeobotany and Palynology*, **59**, 133–143, copyright 1989, with permission from Elsevier Science; Fig. 2.10d: From Labandeira, C.C., Dilcher, D.L., Davis, D.R. and Wagner, D.L. (1994) 97 million years of angiosperm-insect association: paleobiological insights into the meaning of coevolution. *Proceedings of the National Academy of Sciences USA*, **91**, 12278–12282, copyright 1994 National Academy of Sciences, USA; Fig. 2.10f: From Crane, P.R. and Jarembowski, E.A. (1980) *Journal of Natural History*, **14**, 631, with permission from

Taylor & Francis Ltd (*http://www.tandf.co.uk/journals*); Fig. 2.10g: Reprinted from Hickey, L.J. and Hodges, R.W. (1975) Lepidopteran leaf mine from the Early Eocene Wind River Formation of northeastern Wyoming. *Science*, **189**, 718–720; Fig. 2.11c,d: from Labandeira, C.C. and Phillips, T.L. (1996) A Late Carboniferous petiole gall and the origin of holometabolous insects. *Proceedings of the National Academy of Sciences USA*, **93**, 8470–8474, copyright 1996 National Academy of Sciences, USA; Fig. 2.12f: From Genise, J.F. (1995) Upper Cretaceous trace fossils in permineralized plant remains from Patagonia, Argentina. *Ichnos*, **3**, 287–299, copyright OPA (Overseas Publishers Association) N.V., with permission from Taylor & Francis Ltd; Fig. 2.13c: Reprinted with permission from Ren, D. (1998) Flower-associated Brachycera flies as fossil evidence for Jurassic angiosperm origins. *Science*, **280**, 85–88, copyright 1998 American Association for the Advancement of Science; Fig. 2.13g: Reprinted with permission from Crepet, W.L. and Taylor, D.W. (1985) Diversification of the Leguminosae: first fossil evidence of the Mimosoideae and Papilionoideae. *Science*, **228**, 1087–1089, copyright 1985 American Association for the Advancement of Science; Fig. 2.13i: From Poinar G.O., Jr. (1992) Fossil evidence of resin utilization by insects. *Biotropica*, **24**, 466–468, copyrighted 1992 by the Association for Tropical Biology, PO Box 1897, Lawrence, KS 66044-8897, reprinted by permission; Fig. 2.14: Reprinted with permission from Labandeira, C.C. (1998) How old is the flower and the fly? *Science*, **280**, 57–59, copyright 1998 American Association for the Advancement of Science; Fig. 2.15a,b: Reprinted with permission from Edwards, D., Selden, P.A., Richardson, J.B. and Axe, L. (1995) Coprolites as evidence for plant-animal interaction in Siluro-Devonian terrestrial ecosystems. *Nature*, **377**, 329–331, copyright 1995 Macmillan Magazines Limited; Fig. 2.15e: Reprinted from Kerp, J.H.F. (1988) Aspects of Permian palaeobotany and palynology. X. The west- and central European species of the genus *Autunia* Krasser emend. Kerp (Peltaspermaceae) and the form-genus *Rhachiphyllum* Kerp (callipterid foliage). *Review of Palaeobotany and Palynology*, **54**, 249–360, copyright 1988, with permission from Elsevier Science; Fig. 2.15p,q: Reprinted from Crepet, W.L. (1996) Timing in the evolution of derived floral characters — Upper Cretaceous (Turonian) taxa with tricolpate and tricolpate-derived pollen. *Review of Palaeobotany and Palynology*, **90**, 339–359, copyright 1996, with permission from Elsevier

Science; Fig. 2.15r: From Grimaldi, D.A., Bonwich, E., Dellanoy, M. and Doberstein, W. (1994) Electron microscopic studies of mummified tissues in amber fossils. *American Museum Novitates*, **3097**, 1–31, courtesy of the American Museum of Natural History; Fig. 2.20d: Reprinted from Thulborn, R.A. (1991) Morphology, preservation and palaeobiological significance of dinosaur coprolites. *Palaeogeography, Palaeoclimatology, Palaeoecology*, **83**, 341–366, copyright 1991, with permission from Elsevier Science; Fig. 2.20g: Reprinted from Nambudiri, E.M.V. and Binda, P.L. (1989) Dicotyledonous fruits associated with coprolites from the Upper Cretaceous (Maastrichtian) Whitemud Formation, southern Saskatchewan, Canada. *Review of Palaeobotany and Palynology*, **59**, 57–66, copyright 1989, with permission from Elsevier Science; Fig. 2.21: Illustration by Carol A. Abraczinskas and Paul C. Serono. Reprinted with permission from Sereno, P.C. (1999) The evolution of dinosaurs. *Science*, **284** (June 25), 2137–2147, copyright 1999, American Association for the Advancement of Science; Fig. 2.22: Reprinted with permission from Farrell, B.D. (1998) 'Inordinate fondness' explained: why are there so many beetles? *Science*, **281**, 555–559, copyright 1998, American Association for the Advancement of Science; Fig. 4.3: From Van Soest, P.J. (1996) Allometry and ecology of feeding behavior and digestive capacity in herbivores: a review. *Zoo Biology*, **15**, 455–479, copyright © 1996, reprinted by permission of Wiley-Liss, Inc., a subsidiary of John Wiley & Sons, Inc.; Fig. 5.12: from Benkman, C.W. (1999) The selection mosaic and diversifying coevolution between crossbills and lodgepole pine. *American Naturalist*, **153**, S75–S91, with permission from the University of Chicago; Fig. 6.3: From Møller, A.P. (1995) Bumblebee preference for symmetrical flowers. *Proceedings of the National Academy of Sciences USA*, **92**, 2288–2292, copyright 1995 National Academy of Sciences, USA; Fig. 6.4: reprinted by permission from Dornhaus, A. and Chittka, L. (1999) Evolutionary origin of bee dances. *Nature*, **401**, 38, copyright 1999 Macmillan Magazines Ltd; Fig. 6.6: Reprinted with permission from Lewis, A.C. (1986) Memory constraints and flower choice in *Pieris rapae*. *Science*, **232**, 863–865, copyright 1986 American Association for the Advancement of Science; Fig. 6.9 and Plate 6.1: From Schemske, D.W. and Bradshaw, H.D. (1999) Pollinator preference and the evolution of floral traits in monkeyflowers. *Proceedings of the National Academy of Sciences USA*, **96**, 11910–11915, copyright 1999 National Academy of Sciences, USA; Fig.

6.10: Reprinted by permission from Schiestl, F.P., Ayasse, M., Paulus, H.F. et al. (1999) Orchid pollination by sexual swindle. *Nature*, **399**, 421–422, copyright 1999 Macmillan Magazines Ltd; Fig. 6.11: Reprinted with permission from Temeles, E.J., Pan, I.L., Brennan, J.L. and Horwitt, J.N. (2000) Evidence for ecological causation of sexual dimorphism in a hummingbird. *Science*, **289**, 441–443, copyright 2000 American Association for the Advancement of Science; Figs 6.12 and 6.13: Reprinted by permission from Nilsson, L.A. (1988) The evolution of flowers with deep corolla tubes. *Nature*, **334**, 147–149, copyright 1988 Macmillan Magazines Ltd; Fig. 6.15: From Hodges, S.A. (1997) Floral nectar spurs and diversification. *International Journal of Plant Sciences*, **158**, S81–S88, with permission from the University of Chicago; Fig. 6.16: Reprinted by permission from Armbruster, W.S. and Baldwin, B. (1998) Switch from specialized to generalized pollination. *Nature*, **294**, 632, copyright 1998 Macmillan Magazines Ltd; Fig. 7.17: From Dinerstein, E. (1986) Reproductive ecology of fruit bats and the seasonality of fruit production in a Costa Rican cloud forest. *Biotropica*, **18**, 307–318, copyrighted 1986 by the Association for tropical Biology, PO Box 1897, Lawrence, KS 66044-8897, reprinted by permission; Fig. 8.1 and Table 8.2: From Huxley, C.R. and Cutler, D.F. (1991) *Ant–Plant Interactions*. Oxford: Oxford University Press, © Oxford University Press 1991; Fig. 8.14: From Brown, M.J.F. and Human, K.G. (1997) Effects of harvester ants on plant species distribution and abundance in a serpentine grassland. *Oecologia*, **112**, 237–243, fig. 1a.

Geochronological perspective

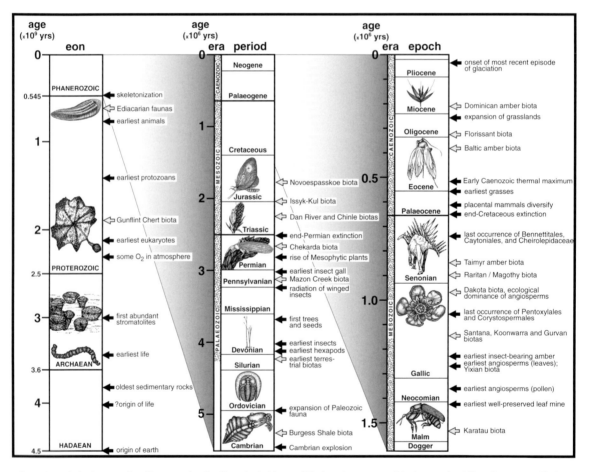

A geochronological perspective of important fossil evidence in the history of life, focusing on events (black arrows) and biotas (grey arrows) in the development of plant–animal associations. The documented terrestrial record begins at approximately 425 Ma, during the Late Silurian (stippled vertical bar), and is subsequently represented by sporadic, well-preserved, and often diverse deposits (*Lagerstätten*), including some insect-bearing ones which are indicated.

Part 1 Introduction

Chapter 1

Species interactions and the evolution of biodiversity

Peter W. Price

1.1 The importance of plant–animal interactions

The majority of terrestrial organisms fly. Conquest of the air has been spectacular, contributing essentially to the evolution of the richest floras and faunas ever recorded on land. The evolution of propelled and passive flight, and their consequences, may well be regarded as the most creative force in the development of biodiversity. Most plants fly at one stage of their life cycle or another, as pollen or as seeds or both. Spores of ferns and fungi fly. Pollen, spores and seeds are carried on the wind by a multitude of winged animals: insects, birds, bats and perhaps pterosaurs in their day. Flight has promoted flight, for the flying insects were a strong inducement for cursorial animals to leap or flap appendages, eventually yielding flying dinosaurs, the birds, flying mammals, the bats and the pterosaurs. Flight is such a spectacle, such a wonder, that it captivates and inspires.

Flight of plants and animals is perhaps the most fundamental element necessary for the understanding of such rich floras and faunas as exist today, and for understanding plant and animal interactions. For the vast majority of terrestrial organisms exist in trophic systems based on plants, be they the plants themselves, herbivores, carnivores, pollinators, frugivores or granivores (Table 1.1). And as we climb the trophic ladder, species richness increases by orders of magnitude. A plant species, such as an oak, birch or willow, may be host to 200–300 insect herbivore species. Each herbivorous insect may be utilized by 10–20 carnivores, either predators or parasites. The plant provides both food and habitat for the associated fauna and many microhabitats are available for colonization (Fig. 1.1). Given the great richness of animals associated with each plant species, plus about 265 000 species of described higher plants (Fig. 1.2a), we can begin to appreciate why there are so many animals associated with plants. Includ-

ing undescribed species, there may be 10–100 million species of all kinds living today, over half of them insects, of which 99.5% can fly in the adult stage. Adaptive radiation of some insect groups has been extraordinary, with flight and close relationships to plants a common characteristic: moths and butterflies (Lepidoptera) and flies (Diptera) with some 160 000 and 125 000 species per order respectively; sawflies and wasps (Hymenoptera) with about 150 000 species and the beetles (Coleoptera) reaching 350 000 species (Fig. 1.2b). Add to the insects about 9000 species of birds and 1000 bat species, together making up 80% of the warm-blooded vertebrates, and we see that conquest of the air has been an evolutionary 'success' story of extreme proportions, if we evaluate success by the extent of the adaptive radiations involved. And let us not forget the pterosaurs with at least 20 families of rhamphorynchoid and pterodactyloid species in the fossil record (Wellnhofer 1991).

The basis for the spectacular radiations of animals on earth today is clearly the resources provided by the plants. They are the major primary producers, autotrophically energizing planet Earth. Regal (1977) made a convincing case for the dominance of flowering plants (angiosperms in Fig. 1.2c) based on mutualistic relationships between plants, their pollinators and seed dispersers.

1 Pollination by animals that fly permits individual plants to outcross with others of the same species, even when plants are very patchy and widely dispersed. Wind pollination is ineffective under such conditions. (Even though air-borne pollen travels a long way, precision is lacking; cf. Jackson & Lyford 1999.)

2 Long-distance seed dispersal by birds and bats resulted in occupation by plants of widely dispersed patches, making plant species hard to find by herbivores and granivores.

3 Such reliable long-distance dispersal resulted in the evolution of specialization of plant species to particular

Table 1.1 Plant resources available to animals and their kinds of exploitation.

Resources available	Name of exploiter
Living plant or plant parts in general	Herbivore (syn. phytophage, vegetarian)
Shoots	None
Leaves	Folivore
Buds	None
Flowers	Florivore
Nectar	Nectarivore
Anthers/pollen	Anthophage
Carpel/fruit	Frugivore
Seeds	Granivore
Spores	None
Cones	None
Wood/xylem	Xylophage
Cambium	None
Bark/cortex/periderm	None
Roots/rhizomes	Rhizophage
Tubers, corms, bulbs	None
Sap (phloem and xylem)	None
Exudates/oozes	None
Dead plant material	Saprophage (syn. detritivore, decomposer)

microhabitats, again reducing herbivore, granivore and parasite pressure.

4 The new ecological relationships of flowering plants resulted in colonizing species with population structures conducive to rapid evolutionary change, divergence of lineages, speciation and impressive adaptive radiation. Impressive diversification is evident in life forms, growth strategies, physical and chemical defences, as well as habitat utilization. Such a population structure, in which colonization and extinction rates may be high and populations small, creates cradles of evolution. Herbs have evolved much more rapidly than shrubs and trees, with much adaptive radiation attributable to chromosomal evolution involving polyploidy and aneuploidy. Herbs are often annuals or short-lived perennials, occupying patchy, disturbed and transient habitats. In such small, ephemeral populations chromosomal changes can become fixed rapidly and speciation results, such that rates of chromosomal evolution and speciation are positively correlated (Fig. 1.3).

This is the fascination of plant and animal relationships: the richness of species, the spectacular variety and complexity of interactions, the beguiling loveliness of evolu-

tionary creations, dynamic evolutionary processes and the quest for understanding. We hardly require more justification to study plant–animal interactions, yet there are many other reasons. We *need* to know. The fields of agriculture, forestry, horticulture, conservation, environmental planning and management, and biodiversity studies all promote and depend upon the understanding of the plants and animals. Plant husbandry is fundamental to the existence and contentment of humankind, and many species of animals are engaged in the life of plants. From the biblical plagues of locusts to the pressing environmental issues of today, humans, plants and animals have been inextricably involved with each other and will continue to be so.

1.2 Biodiversity

Biodiversity may be viewed and evaluated in many different ways, which accumulate to establish the extreme richness of the phenomena involved in plant–animal interactions. The number of species, the roles they play, the kinds of interactions and their results, trophic web structure and the evolutionary mechanisms involved, all constitute aspects of biodiversity studies. Genetic diversity within and among species, phenotypic variation, chemical variation and associated symbionts all contribute to the diversity of life. Habitat heterogeneity, landscape complexity, geographical variation and most things geological are all encompassed in the understanding of biodiversity.

1.2.1 Numbers and percentages

Starting with the simplest gauge of diversity, we can estimate the number of species on this globe (Fig. 1.2). We note immediately the domination of higher plants and insects on this scale, even when all marine organisms are included. Excluding algae, fungi and microbes, Strong et al. (1984) estimated that 22% of the biota is composed of green plants, 26% of herbivorous insects and 31% of insects that are carnivores or saprophages. Of this 31%, about 20% are carnivores and 11% are saprophagous if we use estimates based on the well-studied fauna of the British Isles (cf. Price 1997). The 'other animal' category in Fig. 1.2 includes on-land nematodes, mites, other non-insectan arthropods, amphibians, reptiles and birds, many of which feed on plants directly as herbivores, nectarivores, pollen feeders, frugivores, granivores and

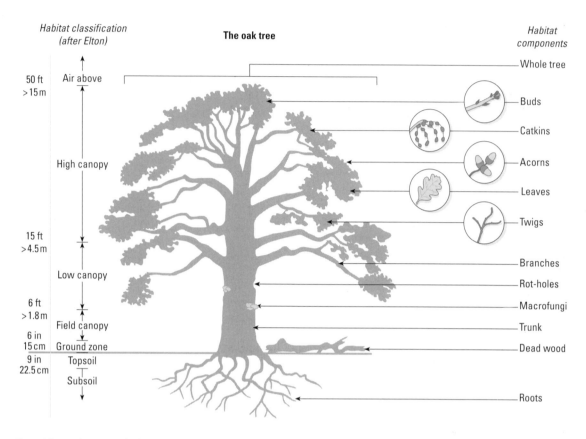

The oak tree

Habitat classification
(after Elton)

Habitat
components

50 ft
>15m — Air above

High canopy

15 ft
>4.5m

Low canopy

6 ft
>1.8m

Field canopy

6 in
15cm — Ground zone

9 in
22.5cm — Topsoil

Subsoil

Whole tree
Buds
Catkins
Acorns
Leaves
Twigs
Branches
Rot-holes
Macrofungi
Trunk
Dead wood
Roots

Figure 1.1 An oak tree in England provides a rich diversity of microhabitats and food resources for animals. The habitat classification follows Elton and Miller (1954). (From Morris 1974.)

saprophages. Others depend upon plants for cover, nesting sites and foraging sites, for much food is available among the herbivores. Predaceous and parasitic insects feed on other insects predominantly, as do many bird and bat species, and as the young and smaller pterodactyls and dinosaurs almost certainly did. And let us not forget the massive adaptive radiation of the ornithischian dinosaurs—the 'bird-hipped croppers' (Sereno 1999, p. 2139), specialized for processing plant food. Just a few familiar groups include (with numbers of genera treated in Norman 1985) the ceratopsids (21), hadrosaurs (17), iguanodontids (11), ankylosaurs (8) and stegosaurs (11). And 'the second great radiation of dinosaurian herbivores'—'the long-necked titans' (Sereno 1999, p. 2140) was made up of the sauropods, including the diplodocids (7 genera), brachiosaurids (6) and titanosaurids (5) (cf. Chapter 2).

All this constitutes the staggering scope of the subject addressed in this book. Well over 90% of energy in terrestrial systems is fixed by autotrophic green plants (the remainder by algae and bacteria), and almost all terrestrial animals depend on autotrophic production, either directly as herbivores or saprophages, or for shelter and microhabitats, or indirectly as predators and parasites utilizing the second trophic level of herbivores.

1.2.2 Diversity of interactions

Here we are entering into the nature of interactions in trophic webs which compound the appreciation of biodiversity. Even in the simplest systems we can observe autotrophic plants, herbivores and carnivores involving four trophic levels (Fig. 1.4). Among the carnivores are parasites, in which larvae live parasitically on the herbivore

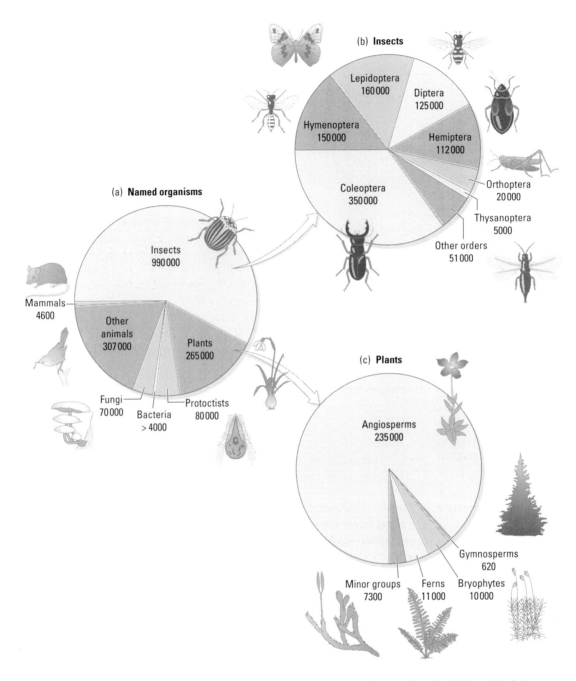

(a) Named organisms

Insects
990 000

Mammals
4600

Other
animals
307 000

Plants
265 000

Fungi
70 000

Bacteria
> 4000

Protoctists
80 000

(b) Insects

Lepidoptera
160 000

Diptera
125 000

Hymenoptera
150 000

Hemiptera
112 000

Coleoptera
350 000

Orthoptera
20 000

Thysanoptera
5000

Other orders
51 000

(c) Plants

Angiosperms
235 000

Gymnosperms
620

Minor groups
7300

Ferns
11 000

Bryophytes
10 000

Figure 1.2 Herbivores and flowering plants constitute a very large proportion of all species. (a) The estimated number of living species of plants, animals, fungi, protoctists and bacteria described and named. Note the dominance of the angiosperms or flowering plants and the large number of insect species. (b) Seven insect orders whose members are herbivores to a large extent make up over 90% of all described insects. (c) The angiosperms (flowering plants) contain a very large proportion of all plant diversity. (Data compiled from Hawksworth & Kalin-Arroyo 1995; Koomen et al. 1995.)

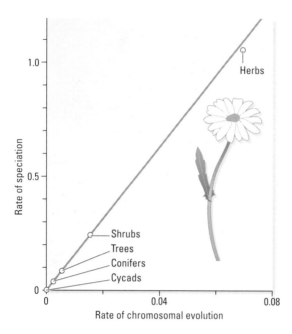

Figure 1.3 The relationship between rate of chromosomal evolution (mean net increase in number of karyotypes per lineage per million years) and the speciation rate (mean net increase in the number of species per lineage per million years) in groups of plants from cycads to angiosperm herbs. (From Levin & Wilson 1976.)

host, but adults are freely flying insects (usually called parasitoids in the literature), and predators, including spiders and the top predator, an anolis lizard. Endless variations on this theme exist in nature. Such trophic webs based on a single plant species may be called **component communities**, usually acting as an element in a **compound community** of several to many plant species and the trophic webs they support, and interactions among them (Root 1973). The web illustrated in Fig 1.4 is typical of many in portraying the narrow perspective of the system under study and not the full diversity of feeding relationships centred on a plant species. As in Fig. 1.1, many resources are provided by any plant species and most are utilized in one way or another. Hence, complete food webs are very rich and complex in nature.

If we study a component community, and a fuller diversity of feeding relationships, we apprehend a mind-boggling diversity of interactions which reach across the landscape, bewildering us with nature's secrets revealed (Table 1.2). A fine example of the expansive nature of interactions has been unravelled by Gilbert (1980) and as-

sociates in the neotropical forests of Costa Rica. Structure is based on plant species in families characterized by similar chemical ecology and similar sets of herbivores and mutualists (Fig. 1.5). We see pollinators involving birds, bats, moths and bees, and seed dispersers including birds, bats and ants. Ants visit extrafloral nectaries, acting as 'bodyguards' for plants. Gilbert (1980, p. 23) coined the term *keystone mutualist*, for 'organisms, typically plants, which provide critical support to large complexes of mobile links' — the mobility provided by flying insects, birds and bats. For example, at La Selva, the canopy tree, *Casearia corymbosa* (Flacourtiaceae), provides for 22 frugivorous bird species which branch out to many other interactions in this wet tropical Atlantic forest. In the same locality most hummingbirds depend on *Heliconia* species (Musaceae) as nectar sources at some time of the year, but the hummingbirds pollinate many other plant species in addition. Thus, a group of *Heliconia* species in a plant community act as keystone mutualists.

Even if we consider only mammals, surely a depauperate taxon relative to most others, the range of species and diversity of interactions is richness itself. In most tropical forest habitats we see examples of mammals that act as browsers, frugivores, omnivores, carnivores (including insectivores), myrmecophages (anteaters) and flower feeders acting as pollinators. In Guatapo National Park, Venezuela, excluding bats and aquatic mammals, 29 species of mammal coexist and interact (Fig. 1.6). In this community, which mirrors many others, the herbivorous species, including browsers, granivores and frugivores, constitute most of the biomass, followed by a group of species combining herbivory with carnivory. At the top of the food web obligate carnivores, including anteaters, constitute only 3% of the mammalian biomass per unit area. The same kind of pyramid of biomass is seen in all terrestrial systems, indicating the inevitable limitation of food supply up the trophic system from autotrophs to top carnivores. Even so, the richness of interactions increases up the trophic ladder, for the top carnivores will hunt and consume almost any mammal species in the lower trophic levels.

1.2.3 Multitrophic level interactions

The role that plants play in the lives of carnivores may not be intuitively obvious, although in all cases effects up the trophic system are strong. Plant distributions define the location and abundance of herbivores, and the kinds of

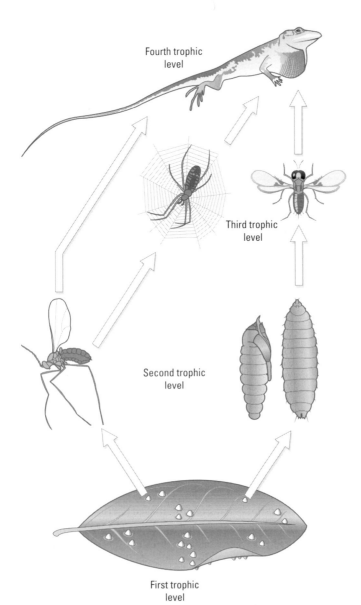

Fourth trophic
level

Third trophic
level

Second trophic
level

First trophic
level

Figure 1.4 A simple food web on small islands in the Bahamas, based on sea grape, *Coccoloba uvifera*, a cecidomyiid gall inducer, with adult, pupa and larva illustrated in the second trophic level, a predatory spider, a parasitoid wasp utilizing larvae in galls and a top predator, an *Anolis* lizard, which feeds on members of the second and third trophic levels. (Based on information in Spiller & Schoener 1990, 1994.)

herbivore, and often the kind of defences that herbivores evolve along with defence against predators. If we consider only chemically mediated interactions in trophic systems, there is a rich variety (Fig. 1.7). Plant families tend to be characterized by a particular kind of chemical defence, such as alkaloids in the potato family (Solanaceae, including potatoes, tomatoes, peppers, ground cherry, etc.), cardiac glycosides in the dogbane and milkweed family (Apocynaceae), mustard oils or glucosinolates in the Brassicaceae (cabbages, mustards, turnip, cresses, etc.), phenolic glucosides in the willows and poplars (Salicaceae) and tannins in many trees (e.g. Fagaceae, with beech, oak and chestnut; Ulmaceae, with the elms; Juglandaceae, with walnut, hickory, pecan, etc.; Corylaceae, with hazels, hornbeams, birches, and alders). The biodiversity of chemicals in plants is extraordinarily rich. Such phytochemicals often inhibit feeding by some herbivores (interaction 3 in Fig. 1.7), but not all, for some

Table 1.2 Consequences of exploitative mechanisms involving interacting species.

Outcomes of interactions	
Mutualism (++)*	Both species benefit from an association
Symbiosis (++, +–, or +0)	Close association between two species: living together
Antagonism (+–, ––)	One species has a negative effect on the other species (parasitism, predation) or both species exert a negative effect (competition)
Amensalism (–0)	One species has a negative effect on the other but there is no measurable reciprocal effect (e.g. highly asymmetric competition)
Competition (––)	Exploitation of a common and limiting resource by two or more species
Commensalism (+0)	One species benefits from an association without any benefit to the other species
Life habits	
Parasitism (+–)	Individuals of one species live in or on a living host, sapping the host's resources for a prolonged period, and exerting a negative but usually non-fatal effect on the host
Predation (+–)	Individuals of one species kill and eat individuals of another organism, the prey species
Inquilinism (+0, +–)	One species enters the domicile of another habitually, with or without causing damage to the host species
Browsing (+–)	Eating tender plant shoots, twigs and leaves of woody plants
Grazing (+–)	Feeding on growing herbage such as grass, nibbling or cutting at surface growth while passing across a patch of vegetation
Cropping (+–)	Cutting off the upper or outer parts of a plant
Exophytic	External feeding on a plant
Endophytic	Internal feeding on a plant
Obligate	A necessary relationship for the existence of a species or individual
Facultative	A relationship that may be used but is not essential

*+ – 0 symbols denote: a beneficial relationship (e.g. ++ is beneficial for both interacting species), a negative effect (e.g. – – is deleterious for both interacting species), or no effect (0).

herbivores specialize on particular plant families and are attracted to volatile components on the surface, which may also act as oviposition cues (interaction 1). Plant volatiles may also attract herbivores (interaction 4) or repel them on associated plants in the community (interaction 5). Moving up the trophic levels, carnivores in the system may be attracted (interaction 16) or repelled (interaction 17) by volatiles from plants, just as odours from herbivores may attract (interaction 11) or repel carnivores (interaction 13). And herbivores ingesting toxic compounds to which they are adapted, such as nicotine from tobacco or cardiac glucosides from milkweeds, may well become toxic to enemies. Toxic phytochemicals may even be stored in tissues or sequestered, as with caterpillars of the monarch butterfly. The stored cardiac glucosides pass into the pupa and adult and act as emetics in predatory birds. So unpleasant is the bird's experience, coupled with the colourful pattern on larvae and adults, that once it is suffered, the predator remembers to avoid similarly patterned butterflies. However, toxins may pass further up the food web, for parasites may sequester phy-

tochemicals as they feed on host tissues. Thus, larvae of the tachinid fly *Zenillia adamsoni* feeding in monarch butterfly larvae, sequester cardiac glucosides and presumably become unpalatable to their predators on the fourth trophic level (interaction 26). Clearly, plant-animal interactions are both direct and indirect and ramify throughout the trophic system. Such multitrophic-level interactions are ubiquitous and important both for the understanding of natural interactions and for effective management of landscapes dominated by humans (agriculture, horticulture, forestry, biological control of weeds and herbivorous pests, conservation, biodiversity studies, animal husbandry, etc.).

1.2.4 Diversity of microbial associates

Another aspect of biodiversity is the richness of microbial mutualists involved with food-web interactions. We have already remarked upon the importance of mutualisms among larger animals, from insects to birds and bats. But for almost every organism, symbiotic associations with

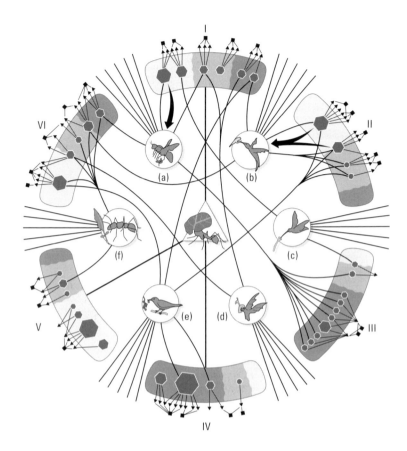

Figure 1.5 A schematic simplified representation of interactions in neotropical forest, as in Costa Rica, emphasizing organizing influences of phytochemical associations and many examples of mutualism. Solid hexagons are plant species, and size of hexagons indicates relative abundance. The plants are grouped in chemically distinct higher taxa, shown as six envelopes, I–VI. In each envelope light grey indicates plants living in successional microhabitats, medium grey in forest understorey and dark grey in forest canopy. On the plants, specialized herbivores adapted to specific phytochemicals are shown as triangles on the outer rim of envelopes, and their specific parasites are indicated as solid boxes. Generalist herbivores are represented by a large triangle at the figure's centre in which a leaf-cutter ant is illustrated (*Atta* species). Mobile links between plant species important in pollination, seed dispersal or plant protection are represented in six circles labelled (a) to (f). Thin lines connect circles to the plants with which the mutualists interact. Many other links to plants not in the figure are indicated by outward-radiating lines. Heavy arrows move from keystone mutualist plants to the taxon that acts as a link among plant species. Plant envelope I contains members of the Solanaceae, or potato family, characterized by alkaloid phytochemicals. One canopy member is pollinated by hummingbirds (b) and fruits are dispersed by other birds (e). Bats (d) pollinate one plant species and disperse seeds of an understorey species. Other plant groups represented are II *Heliconia* (Musaceae), III orchids (Orchidaceae), IV canopy-emergent trees in various families, V grasses, VI passionflower vines in the family Passifloraceae characterized by generally toxic cyanogenic phytochemicals. (From Gilbert 1980.)

bacteria, fungi or protozoa, or combinations of these, are critical, many of them mutualistic. An evolutionary perspective on a grand scale reveals that major breakthroughs in increasing complexity of organisms and interactions have been associated with the acquisition of microbial associates, perhaps initially parasitic but becoming mutualistic eventually (Fig. 1.8). The evolution of the eukaryotic cell, the origin of photosynthesizing protists, the col-

onization of land leading to terrestrial plants, and the utilization of such plants by animals are all major steps mediated by microbial associations. And in the present, take any terrestrial biota and a multitude of microbial associations mediate the interactions in trophic systems (Fig. 1.9). The large artiodactyls, such as antelopes, deer, bison, sheep and goats, all house a fermentation chamber, the rumen, in which bacteria and protozoa with cellulases

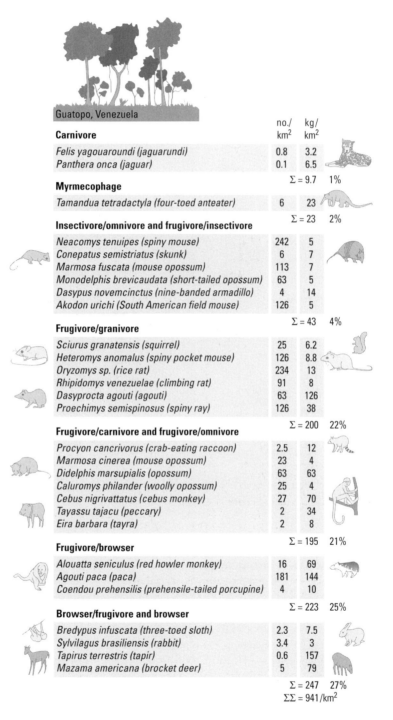

Guatopo, Venezuela

	no./ km²	kg/ km²
Carnivore		
Felis yagouaroundi (jaguarundi)	0.8	3.2
Panthera onca (jaguar)	0.1	6.5
	Σ = 9.7	1%
Myrmecophage		
Tamandua tetradactyla (four-toed anteater)	6	23
	Σ = 23	2%
Insectivore/omnivore and frugivore/insectivore		
Neacomys tenuipes (spiny mouse)	242	5
Conepatus semistriatus (skunk)	6	7
Marmosa fuscata (mouse opossum)	113	7
Monodelphis brevicaudata (short-tailed opossum)	63	5
Dasypus novemcinctus (nine-banded armadillo)	4	14
Akodon urichi (South American field mouse)	126	5
	Σ = 43	4%
Frugivore/granivore		
Sciurus granatensis (squirrel)	25	6.2
Heteromys anomalus (spiny pocket mouse)	126	8.8
Oryzomys sp. (rice rat)	234	13
Rhipidomys venezuelae (climbing rat)	91	8
Dasyprocta agouti (agouti)	63	126
Proechimys semispinosus (spiny ray)	126	38
	Σ = 200	22%
Frugivore/carnivore and frugivore/omnivore		
Procyon cancrivorus (crab-eating raccoon)	2.5	12
Marmosa cinerea (mouse opossum)	23	4
Didelphis marsupialis (opossum)	63	63
Caluromys philander (woolly opossum)	25	4
Cebus nigrivattatus (cebus monkey)	27	70
Tayassu tajacu (peccary)	2	34
Eira barbara (tayra)	2	8
	Σ = 195	21%
Frugivore/browser		
Alouatta seniculus (red howler monkey)	16	69
Agouti paca (paca)	181	144
Coendou prehensilis (prehensile-tailed porcupine)	4	10
	Σ = 223	25%
Browser/frugivore and browser		
Bredypus infuscata (three-toed sloth)	2.3	7.5
Sylvilagus brasiliensis (rabbit)	3.4	3
Tapirus terrestris (tapir)	0.6	157
Mazama americana (brocket deer)	5	79
	Σ = 247	27%
	ΣΣ = 941/km²	

Figure 1.6 The various kinds of mammals and their feeding relationships in the tropical rain forest of Guatopo National Park, Venezuela. Note the large component of herbivores by weight in the community and the very small biomass of strict carnivores, the cats and anteaters. (From Eisenberg 1980.)

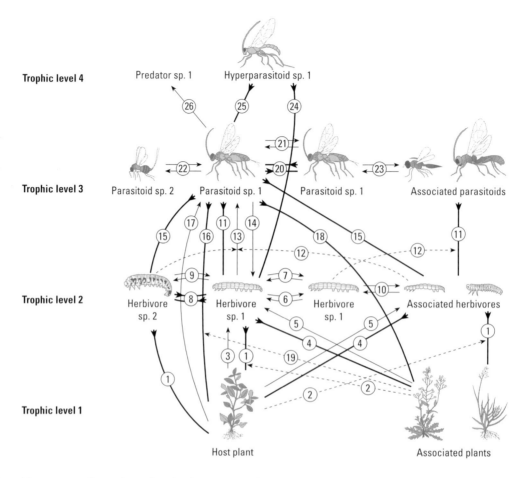

Trophic level 4

Predator sp. 1   Hyperparasitoid sp. 1

Trophic level 3

Parasitoid sp. 2   Parasitoid sp. 1   Parasitoid sp. 1   Associated parasitoids

Trophic level 2

Herbivore sp. 2   Herbivore sp. 1   Herbivore sp. 1   Associated herbivores

Trophic level 1

Host plant   Associated plants

Figure 1.7 A composite of known chemically mediated interactions in a four-trophic-level system of plants, herbivores and carnivores. Linkage emphasizes a component community based on one plant species but associated plants and trophic dependents also interact. (From Price 1981.)

digest refractive plant material. Beetles carry fungi to a new breeding site, inoculating the tree with either a food supply, in the case of ambrosia beetles, or pathogenic killers, such as blue-stain fungi which render the tree suitable for breeding in. Gallflies also inoculate the host plant with fungi on which the larvae feed within the gall. And at the third trophic level herbivore resistance to parasitic larvae is suppressed by inoculated virus, or bacteria render the root-feeding host larva toxic to species other than the inoculating parasitic nematode. Again we observe in fascinating detail the complexity of food webs, layer upon layer of interactions permitting life to burgeon in such a diversity of necessary associations.

These mutualistic associations have been the cornerstones and building blocks of biotic diversity as depicted

in Fig. 1.8. Certainly the new entity arises in a saltational manner, a jump from one species to two species living together. As such, saltational events can multiply until what we call species are actually more like ecosystems with many genomes represented. The eukaryotic cell is an obvious example. The origin of green protists and land plants provides more examples. Additional symbioses in early plant groups is well documented: bryophytes (Boullard 1988; Kost 1988), pteridophytes (Boullard 1979) and lichens (Hawksworth 1988). The rumen in the artiodactyls, the caecum of lagomorphs and perissodactyls, and the paunch of termites are all fermentation chambers supporting a rich brew of microbial symbionts. Arboreal folivores of diverse mammalian taxa all depend on foregut or hindgut fermentation (Cork & Foley 1991):

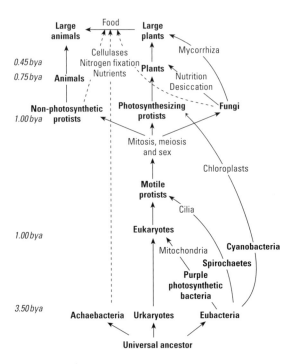

Figure 1.8 A sketch of life going back 3.5 billion years or so, emphasizing the development of symbiotic relationships which created dramatic breakthroughs in the complexity of organisms and interactions. Names in bold type are taxa of organisms or groups of taxa. Solid lines with arrows show links between evolutionary steps, and the contributions made by the taxa in the evolution of biotic complexity are shown in light lettering. Dashed lines indicate the utilization of members of early kingdoms to enable the utilization of plants by large animals such as insects and mammals. For example, the Archaebacteria have provided methanogens critical as mutualists in the ruminant gut, and protozoa and bacteria are essential in the breakdown of cellulose in the termite diet. Note that our understanding of the basal phylogeny of life and the timeline is currently in rapid development (Doolittle 1999) and subject to modification, thus the figure should be used primarily to consider ecological relationships between groups. (From Price 1991a.)

Australian marsupials, New and Old World primates and edentates, the last being the tree sloths. Insectan herbivores may house up to seven species of mutualistic symbiont (Buchner 1965), as in the leaf hoppers (Homoptera: Cicadellidae) and it is hardly credible that such extensive radiation of leaf hoppers, aphids, tree hoppers and jumping plant lice could have occurred in the absence of mutualistic microbes (see also Moran & Telang 1998).

The consequences of mutualistic associations are evident in the millions of eukaryotic species, the adaptive

radiations of herbivores, and are probably involved in any animal feeding on specialized diets, such as nectarivores, pollen feeders, frugivores, granivores and sap and ooze feeders. Add to all these linkages, microbial associations of other kinds, and we may wonder if plant-animal interactions could exist at all without microbial intermediaries.

Biodiversity studies not only involve counting the number of species present in biotas, but they necessarily encompass efforts to understand how the species survive and interact. Just a few examples of the ways in which nature works have been discussed in this chapter and many more views and instances will follow. The complexities are bewildering but fascinating, and one of the most intriguing avenues of investigation is the evolutionary basis for the diversity of life on earth.

1.3 The creation of biodiversity

Obviously, evolution is a very creative process. We can ask about the major episodes in the development of terrestrial life and the major processes involved. To frame our view, focusing on the plant and animal interaction, let us consider first the steps resulting in the evolution of the human species, a topic dear to our hearts. For this scenario illustrates the magnitude of the long-term consequences of apparently simple interactions initially at three trophic levels involving plants, herbivores and carnivores. The most plausible scenario of the origin of primates, although speculative and hard to test, has been called the visual predation hypothesis (Cartmill 1972, 1974a, b). The key was to discern how and why primates diverged from all the other lineages which colonized trees, which retained claws and large snouts, and lacked prehensile appendages, binocular vision and associated characters: opossums in the New World, possums and koalas in Australia, tree squirrels, raccoons and civets. Cartmill argued that the unique traits that humans share with other primates, including binocular vision, a short snout and prehensile appendages, developed among predators in the canopy of woody plants utilizing prey with effective escape responses. The prey were insects feeding at the extremities of branches where rich new growth prevailed. Three basic traits were required in this precarious hunting arena:

1 Extraordinary agility on thin tree limbs, requiring prehensile hands.

2 Excellent binocular vision for visually hunting active, alert and highly mobile insects.

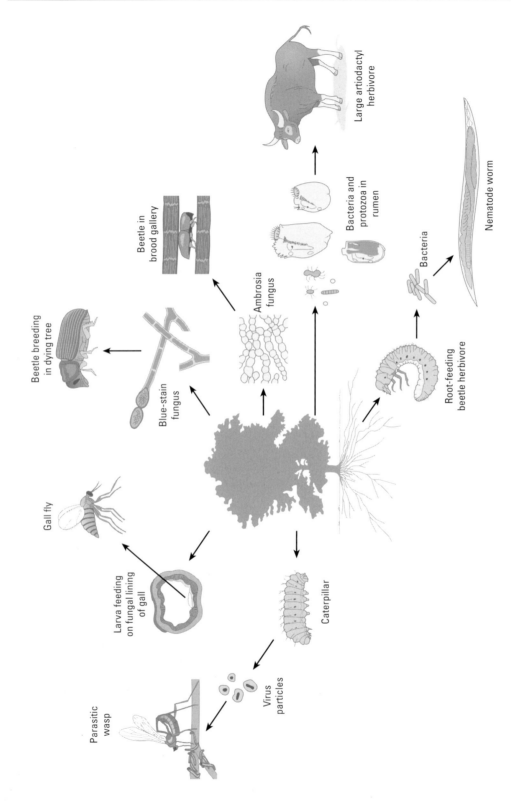

Figure 1.9 A composite sketch illustrating many plant–animal interactions mediated by symbiotic bacteria, protozoa and fungi, as well as cases in which small carnivores utilize viruses and bacteria to ensure access to the host.

3 Both binocular vision and prehensile hands for the accuracy of the final strike and capture of prey.

Increase in visual-field overlap enables compensation for a prey's evasive movements, and prehensile hands ensure a firm grasp of the prey. These traits paved the way for the evolution of the human species. Here, then, we perceive the kinds of long-term consequences of three trophic-level interactions. The evolution of woody plants created new food resources for herbivores, but these fed in precarious sites for the heavier mammalian predators. Plant architecture, coupled with the presence of nutritious and abundant insect food, dictated a path of evolution into the primate stock, starting with agile predators and insectivores—and ending unfortunately in our own species, poorly adapted both for arboreal life and bipedalism.

1.3.1 Colonization of land

On a much broader scale in the terrestrial environment, the adaptive radiation of plants created a cornucopia of habitats and food sources (Table 1.1) with a consequent even richer radiation of herbivores and detritivores, followed closely in time, no doubt, by carnivore radiations. Phases in the colonization of land are covered in Chapter 2, with adaptive radiations based on increasingly complex associations of organisms. For example, many invertebrate taxa must have colonized dead and dying plant material, perhaps depending on fungal, bacterial and protozoan decay organisms to break down refractive cell walls, liberating nutrients. An illustration of this kind of interaction and the extreme diversification that resulted concerns the radiation of scarabaeoid beetles stemming from early stock feeding in humus and on fungi (Fig. 1.10). The group diverged into new adaptive zones still closely related to humus and fungi, colonizing decaying logs (e.g. Passalidae, Lucanidae), nests of termites and ants (e.g. Ceratocanthidae, Valginae), plant roots (e.g. Pleocomidae, Dynastinae), dung and carrion (e.g. Aphodiinae, Scarabaeinae), and eventually feeding on leaves and flowers of living plants (e.g. Melolonthinae, Rutelinae, Cetoniinae). This remarkable group numbers some 29 000 beetle species, with over half the species closely associated with living plants, feeding on roots, leaves, flowers, saps and oozes (Melolonthinae, Dynastinae, Rutelinae, Osmoderminae, Cetoniinae). The subfamily Melolonthinae alone, with larvae feeding on humus, roots or dung, and adults utilizing leaves or flowers, if they feed at all,

numbers about 10 000 species. Clearly, the diversity of plant species, their phytochemicals and the habitats they occupy, provide an extraordinarily diverse template on which plant-related taxa diversify. The origin of new species, or speciation, is spectacular.

1.3.2 Speciation

Speciation itself may be driven by different kinds of circumstances. Species are recognized as groups of populations that interbreed, but which are reproductively isolated from other such groups. Reproductive isolation is a property of species resulting in the loss of fertility when hybridization occurs, keeping gene pools more or less discrete. Such reproductive isolation and speciation may result from populations becoming separated by a geographical barrier, such as a mountain range, so that they evolve independently. They diverge in characters, which may include traits involved with courtship, mating and fertility, resulting in reproductive isolation and new species. This form of speciation, called **geographic** or **allopatric speciation**, may not involve important plant–animal interactions, but another form of speciation places such interactions in a central role. (For more discussion of modes of speciation see texts on evolution such as Price 1996, Ridley 1996, Futuyma 1998 and Howard & Berlocher 1998.)

Speciation within the normal cruising range of a population, that is, in a local geographical area, is referred to as **sympatric speciation**. The difficulty for many biologists is to understand how reproductive isolation can evolve in one locality without a geographical barrier. However, with about 500 species in a leafhopper genus such as *Erythroneura* (Homoptera), all of which feed on plants, it is hard to imagine how there could be so many geographical barriers involved in speciation and how the process could occur relatively rapidly, such that species had not diverged into different genera. The conundrum is resolved when we recognize that many species are specialized to associate with a single host plant species. An accidental shift from one plant species to another could result in isolation between parental and derived populations, eventually resulting in new species. Volatile cues from plant hosts may change, the phenology of development of plant parts and habitat may also change, tending to reduce contact between populations on each plant species. As individuals adapt to a new host plant species, reproductive isolation may evolve, resulting from a behavioural preference for

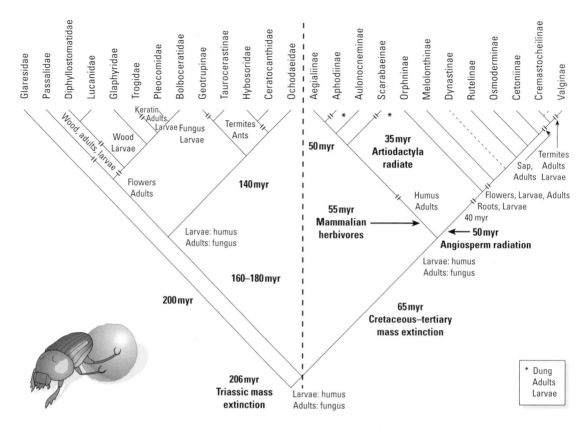

Figure 1.10 A phylogeny of the scarabaeoid beetles starting with larvae feeding in humus and adults feeding on fungi. The group branched into two clades which radiated into similar sets of adaptive zones. Note that the right-hand clade followed the adaptive radiations of the mammalian herbivores and the angiosperms, becoming rich in species exploiting mammalian dung and green plants. (From Scholtz & Chown 1995.)

the new host, or shifted phenology that synchronized better with the new host and adaptation to a new micro-habitat. Such processes may be repeated many times, even in one region and quite rapidly, resulting in the many closely related species we observe feeding on related host plant species (cf. Howard & Berlocher 1998 for extended discussions of sympatric speciation).

1.3.3 Coevolution

The general kind of concordance between insect herbivore groups and plant groups has long been noted by plant pathologists and ecologists alike. Mode (1958) coined the term **coevolution** for reciprocally induced evolutionary change between two or more species or populations, using as his example the interaction of the rust fungus *Melamp-*

sora lini and its host, the flax *Linum usitatissimum*. Ehrlich and Raven (1964) soon popularized the concept of coevolution based on associations of butterflies and plants (Fig. 1.11). Apparently the diversity of living organisms has been promoted by a kind of arms race between antagonistic associations of plants and herbivores. A taxon of herbivores creates selective pressures for plant defences. With new defences, the plant taxon emerges free of herbivores until a new group of herbivores breaks through the defensive barrier, causing selection for new defences. Each breakthrough may result in a new group of plants with new forms of defence, associated with a new taxon of herbivores adapted to these new defences. As we see in Fig. 1.11, paired associations of butterfly taxa and toxic plant families are evident. For example, a subfamily, the Pierinae, including the cabbage white butterfly, is

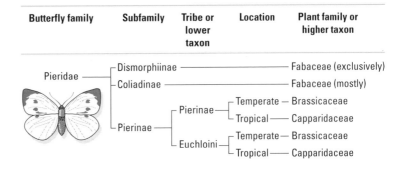

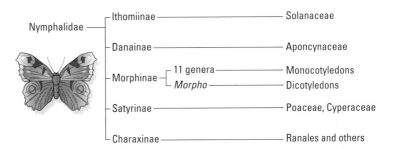

Figure 1.11 Some examples of relationships among butterfly taxa and host plant taxa discussed by Ehrlich and Raven. Butterfly taxon names have been retained as used by Ehrlich and Raven (1964) for easy reference to their text, although the systematics of some groups have since been modified. (Adapted from Price 1996.)

found on the cabbage family, Brassicaceae, in the temperate zones and Capparidaceae in the tropics, both families with toxic mustard oils or glucosinolates. In the Nymphalidae, the Ithomiinae feed on the potato family, Solanaceae, spiked with toxic alkaloids, and the Danainae feed on the dogbanes and milkweeds, Apocynaceae containing cardiac glucosides. Escalation in the arms race appears to generate a great diversity of phytochemicals, herbivores adapted to utilize them and, perhaps, new plant and animal taxa. The actual mechanisms involved, and questions about whether insects really select for toxic chemicals in plants, are still debated and the validity of broad scale coevolution remains in question. As we will see for the *Rhagoletis* fruit flies and as Feinsinger (1983, p. 307) noted for flower and bird relationships, 'Most plants and pollinators move independently over the landscape, not in matched pairs.' Nevertheless, the general scenarios noted by Ehrlich and Raven (1964) remain to be explained mechanistically while making an important contribution to the appreciation of biotic diversity and perhaps its generation.

An example of pattern consistent with the Ehrlich and Raven hypothesis, but lacking a mechanistic explana-

tion, concerns a New World plant group akin to those producing frankincense and myrrh in the Old World. In resin canals ramifying in stems and leaves various compounds circulate, often toxic to herbivores, including terpenes such as pinenes, camphene, phellandrene and limonene (Becerra 1997). Flea beetles (Coleoptera: Chrysomelidae: Alticinae) in the genus *Blepharida* feed on these *Bursera* host plants, with related beetles utilizing plants with related phytochemistry. Still, the question remains of how reciprocally selected these chemically based patterns are.

Nevertheless, the field of coevolution has advanced well beyond the initial vision of Ehrlich and Raven. The kinds of coevolution have been classified by Thompson (1989, 1994) (Table 1.3), some excellent examples are provided by Thompson and associates (1999a) and the future development of the field is discussed in a recent review (Thompson 1999b).

1.3.4 Cospeciation of plants and animals

As we saw earlier in this chapter, plants occupying small and patchy microhabitats have speciated rapidly (Fig. 1.3)

Table 1.3 Kinds of coevolution recognized by Thompson (1989, 1994).

Gene-for-gene coevolution. Cases in which specific, mutually selected genes define the relationships between two species. *Example:* Relationships between pathogenic fungi and host plants are dictated by the presence or absence of virulence or avirulence genes in fungi and resistance or susceptible genes in the host plant.

Specific coevolution. Species have special traits adapted for living with another species, matched by reciprocal traits of the cohabiting species, without any genetically based evidence for traits. *Example:* Mutualisms between ants and acacias in which the plant provides inflated thorns as nesting sites, plus protein and carbohydrate sources, and the ants provide protection as bodyguards against herbivores and competing plants, and against fire.

Guild coevolution. Cases in which a group of species, acting as a guild, interact with a group of other species, resulting in broadly adapted sets of traits mutually induced by each species group. *Example:* Blood parasites of vertebrates in general have probably selected for escalated sophistication of the immune system, which in turn has selected for such stratagems as rapid change in coat antigens and hypermutation in the parasites.

Diversifying coevolution. Cases in which highly specific interactions result in reproductive isolation and speciation, producing sometimes large taxa of associated interacting species. *Example:* High specificity of pollinators when they are also herbivores in fruits, such as the fig wasps, coupled with host plant adaptations that limit access to would-be parasites (thieves) in the system, probably resulted in reproductive isolation between populations as small amounts of variation modified pollinator-host relationships.

Escape-and-radiation coevolution. The scenario envisioned by Ehrlich and Raven (1964) in which a plant taxon escapes from attack by a herbivore group when a mutation results in novel defensive chemicals. This is followed by evolution of a herbivore taxon to utilize the newly defended plant group, resulting in radiation of species in both plant groups and herbivore groups. *Example:* Figure 1.11.

making an important contribution to the dominance of the angiosperms in floras in most parts of the world today. As such plants radiate over landscapes and geographical regions, speciating on the way, they are likely to be tracked by members of the associated fauna of specialists. The hops and jumps in host plant distribution will have created many instances of allopatric speciation, with host plants and specialist animals speciating together and in parallel: **cospeciation** and **parallel cladogenesis** or **phylogenetic tracking**. While such phylogenetic congruence has been recorded for many groups of animal hosts with their animal parasites (e.g. Brooks & McLennan 1993) evidence suggests that plant-dwelling organisms are not so tightly constrained. 'Cases of strictly parallel diversification are rare' (Farrell et al. 1992, p. 41).

More commonly, herbivorous species of insects are loosely associated with a larger plant taxon such as a genus or family, but dramatic shifts among plant taxa are common. A case in point is in the fruit flies of the genus *Rhagoletis* which includes the notorious apple maggot, *R. pomonella* (Fig. 1.12). We see some species grouped on Rosaceae, others on Ericaceae, still others on Juglandaceae and many isolated species on very different plant families: Caprifoliaceae, Cornaceae, Saxifragaceae, Berberidaceae and even the coniferous Cupressaceae. One feature in

common among hosts is the presence of fleshy fruits in which the larvae feed. Another is the coexistence of hosts in similar habitats so that chance shifting from one host to another among ecologically associated plants has a high probability in space, given enough time. Thus, *Rhagoletis* flies and most other herbivores appear to be ecological opportunists in evolutionary time, shifting among hosts within habitats with equivalent resources for breeding, and speciating in the process.

The unusual case of almost complete congruence between phylogenies of host plants and insect herbivores may provide some clues about where to find strict phylogenetic tracking (Fig. 1.13). Beetles in the chrysomelid genus *Phyllobrotica* have speciated on several basal lineages of the Lamiales, especially on *Stachys*, the hedge nettle (Lamiaceae, the mints) and *Scutellaria*, the skullcap, in the same family. We enjoy the Lamiaceae for their aromatic compounds (e.g. mints, sage, thyme, savory, marjoram), but they act as repellents and toxins to unspecialized insects, and attractants and ovipositional and feeding stimulants for specialists. *Phyllobrotica* beetles are such specialists, which may also depend on host-plant compounds for defence against predators (Farrell & Mitter 1990; Mitter et al. 1991). 'The iridoids catalpol and aucubin, present in *Scutellaria* and elsewhere, are

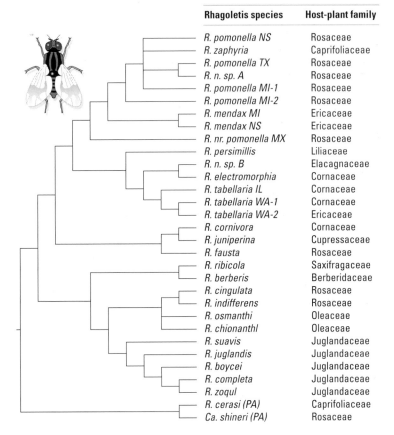

Rhagoletis species	Host-plant family
R. pomonella NS	Rosaceae
R. zaphyria	Caprifoliaceae
R. pomonella TX	Rosaceae
R. n. sp. A	Rosaceae
R. pomonella MI-1	Rosaceae
R. pomonella MI-2	Rosaceae
R. mendax MI	Ericaceae
R. mendax NS	Ericaceae
R. nr. pomonella MX	Rosaceae
R. persimillis	Liliaceae
R. n. sp. B	Elacagnaceae
R. electromorphia	Cornaceae
R. tabellaria IL	Cornaceae
R. tabellaria WA-1	Cornaceae
R. tabellaria WA-2	Ericaceae
R. cornivora	Cornaceae
R. juniperina	Cupressaceae
R. fausta	Rosaceae
R. ribicola	Saxifragaceae
R. berberis	Berberidaceae
R. cingulata	Rosaceae
R. indifferens	Rosaceae
R. osmanthi	Oleaceae
R. chionanthl	Oleaceae
R. suavis	Juglandaceae
R. juglandis	Juglandaceae
R. boycei	Juglandaceae
R. completa	Juglandaceae
R. zoqul	Juglandaceae
R. cerasi (PA)	Caprifoliaceae
Ca. shineri (PA)	Rosaceae

Figure 1.12 Estimated phylogeny of *Rhagoletis* fruit flies and relatives and the host-plant families utilized, illustrating limited concordance between related groups of *Rhagoletis* and the plant families on which larvae feed. This particular tree is a subset of a phylogeny based on mitochondrial DNA sequence data analysed using maximum parsimony. Capital letters after *Rhagoletis* species names denote states or provinces of origin: NS, Nova Scotia; TX, Texas; MI, Michigan; MX, Mexico; WA, Washington; PA, Pennsylvania. The outgroup *Carpomya shineri* (at bottom) is placed in a separate genus, but grouped with a *Rhagoletis* species. (Based on Smith & Bush 1997.)

selectively sequestered by larvae in several lepidopteran families and by *Dibolia* leaf beetles . . . These chemicals have been demonstrated to play a major role in the defensive strategies of several of these herbivores . . . A similar defense for *Phyllobrotica* is suggested by the apparently aposematic yellow-orange and black adult markings and by the bitter taste of the adult beetles' (Farrell & Mitter 1990, p. 1399). In addition, larvae feed endophagously, boring in the roots of plants. Such intimate associations between host plant and insect may set a **phylogenetic constraint** on the host-plant range utilized by the insect, a kind of straitjacket leaving little room for manoeuvring in evolutionary time.

1.3.5 Adaptive radiation

Adaptive radiation may be defined as 'the rise of a diversity of ecological roles and attendant adaptations in

different species within a lineage' and 'one of the most important processes bridging ecology and evolution' (Givnish 1997, p. 1). It results in large-scale increases in biodiversity involving speciation and the other processes treated in the section on the creation of biodiversity. The most celebrated cases of adaptive radiation have been described on oceanic archipelagos—Darwin's finches in the Galapagos, honeycreepers and *Drosophila* on the Hawaiian Islands. But habitat islands and 'archipelagos' on mainlands are also important, for example, in the radiation of cichlid fishes in the rift-valley lakes of East Africa (e.g. Reinthal & Meyer 1997), the adaptive shifts in the Pontederiaceae which use fresh-water habitats, as in the water hyacinth (Barrett & Graham 1997), and the terrestrial columbines (Hodges 1997a). In the last case, rapid radiation has involved the evolution of high diversity in floral morphology and colour that relate to different pollination systems. These involve bees and hawkmoths with

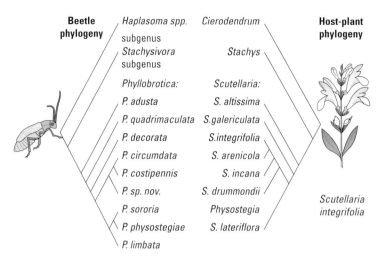

Figure 1.13 The phylogeny of the chrysomelid beetle genus *Phyllobrotica* and its sister genus *Haplasoma*, coupled with the phylogeny of host plants. Note the close resemblance of the phylogenies. The only major shift to a new plant group is by *P. physostegiae* onto *Physostegia*, although the host plant is also a member of the Lamiaceae, as is *Scutellaria*. (From Farrell & Mitter 1990.)

various proboscis lengths, as well as hummingbirds. The author notes that in other taxa the evolution of nectar spurs is generally related to extensive adaptive radiation relative to a sister group without this general trait.

As in the columbines, it is frequently the plant–animal interaction that drives adaptive radiation in one way or another. The case of oncidiine orchids is particularly interesting and instructive (Chase & Palmer 1997), and may involve escalations named **leapfrog radiation**, from the first vegetative shifts in response to new habitats, followed by new associations with pollinators. The oncidiine orchids are obligate twig epiphytes, and in these harsh temporary and precarious habitats many vegetative forms have developed in relation to the variety of habitats colonized (Fig. 1.14)—a radiation in vegetative form. A second round of diversification resulted from multiple pollinator relationships and the kinds of floral rewards available (Fig. 1.15). Some are pollinated by wasps or halictid bees, others by birds or butterflies, and still others by euglossine bees. The specificity of pollinators of orchids has been long recognized and invoked in the development of reproductive isolation between populations, generating new species rapidly and extensively. This kind of leapfrog radiation is akin to Ehrlich and Raven's view of an escalating arms race between plants and herbivores, but with a more mechanistic explanation.

We may then explore the effect of plant species' adaptive radiation on plant-related animals such as herbivores. Mitter et al. (1988) undertook a study of adaptive zones, asking if herbivory promoted insect diversification rela-tive to exploitation of other zones, such as carnivory or saprophagy. In 11 out of 13 tests using sister group comparisons, herbivorous insect taxa were much more diverse than a sister group, and overall herbivores were three times more speciose than sister groups (329 305 species versus 107 021 species). Evidently the need for specialization in herbivores, coupled with the archipelago-like distribution of many host-plant species, appears to have contributed greatly to the adaptive radiation of herbivorous groups.

A final question may be addressed to the circumstances that fostered diversification of insect groups initially, as the land plants colonized solid ground. A most heuristic view was developed by Hamilton (1978), who noted that many primitive insects are associated with rotting trees as well as other decaying plant material. In the close confines under bark, social systems become modified, close associations with microbial associates develop, flight is fostered, involving colonization of widely disjunct habitats, and wing polymorphism evolved many times. Rich plant resources offered a wide range of foods in sequence, starting with yeasts, fungi, spores and bacteria on which insects and mites could feed, providing food for higher trophic levels. We saw how the scarabaeoid beetles radiated from these kinds of beginnings earlier in this chapter. In addition, primitive plants would have provided resources and the basis for the adaptive radiation of groups like the Lepidoptera, ultimately numbering some 160 000 species. The most primitive Lepidopteran family, Micropterygidae, have adults feeding on pollen and

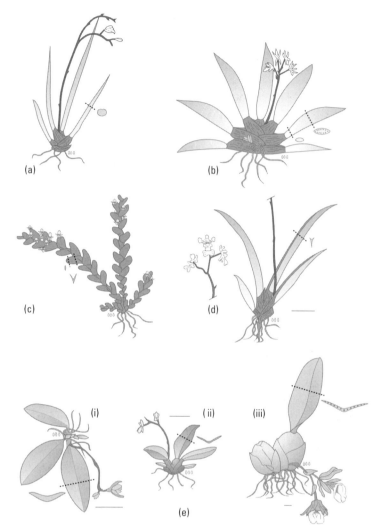

Figure 1.14 Representative growth forms of oncidioid orchids adapted to living at the tips of woody stems in humid cloud and rain forests around the world (with pollinators given in parentheses below). The five major growth habits are illustrated. (a) Type I, small pseudobulbs with apple-shaped leaf cross-sections (butterfly pollinated). (b) Type II, well developed pseudobulbs, with lens-shaped leaf cross-section (pollinated by euglossine bees). (c) Type III, no pseudobulbs, leaf with basal deep V cross-section and distal lens-shaped cross-section (pollination?). (d) Type IV, psygmoid seedlings, sympodial growth, with Y-shaped leaf cross-section (pollination by reward to bees or by deceit). (e) Type V, standard type, generally with large pseudobulbs and broadly V-shaped leaf cross-sections (diverse pollinators including halictid bees, euglossine bees, oil-collecting bees, hummingbirds and other birds, and no-reward flowers). Scale bars equal 1 cm. (From Chase 1986.)

larvae feeding on bryophytes. Hamilton noted the ease with which a lineage could shift diet, from fungal mycelium to spores to pollen. The earliest group of Hymenoptera, represented by the genus *Xyela*, feed on microspores scattered in forest litter, but they are now the basal link to all 10 000 species of plant-associated sawflies and relatives, from which all other groups of hymenopterans are derived: parasitoid wasps, ants, bees and wasps.

From small beginnings coupled with adaptive innovations in feeding habits, habitat use and new associations with other organisms, terrestrial life has burgeoned into the richest display of life forms ever recorded on this planet. If we can document this diversity of species and interactions, explain the mechanisms by which they have arisen, and conserve all for future generations, then we will have completed a monumental task.

1.4 Opportunities and pitfalls

We have seen repeatedly in this chapter that plants and animals have interacted in ways that have fostered the creation of tremendous diversity on this earth. And the opportunities for the further development of this field are equally rich but nevertheless challenging. Brief coverage of the methods of study used in developing the field is worth considering, for the methodologies are the bases

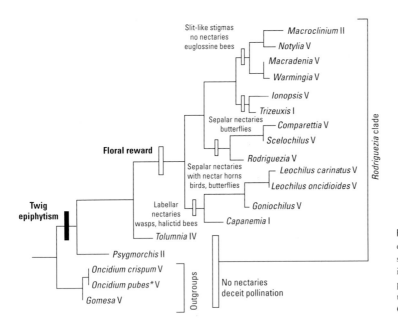

Figure 1.15 A phylogenetic hypothesis of the evolution of a clade of oncidioid orchids, starting with twig epiphytism and radiating into different growth habits, floral traits and pollination syndromes. Next to each genus is the habit type illustrated in Fig. 1.14. (From Chase & Palmer 1997.)

of continued progress and the diversity of approaches is worth noting.

1.4.1 Methods of study

Of course, field studies are fundamental and central to the developing field. Scientists who spend long hours in the field — observing, collecting, recording — provide the grist of natural history, novel insights and general knowledge. Observing relationships, behaviours, habits and habitats are basic (cf. Figs. 1.1, 1.4–1.15). Collecting provides the samples and records needed for herbaria and museums, botanic gardens and arboreta, on which many other studies are based. And the written and published record keeps all informed and in tune with progress. Field studies include those on living organisms and the search for fossils, so that neontology and palaeontology are intertwined in the unravelling of plant and animal relationships, evolutionary trends, phylogenetic relationships and comparative biology.

With the basis of field studies lasting decades and centuries, we have amassed large collections of living and extinct organisms. These form the key to a suite of sciences essential to the analysis of plant-animal interactions. Systematics provides the identity, Latin binomials and

phylogenetic relationships among organisms (cf. Figs. 1.8, 1.10–1.13, 1.15). It is basic to estimates of biodiversity (cf. Fig. 1.2), the correct cataloguing of species, the analysis of adaptive shifts in lineages and adaptive radiations, the practice of agriculture, forestry and horticulture, the conservation of organisms, and informed management of all kinds of landscapes. We have seen the importance of systematic studies in this chapter, using comparisons of morphological traits, electrophoretic analysis of proteins and molecular genetics. As an additional example, evolutionary trends in the ecology of New World monkeys were analysed by Horovitz and Meyer (1997) using nuclear and mitochondrial DNA and morphological data. These analyses were coupled with extensive observations on habitats used, modes of locomotion and foods eaten, including leaves, fruits, exudates, nectar, insects and other animals. The chemical systematics of plants provided the groundwork for chemical ecology, studying the chemically mediated interactions among organisms involving defences, attractions, behavioural responses, trail odours, marking pheromones and many other associations (cf. Fig. 1.7). Plantings of species, ecotypes or geographical representatives allow comparative studies on associated organisms and provide material for genetic studies; for example, broad scale studies on the evolutionary potential of species and phylogenetic relationships are accelerated by relevant living collections, as illustrated by the results shown in Figs. 1.3, 1.14 and 1.15. The comparative method in biology is of fundamental importance in the field of plant and animal relationships. The bigger pictures of nature and the discoveries of patterns in nature depend upon the comparison of traits: morphological, chemical, molecular, behavioural and ecological, relative numbers of species, genera and higher taxa. Most figures in this chapter illustrate results of the comparative method. The approach is so fundamental that we take it for granted on the one hand while forgetting it totally in many studies that would be stronger if the method were used. There is more about this in the next section.

A major approach in comparative biology involves **phylogenetic analysis**, which develops hypotheses on the development of evolutionary relationships from a common ancestor to a derived group of species or **clade** (Hillis et al. 1996; Harvey & Nee 1997). Many kinds of traits may be used for estimating evolved relationships, with morphological and molecular characters most frequently used. Other characters may include habitat use, life history, pollination syndrome and feeding preferences, but all have some connection to morphology. We will use a real example to follow through the steps involved in phylogenetic analysis, using a group of orchids analysed by Chase and Palmer (1997) and discussed earlier in this chapter (Figs. 1.14, 1.15).

Many orchid groups are composed mostly of epiphytes on major branches and stems of trees, a habitat use that appears to be **ancestral**, that is, **plesiomorphic**, relative to smaller and apparently **derived** or **apomorphic** groups that are adapted to terminal small twigs in the canopy. These derived species became obligate twig epiphytes, well represented in the oncidiine orchids, with many unusual life-history characters, including hooked seeds, flattened seedlings and short life cycles (6 months from germination to flowering versus 2–4 years in ancestral species). These characters set the clade of derived obligate twig epiphytes apart from the ancestral species living on large limbs. The *Rodriguezia* clade of twig epiphytes consists of 15 species in 14 genera, indicating major morphological change involved with speciation events, with most species representing new genera (Fig. 1.15).

The phylogenetic analysis starts by identifying a group of species that are closely related to the species whose internal relationships you want to infer. We assume that they reflect to some extent the traits of the common ancestor of the group under study, thus the ancestral states. The species with these plesiomorphic characters then form an **outgroup** to help analyse how a derived group such as obligate twig epiphytes have radiated. Outgroups are indicated in Fig. 1.15 as *Oncidium* and *Gomesa* species typical of the large-branch epiphytes. Then, major shifts from this outgroup are mapped onto the most general traits assumed to be the most basic in the new clade, represented by twig epiphytism. The next new trait to develop is a floral reward, and from here we see radiation into different types of floral reward: specialized for wasps and halictid bees, or birds and butterflies, or only butterflies, or for euglossine bees. In this way a phylogeny is developed, using traits to discern ancestral and derived characters and mapping these traits on the phylogenetic relationships.

Such comparing and contrasting among apparently related species is formalized into a phylogenetic hypothesis for a clade as illustrated in Fig. 1.15. The hypothesis developed using morphological and life-history characters may be compared with an independent phylogeny using molecular traits. If the phylogenies are closely similar, or **congruent**, the general hypothesis is strengthened. Such congruency was found for the twig epiphytes involving pollination syndromes and the associated floral traits, but not for vegetative traits and molecular traits, reinforcing the precautionary tactic of using as many independent derivations of phylogenies as possible.

1.4.2 Special and general approaches

The special-case approach (the idiographic approach) to research involves studying individual cases without regard for the big picture. It is an important form of science because accurate and factual information on particular cases is essential before broader views and patterns can be detected. The idiosyncrasies of a certain interaction are often fascinating and we devote attention to the details of particular cases because the science of natural history is itself a charming occupation. The urge for a reductionist approach, reducing investigation to smaller and smaller units, is often appealing—detail is what we need, and the more detailed we become, the more we reach into fields beyond the focal study of plant–animal interactions: phytochemistry, physiology, molecular biology, etc. This all adds up to good, creative, captivating science.

That said, we should debate how science progresses. Certainly special-case approaches are essential, but they are not enough, not half enough. Surely the core of a science lies in the scientific theories developed and the emphasis devoted to the generation of theory. And here I mean that a **scientific theory** is a factually based, mechanistic explanation for broad patterns in nature. This is exactly what the theory of evolution is. But we have no such central, unifying theory in ecology to work with. The term 'factually based' means founded on real natural history and the real facts of nature, hence the importance of special approaches. The term 'mechanistic explanation' means the processes resulting in a phenomenon are understood: we are not developing laws of nature but the explanation of why laws exist. And the term 'broad patterns in nature' means that we are searching for *general* phenomena in nature—the general or holistic approach (the nomothetic approach).

Generalizable facts and mechanistic explanations are the basis of scientific theory. The broader the reach the better, for the more applicable, useful and predictive the theory becomes. But in the biological sciences, and especially in ecology, we cannot expect a monolithic theory that explains all. Rather, **pluralistic theory** is essential for, as in physics, starting conditions will affect the outcomes (cf. MacArthur 1972; Schoener 1986). For example, a lineage *A*, with a certain set of life-history and behavioural characters, is likely to express an ecology different from a lineage *B* with different traits. The evolutionary starting points *A* and *B* differ, therefore the ecologies of *A* and *B* will differ. But lineages *A* and *C* may have converged in traits, resulting in our expectations for similar ecologies, as exemplified by species *A*. We can keep adding lineages to categories equivalent to life-history types *A*, *B* and others, making the generalizations broader and the predictive power stronger—broadening the holistic approach. For examples in insect herbivore population dynamics, see Price 1994, 1997, Price et al. 1990, 1995. Broad patterns with mechanistic explanations are also illustrated by Figs. 1.3, 1.6, 1.8, 1.10, 1.13–1.15. Of course, the explanations for pattern may be debated, as with Ehrlich and Raven's (1964) scenario for coevolution between plants and herbivores (cf. Thompson 1994), but without the discovery of pattern we have little to debate. And the debate or dialectic is an essential component of theoretical development.

After many decades of intensive study, the field of plant and animal interactions needs some emphasis on the identification of patterns and the development of factually based theory. 'The concept of pattern or regularity is central to science. Pattern implies some sort of repetition, and in nature it is usually an imperfect repetition. The existence of the repetition means some prediction is possible—having witnessed an event once, we can partially predict its future course when it repeats itself' (MacArthur 1972, p. 77). 'Next we present a unified account of the various reasons one area has more species than another. Our effort is not to understand all cases of different numbers of species occupying different places. That would be too complex to be rewarding. Rather, our aim is to select certain cases that reveal interesting things about the mechanisms involved' (MacArthur 1972, pp. 77–78).

MacArthur's masterful, broad-scale view of nature, exemplified in the theory of island biogeography (MacArthur & Wilson 1967), can be emulated in the field of plant-animal interactions. We have banks replete with special-case studies which can be synthesized to fill vaults of factually based theory. What are the patterns in the chemical ecology of plant-animal interactions and where is the theory? What generalizations have been reached on multiple trophic-level interactions? What broad-scale predictions are available for plant and animal mutualistic interactions? The great challenge for this field is the detection of broad-scale patterns, the development of theory and the nurturing of generally more predictive science. A more even balance between special cases and reductionism, and holistic, synthetic developments should result in more rapid advances in knowledge. As Brown (1995, p. 11) noted regarding his broad-scale view, macroecology: 'Macroecology is self-consciously expansive and synthetic. In this regard it does differ philosophically from much of traditional ecology, which I would characterize as becoming increasingly reductionist and specialized.'

1.4.3 Applications

An enriching force in the study of plant and animal interactions is derived from many kinds of investigation related to solving real human problems. Pollination of crop plants, protection of plants against insects and other animals, the biological control of weeds and insects, and conservation initiatives all feed into the general pot, or smorgasbord, of the science. But I am repeatedly struck, when asked to address such areas as an ecologist, how little we have to offer in return in terms of broad principles or concepts helpful to the practical, problem-related scientific issues of our day. For, in terms of knowledge based on real life in nature, we are about as empirically based as the problem-solving aspects of our science. We generate and answer questions idiosyncratically in the main.

This is not a dismal picture nor a dismal end to this introductory chapter. Our sciences of plants and animals are young and vibrant, the growth of knowledge is extra-ordinary and the rate of discovery increases. Recognizing the opportunities and pitfalls fosters growth and the refinement of emphases and initiatives. And we can still learn from Darwin's example — first the great natural-ist, then the detector of pattern, then the formulator of mechanisms, and finally the creator of the central theory for life.

Chapter **2** **The history of associations between plants and animals**

Conrad C. Labandeira*

2.1 Introduction

Ever since the greening of a strip of equatorial coastline that fringed tropical oceans approximately 425 Ma, plants and animals have been extending their ecological reach on land. Today their terrestrial representatives constitute the overwhelming bulk of macroscopic diversity on the planet, and occur in profusion in virtually all terrestrial habitats. The earth's land surface today is fundamentally an abode for vascular plants and five major groups of macroscopic animals: nematodes, gastropods, onychophorans, arthropods and vertebrates. Two subgroups of this menagerie—seed plants and insects—contribute the preponderance of this diversity, and overwhelmingly outnumber in species all other groups combined. Virtually all terrestrial life is affected in some way by their myriad associations.

The origin and evolution of associations between vascular plants and animals has a rich but undervalued fossil history. This history has been integrated into biological and palaeobiological perspectives, each of which has explored the deep historical dimension of vascular plant and animal associations. The empirical data underlying both of these approaches will be interpreted in this chapter within a context of such theoretical questions as the initial launching of herbivory on land; the origin and early expansion of major ecological associations; the establishment of modern associations and the survival of ancient relationships; and the roles of taxonomic extinction, ecological convergence and escalation in long-term patterns of herbivory. Both palaeobiological and biological approaches have unique strengths and liabilities that include the varied uses of uniformitarianism, the quality of fossil evidence for inferring past associations, the role of combining phylogenetic and ecological data

in estimating the duration of associations, and limitations on the use of modern ecological analogues for correctly interpreting fossil material. In this context I have chosen three ecomorphological units—functional feeding groups, dietary guilds and mouthpart classes—as yardsticks to anchor much of the discussion of the fossil record. Emphasis will also be placed on results from varied phylogenetic studies for inferring the history of associations between plant hosts and their animal herbivores. A glossary of technical terms is shown in Box 2.3 at the end of the chapter (p. 73).

2.2 Why study plant–animal associations of the past?

The evolutionary history of biodiversity is essentially a chronicle of species interactions; species do not occur in splendid isolation. This precept suggests that an understanding of the history of life and the generation of biodiversity must include an examination of when, how and why organisms have associated during geological time. By contrast, the overwhelming effort in palaeobiology and biology has concentrated on examinations of isolated species in time and space. As indicated by Price in Chapter 1, there has been a growing realization that comprehending associations locked within a biota is considerably more than the sum of constituent species lists, and alternatively requires placing associations in an evolutionary context (Price 1997). Early efforts at unravelling these associations focused on gross correlations between the taxonomies of plant hosts and their insect herbivores (Ehrlich & Raven 1964), followed by more refined hypotheses (e.g. Funk et al. 1995), demonstrating that cycles of adaptations of plant defence and reciprocal counter-adaptations of herbivore attack resulted in the long-term

*This is contribution 69 of the Evolution of Terrestrial Ecosystems Consortium at the National Museum of Natural History.

improvement of individuals towards coping with decreasing odds of survival. Curiously, this research did not influence palaeobiology.

By contrast, the escalation hypothesis of Vermeij (1987) was proposed from an evaluation of the fossil record of marine invertebrates. Vermeij's formulation describes escalation as the increasing adaptiveness of lineages through time resulting from successively higher levels of selection pressure by predators on prey. It differs from various coevolutionary hypotheses by (1) a focus on the adaptation of evolving species to a changing environment rather than through interbiotic associations, and (2) the primacy of antagonistic associations such as predation over less antagonistic associations such as mutualism. The escalation hypothesis was a logical extension of the Red Queen hypothesis proposed earlier (Van Valen 1973), in which an adaptation in one species affected the relative non-adaptedness of other species, resulting either in the simultaneous improvement of interacting species or their marginalization through extinction or relegation to safer marginal environments.

Until recently, palaeobiology never developed a methodology to extensively test trends in the fossil record resulting from patterns seen in modern plant–insect associations. Rather than investigation of mutualisms or other modes of accommodation, overt predation has been the centre of study. Much of this reluctance is attributable to criticisms regarding a poor fossil record, the inability of fossils, if found, to illuminate relevant questions of phylogeny, or perceived limitations in the interpretation of evidence from plant–insect associations (Farrell & Mitter 1993; Shear & Kukalová-Peck 1990). Nevertheless there is an informative and interpretable record that has been documented and used productively. This is particularly true for well-preserved deposits that provide unique windows into the past (Crepet 1996; Labandeira 1998a). In particular, examination of fossil plant–insect associations is essential for testing hypotheses regarding the macroevolutionary dynamics of plants and their insect associates. These data are the source for patterns, especially temporal trends, that can support or refute hypotheses originating from modern ecological studies (Coley 1999).

2.2.1 Associations and time

An association is an ecological relationship that occurs indirectly or directly among two or more participants that use a resource in a repeatable or otherwise stereotyped or predictable way. Associations range from antagonistic interactions such as predation or parasitism, to neutral relationships such as commensalism, to the collectively beneficial arrangements of various mutualisms. Feeding or feeding-related processes constitute the bulk of known relationships between plants and animals, and thus an association is conceived as an available food resource 'niche' that is variably occupied by different hosts and consumers in time and space. An instructive example are marattialean ferns and their insect herbivores, both of which underwent a complete taxonomic turnover during the 300 million years since the Late Pennsylvanian (see p. xii for a time scale). During the Late Pennsylvanian almost all the fundamental ways that insects consume land plants were established — chewing, piercing and sucking, galling, boring, and consumption of spores and prepollen (Labandeira 1998a). Only leaf-mining was apparently absent. These basic feeding styles, or functional feeding groups, also occur in the modern descendants of marattialean ferns, albeit by herbivores from lineages that have evolved more recently than the late Palaeozoic. In a similar context, Lawton and colleagues (1993) provide an example from the modern herbivores of the cosmopolitan bracken fern, in which continental geography rather than geological time is examined. They conclude that the current insect herbivores of bracken are highly variable globally, often with phylogenetically disparate insect taxa occupying the same functional feeding group on different continents, and in some regions particular functional feeding groups may be occupied by multiple taxa or even remain vacant. The salient point is that while plant hosts and their varied insect herbivores evolve and are constantly replaced in time and space, their associations nonetheless remain constant. A Palaeozoic palaeodictyopterid insect imbibing vascular tissue sap from a marattialean tree fern is functionally playing the same role as an aphid today feeding on the same tissues in an angiosperm (Labandeira & Phillips 1996). An appropriate analogy is a soccer team in which the positions remain the same but players occupying these positions are periodically rotated, retired and recruited.

2.2.2 Theoretical issues about the origin and evolution of associations

Four basic questions have been identified that are of interest to palaeobiologists and biologists. All are united by explicit reference to geological time, which provides the chronometer by which long-term trends and processes are calibrated and ultimately interpreted.

2.2.2.1   How were plant–animal associations launched during the Palaeozoic?

Three hypotheses have been proposed for assessing how ecological units, such as functional feeding groups, dietary guilds and mouthpart classes, expand in macro-evolutionary time (Strong et al. 1984). The first hypothesis, the ecological saturation hypothesis (ESH), advocated by palaeobiologists, maintains that the total number of ecological positions, or roles, has remained approximately constant through time after an initial exponential rise (Fig. 2.1a). Thus taxa enter and exit the ecological arena of the local biological community (Box 2.1), but their associations or roles remain virtually level. By contrast, the expanding resources hypothesis (ERH) is favoured by biologists and states that there is a gradual increase in food resources and availability of niches through time, such that the current high level of exploitation is a geologically recent phenomenon (Fig. 2.1b). Last, the intrinsic trend of diversification hypothesis (ITDH), suggested by Schoonhoven et al. (1998), holds that the long-term patterns of ESH and ERH vary among groups of organisms (Fig. 2.1c). This view would imply that the proportion of occupied ecological roles has a globally disjunct pattern according to group, time and space. Of these, the data currently favors ESH, if one assumes that the ecological clock was set during the Pennsylvanian and the previous fossil record is too poor for analysis. The ITDH could be evaluated on a group-by-group basis, if the temptation is avoided to time-average the proportion of occupied ecological roles across all groups worldwide.

2.2.2.2   What is the role of extinction?

The fossil record of life on land is replete with major and minor episodes of mass extinction that correspond to rates of taxal mortality significantly above the background level (Raup & Sepkoski 1982). Major events include the two devastating ones at the ends of the Permian and the Cretaceous Periods, and less intense events such as the ends of the Middle Pennsylvanian, Late Triassic and Eocene Epochs (Boulter et al. 1988; Labandeira & Sepkoski 1993). Thus a basic question regarding the extinction record of major terrestrial organisms is whether the post-event evolutionary clock was reset, either by plants

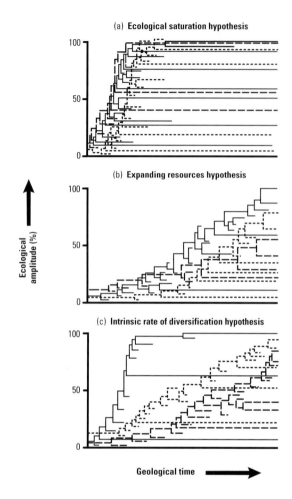

Figure 2.1 The three principal hypotheses for the ecological origin and development of plant–insect associations. Ecological amplitude is a measure of the amount of occupied ecological space, calibrated to the present as equalling 100%. Proxy measurements of ecological amplitude, relevant to plant–insect associations, include functional-feeding groups or dietary guilds or mouthpart classes (see Figs 2.6, 2.16 and 2.17). The ecological saturation hypothesis (a) has been principally advocated by palaeobiologists, and has been supported by Raup (1972) and Behrensmeyer et al. (1992). The expanding resources hypothesis (b) has been mainly propounded by biologists, including Strong et al. (1984). The intrinsic rate of diversification hypothesis (c) is group-specific and has been proposed by Schoonhoven et al. (1998).

becoming unavailable for consumers or the absence of animal associates that can contribute to plant survival through mutualisms such as pollination, seed dispersal or defence from herbivores? Also, what are the consequences for species and their associations when entire

Box 2.1 Plant–animal associations in the context of past biological communities

Biotic communities are local assemblages of trophically linked species, some members of which enter into highly intimate associations. These communities, as well as their myriad constituent plant–animal associations, possess distributions in space and time (Fig. 1) that include extinction (a), re-evolution (b), coalescence (c), complex patterns of extinction, coalescence and persistence (d), persistence (e) and relatively recent origination (f). A uniformitarian approach — where 'the present is the key to the past' — works best for the more recent portion of the biota whereas more ancient examples are likely to have occurred under processes and conditions without recent analogues, necessitating non-uniformitarian approaches.

Although communities evolve, certain highly associated or coevolved constituents may indeed persist unchanged for long geological stretches (Labandeira et al. 1994; Wilf et al. 2000), even though there is evidence that some commu-

nities of today are not analogous to communities even of the very recent past. An intimately coupled association between a host-plant lineage and its insect herbivore lineage may provide a rare opportunity of persistence in the face of species turnover within a community. By contrast, ecologically malleable or geologically fleeting associations, such as ungulate grazers and grasses, are uninformative, if for no other reason than taxal turnover rates are high when compared to plants and their insect taxa (Labandeira & Sepkoski 1993).

Some of the best evidence for modern plant–arthropod associations in the fossil record involves well-preserved angiosperm leaves in Caenozoic deposits with highly stereotyped insect damage, particularly leaf mines and galls (see p. xii). For leaf miners, Opler (1973) has documented modern, genus-level associations for both host plant and insect leaf miners that extends to the early Miocene (20 Ma). Another

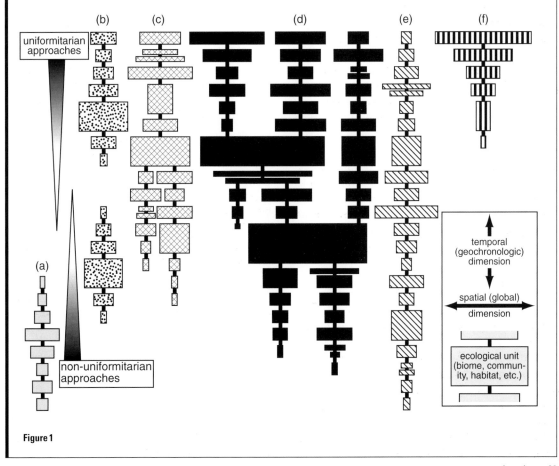

Figure 1

continued on p. 30

Box 2.1 *continued*

direct type of evidence was provided by O'Dowd et al. (1991), who isolated mites and their domatia on angiosperm hosts from the middle Eocene of Australia. Opting for a more indirect approach, Moran (1989) combined biogeographical and palaeobotanical data with successful host transfer experiments between North American and Chinese aphids, indicating that the aphid/sumach/moss life cycle was established by the early Eocene (48 Ma). She concluded that complex multispecies life histories may result from ancient evolution-

ary commitments, allowing a role for historical constraints. However, these older associations are more difficult to track because they are rarer in the fossil record and in the modern biota (Farrell & Mitter 1993) and because increased resolution and phylogenetic scope is needed to detect a weaker signal among extant lineages. In most instances the biological communities in which these older associations occurred are gone or otherwise transformed (Fig. 1), indicating that when they persist today, they occur as relics.

---

communities are devastated? Tentative data indicate that major floral extinction results in a trophically cascading response that affects insect herbivores. This occurred during the Middle to Late Pennsylvanian extinction (Labandeira 1998a), dramatically during the terminal Permian event (Labandeira & Sepkoski 1993) and there is evidence now for a significant turnover in insect herbivore types at the Cretaceous–Tertiary (K/T) boundary (Labandeira et al. 2002), based on plant–insect associational data where previously no extinction was detectable from family-level analyses of body fossils (Labandeira & Sepkoski 1993; Fig. 2.2). Given the taxonomic turnover of vascular plants and herbivorous insects and yet the survival of persistent ecological associations, the phenomenon of ecological convergence is an important long-term pattern, as exemplified by marattialean ferns and their herbivores mentioned above. However, other options are possible, such as the geographical restriction of associations to small refuges and their subsequent expansion and colonization of much larger regions when conditions become favourable.

2.2.2.3 *Do animals track the environment and plants during geological time?*

One of the most pronounced patterns in the distribution of organisms is their overall greater diversity in the tropics than at higher latitudes (Blackburn & Gaston 1996). This pattern is particularly dramatic for insects and their vascular plant associates, and has been documented for bulk defoliation values for local floras or single plant species (Coley & Barone 1996). Insect herbivores typically demonstrate greater removal rates of leaf area in the tropics when compared to higher latitudes. For example, leaf-mining diversity is elevated in tropical regions where centres of diversity exist for several groups (Powell et al.

1999), a trend that parallels wood-boring insect diversity based on species distribution lists. By contrast, galling insects peak at warm-temperate to subtropical xeric latitudes where they are associated with dry, sclerophyll vegetation (Fernandes & Lara 1993). These latitudinal trends are mirrored by altitudinal transects, although notable exceptions are aphids (Dixon 1998) and bees (Michener 1979), which have their greatest diversities at temperate latitudes.

These latitudinal patterns are associated with important trends in climate, including temperature, precipitation and seasonality. Such trends also can be tracked by keeping latitude stationary and varying temporal context such that changes in global climate can be observed vertically in a stratigraphic section containing a taxonomically characterized and well-preserved record of plant fossils. Using this approach, Wilf and Labandeira (1999) studied the Early Caenozoic Thermal Maximum in the Greater Green River Basin of Wyoming, USA, during an interval of a 7 °C to 9 °C rise in mean annual temperature and representing a shift from temperate to subtropical vegetation during the latest Palaeocene (56 Ma) to Early Eocene (53 Ma) interval. Major vegetational changes were documented during this period, as well as differences in the intensity and spectrum of insect-mediated damage on each plant species in before and after floras (Fig. 2.3). A significant trend was the increased intensity and spectrum of insect damage on poorly defended, fast-growing species, such as members of the birch family. This indicated the pre-eminence of a herbivore-accommodating, weed-like defence strategy under the subtropical conditions of the early Eocene, a pattern that was absent during the much cooler latest Palaeocene. This study demonstrates that the fossil record can be used to test long-term predictions of hypotheses based on ecological studies of extant organisms.

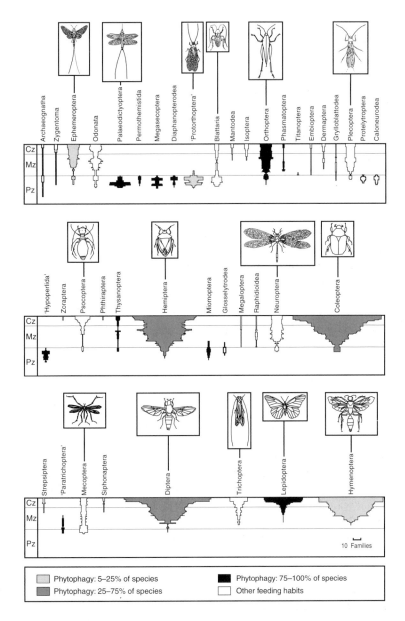

Figure 2.2 Geochronological changes in family-level diversity of major insect clades, categorized by phytophagous dietary preferences. (Modified from Labandeira and Sepkoski 1993.)

2.2.2.4 Applying inverse uniformitarianism: is the past the key to the present?

Catastrophism holds that geologically rapid, global upheavals have been the major influence on the history of life. By contrast, uniformitarianism, which has largely supplanted catastrophism, embodies two distinctive aspects (Gould 1965). One is methodological uniformitarianism, in which the fundamental laws of the modern physical world apply to the past, and the other is substan-tive uniformitarianism, which states that the rates and conditions of current existence are also applicable to the past. Of these two distinct concepts, methodological uniformitarianism is largely unassailed, but substantive uniformitarianism has not withstood close scrutiny (Ager 1993). Indeed, multidisciplinary evidence from various geological disciplines, particularly those applied to the earlier part of the fossil record, indicate that the more ancient the ecosystem, the less it resembles the present. Consequently uncritical application of present-day material

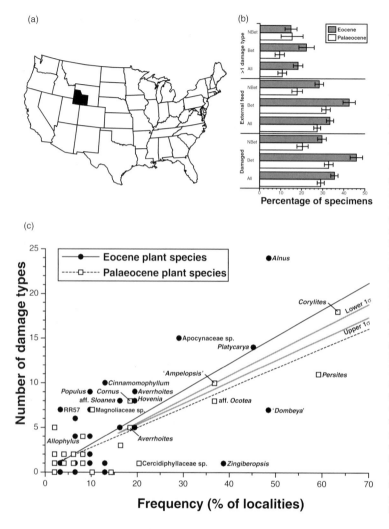

Figure 2.3 A study by Wilf and Labandeira (1999) showing the response of plant–insect associations to the Early Caenozoic Thermal Interval (ECTI), an interval of elevated global warming. A greater variety of insect damage per host species and increased attack frequencies characterize early Eocene plants, occurring under considerably warmer conditions, as compared to earlier late Palaeocene plants. In addition, herbivory is elevated for both late Palaeocene and early Eocene members of the Betulaceae (birch family), a group of fast-growing plants with minimal herbivore defences. (a) Sampling areas are indicated by black polygons. (b) Damage census data for Late Palaeocene and Early Eocene leaves, illustrating the greater percentage of specimens with damage from early Eocene sites. From bottom to top: leaves with any insect damage, leaves externally consumed, and the percentage of damaged leaves bearing more than one damage type. These categories are each analysed separately for all leaves (All), Betulaceae only (Bet) and taxa other than Betulaceae (NBet). (c) Diversity of insect damage for each plant-host species (vertical axis), plotted against the percentage of localities (49 late Palaeocene and 31 early Eocene) at which species occur (horizontal axis). Each data point is one species; many data points overlap at the lower left; survivors are plotted twice. Grey lines show divergence of 1 standard deviation (68%); confidence intervals are for two regressions. This graph shows the higher percentage of localities with more damage types from the early Eocene.

conditions and rates of change to the distant past may not be valid, and the maxim 'the present is the key to the past' may be inappropriate in many instances.

In the context of long-term development of ecosystems and species interactions, the role of very different past conditions for establishing general adaptation and the successful early incumbency of a group is an important guide for understanding its present condition. This grants legitimacy to an inversion of the uniformitarian principle in that 'the past is the key to the present.' As a result, the extent to which substantive uniformitarianism can be extended to the past is debatable. It is generally agreed that the further one delves into the past, eventually a threshold is reached whereby extension of modern conditions be-

comes unreliable (Box 2.1, Fig. 1). It should be stressed that the biological context of plant–animal associations is the local ecological community, which is a spatiotemporally interacting group of organisms that is governed both by external changes in the physical environment and the evolving internal biological integration of its constituents. Depending on the extent of species-level turnover that has occurred, these communities are amenable to greater or lesser application of uniformitarian principles.

2.3 The geochronological context

An appreciation of the history of plant–animal associations involves knowledge of the evolution of terrestrial life. This includes how life colonized the land, processes of fossilization, and biases in the preservation of organisms and past environments. While the terrestrialization of life is a relatively recent event, it has parallels with much earlier episodes of the diversification of marine life, which also resulted in the rapid colonization of vacant but habitable space.

2.3.1 The geological time-scale and important events

The Earth is about 4.6 billion years old, and life originated soon thereafter, according to diagnostic carbon isotopic signatures in sedimentary rocks dated at 3.8 billion years (see p. xii for time scale). The earliest microfossils are 3.5 billion years old, and macroscopic biotic structures resembling prokaryotic algal accumulations occur at 3.0 billion years. Eventually free oxygen accumulated in the atmosphere, reaching levels that became metabolically threatening to prokaryotic life, although mechanisms were evolved to ward off the initial biocidal effect of an oxidative environment. The earliest eukaryotes are algae documented from rocks of 2.2 billion years, for which distinctive geochemical evidence is present by 1.7 billion years; by contrast, the earliest known identifiable animal occurs in 0.9 billion-year-old sediments. The distinctive multicellular Ediacarian fauna occupied shallow marine environments worldwide, occurring from 0.565 to 0.545 billion years ago, immediately prior to the metazoan diversification event during the earliest Cambrian. Multicellular, skeletonized animals, including arthropods, mark the beginning of the Palaeozoic Era. The earliest tangible evidence for terrestrial life is Late Silurian, although suspiciously terrestrial fossils are known earlier from marginal marine deposits.

The earliest terrestrial biota is Late Silurian (approximately 425 Ma) from the United Kingdom and intriguingly contains evidence for plant–animal associations from coprolites deposited by an animal consuming spores and vegetative tissues (Edwards et al. 1995). More recent Early to Middle Devonian biotas that contain insects and their close relatives, as well as evidence for associations with plants, are documented from the Rhynie Chert of Scotland, at Gaspé, Quebec, and at Gilboa, in New York state (Shear & Kukalová-Peck 1990). A 55-million-year gap occurs during the Middle Devonian to the Mississippian/Pennsylvanian boundary during which terrestrial floras are modestly represented but the arthropod record

essentially ceases, only to resume in profusion in the Early Pennsylvanian. During this interval the earliest tetrapods are documented from Greenland in strata of latest Devonian age. A smaller gap, though with very limited arthropod but modest vertebrate occurrences, occurs from the Late Permian to the Middle Triassic, straddling the end-Palaeozoic extinction. The Palaeozoic land biota consisted of clades that are extinct or highly diminished in diversity, by comparison to Mesozoic biotas, which typically harbour precursors of modern lineages. The K/T mass extinction, less severe than that at the end of the Palaeozoic, was responsible for the demise of some important higher-level groups such as rudistid clams, ammonites, dinosaurs, and many seed-plant and insect lineages. Since the expansion of angiosperms during the Middle Cretaceous, animal herbivory has continued unabated, marked by an end-Cretaceous extinction, subsequent reduction in Palaeocene associations, and a rebound during the latest Palaeocene to Early Eocene prior to the diversification of mid-Caenozoic grasslands.

2.3.2 Taphonomic filters and fossil deposits

Taphonomy is the study of the physical, chemical and biotic events that affect organisms after death, including pre-burial, burial and post-burial processes that transform the original living community into an entombed death assemblage that may be encountered by palaeobiologists many aeons later. The fidelity to which the preserved assemblage actually resembles the source community is an issue in discussions of the quality of the fossil record (Behrensmeyer et al. 1992).

Terrestrial *Lagerstätten* deposits occur episodically in the fossil record and are spatiotemporally separated from each other by extensive barren strata (Box 2.2). Nevertheless they represent particularly informative glimpses into once-living, diverse assemblages, and constitute most of our knowledge of many organismic groups and their past environmental settings. They provide virtually all of the documentation for associations between plants and animals. Most *Lagerstätten* are fine-grained, stratified deposits that contain two-dimensional compression fossils that originate from lake, deltaic or river-associated environments. Permineralized deposits, featuring three-dimensional impregnation of fossils typically by carbonate, silica or phosphate, are rarer but include fossils such as coal balls and petrified wood. Amber deposits, like compressions in shales and other fine-grained sedimentary

Box 2.2 *Lagerstätten* and modes of preservation

Through centuries of fossil exploration it has become clear that there are two basic, albeit intergradational, modes of fossil preservation. The most commonly encountered mode is that of fossil deposits whereby organisms with resistant hard parts survive, mostly skeletonized invertebrates in marine deposits and bone-bearing vertebrates in terrestrial deposits. Plant leaves and insect wings as compression or impression fossils are typically encountered in such deposits of mediocre preservation (Fig. 1, *right*).

A very different and special mode of preservation is known as a *Lagerstätte* (plural, *Lagerstätten*) which is an extraordinarily well-preserved fossil deposit that merits considerable scientific interest. Many deposits with superb preservation of plant–animal associations – such as the post-Devonian grey arrows in the time scale on p. xii – are *Lagerstätten* characterized by anatomically spectacular detail resulting from entombment under special conditions of rapid burial and chemical sealing, followed by minimal subsequent decay, usually under anaerobic conditions. Such fossils are characterized by greater than usual resolution of fossil anatomy (Fig. 1, *left*). Postmortem alteration within more typical deposits includes chemical changes of fossil skele-

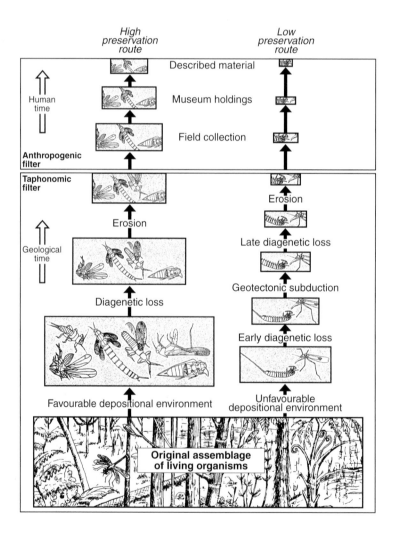

Figure 1

continued on p. 35

tons by mineral replacement or removal by ground water, mineralic phase recrystallization under conditions of elevated temperature or pressure, and sediment compaction. A major source of taphonomic loss is erosion from initial deposition in areas of high elevation that have become uplifted. Degradation of the fossil record occurs in geotectonically active regions where there is faulting, folding or magnetic intrusion.

Geographically, much of the terrestrial fossil record has been obliterated by tectonism and erosion. Geotectonic subduction and associated magmatism at leading edges of continental plates, where they abut with the denser crust of oceanic plates, has resulted in physical removal of transported terranes and the diagenetic alteration of remaining land

masses such that fossil biotas are unrecognizable. For areas of montane uplift in continental interiors there has been accelerated erosion of the sedimentary record. Ironically, major crater-impact structures that destroy the previous sedimentary record also act as subsequent, long-term sediment traps, often containing spectacular *Lagerstätten* (Crepet 1996). By contrast, oceanic islands are geologically ephemeral and undergo either geotectonic subduction or eustatic subsidence and become inundated (Wagner & Funk 1995). The geologically permanent continental centres of lowland equatorial latitudes are frequently unavailable for exploration because of the destructive effect of vegetation on surface exposures. Similarly, present-day polar latitudes are covered by permanent ice caps and are inaccessible.

---

rocks, are also important, although the oldest dated amber with insect inclusions is approximately 125 Ma, representing only the last 30% of terrestrial plant and arthropod history, useful mostly for the earlier evolution of extant clades. Lake deposits, by contrast, have the greatest geochronological persistence, extending about 280 million years, into the Late Palaeozoic, and are critical for discerning premodern plant–animal associations. Special trap deposits such as fissure fills and asphalt pools are unique ephemeral environments generally confined to the Late Caenozoic (Labandeira 1999).

2.4 Approaches towards the study of past associations

The use of two types of approaches — phylogenetic versus ecological, and palaeobiological versus biological — collectively demonstrate much promise for integrating complementary ways for understanding the origin and evolution of plant–animal associations.

2.4.1 When phylogeny matters: mapping ecological attributes onto clades

A central goal of biology is to establish as accurately as possible an explicit relationship among organisms, using methodology that minimizes ad hoc assumptions and is consistent with an understanding of their molecular and morphological structure and occurrence in the fossil record (Funk & Brooks 1990; Fox et al. 1999). A cladogram that is well corroborated by available and diverse

evidence can be considered as a hypothesis of organismic relationships (Funk & Brooks 1990), but equally as a starting-point for testing ecological concepts derived from other, fundamentally different, types of data (Grandcolas 1997). For plant–insect associations, ecologically coupled and putatively related herbivore taxa can be mapped onto a phylogeny of their constituent, uniquely derived (monophyletic) plant hosts (Brooks & Mitter 1984). The degree to which there is a phyletic association between a monophyletic clade of host plants and a monophyletic or otherwise clade of their insect herbivores can be evaluated statistically for congruence. Degrees of coordinated association between host plants and their insect herbivores have been documented, ranging from parallel cladogenesis approaching one-for-one matches between host and herbivore (Roderick 1997; Farrell & Mitter 1998) to looser arrangements (Janz & Nylin 1998) to diffuse associations where weak or anecdotal evidence supports occasional associations between related host plants and their related insect herbivores (Armbruster 1992). Typically, estimates for the divergence of radiations of plant or herbivore clades are pegged to a molecular clock, sometimes calibrated to relevant fossil occurrences of plant hosts or insect herbivores, or are linked to biogeographic events (Funk et al. 1995; Janz & Nylin 1998).

Less resolved phylogenetic data, without explicit congruence of host and herbivore clade topology at the species level, can be used for ascertaining the degree of relationship between plants and their presumed animal associates. There are two common techniques that have exploited comparative ecological or temporal data to determine whether one of two or more clades have promoted taxonomic diversification (Mitter et al. 1988). The first technique is whether a key ecological trait, usually a synapomorphy, which was unique to the early phylogeny of one of two sister-groups, has also been associated with significantly greater taxonomic diversity (Coddington 1988). Examples of such a sister-taxon comparison is the study by Mitter and colleagues (1988) which demonstrated, in an overwhelming number of pair-wise contrasts, that herbivorous clades are more diverse than their non-herbivorous sister-clades, and the examination by Dodd and colleagues (1999) that biotically pollinated angiosperm families are more diverse than their abiotically pollinated sister-clades. Also illustrating this method is the study by Farrell et al. (1991), who concluded that increased taxonomic diversity occurs in those plant sister-lineages that singularly possess latex-canal defences to resist insect herbivory. An opposite conclusion was obtained by Wiegmann et al. (1993) when examining whether carnivorous parasitic insects were more diverse than their non-parasitic sister-lineages.

The second technique determines if there is a preferred pattern of appearance in the fossil record of specific related or unrelated clades possessing predicted ecological characters associated with increased taxonomic diversity. An application of this method is the appearance of certain insect-related floral structures in several angiosperm lineages during their Cretaceous ecological expansion, indicative of the acquisition of particular pollination styles or 'syndromes' (Crepet & Friis 1987). On balance, these clade comparisons for detection of the presence or timing in the appearance of clade attributes can provide strong evidence indicating whether particular associations, such as generalized herbivory, pollination or seed-dispersal syndromes, have been associated with increased species-level diversification.

2.4.2 Non-phylogenetic data: measures of herbivore impact and specificity

Not all insights on past plant–insect associations are derivable only from phylogenetic analyses of modern taxa. According to Wing & Tiffney (1987: 204), 'it is possible to infer past ecological interactions through morphological analogies with extant organisms and systems. Such inferences can be made independent of systematic affinity, and permit us to escape the potential trap of assuming ecological features and community structure on the basis of the taxonomic affinities of the organisms in question.' These inferences apply to the more distant fossil record, and recognize the validity of methodological uniformitarianism. Accordingly, the fossil record becomes an independent source of direct inference, principally as a repository of animal-mediated damage to fossil plants. These types of records not only extend biological data derived from modern descendant taxa (Moran 1989; O'Dowd 1991), they also provide primary associational data from the fossil record that record the role of plants and insects in past ecosystems. Such data includes the intensity of herbivore attack, levels of host specificity in particular floras, and the temporal trend of herbivore pressure on bulk floras or on component communities of plant lineages (Futuyma & Mitter 1996). Such data, as well as fossil documentation of the spectrum of recognizable damage types (Stephenson 1992; Wilf & Labandeira 1999), have established trends regarding the spectrum and intensity of herbivore pressure through time.

2.4.3 Palaeobiological approaches

The source of primary data for palaeobiology is fossil material whereas for biology it is information from modern organisms. The boundary between these two disciplines typically relies on whether the preponderance of the research is centred on fossil versus extant material, and is occasionally fuzzy. Figure 2.4 is a summary of the published literature, and reveals that palaeobiological approaches are applicable to the entire 425-million-year terrestrial fossil record whereas biological approaches are generally restricted to the past 150 million years, except for the extension of modern mouthpart classes to the entire fossil record and the assignment of Permian dates for the origin of feeding strategies in a few phylogenetic analyses. A perusal of all 12 types of data reveals a few glaring gaps, notably the Middle Devonian to latest Mississippian, the Permo-Triassic boundary interval and the Late Triassic through Late Jurassic. As expected, the fossil record improves towards the present in terms of the frequency of deposits and the greater percentage of plant/animal associations that are recognizably modern. This phenomenon, termed 'the pull of the Recent' when initially described for taxonomic data (Raup 1972), reflects the increase in quality of fossil assignments during the more recent past.

This evidence for the associations between plants and insects, and to a lesser extent mites, has been partitioned into three types of ecological data: functional feeding groups, dietary guilds and mouthpart classes. Each of these three ecomorphological units is defined by the type of relationship existing between the insect consumer and the consumed food. Functional feeding groups are defined according *to how their food is consumed*, and includes types such as external foliage-feeding and leaf-mining. Dietary guilds are delimited by *what type of food is consumed* and consist of herbivores on live tissue and detritivores consuming decaying litter. In some instances entire organisms are consumed, such as many algae, spores and seeds, while in other cases whole organs with multiple tissues are eaten, such as roots, twigs or leaves, and in other examples a single or a few tissues are consumed, such as sap, cambium or secretory gland products. Mouthpart classes, by contrast, are classified by *what type of structural apparatus is used* to consume food. Mouthpart classes have been defined from the results of a phenetic analysis of modern insects and their relatives, in which 34 distinctive clusters were elucidated (Labandeira 1997), almost all of which were recognized independently in classic morphological studies (Chaudonneret 1990). Mouthpart classes are grouped into broader groupings by shared structural features. These three ecomorphological units provide a vital ecological component to the history of insects and plants independent of taxonomy. Each displays considerable ecological convergence in time and space.

2.4.3.1 Types of evidence for arthropod and plant associations

A full appreciation of the fossil associational record requires an evaluation of the five major types of qualitative evidence: plant reproductive biology, plant damage, dispersed coprolites, gut contents, and insect mouthparts. There are two major preservational modes that are relevant to assessments of evidence type. Two-dimensional compressions are good for preserving flat organs, particularly leaves and floral structures, and are ideal for detecting gut contents and plant damage, such as external foliage-feeding and leaf-mines. By contrast, three-dimensional permineralization and transformation into charcoal preserve structures such as twigs and seeds, and are prone to preserve spheroidal coprolites and thickened plant damage produced by piercers-and-suckers, gallers and

borers. Collectively, these five types of evidence range from the direct, 'smoking gun' of gut contents, where the consumer and consumed are typically identifiable, to the more remote and circumstantial evidence of floral reproductive biology and mouthparts, where inferences are based on functional understanding, usually from modern analogues. Plant damage and dispersed coprolites are within the middle of this continuum, and represent high resolution of targeted plants but poor definition of the insect culprits. The spatiotemporal distribution of these types of evidence is highly variable in the fossil record: compressions provide broad-scale data, although less frequent permineralization can document in considerable detail the plant-damage structure, especially histological features important for inferring the plant-host's response to particular types of herbivory (Labandeira & Phillips 1996). Amber has been minimally used in this context.

The fossil history of these five qualitative types of evidence is illustrated in Fig. 2.5. Each horizontal bar represents a *Lagerstätten*, placed in relative geochronological order, and containing multiple but not all of the five types of evidence for plant–insect associations. In such deposits each type of evidence, when present, provides data that links plant–insect associations to the separate plant and insect fossil records. Assembly of this evidence from the available associational, plant and insect fossil records reveals a solution to a larger puzzle that addresses the trophic network within each of these ancient ecosystems.

2.4.3.1.1 Plant reproductive biology (Fig. 2.4f)
Evidence for arthropod associations from plant reproductive biology originates from a wide variety of deposits through time. One of the earliest examples are megaspores possessing long appendages with grapnel-like terminal hooks from the Late Devonian of England (Kevan et al. 1975) and the Mississippian of Egypt, interpreted as structures for dispersal by arthropods. In coal-swamp forests of Late Pennsylvanian age, medullosan seed ferns bore unusually large prepollen, which considerably exceeded the range for wind-dispersed pollen, implicating insects as transport agents (Taylor & Millay 1979). This inference of arthropod transport of prepollen is buttressed by piercing-and-sucking damage of male organs that contain the same prepollen type (Labandeira 1998a) and the probable presence of this same prepollen type in dispersed coprolites. All three disparate types of evidence occur in

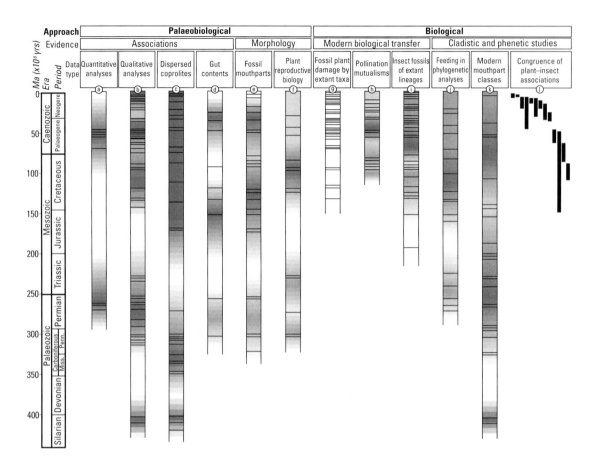

Figure 2.4   Summary of palaeobiological versus biological approaches in examinations of the temporal dimension of plant–animal associations. The types of data and their categorization at the top are a representative sample of the literature. The data of fossil occurrences are horizontal lines within vertical bars. Palaeobiological approaches typically span the terrestrial Phanerozoic record, but there are important earlier Palaeozoic and mid-Mesozoic gaps. By contrast, biological approaches, with the exception of modern mouthpart classes and perhaps feeding inferences in phylogenetic analyses, best describe the Cretaceous and Caenozoic, and are approximately equivalent to the duration of most of the larger extant clades. Abbreviations: Sil. = Silurian, Miss. = Mississippian, Penn. = Pennsylvanian, Neog. = Neogene.

   Palaeobiological approaches (a to f), are given as follows: (a), quantitative analyses; (b), qualitative analyses; (c) dispersed coprolites; (d) gut contents; (e) fossil insect mouthparts; and (f) fossil plant reproductive biology. Biological approaches are partitioned as follows: (g), assignment of highly stereotyped damage in the fossil record to extant taxa; (h), the extension of modern pollination mutualisms by reference to diagnostic plant and insect morphological attributes in fossils; and (i), inferences of plant–insect associations based solely on insect fossils of modern lineages and modern ecological affiliations, biogeographical patterns or other supportive biological data. Assessments from phylogenetic analyses of the time of origin of plant-related feeding attributes (j) are typically approximate, and midpoints have been assigned to time interval estimates from the literature. Data for the geochronological extension of modern mouthpart classes are provided in (k). A brief survey of analyses demonstrating congruent associations between plants and insects constitutes a sample of 13, each of which provides estimates for the origin of subclades (l). Data for each horizontal line within the vertical columns, including deposit, age, locality and literature source, will be found in the Appendix.

the same deposit, the Calhoun Coal of Fig. 2.5e. The same plant bore butternut-sized seeds containing abundant and nutrition-laden megagametophytic tissue surrounded by a thick woody layer; – promising candidates for vertebrate endozoochory, although a credible disperser remains elusive.

Like some seed ferns, some Palaeozoic cycadophytes in somewhat younger, Early Permian deposits possessed leaves that had prominent glandular hairs, considered by some as deterrents to herbivory (Beck & Labandeira 1998). In contrast, closely related or conspecific fructifi- cations bore similar glands among their seeds that have been considered as inducements for pollinating insects. The occurrence of a pollination drop mechanism in many of these seed plants, accompanied by a tubular micropyle leading to a basal pollen chamber, is similar to that in extant conifers (Tomlinson et al. 1991) and cycads (Norstog & Nicholls 1997), and may have been a source of nutrition for surface fluid-feeding and pollinating insects. A modification of this attraction mechanism, involving accumulation of fluid from specialized tissue into a depression encircling the seed apex, has been proposed for the Early Permian conifer *Fergliocladus* (Archangelsky & Cuneo 1987).

The micropyle-pollination drop mechanism also occurred among Mesozoic seed plants, including Late Triassic 'gnetophytes' (Cornet 1996) and Early Cretaceous bennettitaleans. However, it is among Middle Cretaceous deposits from the USA Atlantic Coastal Plain and southern Sweden, representing the early angiosperm radiation, that a variety of flowers and fruits have yielded important discoveries regarding the early history of modern types of pollination. This material consists of small, intact reproductive material preserved as charcoalified structures containing remarkably preserved, three-dimensional anatomy and external structure, frequently with in situ pollen and ovular contents. Beetle, fly, and bee styles of pollination have been recognized during this interval, including mid-Cretaceous pollination by anthophorid bees, as evidenced by the presence of oil-secreting nectaries of guttifer flowers (Crepet 1996). Similar types of analyses from Late Palaeocene to middle Eocene compression material have revealed associations between highly faithful insects and modified angiosperm inflorescences, including bee pollination of leguminous brush blossoms. It is notable that these abundant and distinctive floral features indicate the presence of insect pollination styles in the absence of body-fossil support: for example, only one documented bee body-fossil is known from the Cretaceous.

2.4.3.1.2 *Plant damage* (Fig. 2.4b)
Of all types of evidence for plant–arthropod associations, plant damage has the most extensive fossil record, occurring in the widest variety of *Lagerstätten* deposits from the Early Devonian to the late Caenozoic (Box 2.2). Most functional feeding groups are revealed predominantly by plant damage; for surface fluid feeding, pollination and aquatic feeding the evidence principally originates from plant reproductive structure, gut contents and insect

mouthparts. The earliest occurrences of plant damage are Early Devonian silicified stems of primitive vascular plants from the Rhynie Chert of Scotland (Kevan et al. 1975) and at Gaspé and Campbellton, eastern maritime Canada (Banks & Colthart 1993). During the Late Pennsylvanian of Euramerica, carbonate coal-ball permineralizations predominate, originating from extensive equatorial wetlands that contain tissues damaged by external feeders, borers, gallers and piercers-and-suckers (Labandeira 1998a). Permian carbonate and silica permineralizations containing oribatid mite and insect-sized borings are documented for glossopterid and other seed-plant woods, mostly from Gondwanaland continents (Weaver et al. 1997). Much of the subsequent permineralized fossil record contains insect borings and seed predation in silicified material. Permineralized material in Mesozoic deposits overwhelmingly contains beetle borings in gymnospermous wood. By contrast, a more diverse plant-damage spectrum occurs in Caenozoic deposits, typically in angiospermous woods and seeds, and includes damage by termites, ants and wasps, flies and especially beetles, many identifiable at the family or genus level. The fossil record of permineralized plant damage is largely decoupled from the compression/impression fossil record and each documents the history of different functional feeding groups.

The fossil record of plant damage from compressions and impressions is more persistent and abundant than that of permineralizations. The earliest major deposits are latest Pennsylvanian to Early Permian, representing riparian or estuarine deposits close to sea level or as humid settings marginal to peat accumulations (Labandeira 1998a). All major Palaeozoic plant taxa are represented in these deposits, including sphenopsids, ferns, seed ferns (such as medullosans, glossopterids and gigantopterids), cordaites, conifers and cycadophytes. By the Late Triassic there is a shift towards feeding on advanced fern and seed-fern clades, conifers and bennettitaleans, documented in floras from the south-western USA (Ash 1997) and western Europe (Grauvogel-Stamm & Kelber 1996). Although evidence for plant damage in Jurassic and earlier Cretaceous floras is poor (Fig. 2.4*b*), the record becomes more informative for floras of mid-Cretaceous age where there is considerable evidence for diverse and abundant herbivory (Stephenson 1992; Labandeira et al. 1994) coincident with the ecological expansion of angiosperms. After the K/T extinction, there is depauperate representation of plant damage in Palaeocene floras (Labandeira

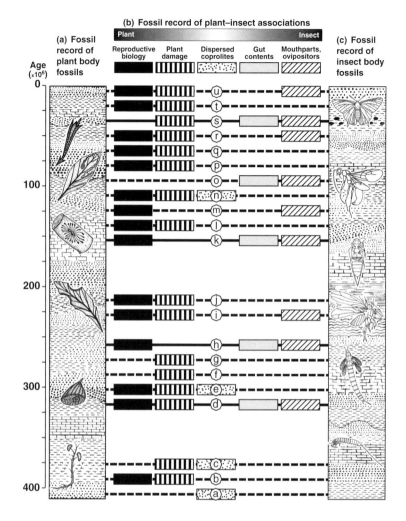

Figure 2.5 Types of evidence for plants, insects and their associations for 21 biotas, a small but representative sample selected from the fossil record and documented from the literature. The fossil record of plant–insect associations (centre panel with links) consists of one to four of the five major categories of evidence for plant–insect associations that have been used to establish direct (solid lines) to indirect links (dashed lines) between plants and insects. These types of evidence range from those centred primarily on the plant (reproductive biology) to those concentrating on both plant and insect (dispersed coprolites) to those focusing on insect structures (mouthparts and ovipositors). These biotas are not a complete inventory but represent most of the best-case examples, placed in geochronological order and approximately pegged to their absolute ages (left).

Each biota is identified by a circled letter and are from oldest to youngest: (a) Downton Castle Formation and Ditton Group, Late Silurian and Lower Devonian; (b) Rhynie Chert, Lower Devonian; (c) Battery Point Formation, Lower Devonian; (d) Carbondale Formation, Middle Pennsylvanian; (e) Mattoon Formation, Late Pennsylvanian; (f) La Magdalena coalfield, Late Pennsylvanian; (g) Waggoner Ranch Formation, Early Permian (Artinskian); (h) Koshelevo Formation, Early Permian; (i) Upper Buntsandstein and Lower Keuper Formations, Middle Triassic; (j) Chinle Formation, Late Triassic; (k) Karabastau Formation, Late Jurassic; (l) Wealden Formation, Early Cretaceous; (m) Yixian Formation, Early Cretaceous; (n) Kootenai Formation, Early Cretaceous; (o) Santana Formation, Early Cretaceous; (p) Dakota Formation, earliest Late Cretaceous; (q) Hell Creek and Fort Union Formations, latest Cretaceous to earliest Palaeocene; (r) Klondike Mountain Formation, middle Eocene; (s) Baltic amber, late Eocene; (t) Passamari Formation, late Oligocene; (u) the Störungszone, Pliocene. See the Appendix for additional locality details and references.

et al. 2002), after which the plant damage record improves considerably from the early Eocene on (Wilf & Labandeira 1999), and continues thereafter in deposits from intracontinental basins (Schaal & Ziegler 1992; Stephenson 1992).

2.4.3.1.3 Dispersed coprolites (Fig. 2.4c)

For sheer quality of preservation and abundance of deposits containing matrix-dispersed coprolites, the Pennsylvanian is the single most important taphonomic window worldwide. Euramerican coal-balls of Pennsylvanian age were deposited as nodular accumulations of plant organs with spectacular histological resolution, revealing peat-litter horizons that contain plant organs and interspersed coprolites with identifiable tissues from major plant taxa (Behrensmeyer et al. 1992). These coprolites range in size from those of oribatid mites (50 to 200 µm in length) to those of large mandibulate insects and perhaps millipedes (10^4 µm in length), and represent detritivorous and herbivorous consumption of almost all accessible plant tissues, including root, stem and foliar parenchyma, epidermis, xylem and phloem, and spores and pollen. There also is evidence for detritivorous wood-boring and coprophagy, principally by mites (Labandeira et al. 1997). The presence of such rich data for inferring Pennsylvanian arthropod diets is attributable to a carbonate matrix forming in situ permineralization which preserves the cellular detail of plants. Unfortunately such a wealth of dietary data is difficult to attribute to particular arthropod culprits since the taphonomic processes that are ideal for retention of plant-tissue detail are apparently inhospitable to the preservation of arthropod chitin.

Arthropod coprolite assemblages are found in other fossil deposits, including some of the earliest known plant communities in the United Kingdom and the littoral of north-eastern North America (Edwards et al. 1995). They are typically found as ellipsoidal, spheroidal or even spiral to somewhat flattened coprolites bearing spores and vegetative tissues. Assemblages of large dispersed coprolites containing dominantly sporangial tissues and lesser amounts of conspecific vegetative tissue are known from the Early Mississippian to Early Permian. After the Palaeozoic, dispersed, frequently pollen-containing, coprolites occur in mid-Cretaceous sediments containing charcoalified flowers and vegetative tissues. In addition, an increasingly rich record of herbivorous dinosaur coprolites is known from the Late Triassic to Late Cretaceous, including finds often with identifiable plant organs

from the Western Interior of North America (Chin & Gill 1996), the United Kingdom and India (Nambudiri & Binda 1989). These coprolites are typically composed of calcium phosphate, although younger coal-like versions are known from mammals at several Caenozoic deposits, including the middle Eocene lake at Messel in Germany (Schaal & Ziegler 1992). Avian and mammalian mummified dung has been recorded in cave deposits from extinct Quaternary herbivores, particularly from biotas such as Hawaii and the Balearic Islands (James & Burney 1997; Alcover et al. 1999), but also continental interiors (Thompson et al. 1980).

2.4.3.1.4 Gut contents (Fig. 2.4d)

Although gut contents are the rarest type of evidence for plant–animal associations in the fossil record, they are the most informative, since the taxonomic identities of both the consumed plant and the herbivore consumer are typically available. Gut contents of plant material are found in both the arthropod and vertebrate fossil records, and provide valuable dietary information that can verify evidence from insect mouthparts, vertebrate dentition and even plant reproductive biology. The most renowned site for gut contents is the Early Permian deposit at Chekarda, in the Ural Mountains of Russia, in which hypoperlid insect taxa previously presumed to be pollinivores based on mouthpart morphology, were subsequently discovered to contain gut contents of predicted seed-plant prepollen and pollen (Krassilov & Rasnitsyn 1999). This site has yielded additional insect taxa from other major clades, including Grylloblattida and Psocoptera, that variously bore monospecific to multitaxal pollen assemblages from seed-fern, cordaite and walchian conifer source plants (Krassilov & Rasnitsyn 1999). Three other sites where insect gut contents have been preserved are Karatau, from the Late Jurassic of south-eastern Kazakhstan, where grasshoppers consumed the pollen of an extinct family of conifers; the Early Cretaceous of Baissa, Russia, featuring xyelid sawfly guts containing pinalean conifer and similar seed-fern pollen; and Santana, from the Early Cretaceous of north-eastern Brazil, in which another xyelid sawfly species consumed apparently angiosperm pollen (Krassilov & Rasnitsyn 1999). Caenozoic insect-gut contents occur in more recent settings, such as the middle Eocene Messel deposit of Germany, late Eocene Baltic amber and early Miocene Dominican amber (Schaal & Ziegler 1992).

Vertebrate gut contents are rarer than those of insects.

The geologically earliest, well-documented example are sizable conifer ovules and associated gizzard stones in a food bolus from the stomach region of a pareiasaur (Munk & Sues 1993). Occasional gut contents have also been found in stomach cavities within dinosaur skeletons, such as conifer needles and unascribed fruits in the Late Cretaceous Canadian ornithopod *Trachodon*. The most spectacular discovery to date has been an abundance of gut contents in numerous plant-associated avian and mammalian taxa from the middle Eocene oil shales at Messel, including foliar remains in the rodent *Ailuravus* and foliar material and grape seeds in the early horse *Propalaeotherium* (Schaal & Ziegler 1992).

2.4.3.1.5 Mouthparts and feeding mechanisms (Fig. 2.4e)
The body-fossil record of animal mouthparts is sporadic but is controlled by the dominant *Lagerstätten* representing a diversity of taphonomic modes. Although geologically more recent amber deposits perhaps offer the best resolution of insect mouthpart structure, other preservational types, including silica permineralization, ironstone nodules and especially fine-grained compression deposits offer valuable data, particularly in the older fossil record (Shear & Kukalová-Peck 1990; Labandeira 1999). In several permineralized and compression deposits mouthpart structure can reveal considerable detail of robust individual elements, but also details of the setation and lobation for less sclerotized elements and the construction of multi-element mouthpart complexes (Labandeira 1997). A related, alternative approach is based on geochronological extensions into the fossil record of modern mouthpart classes. This tack, detailed in Section 2.4.3.4, provides a parallel system of inferring the history of mouthpart types.

Studies of fossil vertebrate chewing have focused on biomechanical analyses, as compared to the more descriptive documentation of arthropod mouthparts that are typically based on modern descendants. Within vertebrates, the relatively early origin of herbivory has been advanced from reconstructions of jaw movement, tooth-shearing planes and dental microwear striations for Permian synapsids and reptiles (Hotton et al. 1997; Sues & Reisz 1998). These have also been applied to a wide variety of herbivorous dinosaurs and other reptile-grade groups (Farlow 1987; Weishampel & Norman 1989). In particular, considerable effort has been directed towards understanding the complex mechanics of ornithopod and ceratopsian chewing (Weishampel & Norman 1989; Sereno 1999) as well as the processing of large quantities

of high-fibre food by tooth-poor sauropods (Bakker 1986; Farlow 1987). The chewing mechanics of early Caenozoic mammalian taxa have been examined for multituberculates, primates and other mammals, although much of this inference relies on research from modern descendants (Rensberger 1986).

2.4.3.2 Insect functional feeding groups

Functional feeding groups can be sorted into 14 basic ways that insects access food (Fig. 2.6). The concept of the functional feeding group was formalized initially from ecological studies of freshwater insects (Cummins & Merritt 1984), although the concept has also been used informally to refer to the varied feeding modes of terrestrial insects (Lawton et al. 1993). Functional feeding groups, like mouthpart classes, are ancient and originated independently multiple times within unrelated insect clades (Cummins & Merritt 1984; Labandeira 1997).

2.4.3.2.1 External foliage feeding (Fig. 2.7)
External foliage-feeders comprise the larval and adult stages of mandibulate insects that consume the entire or partial thickness of live leaf tissue from the outside. Several major subtypes of external foliage-feeding are recognized, namely margin-feeding, characterized generally by semicircular excavations of leaf-margin removal; hole-feeding, whereby interior circular or polygonal portions of the leaf are excised; and skeletonization, which features consumption of a non-marginal part of a leaf with venation remaining, often as a latticework of fine or coarser veinlets. Bud-feeding is a specialized type of external foliage-feeding that is caused by a larva tunnelling through imbricate leaf blades that are folded within a bud, leaving a characteristic pattern of symmetrical holes once the young leaf has unfurled. Free-feeding is an extreme type of external foliage-feeding whereby most of the leaf is consumed, with only major veins and occasional flaps of leaf tissue remaining. These basic types of external foliage-feeding and their geochronologies are illustrated in Fig. 2.6, and are compared to generalized feeding, equivalent to consumption of dead foliar tissue. In well-preserved Cretaceous and Caenozoic angiosperm-dominated floras, there are approximately 30 distinct damage types of external foliage-feeding, ranging from generalized bite-marks on margins to highly stereotyped and often intricate patterns of slot-hole feeding; earlier floras have fewer recognizable types of damage. Most

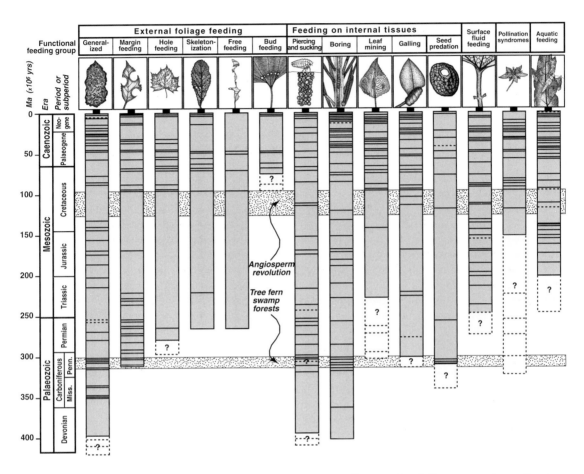

Figure 2.6 The fossil record of plant-associated insect functional-feeding groups. Horizontal lines on each vertical bar represents a datum collected from the literature of fossil plant–insect associations, and to a lesser extent, the fossil record of relevant plant and insect occurrences. Solid horizontal lines are well-supported data; dotted horizontal lines indicate probable or possible occurrences. Darker vertical bars encompassing the horizontal lines represent the geochronological duration of each functional-feeding group; less compelling earlier occurrences are indicated by lighter shading and question marks. Each functional-feeding group is included within a more inclusive feeding category based on terrestrial versus aquatic and internal versus external feeding modes. Pollination syndromes are from plant-reproductive and insect structural features that indicate associations, particularly mutualisms; they include interactional features that transcend the mere consumption of spores and pollen. Included under 'borings' are Early and Late Devonian tunnels and galleries in the 'woody' stroma of *Prototaxites*, a large, enigmatic, lignin-bearing fungus. For delimitation of functional-feeding groups see Coulson and Witter (1984) and Cummins & Merritt (1984). Abbreviations: Miss. = Mississippian; Penn. = Pennsylvanian; Neog. = Neogene. Documentation is available in the Appendix in figure captions for functional feeding groups (Figs. 2.7–2.13 and 2.15).

extant taxa of external foliage-feeders are immatures and adults of almost all species of Orthoptera, Phasmatodea and Lepidoptera, and a lesser percentage of Coleoptera and Hymenoptera. Palaeozoic external foliage-feeders were probably the Protorthoptera, Paratrichoptera and some Hypoperlida (Fig. 2.2).

Although generalized detritivory extends to the Early Devonian, based on dispersed coprolites, there also is evidence for external chewing on stems, based on plant damage with response tissue (Fig. 2.7a,b). Leaves were a subsequent development, and the fossil record

documents margin-feeding of foliage during the Middle Pennsylvanian (Labandeira 1998a). Hole-feeding, skeletonization and free-feeding were present during the Early Permian as well (Fig. 2.7c–h), although earlier origins are likely. Early external foliage-feeders were preferentially targeting diverse seed-ferns with fern-like foliage, large foliose leaves of gigantopterids, and to a much lesser degree ferns and cycadophytes; conifers were virtually free from attack. These functional feeding groups reappear on Middle and Late Triassic vegetation (Grauvogel-Stamm & Kelber 1996; Ash 1997), principally on ferns, but are

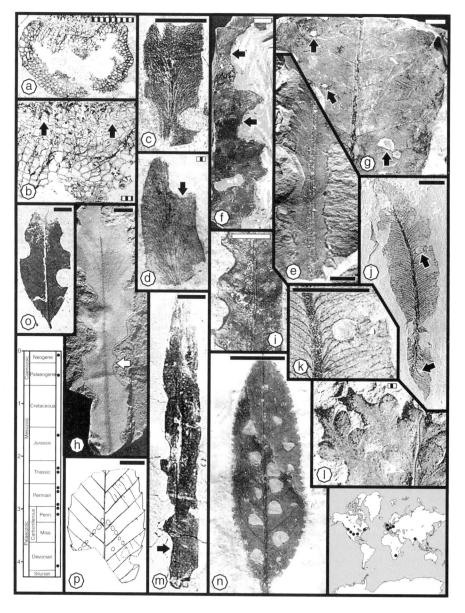

Figure 2.7 The fossil history of insect external foliage-feeding. (a) Transverse section of a permineralized specimen of a Lower Devonian trimerophyte, showing cortical collenchyma and wound response at top; inner tissues missing. (b) Detail of wound response tissue in (a), indicated by arrows, suggesting surface grazing. (c) A pinnule of a Middle Pennsylvanian seed-fern displaying cuspate margin-feeding excisions. (d) A Late Pennsylvanian seed-fern exhibiting a bite mark (arrow) attributable to an external feeder. (e) A Lower Permian specimen of a seed-fern leaf showing extensive, scalloped, margin-feeding (arrow). (f) A Lower Permian cycadophyte displaying extensive margin-feeding (arrows). (g) From the same deposit as (f) is a gigantopterid plant displaying hole-feeding and surrounding necrotic blotches (arrows). (h) Specimen of an Upper Permian seed-fern showing cuspate excavations (arrow) along the leaf margin and extending almost to the midrib. (i) Short leaf segment of a Middle Triassic cycadophyte leaf, showing three serial, cuspate excavations that have projecting veinal stringers. (j) Leaf fragment of the Upper Triassic filicalean fern showing margin- and hole-feeding traces (arrows). (k) Detail of hole-feeding trace in the upper part of (j), exhibiting a reaction rim. (l) Cuspate margin-feeding on a probable seed-fern from the same provenance as (j). (m) A Middle Jurassic cycad leaf exhibiting margin-feeding. (n) An unidentified angiosperm leaf with hemispheric- to deltoid-shaped feeding holes between secondary veins, from the middle Eocene. (o) Damage by an adult leafcutter bee (Megachilidae) on a middle Eocene rosaceous leaf. (p) Insect bud-feeding on a lower Pliocene chestnut leaf. Scale bars for this and succeeding figures: crosshatched = 10 cm, solid = 1 cm, striped = 0.1 cm, dotted = 0.01 cm (100 μ), and back-slashed = 0.001 cm (10 μ). Additional locality data and source references can be found in the Appendix.

poorly documented in Jurassic and Early Cretaceous deposits (Figs. 2.4, 2.6, 2.7i–m). New foliage types of mid-Cretaceous angiosperms were heavily attacked by external feeders, but also leaf-miners and gallers (Stephenson 1992). Bud-feeding can be traced to the Late Cretaceous and so far has only been found on angiosperms (Fig. 2.7p). This intensification and diversification of external foliage-feeding continued until the K/T extinction, after which floras had generalized and lower levels of damage (Labandeira et al. 2002) until the Palaeocene to Eocene transition, after which many modern damage types are recognizable (Fig. 2.7n,p; Wilf & Labandeira 1999).

2.4.3.2.2 *Piercing and sucking* (Fig. 2.8)

Piercing-and-sucking insects possess specialized mouthparts composed typically of one or two pairs of stylets for penetrating plant tissues and include accessory structures such as a muscle-controlled cibarial pump for creating suction for sap uptake. Piercers-and-suckers exhibit a unique combination of feeding on internal tissues while they are stationed external to the attacked plant organ. These invasive feeders typically target one of three vascular plant tissues—xylem, phloem or mesophyll—and they leave a characteristic pattern of internally and externally evident plant damage (Johnson & Lyon 1991). This type of plant damage has been found in Early Devonian deposits, although evidence for this type of feeding is rare and generally requires three-dimensionally preserved permineralized material. Thus there are a limited number of examples from the fossil record, most of which are Devonian (Kevan et al. 1975; Banks & Colthart 1993) and Pennsylvanian in age (Labandeira 1998a; Labandeira & Phillips 1996), with sporadic occurrences thereafter. Compelling evidence for the presence of piercing-and-sucking damage includes stylet tracks in plant tissue, frequently expressed as multiple radiating probes from a point of origin on the surface; the presence of a stylet terminus indicating the target tissue, with or without a feeding pool; disruption of surrounding, often parenchymatous tissue showing abnormal cellular development; and the presence of typically opaque, acellular material sheathing the stylet track or feeding pool (Labandeira & Phillips 1996). Two-dimensional surface views of punctures in compression material are less informative, revealing only a central perforation and a thickened, surrounding callus or other reaction tissue (Fig. 2.8l).

Insect groups responsible for piercing and sucking have changed significantly through time. They range from unknown Devonian microarthropods (Fig. 2.8a–e) to four orders of later Palaeozoic palaeodictyopteroids (Fig. 2.8f–k; Carpenter 1971; Labandeira & Phillips 1996), which in turn were replaced by basal lineages of Hemiptera and Thysanoptera during the mid-Permian. After the demise of palaeodictyopteroids by the terminal Permian extinction, hemipteran and thysanopteran clades extended their ecological reach of feeding styles onto diverse vascular plants (Fig. 2.8l). Currently these two groups are the predominant piercing-and-sucking herbivores, although proturans, springtails and several families of beetles are also represented, many of which feed on fungal hyphae, algal filaments and mosses (Crowson 1981; Verhoef et al. 1988).

2.4.3.2.3 *Boring* (Fig. 2.9)

Arthropod borers construct tunnels through hard plant tissues, especially wood, but also bark, collenchyma and sclerenchyma. Additionally, the hard, chitinous basidiocarps of some fungi are bored in much the same manner as wood. Boring is overwhelmingly done by oribatid mites, termites and holometabolous insect larvae, the latter two of which bear mandibulate mouthparts for macerating or shredding indurated tissue into smaller particles that may be digested and voided as faecal pellets. These pellets are often formed into internal tunnel coatings or packed as undigested frass behind the advancing borer. Some mites and insects are known to possess a special gut microbiota responsible for the digestion of lignin and other structural polysaccharides found in wood and similar tissues that are typically resistant to enzymatic breakdown (Wilding et al. 1989). By contrast, other insect groups form a mutualism by supplying wood fragments and frass as a substrate, and feeding on portions of the resulting fungi for their nutrition. Modern borers are predominantly termites and larval beetles which variously bore through heartwood, cambium and bark (Crowson 1981). The larval wood-borers of other holometabolous insects include fungus gnats and leaf-miner flies among the Diptera, ghost moths, clearwing moths, carpenter worms and cutworms among the Lepidoptera, and the common sawflies, horntails, wood wasps and carpenter ants among the Hymenoptera (Johnson & Lyon 1991), many of which are cambium- or pith-borers that consume softer tissues. Insect borers inhabit wood in live or dead plants whereas oribatid mite borers almost always occur in dead wood and construct tunnels whose diameters are minimally an order of magnitude narrower.

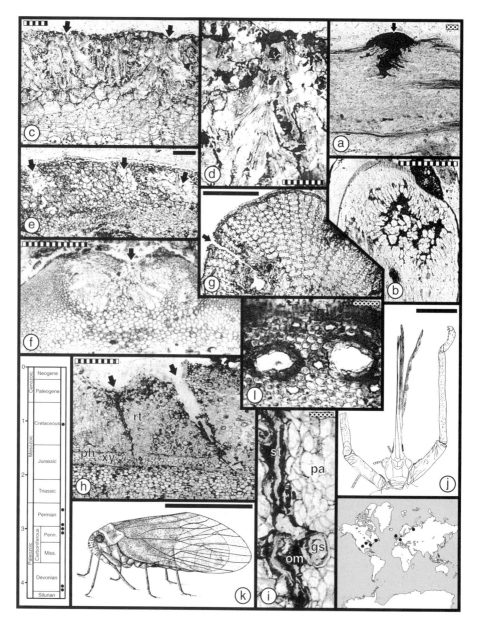

Figure 2.8 The fossil history of piercing-and-sucking. (a) An oblique, longitudinal section of an axis of a Lower Devonian rhyniophyte showing a lesion plugged with opaque material and extending to subjacent vascular tissue. (b) Another Lower Devonian axis in transverse section, displaying hypertrophied cortical cells and associated opaque material. (c) A specimen of the Lower Devonian trimerophyte, exhibiting three sites of piercing (arrows). Below each puncture site is a cone of lysed subepidermal tissue that is floored by unaltered periderm tissue. (d) Detail in (c), showing an enlarged cone of lysed tissue with radiate stylet tracks and damaged epidermal cells at the top. (e) Three piercing wounds on a Lower Devonian trimerophyte, eliciting a light-hued response periderm (arrows). (f) Damage to a Middle Pennsylvanian fern petiole, probably by an insect with stylate mouthparts, showing disorganized tissues enveloping the puncture wound. (g) A stylet track with terminal feeding pool, surrounded by reaction tissue, in a seed-fern prepollen organ from the Late Pennsylvanian. (h) Two stylet tracks targeting vascular tissue (*xy*lem and *ph*loem) of a Late Pennsylvanian marattialean fern. The right track, approximately 3 mm long, is sectioned lengthwise and shows surrounding reaction tissue (*rt*) and a terminal feeding pool. (i) Detail of the left stylet track in (h), showing stylet track (*st*), surrounding opaque material (*om*), penetration of undifferentiated parenchyma (*pa*) and avoidance of large gum-sac cells (*gs*). (j) Head and 3.2 cm-long stylate mouthparts of the Lower Permian palaeodictyopterid insect, *Eugereon*. (k) Reconstruction of the Lower Permian early hemipteran, *Permocicada*. (l) Two stylet probes, with surrounding rims of opaque material, on a Lower Cretaceous cheirolepidiaceous conifer. See Fig. 2.7 for scale-bar conventions; additional locality data and source references can be found in the Appendix.

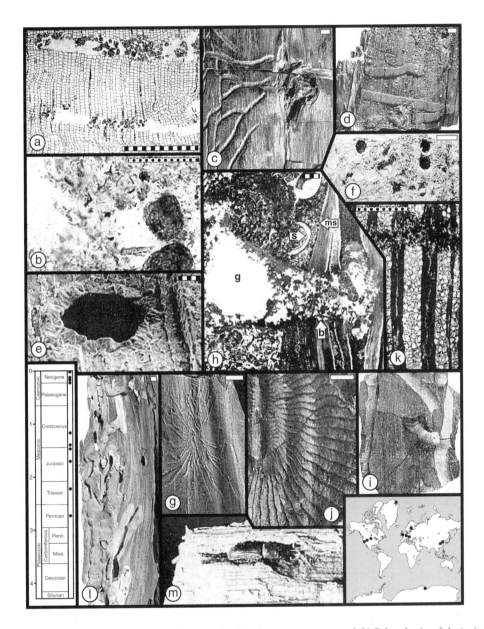

Figure 2.9 The fossil history of borings. (a) Oribatid mite borings in Late Permian gymnospermous wood. (b). Enlarged region of a boring in (a), showing ellipsoidal coprolites and associated undigested frass. (c) Insect (probably beetle) borings in cambium of a Late Triassic conifer. (d) Cambium borings in a conifer as in (c), but fabricated by a different insect species. (e) Scanning electron micrograph of a probable beetle boring in Middle Jurassic coniferous wood, healed by parenchymatous tissue. (f) Late Jurassic beetle borings in gymnospermous wood, assignable to the family Cupedidae. (g) A bark-beetle cambium boring in an unnamed, Early Cretaceous conifer. (h) Probable beetle invasion of the androecium of a Lower Cretaceous bennettitalean, showing consumption of synangial-associated tissues (*s*, synangium), a gallery (*g*) and an exit or entry tunnel (*b*) across the microsporophyll (*ms*) which are bract-like, enveloping structures. (i) A late Oligocene or early Miocene longhorn-beetle boring in unknown wood. (j) A bark-beetle cambium boring in unknown wood of middle Miocene age. (k) An upper Miocene dipteran cambium miner (Agromyzidae) in sycamore wood, displaying tissue damage at the top. (l) A late Pliocene boring of a longhorn beetle in larch (Pinaceae) wood. (m) The bark-beetle *Eremotes* within a boring of unknown Holocene wood. See Fig. 2.7 for scale-bar conventions; additional locality data and source references can be found in the Appendix.

The earliest examples of borers in terrestrial ecosystems are undescribed three-dimensional networks in living tissues of massive lignified fungi from the Early and Late Devonian of eastern North America. These specimens are assigned to *Prototaxites*, now considered a columnar, polypore-like basidiomycete. During the subsequent Mississippian microarthropod-size coprolites are known within plant tissues, but it is from coal-swamp forests during the Pennsylvanian that extensive evidence has been recorded for detritivorous oribatid mite tunnels in a variety of hard plant tissues (Labandeira et al. 1997), particularly cordaite and calamite wood, but also fleshy leaf cushions of arborescent lycopods, other coprolites and seed-fern pinnules. During the Permian there is limited evidence for oribatid mite damage in conifer and seed-fern woods (Fig. 2.9a,b), as well as the earliest appearance of insect-size borings (Weaver et al. 1997). During the Late Triassic to Early Cretaceous various borings assigned to beetles have been documented from a variety of woods and pithy tissues that include conifers, seed ferns and bennettitaleans (Fig. 2.9c–f,h). The earliest bark-beetle damage is probably Early Cretaceous (Fig. 2.9g), and diverse damage has been described from the Late Cretaceous of Argentina. During the Caenozoic, recognizably modern groups are documented, including most of the modern taxa mentioned above (Fig. 2.9i–m; for a review see Labandeira et al. 1997).

2.4.3.2.4 *Leaf-mining* (Fig. 2.10)

Leaf-miners are mobile consumers of soft foliar tissue that do not elicit a major histological response. They are the larvae of holometabolous insects (Connor & Taverner 1997) and rarely mites; the former exhibiting a characteristic ontogenetic pattern beginning with an oviposition site containing an egg either embedded in leaf tissue or laid on the surface. This is followed by larval consumption of a particular tissue layer from which a characteristic frass trail is produced, frequently terminating in an enlarged, terminal chamber used for pupation (Frost 1924). In serpentine miners the frass trail is centrally positioned within a mine that becomes enlarged step-wise as the larva moults and increases in size, and may eventually assume a distinctive trajectory or shape immediately prior to pupation. Blotch mines are seemingly less structured externally, although the frequently obscured frass pattern, the specificity of the plant host and its tissue type, and the mine location on the leaf can be equally stereotyped. Modern leaf-miners comprise the larvae of the four major holometabolous orders: Coleoptera, Diptera, Lepidoptera and Hymenoptera (Hespenheide 1991; Connor & Taverna 1997). Coleopteran leaf-miners occur in seven families, each representing an independent origin (Crowson 1981; Hespenheide 1991). Dipteran leaf-miners also occur sporadically throughout the order, evolved at least seven times, and include more basal members such as crane flies, aquatic midges and root gnats as well as more advanced members such as hover flies, fruit flies, and especially leaf-miner flies (Frost 1924). Within the Hymenoptera fewer than 100 leaf-mining species are known even though this life-habit evolved six times, all occurring within the sawfly superfamily Tenthredinoidea (Connor & Taverner 1997). The Lepidoptera overwhelmingly contains the majority of leaf-mining species, but unlike other orders characterized by multiple originations of relatively derived taxa from external feeders, the mining habit in moths probably originated once and is a primitive feature for the order (Kristensen 1997; Powell et al. 1999). The most basal lepidopteran lineage is similar to Mesozoic forms occurring in deposits as old as the Early Jurassic, whose larvae are mandibulate external feeders. Apparently leaf-mining originated in the next derived clade, the Heterobathmiidae, and was retained plesiomorphically during subsequent major cladogenetic events, although many lineages evolved external feeding as a secondary life-habit (Kristensen 1997; Connor & Taverner 1997).

Leaf-mining may be one of the few herbivore functional-feeding groups that originated during the Mesozoic, as Palaeozoic examples are currently unconvincing either histologically as permineralized sections or as foliar surface compressions with diagnostic features (Labandeira 1998a). Suspect Palaeozoic occurrences include U-shaped interveinal areas on seed-fern pinnules of Middle Pennsylvanian to Early Permian *Macroneuropteris* from several Euramerican localities that could be indicative of mining activity (Fig. 2.10a). Although these necrotic zones resemble blotch mines, no requisite anatomical evidence is present. Similarly, short sinusoidal features on pinnules of the seed-fern *Autunia* may be mines (Müller 1982) although they lack frass trails, discrete step-wise width expansions and terminal chambers. The earliest reliably identifiable leaf-mine was initially attributed to a nepticuloid lepidopteran, originates from the Jurassic/Cretaceous boundary of northern Australia, and occurs on the seed-fern *Pachipteris* (Fig. 10b,c). This occurrence (Rozefelds 1988) suggests that at least one

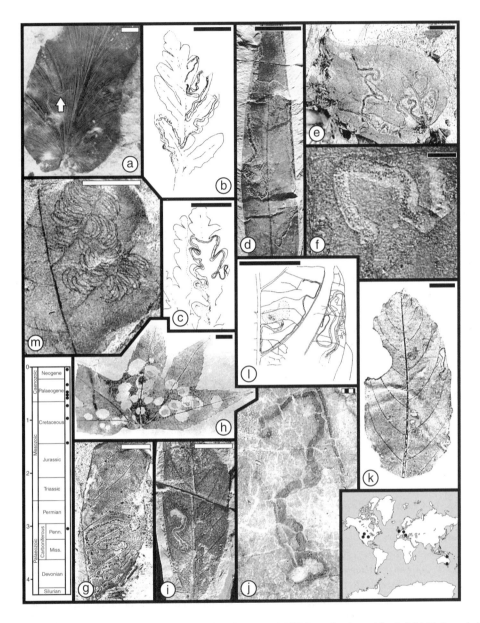

Figure 2.10 The fossil history of leaf-mining. (a) Possible U-shaped blotch mine on a Middle Pennsylvanian seed-fern leaf. (b) A holometabolan serpentine leaf-mine on a latest Jurassic to earliest Cretaceous corystosperm seed-fern. These leaf-mines antedate the earliest documented angiosperms by approximately 15 million years. (c) Another serpentine leaf-mine on a different corystosperm leaf, from the same deposit as (b). (d) A mid-Cretaceous serpentine leaf-mine produced by a moth (Gracillariidae) on a primitive dicot, showing oviposition site, frass trail and pupation chamber. (e) A Late Cretaceous serpentine leaf-mine of a nepticulid moth (Nepticulidae) on a katsura tree leaf (Cercidiphyllaceae). (f) Detail of leaf mine in (e), showing a medial frass trail of particulate coprolites. (g) Portion of an early Eocene mahogany leaflet (Meliaceae) with a serpentine, frass-bearing leaf-mine of a moth (Gracillariidae). (h) Middle Eocene circular leaf-mines of fairy moths (Incurvariidae) on a sycamore-like leaf (Platanaceae). (i) A middle Eocene dicotyledonous leaf bearing a serpentine leaf-mine assignable to the Lepidoptera, probably Nepticulidae. (j) Detail of leaf-mine in (i), showing complete developmental progression from oviposition to pupation. (k) A late Eocene leaf of Lauraceae exhibiting a serpentine lepidopteran mine (Nepticulidae). (l) Camera lucida enlargements of two mines in (k). (m) An upper Miocene dipteran blotch mine (Agromyzidae) on a leaf of Lauraceae. See Fig. 2.7 for scale-bar conventions; additional locality data and source references can be found in the Appendix.

lineage of holometabolous leaf-mining insects was selectively mining seed-fern hosts significantly before the earliest documented angiosperms (Labandeira et al. 1994; see also Powell et al. 1999). During the Middle Cretaceous ecological radiation of angiosperms several leaf-mining lineages of insects became associated with foliar tissues of principally primitive dicotyledonous angiosperms (Fig. 2.10d,e). This expansion of plant/insect associations, at least in North America, was disrupted by the K/T extinction (Labandeira et al. 2002), but slowly rebounded thereafter (Fig. 2.10f–m).

2.4.3.2.5 Galling (Fig. 2.11)

An arthropod gall is an atypically enlarged plant structure that is endophytically induced by a larval or nymphal arthropod, and results in the production of certain nutrient-rich tissues that are eventually consumed. Galls are three-dimensional, conspicuous, often externally hardened structures that can occur on any plant organ and represent the metabolic control of host-tissue production by the encapsulated mite, hemipteroid nymph or holometabolous larva (Shorthouse & Rohfritsch 1992). The galling life-habit originated independently numerous times and includes tarsonemoid and eriophyoid mites; and numerous lineages of Thysanoptera (thrips), Hemiptera, Coleoptera, Diptera, Lepidoptera and Hymenoptera (Shorthouse & Rohfritsch 1992). Insect gallers are highly tissue- and host-specific. Although about 80% of extant galls occur on leaves, the fossil record of galls indicates that the earliest known galls were found on stems or petioles—an ancient feeding mode rooted in Pennsylvanian coal-swamp forests and probably derived from insect borers (Labandeira 1998a).

These earliest, anatomically documented galls occurred on frond petioles of arborescent marattialean ferns (Fig. 2.11c,d) and probably the terminal shoots of calamitalean sphenopsids (Fig. 2.11a,b). Evidence for Permian galls is sparse and unconvincing. Nevertheless there are well-documented examples of spindle-shaped thickenings of branchlets from two taxa of voltzialean conifers from the Middle Triassic of western Europe (Fig. 2.11e). In addition an undetermined gymnosperm from the Late Triassic of the south-western United States bears spheroidal mite-sized galls on leaves (Fig. 2.11f,g). Similar but somewhat larger mite-sized, foliar galls have been recorded on a Middle Jurassic bennettitalean leaf (Fig. 2.11h). After a hiatus of about 65 million years, galls reappear in the fossil record, but now on angiosperms, and

sparingly from the earlier Middle Cretaceous of the eastern United States, later becoming abundant on various taxa from the Dakota Formation of Kansas and Nebraska. Foliar galls from a diversity of Dakota angiosperms occur on midribs, other major veins, petioles and non-veinal leaf-blades, and thus present a modern spectrum of foliar gall types. Subsequent Cretaceous and Caenozoic gall types are elaborations of this theme (Fig. 2.13i–m; Scott et al. 1994), although there is an apparent extinction of certain gall types at the K/T boundary. During the Caenozoic distinctive gall morphologies are occasionally discovered that are very similar to modern gall taxa and their hosts, including the gall wasp *Antron* on oak, and the gall midge *Thecodiplosis* on bald cypress (Fig. 2.11j,m). Fossil galls are found in compression and permineralized deposits (Larew 1992), rarely in amber, and are best preserved in carbonate-permineralized coal-balls where histological detail is similar to the anatomical sections of modern galls (Fig. 2.11c,d).

2.4.3.2.6 Seed predation (Fig. 2.12)

Insects preying on seeds typically penetrate hardened tissues to reach target food reserves, particularly endosperm or other analogous tissues essential for sporophyte survival. Palaeodictyopteroids have been implicated as Palaeozoic seed predators (Shear & Kukalová-Peck 1990), although present-day seed predators include seed bugs, seed beetles, weevils, seed-chalcid wasps and oecophorid moths. Ants are probably the foremost plant associates that use seeds mutualistically, only serving a role in seed dispersal. The fossil record of seed predation extends to the Middle Pennsylvanian in the form of punctured spores and three-dimensional seeds of seed-ferns bearing plugs indicating the presence of a borer (Fig. 2.12a,d), and especially cordaite seeds with circular drill holes that correspond to the width of certain co-occurring palaeodictyopteran beaks (Fig. 2.12b,c). Gingkophyte seeds were used to construct caddisfly cases in Early Cretaceous lake deposits of Mongolia (Fig. 2.12e), and borer damage in dicotyledonous seeds from the Late Cretaceous of Argentina are similar to modern seed-beetle predation (Fig. 2.12f). The Caenozoic has produced a richer record of seed predation, including small holes in seeds of the citrus and legume families (Fig. 2.12g–j).

2.4.3.2.7 Surface fluid-feeding (Fig. 2.13)

Whereas pollinators consume fluids associated with flowers and are involved in pollen transfer, surface fluid-

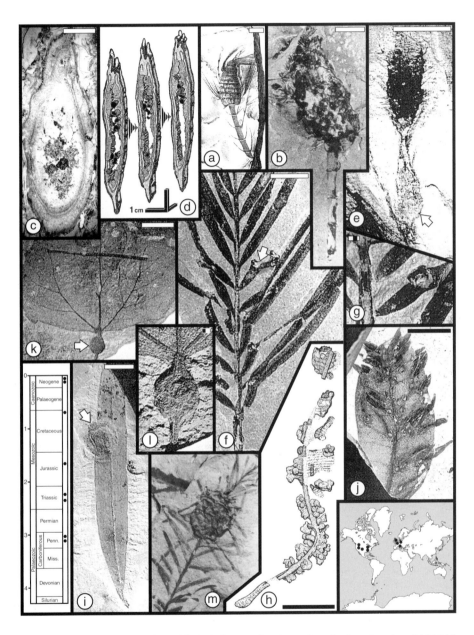

Figure 2.11 The fossil history of galling. (a) A probable Middle Pennsylvanian gall on the terminus of a sphenophyte shoot. (b) A Middle Pennsylvanian 'fructification' reinterpreted as a probable stem gall. (c) A Late Pennsylvanian gall of a holometabolous insect on the rachis of an arborescent marattialean fern. Note barrel-shaped coprolites and frass in the central lumen, and surrounding tufts of hyperplasic and hypertrophic parenchyma. (d) Three-dimensional reconstruction of the same gall type in (c), from the same deposit. (e) An aborted cone of a Middle Triassic voltzialean conifer showing a basal stem expansion interpreted as a gall. (f) Round, oval and deltoid leaf galls on an enigmatic Upper Triassic gymnosperm, expressed as swellings typically occurring about 1 to 1.5 cm from the base of the leaf. (g) Enlargement of a gall in (f) (arrow), showing a deltoid shape and extension beyond the leaf margin. (h) Abundant, bulbous galls on the leaf of a Middle Jurassic bennettitalean, occurring in clusters and preserved three-dimensionally. (i) A large spheroidal gall on a Late Cretaceous angiosperm leaf, similar in form to those produced by gall wasps (Hymenoptera, Cynipidae).(j). Cynipid spindle galls on the leaf of a middle Miocene oak. These distinctive galls are most similar to those of the extant gall wasp *Antron clavuloides*, which parasitizes oaks. (k) A petiolar gall, attributed to a gall aphid, on an upper Oligocene cottonwood leaf. (l) Enlargement of the gall in (k), showing a characteristic stem expansion. (m) A cone-mimicking gall of a gall-midge dipteran on a middle Miocene bald cypress, a swamp-dwelling conifer. See Fig. 2.7 for scale-bar convention; additional locality data and source references can be found in the Appendix.

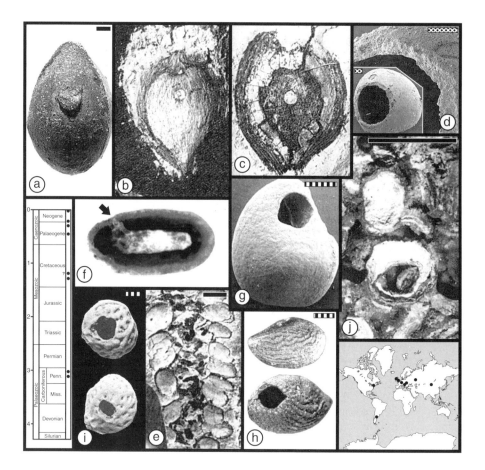

Figure 2.12 The fossil history of seed predation. (a) A sandstone cast of a Pennsylvanian seed displaying a plug infilling a presumptive hole in the seed coat. (b) External surface of a Middle Pennsylvanian cordaitalean seed. The culprit is probably a palaeodictyopterid insect in the same deposit possessing a beak width matching the hole diameter. (c) A similarly bored seed from the same deposit, in longitudinal section. (d) A boring in a Middle Pennsylvanian lycopsid megaspore, showing the entire spore (inset) and a detail of the bored margin. (e) A caddisfly case of gingkophyte seeds from the Lower Cretaceous. (f) Longitudinal section of a permineralized dicotyledonous seed or fruit from the probable Upper Cretaceous, showing a boring assigned to the trace-fossil genus *Carpoichnus*, with exit hole at arrow. (g) A middle Eocene seed of the citrus family (Rutaceae) with an insect exit hole. (h) Pristine (top) and bored (bottom) seeds of another type of citrus family seed, from the early Oligocene. (i) Bored early Miocene stone fruits of a hackberry fruit (Ulmaceae), assigned to the trace-fossil genus *Lamniporichnus*. This damage is typical of weevils. (j) Additional damage of another hackberry fruit, from a Pleistocene hot-spring deposit. See Fig. 2.7 for scale-bar conventions; additional locality data and source references can be found in the Appendix.

feeders have a broader range of imbibed fluids that are produced by plants and many are not necessarily involved with pollination. Common foods for surface fluid-feeders originate from all major vascular plant groups and include floral and extra-floral nectaries, sap flows, exposed subdermal tissue and resulting fluids, the products of guttation, honeydew from sap-sucking insects and deliquescing fungal fruiting bodies (Fahn 1979; Setsuda 1995). Surface fluid-feeders are a diverse assemblage of insects from overwhelmingly adult stages that bear a variety of mouthpart mechanisms for imbibing fluid food. The most common types of fluid uptake are sponging in caddisflies and flies, siphoning in lepidopterans and beetles, and various combinations of lapping, sponging and siphoning among those hymenopterans bearing a retractile maxillolabial apparatus (Chaudonneret 1990; Labandeira 1997; Jervis & Vilhelmsen 2000). Surface fluid-feeders are traceable to the Early Triassic, coincident with the early radiation of basal Diptera and other holometabolous insects (Wootton 1988; Krzeminski 1992), although they probably have Palaeozoic roots (Fig. 2.6). Most of the evidence for surface fluid-feeding

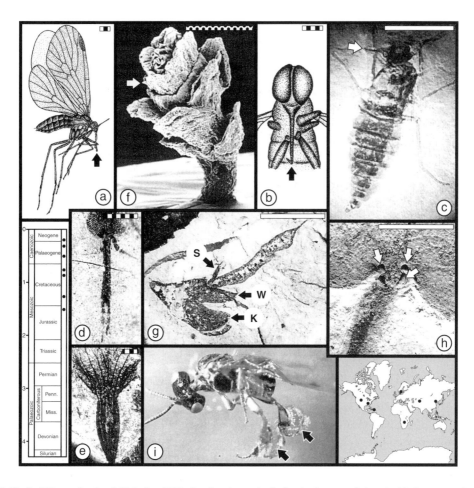

Figure 2.13 The fossil history of surface fluid-feeding. (a) The Late Jurassic scorpionfly *Pseudopolycentropus latipennis*, with elongate mouthparts (arrow), redrawn from a camera lucida sketch in Novokshonov (1997). (b) Head and mouthparts (arrow) of the tanglevein fly *Protnemestrius rohdendorfi*, of the family Nemestrinidae, from the same provenance as (a). (c) Another tanglevein fly *Florinemestrius pulcherrimus*, from the Early Cretaceous of western Liaoning, China. Note elongate mouthparts (arrow) which are similar in form to extant nectar-feeding nemestrinids. (d) Head and mouthparts of the mid-Cretaceous cranefly *Helius botswanensis* (Tipulidae), some extant descendants of which feed on flowers. (e) From the same deposit as (d) is a funnel-shaped flower consisting of fused petal bases and a deep throat, indicative of insect pollination. (f) Scanning electron micrograph of a small charcoalified flower showing a nectary disk (white arrow) above the region of petals, from the Late Cretaceous. (g) A showy, bilaterally symmetrical, papilionoid flower (Fabaceae) from the Palaeocene to Eocene boundary, exhibiting an upper banner petal (*s*), two lateral wing petals (*w*) and bottom keel petals (*k*), associated with bee pollination. (h) Extra-floral nectaries (arrows) at the junction of the leaf blade and petiolar base of a cottonwood (Salicaceae), from the uppermost Eocene and suggestive of an ant/plant interaction. (i) A worker of the early Miocene stingless bee *Proplebia dominicana* (Apidae) bearing conspicuous resin balls (arrows) attached to hind-leg corbiculae. See Fig. 2.7 for scale-bar conventions; additional locality data and source references can be found in the Appendix.

comes from insect mouthparts of body fossils from compression deposits (Fig. 2.13a,b–d) although various plant secretory tissues, including nectaries, floral oil glands and resin-secreting glands, suggest their presence (Fig. 2.13e–i).

2.4.3.2.8 *Pollination* (Figs 2.14 and 2.15)

The present associations between animals and plants are perhaps nowhere highlighted more than by the varied pollination mutualisms which mediate the advantages of outcrossing in plants with the availability of accessible food resources to insects. These plant products include pollen, nectar and oils for nutrition, but floral structures also serve as mating sites and shelter (Proctor et al. 1996). Extant pollinators of angiosperms include members from virtually every major insect order, but notably

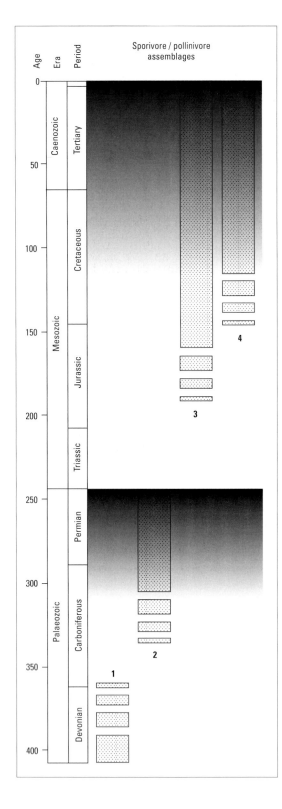

Thysanoptera, Coleoptera, Lepidoptera, Diptera and Hymenoptera. The mutually beneficial reward system that is the essence of pollination can be extended to the mid-Mesozoic, with the best examples being bennettitaleans and especially cycads, based on a variety of fossil and modern evidence (Norstog & Nicholls 1997; Windsor et al. 1999). In fact, the presence of extant cycad–beetle pollination associations, coupled with the fossil histories of cycads and their pollinators, indicate that the origin of some host-specific mutualisms occurred independent of and before the appearance of flowering plants (Farrell 1998; Windsor et al. 1999), particularly from the perspective that modern cycad families extend to the interval from the Late Jurassic to the Middle Cretaceous (Kvacek 1997; Norstog & Nicholls 1997; Artabe & Stevenson 1999).

The fossil record reveals four distinctive, temporal assemblages of spore and pollen consumption based on taxonomic affinities of source plants and their spore and pollen consumers (Labandeira 1998b; Fig. 2.14). The earliest assemblage consists of Late Silurian to Early Devonian spore-rich, dispersed coprolites (Fig. 2.15a,b), although there is no direct evidence of trophically linked arthropods. The second assemblage, Middle Pennsylvanian to Permian in age, consists of orthopteroid and hemipteroid insects associated with pteridophytes and early seed plants, evidenced by dispersed coprolites and gut contents (Fig. 2.15c–h. Evidently this assemblage was extinguished by the end-Permian mass extinction and was succeeded by the third phase during the earlier Mesozoic (Fig. 2.15i–m), dominated by advanced seed-plant hosts and orthopteran and basal holometabolan lineages (Krassilov & Rasnitsyn 1999). Present-day survivors include cycads and their weevil pollinators (Norstog & Nicholls 1997). The fourth, most recent assemblage (Fig. 2.15n–r) began during the mid-Cretaceous and comprises angiosperms and more derived holometabolan lineages (Crepet & Friis 1987). Assemblages 3 and 4 are associated with pollination 'syndromes', and it is likely that

Figure 2.14 The four distinctive assemblages of fossils representing insect consumption of spores, pollen or nectar, based on a variety of evidence, including that illustrated in Figures 2.15 and 2.17. Assemblages 3 and 4 include prominent pollination mutualisms; assemblage 2 is probably associated with pollination mutualisms; and assemblage 1 lacks known arthropod culprits. The presence and intensity of background shading indicates the probable duration and pervasiveness of pollination. (Modified from Labandeira 1998b.)

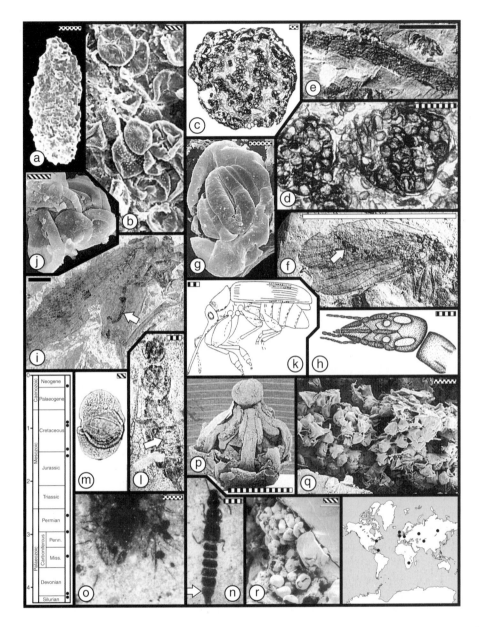

Figure 2.15 The fossil history of spore and pollen consumption, and pollination. (a) An elliptical, somewhat flattened coprolite containing plant cuticle and mostly spores, from the Late Silurian. (b) Contents of a Lower Devonian coprolite containing abundant vascular plant spores. (c) Sporangial fragments and isolated spores of the Mississippian-age coprolite *Bensoniotheca*, attributed to a lyginopterid seed-fern. (d) Two relatively intact sporangia in a *Bensoniotheca* coprolite from the same provenance as (c). (e) The latest Pennsylvanian or earliest Permian coprolite *Thuringia*, consisting of the digested remnants of a seed-fern prepollen organ. (f) The hypoperlid insect from the Lower Permian, containing a plug of pollen (arrow) preserved in its gut. (g) Detail of a *Lunatisporites*-type pollen grain extracted from (f), attributable to a voltzialean conifer. (h) Reconstruction of the head and mouthparts of another hypoperlid insect with pollinivorous habits from the same provenance as (f). (i) A Late Jurassic grasshopper with an arrow indicating a bolus of pollen in its intestine. (j) A cluster of digested *Classopollis* pollen from a cheirolepidiaceous conifer, extracted from the gut of (i) below. (k) A nemonychid weevil from the same provenance as (i), showing a long, decurved rostrum. (l) A Lower Cretaceous xyelid sawfly bearing well-preserved gut contents of pollen (arrow). (m) A pollen grain of *Pinuspollenites*, from a pinalean conifer, found in the sawfly gut in (l). (n) A rove beetle from the Late Cretaceous, displaying pollen transportation (arrow). (o) Enlargement of terminal abdominal region of (n), showing pollen grains trapped among hairs. (p) Lateral view of an Upper Cretaceous charcoalified flower belonging to the Ericales/Theales angiosperm complex. (q) Contents of an anther from the same taxon as (p), displaying the characteristic network of viscin threads, implying a pollinator mutualism. (r) A clump of two types of pollen from the external abdominal surface of a lower Miocene stingless bee, revealed by scanning electron microscopy. See Fig. 2.7 for scale-bar conventions; additional locality data and source references can be found in the Appendix.

Assemblage 2 was too, based on circumstantial evidence such as plant reproductive structure mentioned previously (Labandeira 1998a). Compression-type preservation provides the best evidence for pollination, although this mutualism is the most difficult to demonstrate of all functional feeding groups because of the very indirect nature of the evidence for plant entomophily and insect pollen transfer.

2.4.3.2.9 Aquatic feeding

Three distinctive modes of insect feeding in aquatic ecosystems—scraping, shredding and filtering—each represent a different solution to acquiring plant food that has a varied size-range and spatial distribution within freshwater habitats. Scrapers remove substrate-attached algae and associated material by specialized mouthparts that often include blade-like mandibles (Cummins & Merritt 1984). Scrapers that graze on algal film surfaces of mineralic and organic surfaces include Ephemeroptera, Hemiptera, Trichoptera, rare Lepidoptera, and especially Coleoptera and Diptera, most members of which feed in essentially in similar ways (Cummins & Merritt 1984). Shredders comminute dead and live plant material, including wood, in much the same way as terrestrial external foliage-feeders and borers, typically ingesting plant tissues as coarse particulate organic matter. Shredders are principally represented by members of Trichoptera, Coleoptera, Diptera and subordinately Plecoptera and Ephemeroptera. Filterers collect suspended particulate matter, including phytoplankton, by active sieving or passive screening by means of labral fans or mouthpart-associated brushes. Filterers prominently include larval Diptera, but are found sporadically in other aquatic insect groups. These three modes of aquatic feeding are the principal ways of incorporating dead and live plant material into higher trophic levels within aquatic ecosystems. Much of this material exists in the form of particulate plant detritus, and algae and microbes as part of substrate-attached films.

Insect aquatic feeding has a Permian origin, based on the inferred presence of aquatic immature stages of terrestrial adults recovered from several deposits. Most of these lineages with Permian representatives presumably had aquatic immatures that were predaceous in lotic environments, including Odonata, Megaloptera and Coleoptera (Kukalová-Peck 1991; Wootton 1988). However, there is also sparse fossil evidence for contemporaneous ephemeropteran and plecopteran naiads as detritivores (Kukalová-Peck 1991). Evidence for elevated detritivory

and herbivory in aquatic ecosystems increased during the Middle to Late Triassic, marked by the radiation of nematocerous Diptera (Shcherbakov et al. 1995), including lineages with aquatic stages, based on information from Early Jurassic occurrences, modern ecological associations and rarely preserved Triassic larvae. This interval documents the earliest known appearance of scrapers and shredders (Fig. 2.6). These early aquatic faunas were preserved in Middle to Late Triassic lacustrine faunas from the eastern United States, France and central Asia (Krzeminski 1992; Shcherbakov et al. 1995). They signal the early invasion and trophic partitioning of lotic habitats, including the neuston zone, water column and benthos, which was followed by expansion into lentic habitats during the Early Jurassic (Wootton 1988). This pattern of increasing food exploitation and partitioning, including the use of aquatic macrophytes, wood, phytoplankton and substrate-encrusting algae, continued into the Cretaceous (Ponomarenko 1996), during which the modern spectrum of aquatic feeding styles was established.

2.4.3.3 Insect dietary guilds

Dietary guilds are defined by what an insect consumes and are more phylogenetically labile than a functional feeding groups. Particular types of herbivory have originated from a few to undoubtedly hundreds of times among unrelated insect lineages. For example, the consumption of pollen has been considered a plesiomorphic feature in certain holometabolous insects because of shared, functionally similar mouthpart features in basal Coleoptera, Hymenoptera, Lepidoptera and perhaps related clades (Vilhelmsen 1996). However, pollinivory, with no implication for pollination, also occurs sporadically among numerous phylogenetically disparate groups of recent insects. This evidence is based principally on gut contents, observations of pollen extensively investing surfaces of mouthpart elements and functional morphology. Pollinivores include Collembola, katydids, three families of thrips, several lineages of beetles, several lineages of flies, bees, sphecid and vespid wasps, even two families of butterflies, and other groups not typically associated with pollen consumption (Proctor et al. 1996; Gilbert & Jervis 1998). As discussed previously, fossil pollen-feeders extend this list to orthopterans, hypoperlids, grylloblattodeans, booklice and xyelid sawflies. Other diets can be cited, such as meristematic stem tissue (cambium) or seeds, reinforcing a view that dietary convergence is ubiquitous and results from the opportunistic nature

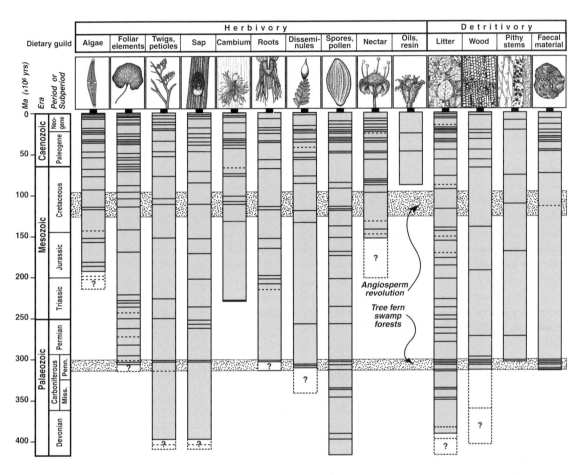

Figure 2.16 The fossil record of plant-associated insect dietary guilds. Horizontal lines on each vertical bar represents a datum collected from the literature of fossil plant–insect associations, and the fossil record of relevant plant and insect occurrences. Solid horizontal lines are well-supported data; dotted horizontal lines indicate probable or possible occurrences. Darker vertical bars surrounding the horizontal data lines represent the geochronological duration of each dietary group; less compelling earlier occurrences are indicated by lighter shading and question marks. Dietary guilds are consolidated into herbivory or consumption of live plant tissue (left), and detritivory or consumption of dead plant tissue (right). Herbivory of spores and pollen refers to evidence for consumption only, and excludes other types of evidence for pollination syndromes detailed in Fig. 2.8. The category 'disseminules' refers to fruits, seeds and analogous reproductive structures. The consumption of wood includes the lignified and hardened tissues of the polypore-like basidiomycete during the Devonian. For delimitation of dietary guilds see Slansky and Rodriguez (1987) and references there. Abbreviations: Miss. = Mississippian; Penn. = Pennsylvanian; Neog. = Neogene.

of insect feeding as exploitable food resources become available in time and space (Simpson & Neff 1983; Lawton et al. 1993).

The history of arthropod feeding on plants began during the Late Silurian to Early Devonian, characterized by coarse dietary subdivision of food into its three elemental forms: solid, particulate and fluid (Fig. 2.16, bottom). Litter detritus and live solid food like stem tissues were undoubtedly eaten by several types of microarthropods, including collembolans, archaeognathans and mites, and possibly myriapods such as millipedes and arthropleurids.

In addition there was consumption of woody fungal tissues by unknown arthropod borers. Evidence for fluid-feeding on plant sap originates from plant damage interpreted as stylet-track damage (Banks & Colthart 1993; Labandeira & Phillips 1996) and associated wound responses (Kevan et al. 1975). Lastly, the oldest of these most fundamental of feeding modes—consumption of particulate matter—is documented from the Late Silurian of England (Edwards et al. 1995) and is revealed by spore-packed spindle-shaped coprolites that indicate the wholesale consumption of spores and perhaps entire sporangia by undetermined arthropods. Thus the chewing of plant organs, the piercing of tissues for sap consumption, sporivory and the boring of hard fungal tissues are the most ancient arthropod diets on land.

By the close of the Pennsylvanian, the expansion of arthropod herbivory had invaded all plant organs and virtually all plant tissues (Fig. 2.16, lower horizontal bar). This expansion of dietary breadth provided a modern cast to the spectrum of insect diets. The best qualitative data comes from the Late Pennsylvanian Calhoun Coal Flora of Illinois. Targeted tissues included xylem and phloem, petiolar parenchyma, root epidermis, sporangial and pre-pollen organ tissues, and wood (Labandeira 1998a). The first evidence for faecal consumption (coprophagy) is also present, with oribatid mites implicated (Labandeira et al. 1997). By the Late Triassic, cambial tissues were added to this inventory but the invasion of lotic and probably lentic aquatic ecosystems by consumers of substrate-attached and suspended algae was also important (Wootton 1988, Ponomarenko 1996). Later, nectarivory became a staple for surface fluid-feeding and pollinating insects, undoubtedly linked to the diversification of advanced seed plants that included the presence of extra-floral nectaries and a transfer in function of the pollination drop mechanism. Last, with the ecological expansion of flowering plants during the mid-Cretaceous (Fig. 2.16, upper horizontal bar), there is documentation for the consumption of oils, supported by highly distinctive floral features in plant taxa related to modern guttifer and malpighia families (Crepet 1996) that are often pollinated today by bees. In summary (Fig. 2.16), while the overwhelming bulk of the 14 plant-associated diet types was in place during the Late Pennsylvanian, it was followed by the addition of 4 novel diet types during the Mesozoic in conjunction with the establishment of freshwater ecosystems and the diversification of advanced seed plants.

2.4.3.4 Insect mouthpart classes

Mouthpart classes, the third ecomorphological unit that will be considered, occur as body fossils and thus provide a direct record of the feeding apparatus insects have used to consume food. While there are some spectacular examples of insect body fossils from *Lagerstätten* that bear well-preserved mouthparts, the historical record of mouthparts is sporadic. However, a complementary method was devised to infer the fossil history of insect mouthpart classes (Labandeira 1997).

Six conclusions have come from a study of the geochronological history of insect mouthparts (Labandeira 1997). The first is that there are 34 fundamental, modern mouthpart types, although subsequently 2 additional (and plant-associated) mouthpart types have been recognized in the Palaeozoic (Fig. 2.17). Of these mouthpart types, about half are associated with plants in major ways. Second, about half the mouthpart classes have originated multiple times when evaluated at coarser taxonomic levels, some several times, thus providing a major role for morphological convergence in the generation of mouthpart morphology and innovation (Cummins & Merritt 1984; Labandeira 1997). Third, there are moderate to strong associations between mouthpart class and general dietary features, such as the form of the food and dietary preferences. Fourth, whereas taxonomic diversity is best characterized as a semi-exponential rise with constant and ever-increasing accretion of taxa towards the present, mouthpart class diversity is logistic and has essentially stabilized since the mid-Mesozoic (Fig. 2.18). Because of this dynamic decoupling, mouthpart data tend to support ecological saturation rather than expanding resources (Labandeira 1997), although there is a modest time lag for mouthpart class accumulation into the earlier Mesozoic. Similar data from functional feeding groups and dietary guilds independently buttress this conclusion. Fifth, a recurring pattern has been the serial derivation of fluid-feeding from generalized mandibulate insects, principally through unique co-optations of mouthpart regions and associated elements to produce haustellate structures such as the lepidopteran siphon, the dipteran sponging labellum and the hymenopteran maxillolabiate apparatus (Chaudonneret 1990; Jervis & Vilhelmsen 2000). Last, there are five major phases present in the temporal distribution of mouthpart classes, each associated with the relatively rapid origination of structural novelty.

From what is known of a sparse Late Silurian to Middle Devonian arthropod record and the phylogenetic relationships among extant basal lineages of hexapods, four or perhaps five ancient mouthpart classes characterize the first phase of mouthpart evolution (Fig. 2.17: *1–4, ?22*; Labandeira 1997). Hexapods and possibly some myriapods possessed these 4 or 5 mouthpart types that were collectively responsible for chewing and piercing dead and live plant tissue, as well as boring through hard fungal tissues. By the Late Pennsylvanian, the second phase of mouthpart evolution was underway, through the addi-

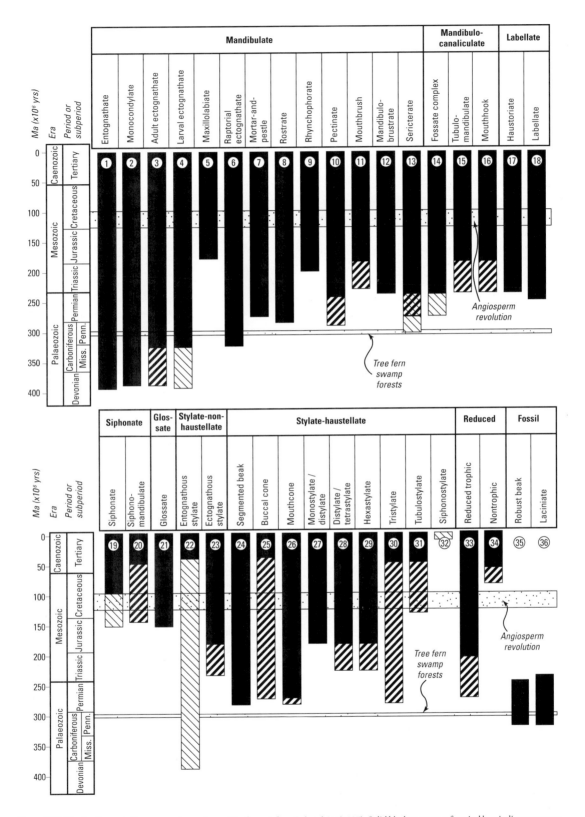

Figure 2.17 The fossil record of insect mouthpart classes. The figure is from Labandeira (1997). Solid black segments of vertical bars indicate presence of a mouthpart class evidenced as body fossils in well-preserved deposits; heavy slashed segments indicate presence based on sister-group relationships when one lineage of a pair occurs as fossils and the sister lineage, whose modern representatives bear the mouthpart class in question, is inferred to have been present. The lightly slashed segments indicate more indirect evidence for presence, such as trace fossils, and the occurrence of a mouthpart class in one life-stage of a species (e.g. adult) when the mouthpart class of interest is inferred to have been present in another life-stage that lacks a fossil record (e.g. larva). Abbreviations: Devon. = Devonian, Carbonif. = Carboniferous, Miss. = Mississippian, Penn. = Pennsylvanian, Perm. = Permian, Trias. = Triassic, Caenoz. = Caenozoic.

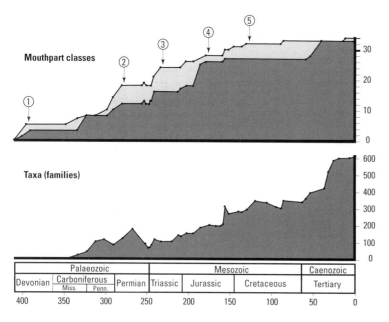

Figure 2.18 Comparison of mouthpart class diversity (upper panel) and family-level taxonomic diversity (lower panel) since the Early Devonian (from Labandeira 1997). Data for both graphs are resolved at the stage level, shown at the bottom sequentially as data points but included in the upper panel to reflect diversity changes. For mouthpart diversity, the darker pattern indicates strong evidence for presence (equivalent to the black portions of bars in Fig. 2.17) and the lighter pattern indicates less reliable evidence (corresponding to slashed patterns of bars in Fig. 2.17). Arrowed numbers refer to the five phases of mouthpart class diversification discussed in the text. Mouthpart data are an updated version of Labandeira (1990); taxonomic data are from Labandeira and Sepkoski (1993).

tion of several mouthpart classes involved in herbivory, including two Palaeozoic-only classes involved in fluid-feeding and spore and pollen consumption, and probably the mouthparts of larval holometabolans (Fig. 2.17: *7, 8, 10, 14, 24–26, 30, 33*; Labandeira 1997, 1998a). By contrast, the third phase of the Early Permian marked a great increase in mouthpart innovation, represented by sap-feeding mouthpart classes from the hemipteroid radiation, and early occurrences of a diverse suite of mouthparts borne mostly by early holometabolous lineages. The presence of morphologically distinct and ecologically separate larval and adult stages associated with holometabolous development resulted in the generation of 7 mouthpart classes (Fig. 2.17: *4, 11–16*). This major structural innovation offered for the first time the presence of 2 distinctive mouthpart classes and feeding strategies encompassed by the same biological species.

The fourth and most expansive phase is attributable mostly to the radiation of basal groups of Diptera and Trichoptera during the Late Triassic to Early Jurassic (Shcherbakov et al. 1995). These groups provided new mouthpart classes involved in the invasion of freshwater habitats by larvae (Fig. 2.17: *5, 9, 11, 12, 15–18, 23, 27–29*), and the increasing reliance on exposed and tissue-enveloped fluid food by adults. The emergence of advanced lineages of seed plants was exploited by highly stereotyped mouthparts borne by adult weevils and small

parasitic wasps, involving the chewing of hardened tissue and the extraction of surface fluids respectively. Stylate penetration of vascular plant tissues and algal and fungal filaments by several independent lineages of small beetles presumably occurred during this interval. Mouthpart types appearing during the fifth phase, from the Late Jurassic to Early Cretaceous, were involved in fluid-feeding on various plant exudates, including nectar (Fig. 2.17: *19–21, 31, ?34*). When expressed as a diversity curve spanning the past 400 million years, there is a linear but stepped rise in mouthpart class diversity from the Early Devonian to the Early Jurassic, where it reached a plateau, followed by only a few subsequent additions (Fig. 2.18). Thus virtually all basic mouthpart innovation, including plant-associated mouthpart classes, was established prior to the angiosperm ecological expansion during the Middle Cretaceous, suggesting that mouthpart classes are attributable to basic associations with seed plants, or vascular plants of the more remote past, rather than the relatively late-appearing angiosperms (Labandeira & Sepkoski 1993).

2.4.3.5 Quantitative analyses

Quantitative analyses of herbivore damage of leaf floras is currently the only way of measuring the intensity of herbivory through time, provided that analysed assemblages are well preserved, sufficiently diverse, abundant and comparable to analogous modern studies. An essential goal in such quantitative analyses is evaluation of the amount of damage inflicted by insects on past floras, as measured by the amount of herbivorized surface area and the frequency of attack of leaves. Site-specific, time-averaged floras, a veritable given in the fossil record, are ideal for providing bulk values for the long-term examination of herbivory in the fossil record and for comparisons with modern floras (Coley & Barone 1996). Although the application of this approach to the fossil record is still in its infancy (Fig. 2.4a), it shows promise for resolving such basic questions as the origin of substantive herbivory during the late Palaeozoic in a variety of environments (Labandeira 1998a). Thus vegetation from an identifiable and geochronologically persistent environment, preserved as taphonomically equivalent deposits, can be tracked vertically through time regardless of taxonomic turnover in the resident host plants or insect herbivores. Variations in foliar attack frequencies and levels of herbivore leaf-area removal, host specificity and insect damage-type spectra provide the basic data for deducing the role of insect herbivores in plant community evolution. A more focused approach can be applied to the component community of a particular plant taxon or lineage through time, in which changes in herbivore type, feeding intensity and plant-host specificity, in response to long-term environmental perturbations, is tracked through time. This latter approach could provide rare, crucial data for evaluating the long-term origin, persistence or extinction of insect associates of particular plant taxa. Many associations between plants and insects are evolutionarily very conservative and ancient (Funk et al. 1995; Futuyma & Mitter 1996; Farrell 1998). If host-plant damage patterns of these ancient lineages are recognizable, a direct palaeobiological dimension can supplement various congruence studies that establish estimates of the origin times for plant-hosts and their insect herbivores.

2.4.3.6 Substrates for insect oviposition and shelter

Although mouthparts are the primary structure by which insects interact with plants, ovipositors also are important (Fig. 2.19). Plant-piercing ovipositors for the insertion of eggs into plant tissues occur in Orthoptera, Hemiptera, Coleoptera, Lepidoptera and Hymenoptera, among others, and are frequently characterized by lateral compression of valves and ridged or sawtooth ornamentation. Ovipositors also are present in exclusively predaceous

dragonflies and in exclusively fossil groups such as the Palaeodictyopteroidea and Hypoperlida (Fig. 2.19a,b), which were predominantly used for penetrating plant tissues. Damage to stem and foliar tissues can be externally distinctive (Fig. 2.19c–j), resulting in a characteristic pattern of linear to crescentic rows of ovate, highly patterned, slit-like scars that elicit a reaction tissue response. The evidence for fossil oviposition originates from plant damage (Grauvogel-Stamm & Kelber 1996) and ovipositor structure (Kukalová-Peck 1991).

Arthropods have used plants extensively for shelter probably since the Early Devonian (Kevan et al. 1975). In many instances the use of a domicile is synonymous with endophytic feeding modes such as leaf-mining, galling, and boring, whereas in other contexts shelter is provided externally by insects such as rollers, tiers and tent-makers, or by case-makers such as bagworms, shield-bearers or caddisflies. These latter types of shelter are seldom identified in the fossil record, except for caddisfly cases fabricated from plant materials, which occur as trace fossils from the later Mesozoic of Eurasia, and in such Caenozoic deposits as the Miocene Latah biota of the Pacific Northwest. Ant–plant mutualisms are poorly documented in the fossil record, and there are no documented cases of aperturate inflated thorns, galleried fern tubers or excavated pithy stems serving as ant domiciles (see also Mueller et al. 1998). Extra-floral nectaries on the petiole of an Oligocene cottonwood leaf (Fig. 2.13h), almost identical in position and structure to ant-tended nectaries on an extant descendant, is the only example indicating past ant–plant mutualisms. Mite leaf domatia, however, have been found in Eocene deposits of Australia, with entombed mites residing in vein-axil pouches (O'Dowd et al. 1991). Last, a surprising pattern is the depauperate representation of sheltered insect eggs on plant surfaces, though they have been described from Permian *Glossopteris* leaves, Middle Triassic horsetail stems and Jurassic fern fronds (e.g. Grauvogel-Stamm & Kelber 1996).

2.4.3.7 The fossil record of vertebrate associations with plants

The amount of live plant tissue assimilated by arthropods is significantly greater than that of vertebrates in virtually all biomes except grasslands (Crawley 1983). The fossil evidence indicates that this arthropod dominance has probably been the case since the establishment of the earliest terrestrial ecosystems. In fact, it was not until the lat-

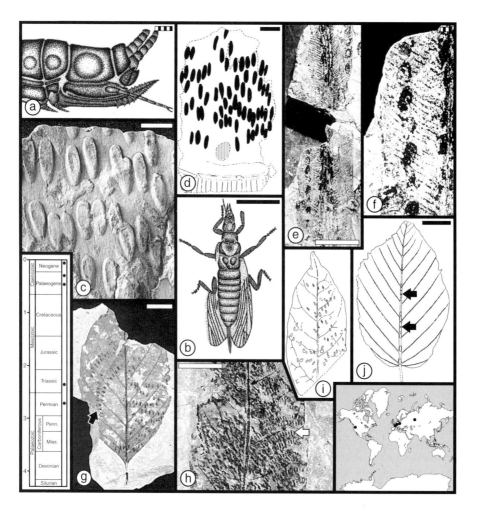

Figure 2.19 The fossil history of oviposition. (a) Reconstruction of the terminal abdominal region of a female diaphanopterodean from the Lower Permian, illustrating a vertically compressed, sawtooth ovipositor used for inserting eggs into plant tissues. (b) A pollinivorous hypoperlid insect from the same provenance as (a). Note the pronounced, flattened ovipositor that was probably used for the endophytic insertion of eggs. (c) Elongate-oval oviposition scars on a Middle Triassic horsetail. (d) Camera lucida drawing of a leaf sheath from another Middle Triassic horsetail species, showing dense, elongate-oval oviposition scars arranged in a zigzag pattern. (e) Probable insect eggs in or on a Middle Triassic cycadophyte leaf. (f) Enlargement of oval oviposition scars in (e) linearly and obliquely placed between the midrib and margin. (g) Oviposition scars of odonatan eggs, inserted as eccentric arcs (arrow) on a middle Eocene alder leaf (Betulaceae). (h) Oviposition scars on an unidentified middle Eocene leaf, similar to and approximately contemporaneous with those of (g), and presumably produced by a odonatan. The arrow refers to an arcuate row of scars. (i) An upper Miocene angiosperm leaf probably of the walnut family, showing odonatan oviposition scars typical of the Coenagrionidae. (j) A middle Oligocene leaf of hornbeam (Betulaceae) exhibiting petiolar oviposition scars (arrows) typical of the odonatan family Lestidae. See Fig. 2.7 for scale-bar conventions; additional locality data and source references can be found in the Appendix.

est Devonian that vertebrates emerged on land (Marshall et al. 1999), for which evidence indicates obligate carnivory. The radiation of early amphibians and reptiles that started during the Mississippian continued into the Pennsylvanian (Behrensmeyer et al. 1992). Direct evidence for vertebrate herbivory does not occur until the latest Penn-

sylvanian to earliest Permian (Weishampel & Norman 1989; Sues & Reisz 1998), about 100 million years after it appeared among mid-Palaeozoic arthropods.

Compared to arthropods, vertebrates are large, typically occur in low abundance in deposits, and possess a very different mechanical apparatus for processing food. For

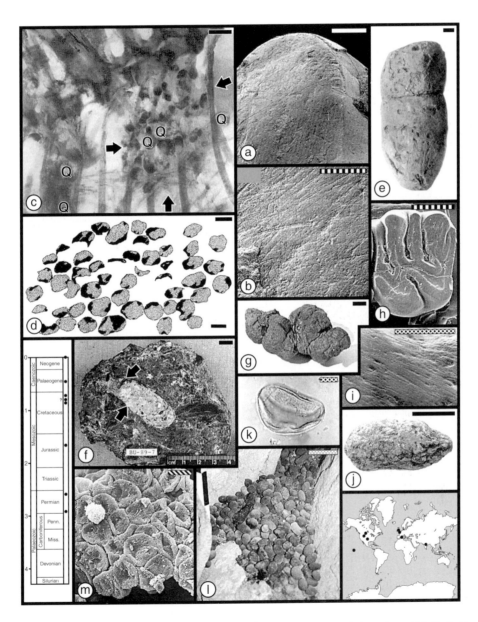

Figure 2.20 The fossil history of plant–vertebrate associations. (a) A crown of a 'molariform' cheek tooth from *Diadectes*, a high-fibre herbivore from the latest Pennsylvanian. (b) Detail of striations in (a) on the wear facet parallel to the long axis of the jaw. (c) Gut contents of the Upper Permian pareiasaur *Protorosaurus speneri*, illustrating a food bolus (delimited by arrows) that contains conifer ovules (dark circular or hemispherical structures) and quartzose gizzard stones (*Q*). (d) A cluster of small, loose pellet-like coprolites attributed to a Middle Jurassic ornithopod dinosaur, each containing abundant bennettitalean leaf cuticle. (e) A Late Cretaceous, herbivorous dinosaur coprolite, showing constrictions and resembling the segmented faeces of extant herbivorous mammals. (f) A Late Cretaceous, herbivore coprolite with two back-filled burrows (one at upper arrow) within a dark groundmass composed of comminuted xylem fragments, and the other light-coloured (lower arrow) packed with intermixed sediment and dung. Both burrows are inferred to have been made by dung beetles (Scarabaeidae), based on burrow patterns. (g) A serpentine dinosaur coprolite displaying conifer twigs, from the Upper Cretaceous. (h) An upper molar of the Eocene rodent *Thalerimys headonensis*, a herbivorous browser; note the broad shelf for seed grinding. (i) A rodent from the late Eocene, with subparallel microwear scratches and pits on upper deciduous molars, suggesting consumption of indurated food. (j) A Holocene coprolite from an extinct folivorous duck of Hawaii. (k) A monolete fern spore from another coprolite from the same locality in (j); ferns were a major component of the diet of this herbivore. (l) A heap of coprolites from an extinct goat-like bovid, found in a Holocene cave. (m) Pollen of the box family (Buxaceae) from coprolites of the bovid mentioned in (l), its major dietary constituent. See Fig. 2.7 for scale-bar conventions; additional locality data and source references can be found in the Appendix.

this reason, the vertebrate fossil record of associations with plants is dominated by coprolites and features of the chewing apparatus (Fig. 2.20). A consequence of large vertebrate size is that consumption of plant organs is frequently complete and not partial as it is among arthropods, leaving minimal evidence from leaves, seeds and other wholly-consumed items. Also, the rarity of vertebrates when compared to arthropods may result in an underestimate of vertebrate importance in their interactions with plants. Lastly, the vastly different and relatively simple construction of vertebrate chewing apparatus, as opposed to arthropod mouthpart complexes, has allowed for inferences in vertebrate food consumption that have stressed the biomechanics of the jaws and associated muscles (Weishampel & Norman 1989), microstructural details of tooth surfaces (Collinson & Hooker 1991) and digestive physiology (Farlow 1987).

2.4.3.7.1 Late Palaeozoic herbivores

At the end of the Palaeozoic, at 245 Ma, three major amniote clades were established on land: synapsids, parareptiles and reptiles, each constituting a significant diversity of lineages (Sues & Reisz 1998). Seven of these lineages possessed herbivorous Permian members with high-fibre diets: diadectomorphs, caseids, edaphosaurs, dinocephalians and anomodonts within the Synapsida; pareiasaurs within the Parareptilia; and captorhinids within the Reptilia. These forms exhibited overwhelmingly large, barrel-shaped thoraces that apparently housed extensive digestive tracts for fermentative degradation of cellulose-rich plant matter by microorganisms. This feature was associated with a posterior expansion of the skull for the accommodation of extensive musculature for jaws closure in the processing plant material. Depending on the lineage, herbivore innovations in dentition included the modification of molariform teeth into transversely expanded grinding or crushing surfaces (Fig. 2.20a,b), the presence of additional rows of teeth in both the upper and lower jaws, or replacement of the teeth by a keratinous beak (Sues & Reisz 1998). Rare finds of gut contents in some fossil skeletons have provided direct evidence for diets, such as conifer and seed-fern seeds in the Late Permian diapsid reptile *Protorosaurus* (Fig. 2.20c; Munk & Sues 1993). These discoveries, in concert with current understanding of seed-fern reproductive biology, indicate that seeds may have been an important dietary component for some large tetrapods (Tiffney 1986).

Anamniote batrachosaurs, which diversified during the Mississippian, have been considered to be omnivores that supplemented their diet with low-fibre plant material (Hotton et al. 1997). However, the best evidence for the earliest plant-feeding vertebrates are the diadectomorphs from the latest Pennsylvanian of North America and Europe (Sues & Reisz 1998), considered as either amniotes or their close relatives (Laurin & Reisz 1995). These taxa bear jutting incisors and a transversely expanded battery of molariform teeth. The unrelated sail-backed *Edaphosaurus* occurs in deposits of similar age and bears additional rows of blunt teeth in both jaws and a disproportionately large, cylindrical trunk when compared to other coexisting vertebrates, strongly indicating high-fibre herbivory. Other Early Permian reptile lineages bear similar features, although it is during the Late Permian that these lineages evolved herbivory, especially dinocephalian and anomodont diapsids in mesic to xeric environments. For dinocephalian diapsids, considerable remodelling of jaw closure musculature and their sites of insertion allowed for articular flexibility in jaw movement, easing the cropping and powerful processing of fibrous plant matter. In a review of Palaeozoic tetrapod herbivory, Sues and Reisz (1998) concluded that two herbivore types evolved during the Late Pennsylvanian to Late Permian: those bearing higher-crowned and leaf-shaped teeth occurring on the external jaw margin, used for shearing and puncturing plant tissue elements; and those bearing crushing to grinding dentitions where occlusion of horizontal tooth surfaces physically degraded plant material.

2.4.3.7.2 Dinosaurs and Mesozoic herbivory

An interesting aspect of Palaeozoic tetrapod herbivores is that they were uniformly short-necked and short-limbed browsers that cropped plant material within a metre to perhaps 2 metres of the ground surface. This trend continued with surviving diapsid and synapsid lineages from the end-Permian extinctions into the Late Triassic, at which time basal dinosaur lineages began their diversification into virtually all major terrestrial feeding niches (Fig. 2.21). Early members of this radiation were prosauropod dinosaurs with elongate necks and probable tripodal stances that increased the browse height several times. This fundamental shift to high-browsing prosauropods, accentuated in the Jurassic with the emergence of much larger quadrupedal sauropods browsing up to 12 metres (Bakker 1986), occurred in concert with a shift from an earlier flora of mostly seed-ferns to a later

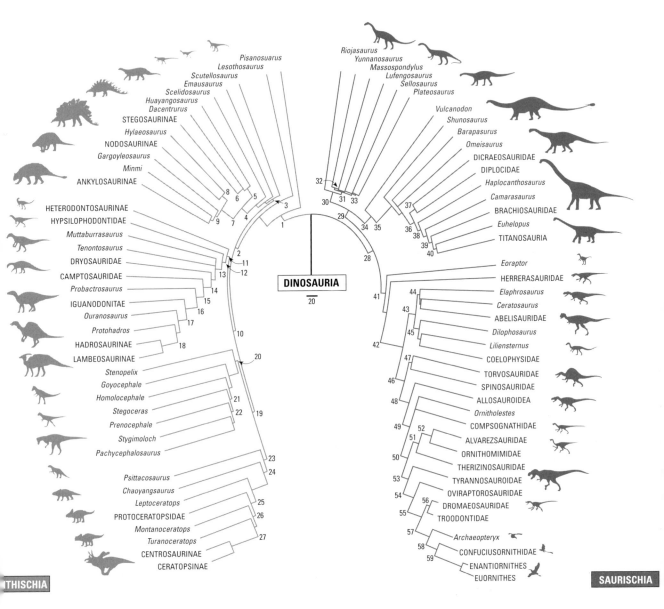

Figure 2.21 Phylogeny of the Dinosauria, showing the relationships among ornithischians (left) and saurischians (right). Plant-feeding clades are the Ornithischia at left and Sauropodomorpha at upper right. For more details, see Sereno (1999), from which this figure is taken. Thickened internal branches are scaled to reflect the number of supporting synapomorphies (scale bar equals 20 synapomorphies). Numbered nodes are as follows, with normal and bold type indicating stem- and node-based taxa respectively: 1, Ornithischia; 2, **Genasauria**; 3, Thyreophora; 4, **Eurypoda**; 5, Stegosauria; 6, Stegosauridae; 7, Ankylosauria; 8, Nodosauridae; 9, Ankylosauridae; 10, Neornithischia; 11, **Ornithopoda**; 12, Euornithopoda; 13, Iguanodontia; 14, **Ankylopollexia**; 15, Styracosterna; 16, **Hadrosauriformes**; 17, Hadrosauroidea; 18, **Hadrosauridae**; 19, Marginocephalia; 20, Pachycephalosauria; 21, **Pachycephalosauridae**; 22, Pachycephalosaurinae; 23, Ceratopsia; 24, Neoceratopsia; 25, **Coronosauria**; 26, Ceratopsoidea; 27, **Ceratopsidae**; 28, Saurischia; 29, **Sauropodomorpha**; 30, Prosauropoda; 31, **Plateosauria**; 32, Massospondylidae; 33, Plateosauridae; 34, Sauropoda; 35, Eusauropoda; 36, **Neosauropoda**; 37, Diplodocoidea; 38, Macronaria; 39, **Titanosauriformes**; 40, Somphospondyli; 41, Theropoda; 42, **Neotheropoda**; 43, Ceratosauria; 44, Ceratosauroidea; 45, Coelophysoidea; 46, Tetanurae; 47, **Spinosauroidea**; 48, **Neotetanurae**; 49, Coelurosauria; 50, **Maniraptoriformes**; 51, Ornithomimosauria; 52, Ornithomimoidea; 53, Tyrannoraptora; 54, **Maniraptora**; 55, Paraves; 56, **Deinonychosauria**; 57, **Aves**; 58, Ornithurae; 59, **Ornithothoraces**. Silhouettes are not drawn to scale.

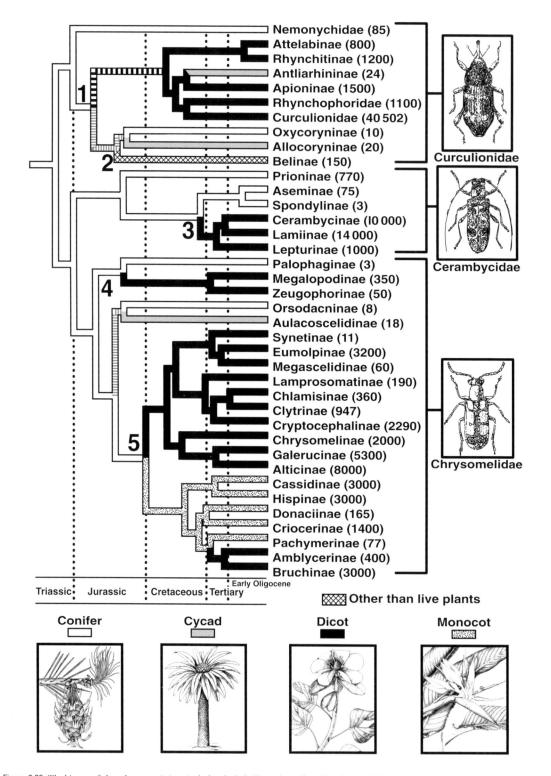

Figure 2.22 The history of plant–host associations in the beetle clade Phytophaga (from Farrell 1998). The Phytophaga consists of weevils, longhorn beetles and leaf beetles (upper right insets), and their seed-plant hosts are characterized broadly (bottom insets) as conifer and cycad gymnosperms and dicot and monocot angiosperms. When plant-host use is mapped onto the phylogeny of major Phytophaga clades, there is a pattern of more primitive beetle lineages using primitive seed plants whereas the more advanced beetle lineages are using advanced seed plants. The five most parsimonious origins of angiosperm associations are numbered. Calibration of each beetle clade's age is provided from fossil occurrences; an estimate of the species-level diversity of each clade is given in parentheses. See Farrell (1998) for additional details.

flora dominated by towering, pole-like conifers. During the Cretaceous the intensified and more specialized cropping of vegetation by low browsers is consistent with angiosperm survival strategies such as rapid regeneration of foliar units, quick recolonization of disturbed habitats and the evolution of seed-dispersal syndromes (Wing & Tiffney 1987).

Dinosaur coprolites offer a direct avenue for inferring herbivory, although their presence in the fossil record is rare. These large, often phosphatized, structures, range from 2 to 10 cm in diameter and up to 15 cm or somewhat more in length (Thulborn 1991). The shapes of dinosaur coprolites are variable, and most have a pinched-off end, formed by a cloacal sphincter muscle, whereas others are flattened, presumably resulting from impact with the ground from a significant vertical drop (Thulborn 1991). The contents of these coprolites range from bennettitalean seeds, seed-fern pollen, gingko foliage and fern fronds among earlier occurrences, to conifer twigs, angiospermous fruits and distinctive angiosperm pollen in later occurrences, some of which have been modified by dung-beetle nest construction (Chin & Gill 1996). A related feature is gizzard-stone concentrations associated with the gut regions of dinosaur skeletons (Bakker 1986), indicating digestive degradation of plants akin to some herbivorous birds. Lastly, gut contents of dinosaurs have been described, such as a Late Cretaceous *Trachodon* containing conifer shoots and abundant small fruits and seeds. These data indicate that Late Triassic to Early Cretaceous dinosaurs typically consumed high-fibre vegetation that relied on physical digestive breakdown, whereas during the Late Cretaceous there was greater specialization, particularly by dentally sophisticated ornithopods on angiosperm foliage and seeds (Wing & Tiffney 1987).

The basic dental modifications for processing high-fibre vegetation that were present during the Permian continued into the Triassic. These feeding modes required considerable oral processing and mastication of food with relatively simple up-and-down (orthal) strokes for slicing, pulping or puncture-crushing of food items in an approximate vertical plane (Weishampel & Norman 1989). In addition, fore-and-aft protraction and retraction for movement of the lower jaw was an efficient mechanism for grinding fibrous food. Orthal puncture-crushers did not survive the Triassic, and some orthal slicers receded into relative obscurity. By contrast, another replacement, the ceratopsians, elaborated slicing through extensive contact surfaces and the possession of a keratinous beak. Orthal pulpers were replaced by stegosaurs, pachycephalosaurids, ankylosaurs and segnosaurs, all of which persisted into the Late Cretaceous; fore-and-aft grinders such as Triassic and Early Jurassic dicynodonts and tritylodonts were succeeded by Late Jurassic to early Tertiary multituberculates. During the Late Triassic to Early Jurassic, prosauropods pioneered the gut processing of vegetation. They were superseded by sauropods during the Middle Jurassic, characterized by slow and heat-generating fermentation of gut compost (Farlow 1987). Other than gut processers, the most significant innovation of Mesozoic food-processing were the transverse grinders (Weishampel & Norman 1989), developed by Late Triassic to Early Jurassic heterodontosaurids, and elaborated especially by the hadrosaurs. For hadrosaurs, there was increased rotation of the upper jaws, shear surfaces were approximately 45° from horizontal, and the dental battery was closely packed with teeth bearing broad contact surfaces (Sereno 1999) — a feature convergent on the transverse chewing stroke of herbivorous mammals.

2.4.3.7.3 Caenozoic patterns of mammalian herbivory

Following the end-Cretaceous demise of dinosaurs (Sereno 1999), there was a Palaeocene interval with low vertebrate diversities that characterized all known ecosystems that have been studied to date. New mammalian clades and some surviving from the Cretaceous diversified during this recovery phase and established associations among plants and other organisms. Although Palaeocene mammalian faunas have a comparatively sparse fossil record, they comprise small-sized members of lineages that gradually increased in body mass. By Eocene times a clearer picture had emerged regarding the associations between taxa stemming from the basal eutherian radiation and modern-aspect plant taxa. One particularly renowned deposit documenting many of these associations is the Messel Oil Shale in Germany (Schaal & Ziegler 1992), of middle Eocene age (48 Ma). This *Lagerstätten* has provided extensive data on superbly preserved gut contents, including histological detail of plant tissues such as leaves, pollen and seeds, in addition to the presence of dental microstructure, dispersed vertebrate coprolites and frequent soft-part anatomy. Less spectacular but coeval deposits from England provide evidence from tooth shape and dental microwear patterns that indicate primate and rodent fruit consumption (Fig. 2.20h,i), including the probable hoarding of nuts (Collinson & Hooker 1991). During this interval there is the earliest evidence for modern mammalian seed-dispersal syndromes (Tiffney 1986), which was preceded during the earlier Eocene by the evolution of relatively large fruits with thick and fleshy outer tissues.

While Palaeocene to middle Eocene mammalian herbivores were dominated by small to medium-sized forms consuming fruit, seeds and leaves, later herbivores were much larger, and invaded the browsing and eventually grazing adaptive zones (Wing & Tiffney 1987). This shift is related to the mid-Caenozoic origin of savanna and grassland biomes concomitant with the ecological spread of grasses. The oldest grasses reliably documented in the fossil record occur at the Palaeocene/Eocene boundary (Crepet & Feldman 1991), although the earliest evidence for a grassland-adapted mammalian fauna is from the middle Oligocene of Mongolia (Behrensmeyer et al. 1992). By the late Miocene, woodland savanna and grasslands were present, exemplified by the diverse Siwalik Fauna on the Indian subcontinent, which included browsing and grazing ungulates. Towards the late Miocene, there was a major worldwide transition from browsers with low-crowned molars to grazers with high-crowned molars. This shift is associated with increasing aridity and an upturn in the abundance of plants from the legume, goosefoot and sunflower families (Behrensmeyer et al. 1992). Although grasses contain considerable silica in their leaves that render folivory an abrasive proposition, the presence of an ever-growing basal meristem probably allowed for a significant increase in herbivore biomass when compared to browsed temperate forest.

Recent technological innovations in the detection of carbon isotopic fractionation in the fossil record has spurred research on determining the diets of later Caenozoic herbivorous mammals (Barrick 1998). This research is based on the fortuitous presence of two basic trajectories by which vascular land plants photosynthetically process the ^{13}C isotope of carbon — the C3 pathway (Calvin Cycle) typical of almost all arborescent plants, herbs and cool-temperate grasses, and the C4 pathway (Hatch–Slack Cycle) characteristic of warm-temperate to tropical grasses. Each of these two suites of plants process ^{13}C such that in C3 plants the ^{13}C isotope is preferentially depleted whereas in C4 plants the same isotope is preferentially enriched. Mammals that ingest assemblages of C3, C4 or a mixed assemblage of plants will deposit in their tooth enamel and bone protein isotope values that will reflect either of these end-member values or an inter-

mediate value for mixed diets. These teeth and bone values can be measured as fossils not only to determine their past diets, but also to assess whether they were grazers or browsers, given supplemental palaeobotanical and palaeoenvironmental knowledge. In addition, recent refinements in measurement have revealed that short-term seasonal to longer-term palaeolatitudinal patterns of herbivory can be ascertained as well. Examples include Cerling and colleagues (1997) who documented the worldwide shift from C3 to C4 plants during the late Miocene (8 to 6 Ma) based on analyses of ^{13}C in the teeth of megafauna from a variety of continents. A more detailed study involved the use of variable distributions of ^{13}C within both C3 and C4 plants, as well as dental microwear striations, to infer that the extinct South American mammal *Toxodon* was not a grazer but fed predominantly on aquatic C4 vegetation in non-forested areas (Barrick 1998).

2.4.3.7.4 Quaternary extinctions

During the Pleistocene (2.65 Ma to 11 000 yr BP), much of the planet underwent severe climatic perturbations from five major episodes of continental and associated alpine glaciation. Continental faunas were considerably reorganized during and after this interval in terms of the dominance and composition of species, at least in North America (Graham 1986). The last reorganization event is associated with significant vertebrate extinction, sea-level rise and major climatic change. This great megafaunal extinction, centred approximately 11 000 yr BP, occurred worldwide, although it was ameliorated in Africa. Much evidence now supports a view that continental species did not respond as cohesive assemblages to these major environmental shifts, but rather individualistically, supported by vertebrate, pollen and generalist insect distributions (Graham 1986; Overpeck et al. 1992; Coope 1995). An important exception to this trend are insects with high host specificity, which responded differently, retaining ancestral plant associations to the present (Labandeira et al. 1994; Farrell 1998; Wilf et al. 2000) or becoming extinct. Herbivorous mammals have less obligate dependence on plant species (Crawley 1983) and thus exhibit greater dietary flexibility during times of major environmental stress. However, there was a great·megafaunal extinction, attributable variously to severe range contraction and environmental degradation or anthropogenic overkill (Thompson et al. 1980; Graham 1986). This extinction event is well illustrated on islands, where severe species reductions occurred during the latest Pleistocene and Holocene (Steadman 1995). Two notable examples of extinctions occurring around 4000 yr BP have been deduced primarily from well-preserved coprolitic and skeletal material in island caves. First is the herbivorous swan-like duck *Thambetochin chauliodous* from the Hawaiian Islands, which fed almost exclusively on ferns (Fig. 2.20j,k); the second is the goat-like *Myotragus balearicus* from the Mediterranean Balearic Islands, which fed dominantly on alkaloid-laden boxwood (Fig. 2.20l,m). Both examples provide evidence of unusual feeding specializations and associated coprolites that demonstrate highly plant-specific diets, rendering these taxa susceptible to extinction during periods of major environmental change.

2.4.4 Biological approaches

The traditional method of using modern biology to understand past plant–animal associations is to assign a known ecological association to an insect-damaged fossil plant that exhibits recognizable and stereotyped insect damage. This procedure of ecological transfer is based on a uniformitarian approach and is typically available for Caenozoic and some Late Cretaceous taxa. The two most prominent uses of this method are identification of stereotyped herbivore damage on plant organs (Fig. 2.4g) and evaluation of suites of distinctively specialized floral characters indicative of unique pollinator presence (Fig. 2.4h). A third use involves the simple presence in *Lagerstätten* of insect body fossils from extant lineages with known ecologies (Fig. 2.4i).

The major alternative way of inferring the history of plant–animal associations using a biological approach is to take advantage of insect or plant phylogeny and identify those nodes which reveal character states that have direct correlates to feeding strategies (Fig. 2.4j). One such type of study is to ascertain the time of origin of particular character states (synapomorphies) on a cladogram that reveal features such as digestive physiologies or alimentary structures. A second methodology employs a phenetic classification of an ecomorphological feature in a group of modern animals of importance to plant–animal associations. The fundamental structural entities of this classification are then mapped onto the phylogenies that bear them, and are subsequently tracked geochronologically by established phylogenetic relationships calibrated by known fossil occurrences (Labandeira 1997). Such studies can provide minimum geological dates for the origin of feeding strategies, including types of herbivory

(Fig. 2.5k). Lastly, assessment of phylogenetic congruence between plant hosts and their animal herbivores, combined with age estimates for the times of origin of their radiations, is also a powerful way of using phylogenies to infer the history of associations (Fig. 2.5l).

2.4.4.1 Transfer ecology from extant descendants of fossil-bearing lineages

Knowledge of modern autecology frequently has been used to infer the life-habits of fossil representatives that have modern descendants (Behrensmeyer et al. 1992). This equivalence is based on three assumptions, namely that a close genealogical relationship exists between the extinct and extant taxa, that there has been minimal ecological change separating the taxa during the intervening time, and that the extant taxon has not been ecologically restricted relative to that of its extinct antecedent (Behrensmeyer et al. 1992). The applicability of transfer ecology is also dependent upon whether the ecologies of documented modern representatives are known well enough, if at all, to comment on the fossil taxa. Perhaps the most serious limitation on the applicability of transfer ecology is whether the time interval separating the past from present taxa is not excessively long so as to invalidate uniformitarian argumentation. In practice, this means that transfer ecology is reasonably sound for lineages that occur in the later Caenozoic, and is applicable to the earlier Caenozoic and later Mesozoic for those lineages that either represent ancient associations or are found in environments and exhibit sufficiently stereotyped morphologies or life-habits for them to be compared to more distantly related extant taxa, but still encompassed taxonomically by the same and more inclusive recent clade. It should be noted that varied evidence indicates that ancient associations between herbivorous insects and their plant hosts should be expected in the fossil record. This prediction is buttressed by studies of genetic variation and physiology that demonstrate pronounced plant-host conservatism in many herbivorous insects (Futuyma & Mitter 1996), the absence of evidence for insect extinction during the Pleistocene glaciations, the most climatically punishing period of the Caenozoic (Coope 1995), and the comparatively long geochronological longevities of insect families when compared to other terrestrial taxa (Labandeira & Sepkoski 1993).

Transfer ecology has been used extensively for body-fossil data, and much less employed for plant–animal associations. In some instances there are excellent matches between a fossil plant–insect association and a particular diagnostic syndrome of herbivore-mediated damage occurring today on the same or a similar host plant (Opler 1973; Wilf et al. 2000). For some palaeoecological data, transfer ecology based on plant–animal associations is more informative than those of the body fossils, since the ecologies are literally preserved in an associational context, such as moth leaf-mines or bark-beetle borings. In fact, such inference from body-fossil morphology or taxonomic identity alone is more difficult, and for many groups such as pollinator and leaf-miner lineages, the trace-fossil record is richer and more ecologically informative than the corresponding body-fossil record.

2.4.4.2 Inferences about feeding from phylogenetic analyses of clades

There are three basic methods for ascertaining the macroevolutionary history of plant–animal associations that employ either a phylogenetic analysis only or a combination of a phylogenetic and phenetic analysis of ecomorphological data. The first method is to locate on a clade those synapomorphies that record relevant feeding characters. The branching sequence of feeding-related characters on a clade can provide data for inferring the relative temporal distribution of feeding strategies. If it is possible to peg these data to a geochronology, then 'absolute' ages also can be inferred. The second method is to map non-phylogenetic ecological attributes onto independently established phylogenies to determine the taxal distribution and times or origin of, for example, feeding types. In these analyses, ecological attributes are logically separate from characters used in phylogeny reconstruction and are thus analysed independently by a variety of techniques or are simply recorded as attributes onto clade taxa. Examples include the evolution of dung use in scarab beetles (Cambefort 1997) and the evolution of the blue and ultraviolet components of colour vision within Hymenoptera prior to the advent of angiosperms (Chittka 1996). An application of this method employs a multivariate phenetic analysis of the ecological data, followed by mapping the resulting phenetic classification onto cladograms or phylograms based on modern taxa (Labandeira 1997), discussed previously. The detection of ecological convergence in attributes is a central goal of such methods.

A third method of investigating the macroevolutionary history of plant–animal associations is analysis of charac-

ter-state evolution within the context of a cladogram, using a variety of techniques. One example is multivariate analyses of phylogenetic and ecological distance matrices to ascertain if and to what extent ecological attributes are associated with phylogeny (Pagel 1999). One of these approaches, used by Gilbert and colleagues (1994) to elucidate the evolution of diet in larval syrphid flies, combines cladistic analysis with ahistorical techniques typically used in ecology. Moreover, depending on the question being asked, one can test for whether phylogeny influences an ecological trait of interest or whether phylogeny and ecological traits co-vary. Since these techniques use phylogenies directly or as part of a combined study, a historical component can now be added to plant–animal associations derived exclusively from modern data. Much of the motivation in such analysis is to search for a common pattern among convergent characters (Coddington 1988) and to evaluate the role of environmental variables and phylogeny in the production of functionally equivalent morphologies. Ecological convergence and the exploration of character change can also be investigated using plant hosts, in addition to the examples of animal herbivores mentioned previously. Such an example is Crisp's (1994) analysis of floral features from a tribe of Australian legumes to determine the origin of bird-pollination syndromes. In summary, analyses of plant- or animal-associated character-state evolution in the context of a cladogram considers ecological convergence as a phenomenon of immense interest demanding explanation, unlike its characterization as a problem in phylogenetic analyses.

2.4.4.3 Phylogenetic congruence of plant hosts and their insect herbivores

The history of observations and theories supporting the idea of mutual phylogenetic association originated along several parallel tracks from 1860 to 1920. During this formative interval, studies on plant and pollinator co-specialization, mimicry (e.g. Darwin 1877), and similarities between co-associated animals and parasite taxa both within and among biogeographic regions, led to a view that specialization could occur among multiple species, causing unidirectional effects on one partner or even bilateral consequences. Although early researchers recognized a conserved phylogenetic pattern of specificity between insects and their host plants, the seminal break occurred in Ehrlich and Raven's (1964) paper in which these phylogenetic constraints were placed in an explicit theoretical context that later became known as escape-and-radiate evolution (Schoonhoven et al. 1998). Ehrlich and Raven's (1964) thesis was based on adversity, borrowed from earlier studies of animals and their host parasites (Thompson 1994), unlike Janzen's (1966) subsequent investigations, which considered the associations between leguminous trees and their ant associates to be true mutualisms. Continued and ever-expanding research in the field of plant–animal associations in general and plant–insect associations in particular—such as the fig and fig wasp, yucca and yucca moth, and milkweed and milkweed beetle systems—has resulted in a large literature and increasing investigation of congruence between host-plant lineages and their associated insect herbivore lineages. An exciting study is Farrell's (1998) analysis of subfamily relationships within the highly diverse and herbivorous clade Phytophaga, consisting of leaf beetles, longhorn beetles and weevils (Fig. 2.22). Farrell's study demonstrated that the most basal and fundamental splits within this clade are ancient associations with pre-angiospermous seed plants (cycads and conifers) of the Jurassic, and that more recently evolved diversification is associated with lower-level taxa radiating in parallel with the angiosperm ecological expansion. Of those congruence studies that have provided time estimates for the origin of such associations, 13 are displayed in Fig. 2.4l. If representative, these studies indicate that there are two clusters for dates of origin: one concentrated in the later Caenozoic, and the other during the later Cretaceous. This bimodal distribution is consistent with independent evidence gleaned from the fossil record, and is attributable to a major spurt of herbivore diversification coinciding with the mid-Cretaceous ecological expansion of angiosperms, succeeded by the extirpation of many specialized associations at the K/T boundary, followed by gradual rebound to Late Cretaceous levels during the earlier Cenozoic (Wilf & Labandeira 1999; Labandeira et al. 2002).

Currently, four major hypotheses account for the origin of plant–insect associations (Schoonhoven et al. 1998; Thompson 1999c). These hypotheses are distinguished principally by the extent of reciprocal genetic selection of one partner on the other, and by the relative timing of events by which a herbivore clade becomes associated with a plant clade (Fig. 2.23). Parallel cladogenesis, or phylogenetic tracking, is an association typically characterized by a one-to-one match between host and herbivore (Fig. 2.23a; Mitter & Brooks 1983). Although there is simultaneous speciation of host and herbivore, it is not caused by coevolutionary genetic feedback. An example is *Nesosydne* planthoppers on Hawaiian silverswords, members of the sunflower family (Roderick 1997). By contrast, sequential evolution (Jermy 1984) indicates that hosts and herbivores lack reciprocal selection and thus do not coevolve. Host lineages undergo phases of diversification independent of herbivore association, and are subsequently, often much later, colonized by herbivores. An example is *Ophraella* leaf beetles on North American hosts of the sunflower family (Fig. 2.23b; Funk et al. 1995). Alternatively, the escape-and-radiate hypothesis (Ehrlich & Raven 1964) invokes a more complex series of events in which mutant plant-host species evolve chemical or physical defences from their herbivores, allowing diversification in enemy-free space. Novel herbivore populations subsequently breach those plant defences, radiate onto the host lineage and colonize individual plant species as descendant herbivore populations become specialized. Later, a plant-host species develops another round of anti-herbivore defence, escaping herbivory but eventually becoming colonized as the process of escape and radiation is repeated. This process is driven by host defences and herbivore counter-defences, and there is not necessarily a one-for-one match between host and herbivore on individual host species. An example is *Tetraopes* longhorn beetles on North American milkweeds (Fig. 2.23c; Farrell & Mitter 1998). Lastly, diffuse coevolution, including diversifying coevolution (Thompson 1994, 1999), is a hypothesis that includes several mechanisms in which herbivore populations specialize on plants based on local processes of reciprocal adaptation, but result in patterns that are not typically phylogenetically congruent. An example is bee pollinators of *Dalechampia* vines (Armbruster 1992), members of the spurge family, in which there is a mixture of limited clade congruence and isolated colonization (Fig. 2.23d). Current evidence favours sequential evolution, although there are several well-established cases of parallel cladogenesis and escape-and-radiate evolution. Diffuse coevolution is difficult to evaluate because of the plurality of mechanisms involved.

2.4.5 Complementarity of both approaches

Palaeobiological and biological approaches both address the same questions regarding the spatiotemporal patterns and processes involved in plant–animal associations.

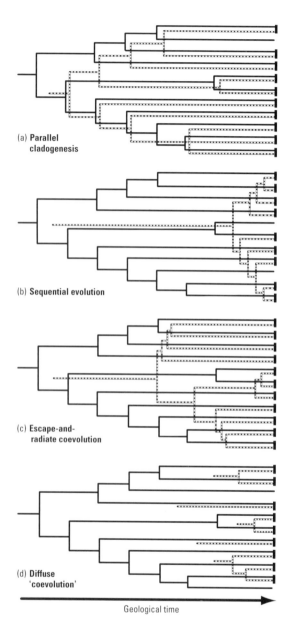

(a) Parallel cladogenesis

(b) Sequential evolution

(c) Escape-and-radiate coevolution

(d) Diffuse 'coevolution'

Geological time

Figure 2.23 Four major hypotheses regarding the evolution of plant–insect associations (Schoonhoven et al. 1998; Thompson 1999d). See text for additional explanation. Examples are idealized, although case-studies may be consulted from the literature indicated. Plant hosts are deployed as black lines, insect herbivores are dashed lines; terminal black rectangles indicate an extant host plant/insect herbivore association. (a) Parallel cladogenesis, or phylogenetic tracking (Mitter & Brooks 1983); for example, *Nesosydne* planthoppers on Hawaiian silverswords (Roderick 1997). (b) Sequential evolution (Jermy 1984); for example, *Ophraella* leaf beetles on North American asteraceous hosts (Futuyma et al. 1995). (c) The escape-and-radiate hypothesis (Ehrlich & Raven 1964); for example, *Tetraopes* longhorn beetles on North American milkweeds (Farrell & Mitter 1998). (d) Diffuse coevolution, including diversifying coevolution (Thompson 1994); for example, bee pollinators of euphorbiaceous *Dalechampia* vines (Armbruster 1992).

direct versus indirect observation, varying ways of inferring deep geological time and a spectrum of methodologies for assessing the history of plant–animal associations.

One context that clouds the distinction between palaeobiological and biological approaches is the quality of evidence. Fossils that include slabs of compression imprints, permineralized tissue or amber can be compared to modern material such as pressed plants in herbaria, insects in alcohol vials or mounted on pins and histological sections of damaged plants. The view that modern material is markedly better-preserved than fossil material is certainly true in the aggregate, but permineralized material, amber and even compressions do offer surprises. For example, permineralized plant tissues from coal-ball floras of Middle and Late Pennsylvanian age often display histological detail similar to microtomed sections of modern embedded plant tissues. Amber contains details of surface ornamentation, setation, genitalia and other key characters that allow taxonomic resolution within a systematic context of extinct and modern taxa. Compression fossils, typically considered as substandard for detection of important diagnostic characters, often present features in *Lagerstätten* that reveal considerable plant-associational data, including mouthpart and ovipositor structure, habitus and important structures such as distinctive hair, surface ornamentation and gut contents. In addition, compression and permineralized material is virtually our only source of ecological data for leaf-mines, galls, borings and other stereotyped feeding patterns, and often provide taxonomically highly-resolved associational data not available from other preservational modes such as

Their distinction principally involves one of scope: palaeobiological approaches have been primarily concerned with the temporal scale of associations variably preserved in the fossil record whereas biological approaches have investigated the spatial wealth and intricacy of associations occurring today. Each approach presents a unique take on reliability of evidence, what constitutes amber. Admittedly these data provide an inadequate inventory of the totality of past associations, but they are robust at higher taxonomic and ecological scales (Figs 2.4, 2.6, 2.16–2.18). Conversely. the extant insect fauna also is poorly understood at lower taxonomic and ecological scales. Estimates of the true species-level abundance of insects indicate that a significant number of modern plant–insect associations are unknown at lower latitudes, and others that are known are typically gleaned from evidence representing either adult or immature phases, but not both. Issues of sampling completeness, variable coverage and attribution of known plant damage to insect hosts afflicts fossil and modern material alike, both of which provide a small representation of true associational diversity in time and space. Perhaps the greater inventory of modern plant–animal associations is balanced by the greater temporal breadth of associations documented in the fossil record.

Increasingly both approaches have supplemented each other and their interface is becoming fuzzy (Fig. 2.4). Biological approaches have gradually relied on the fossil record to calibrate plant–host and animal–associate phylogenies whereas palaeobiologists have begun to test evolutionary and ecological hypotheses based on modern associational data. The fossil record has even been used as primary data for understanding how insect herbivores respond to palaeoclimatically induced vegetational change. This mutual dependence of palaeobiological and biological associational data will only increase in the near future.

Box 2.3 Glossary

assemblage   A group of fossil organisms occurring in a deposit, without any implication of contemporaneous biological association, as in a modern community.

biomechanics   The study of the mechanisms by which organisms function, usually investigated by mathematical modelling of how organismic structure responds to external perturbation.

body fossil   A fossil consisting of its physical remains or its mineralic alteration, rather than the activity of the organism expressed as a trace fossil, such as plant damage or coprolites.

catastrophism   The doctrine that past geological paroxysms, such as relatively sudden climatic change and tectonism, has had a major impact on the course of the history of life.

cladogram   A diagrammatic depiction of branching phylogenetic relationships, which lists key characters (synapomorphies) that define particular taxa.

coal-ball   A globular, concretionary fossil permineralized by carbonate minerals, occurring in coal seams or adjacent strata, and preserving fossil plants, often with superb detail of tissues.

community   A group of organisms representing various species that co-occur in the same spatial habitat and are linked trophically such that they form a web of interdependent associations.

component community   A plant species and all of its dependent organisms, including herbivores, pathogens, fungi and all trophically dependent predators, parasitoids and parasites.

compression   Fossil preservation of an entombed organism on a bedding surface, represented by a carbon film, and often as flattened resistant tissue such as plant cuticle or insect chitin.

convergence   The independent evolution of structural or functional similarity in two or more unrelated lineages that is not based on genotypic similarity; homoplasy.

domatium   A region on a leaf—usually a pit, flap or hollowed area—typically located at the angle of two major veins, serving as lodging for mites or insects.

ecomorphological feature   An organismic attribute characterized by ecological or structural criteria, often contrasted to clades on which such features are frequently mapped or contrasted.

entomophily   The pollination of seed plants by insects, as opposed to other organisms or physical agents such as wind or water.

escalation   The hypothesis that, because of the greater intensity of selection within a habitat through macroevolutionary time, lineages must become successfully better adapted to exist.

geotectonic   Broad, regional processes, related directly or indirectly to the planet's tectonic plates, that deform or otherwise affect the earth's crust.

holometabolous   Possessing a complete developmental transformation in insects, consisting of the separate and distinctive sequence of egg, larval, pupal and adult stages.

impression   A mode of fossilization in which the flattened imprint of an organism forms on a bedding surface, without the preservation of original organismic tissue.

individualistic response   The reaction of organisms to profound ecological change by species-specific abiotic and

continued on p. 74

Box 2.3 *continued*

biological preferences rather than community-wide readjustment.

isotope geochemistry   The study of the relative abundances of isotopes in sediments to determine environmental conditions of the past, including salinity and especially climate.

macroevolutionary   Evolutionary processes viewed from the perspective of geological time, such as the origin of higher taxonomic categories, and attributable either to the prolonged accumulation of microevolutionary processes or the operation of hierarchically different processes during species formation.

megafauna   A large-statured, continental vertebrate fauna associated with the Quaternary, whose extinction around 11 000 BP is associated with climatic deterioration or anthropogenic overkill.

molecular clock hypothesis   The idea that point mutations occur at a sufficiently regular rate to permit the dating of phylogenetic sister-taxa.

pollination drop mechanism   A type of gymnosperm pollination in which a mucilaginous drop exuded from the ovular micropyle traps and detains pollen grains in a pollen chamber, resulting in pollen germination and subsequent fertilization.

prepollen   Fossil pollen grains that function as pollen but morphologically resemble spores in that they bear trilete marks and germinate proximally.

pull of the recent   The inherent bias in the fossil record attributable to greater representation of more recently deposited sediments than those deposited in the more remote past.

radiation   The geologically rapid, evolutionary divergence of lineage members into a variety of adaptive types, frequently with differentiation of taxa in the use of resources or habitats.

syndrome   A controversial concept that defines discrete types of biotic associations, especially pollination and seed dispersal, based on symptoms and structural features that characterize the dispersed plant and the dispersing animal.

time-averaging   The incorporation, either within a deposit or among adjacent successive deposits, of material representing different instantaneous occurrences in time, resulting in the averaging of more discrete and shorter-term evolutionary processes or events.

trace fossil   A type of behavioural evidence for life of the past, consisting of animal-mediated plant damage, coprolites, tracks, trails and burrows that occur in the fossil record.

transfer ecology   The use of the ecological attributes of modern taxa to understand the palaeoecology of phylogenetically related, extinct taxa.

uniformitarianism   The principle that there are inviolable laws of nature that have not changed during the course of time.

Part 2 Mostly Antagonisms

Chapter 3 Plant–insect interactions in terrestrial ecosystems

Sharon Y. Strauss and Arthur R. Zangerl

3.1 Introduction

Herbivory, which is the consumption of plants by animals, encompasses many different types of interactions that differ in their duration and deadliness to the plant. Insect herbivores, like mammals, feed on plants in numerous ways. Seed and seedling herbivory are predatory interactions because herbivores immediately kill individuals in the plant population. Insect herbivores that feed on leaves and other parts of mature plants typically do not cause plant mortality. In the rare cases when they do, it usually requires much time to kill the host plant. Such relationships are closer to parasite–host than predator–prey relationships. Like other parasitic interactions, insect herbivores can feed either externally (leaf-, bud- or flower-feeders) or internally within plants (miners, stem-borers, gall-makers).

Insect herbivores differ from mammalian herbivores in their size, numbers, and the kinds of damage they inflict. Because of their small size, insects often have an intimate, lifelong association with the host plant. Moreover, while their associations are lifelong, often their lives are rather short, predisposing them to rapid rates of evolution. On average, insect herbivores are much more specialized than their mammalian counterparts. For example, the average North American chrysomelid beetle feeds on one or two species of host plant (Wilcox 1979). In an extraordinary effort, tropical ecologist Daniel Janzen collected over 60 000 specimens of Lepidoptera (moths and butterflies) from 725 plant species in Costa Rica. Over half the species collected fed exclusively on one plant species, and many others fed only on a small group of closely related plants (Janzen 1988).

Understanding the relationships between insects and their host plants entails knowledge of biochemistry, ecology, behaviour, physiology and genetics. Insect–plant interactions are important for the ways in which they affect the distribution and abundance of plant species, for their role in energy and nutrient dynamics in ecosystems, and as systems to understand evolutionary mechanisms responsible for much of the diversity of life we see on earth. In this chapter, we describe nutritional and other obstacles that need to be overcome by insects in order for them to use host plants as their sole resource. We then discuss counter-measures plants use to defend themselves against herbivore attack. We also describe how plants and herbivores affect each other's fitness, and how each has evolved in response to these selective pressures. Finally, we explore the ecological ramifications of plant–insect interactions. These include the effects of herbivores on the distribution and abundance of plant species and the role herbivores play in determining the species composition of plant communities. We also describe the consequences of herbivory in ecosystem properties such as nutrient cycling.

3.2 Insect herbivores show many adaptations for acquiring resources from plants

Herbivory as a way of life for insects has existed for millions of years, and plant–insect interactions are common throughout from early on in the plant fossil record; the history of herbivory is discussed in detail in Chapter 2.

One of the most striking features of herbivorous insects is the variety of ways in which different insect feeding guilds harvest their food. Chewing insects include members of the almost entirely herbivorous orders Lepidoptera, consisting of 160 000 named species of moths and butterflies, and Orthoptera, with 20 000 species, including grasshoppers. Other orders that harbour an abundance of herbivorous chewers are the Coleoptera (350 000 beetle species) and the Hymenoptera, comprising 150 000 species, including wasps, bees and ants (Fig. 3.1). These insects possess 'toothed' mandibles, equal to

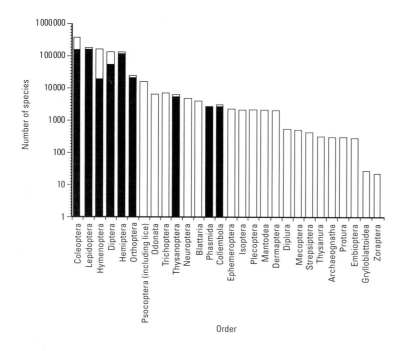

Figure 3.1 Insect orders that largely use live plant tissue are among the richest in species. Insect orders ranked by estimated species number in the world, with the proportion of plant-feeding species indicated by shading. Data for the proportion of herbivorous species give the proportion of herbivores in the fauna of the British Isles (Price 1997); the extent to which this reflects global patterns has yet to be explored.

the tasks of cutting, crushing and/or macerating plant tissues. Most chewing insects feed externally on the plant, others, such as the leaf miners, harvest the tissue layers between the upper and lower leaf epidermis; still others feed inside roots and stems.

Other insects extract their nutrition not by removing chunks of plant tissue but by siphoning fluids from their hosts through specialized tubular mouthparts. These insects belong to the order Hemiptera, consisting of about 112 000 species, the majority of which are plant feeders. One of its most prominent subgroups is aphids that feed on phloem sap. Because this sap is dilute in amino acids, aphids must process large quantities of sugar-rich fluid just to satisfy their amino acid requirement; the bulk of this sugar is of no use to the aphid and is excreted as honeydew. Among the members of the superfamily Cicadoidea, which make their living feeding on xylem sap, are the periodical cicadas. Nymphs of the 17-year periodical cicada spend 17 years feeding underground on root xylem before emerging as adults! However, feeding on the dilute solutions present in xylem vessels does not necessarily dictate a lengthy period of feeding. As their name implies, spittlebugs, members of the same superfamily as the cicadas, are surrounded by a spittle-like froth that they

produce while feeding on xylem sap. These insects develop far more rapidly than the cicadas, completing development within a single season. Other hemipterans, such as the whiteflies, feed neither on xylem or phloem but on the contents of mesophyll cells in leaves.

Gall-making insects manipulate their plant host's tissues to provide themselves with a rich source of nutrients and also a protective shelter (Fig. 3.2). Females of these species lay their eggs in a specific part of their host plant, e.g. the stem, leaf or petiole. Subsequently, the plant tissue surrounding this egg will start growing very rapidly. The growth is called a 'gall' and appears as an abnormal swelling or protrusion of the afflicted plant part. The insect larva within the gall feeds on the highly nutritious inner tissues and eventually emerges from the gall as a fully formed adult. Gall-making insects, which feed predominantly on dicotyledonous angiosperms and on conifers, have evolved many times, and are especially common among wasps, flies, thrips and aphids.

3.2.1 Specialization

The orders of primarily herbivorous insects are the richest in species and contribute tremendously to the biodiver-

Figure 3.2 Gall-making insects modify their host substrate to create sheltered tissues in which their larvae can feed. Galls of different species of gall wasps (Eurytomidae) on scrub oaks in Florida. Affected plant parts (top left, clockwise) include leaf bud, twig, male catkins and fine branch. One larva inhabits each gall. The inhabiting species are *Andriscus quercusfoliatus*, *Callirhytis* sp., *Callirhytis* sp. and *Disholcaspis quercusvirens*. (Photographs by courtesy of Olle Pellmyr.)

sity we see on earth (see Fig. 1.2). In all but two of 13 sister groups (those that are taxonomically related), herbivorous insects had at least twice the number of species contained in non-herbivorous groups (Mitter et al. 1988). As mentioned before, many of these species are relatively specialized in terms of the number of plant species upon which they can or will feed, and this specialization may be linked to increased speciation. If insects switch from one host plant to another as a result of a new genetic mutation, and if mating and reproduction take place on that new plant species, then reduced genetic exchange between insect populations, coupled with natural selection favouring the separation of gene pools, may result in the formation of a new species. Generalists that feed on many plant species would not necessarily experience the limited gene flow that is an important part of the speciation process.

3.2.1.1 Tradeoffs of generalist and specialist habits

There has long been debate over why specialist feeding habits are widespread in herbivorous insects. Intuitively, it seems hard to believe that the fitness of an insect would be greater if it used only a limited number of host plants rather than a large variety. For example, in a bad year, a herbivore dependent on only one or two plant species may have a hard time finding food, whereas a more generalized feeder could use other host-plant species. In an extreme case, a host plant may become extinct and then a specialized herbivore will go extinct with it. Three main arguments exist for the evolution of specialization.

The first argument relies on the old adage that 'a jack of all trades is master of none,' that is, specialists are better at finding and extracting resources from their hosts than are generalists. Selection may fine-tune insect physiology and host-finding to be maximally efficient on a particular host plant. Tests of this hypothesis compare the performance of generalists to specialists on the diet of the specialists, and provide mixed results. For example, in many cases generalist and specialist lepidopterans did not differ in their ability to use the host plant of the specialist (e.g. Scriber & Feeny 1979). In contrast, there are also numerous examples in which specialists do outperform generalists on a shared host (e.g. Montandon et al. 1987; Moran 1988; Cottee et al. 1988). Thus it does not appear that specialization necessarily confers superior ability to utilize hosts.

A more recent version of this idea is that the limited neural ability of insects to process information means that specialists might be more efficient at finding and using host plants than generalists. Specialists searching in a complex environment respond to only a few cues provided by the specific host. In contrast, generalist herbivores must sort through a number of acceptable stimuli from a variety of hosts and decide where to feed and oviposit. This hypothesis has received only two tests, and both support the idea of greater efficiency by specialists. Bernays and Funk (1999) found that specialist aphids were more efficient at every aspect of host location and oviposition than were conspecifics with a more generalized diet. Similar results were found in a comparison between species of specialist and generalist grasshoppers (Janz & Nylin 1997).

A second argument for the evolution of specialization is that the value of some plants as hosts lies more in their ability to protect herbivores from enemies (providing 'enemy-free space') than in their nutritional quality (Jeffries & Lawton 1984; Bernays 1997). For example, chrysomelid beetles have greater survivorship on two species of willow that allow them to produce noxious, predator-repelling defensive compounds than on another species that allows faster and better larval growth but poorer defence against predator (Denno et al. 1990). The argument about enemy-free space seems to work best in explaining why some insects are found on only a subset of the plants upon which they can feed (i.e. on some willows but not others). It is less effective in explaining why total potential diet breadth is still very limited for many herbivorous species (i.e. why these beetles feed only on willows or close relatives but not also on other groups like oaks, maples, etc.).

3.2.1.2 Specialization as a non-adaptive process

Finally, the last hypothesis suggests that specialization is not an adaptive trait but rather an evolutionary dead end. For example, in aphids, many species switch hosts part of the way through the season or alternate hosts. The fundatrix, or founding female, typically feeds on woody plants that are just leafing out at the beginning of spring. As leaves on these woody plants mature, her asexually produced offspring migrate to unrelated herbaceous hosts, where they reproduce asexually throughout the rest of the season. These herbaceous hosts are often nutritionally superior to the woody hosts. At the end of the season, asexual aphids produce winged, sexual adults that mate to produce the fundatrices for the next season. In several groups of aphids the fundatrix stage has been lost, and aphids feed only on the herbaceous hosts (Moran 1988). In these cases, the simplest explanation for the loss of the fundatrix is the overspecialization of the fundatrix stage; that is, selection favoured the loss of the fundatrix as a result of her inability to use the alternative herbaceous host. In a similar way, other herbivorous insects may become overspecialized and may be unable to respond to changing environments that might favour the use of other host-plant species. Over time, superior performance on a single host-plant species may cause adult females to ignore other suitable hosts for oviposition; herbivores may eventually lose the ability to use other hosts (presumably because it is costly to maintain physiological mechanisms that allow an insect to feed on a particular host plant).

The loss of ability to use other hosts is the crux of this argument, and there is some evidence to support this hypothesis. Spider mites originating from tomato plants had decreased fitness on tomatoes after having been reared for many generations on lima bean plants (Fry 1990). *Ophraella* beetles exhibited little or no genetic variation in ability to use host plants of congeners, thus host range is constrained as a result of forces acting over longer evolutionary time scales (Futuyma et al. 1995). Once a herbivore becomes limited to only a small subset of resources, it may then face a greater risk of extinction.

One approach to testing the idea that specialization is a dead end is to ask 'Are specialists always derived from generalist ancestors?' This indirect approach requires that a phylogeny be constructed based on characters that are independent of diet breadth (e.g. relatedness based on DNA sequence similarity) for a large number of species within an insect group. Knowledge of the actual diets is also nec-

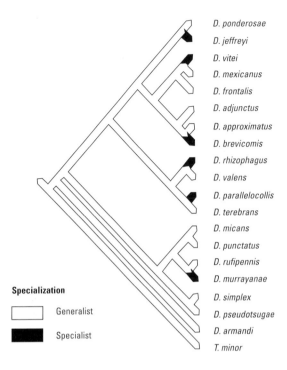

D. ponderosae
D. jeffreyi
D. vitei
D. mexicanus
D. frontalis
D. adjunctus
D. approximatus
D. brevicomis
D. rhizophagus
D. valens
D. parallelocollis
D. terebrans
D. micans
D. punctatus
D. rufipennis
D. murrayanae
D. simplex
D. pseudotsugae
D. armandi
T. minor

Specialization

☐ Generalist

■ Specialist

Figure 3.3 In some instances, more specialized organisms may be derived from ancestors with a wider diet breadth. Phylogram of a group of *Dendroctonus* bark beetles, with species using a small proportion of available conifers indicated by filled branches. In this case, specificity was measured as the proportion of available conifers used as host. Note that the switch consistently appears to have been from a broad to a narrower diet. *Tomicus minor*, another pine-feeding beetle, serves as the outgroup for the tree. (From Kelley & Farrell 1998.)

essary, and perhaps also the possible diets (based on plant distribution) for each insect. One can then determine whether specialists are at the 'tips' of these phylogenies (the youngest groups) and whether they have been derived from generalist feeders. In *Dendroctonus* bark beetles, specialists did tend to be derived in this way, and specialists arose from a generalist feeding habit six times independently (Fig. 3.3; Kelley & Farrell 1998). One caveat in this paper, however, was that 'specialist' was defined according to the *proportion* of potential host plants used by a species. In other words, according to the authors' definition, a specialist living in an area with many coniferous tree species may actually feed on more host-plant species, but a lower proportion of possible hosts, than a generalist utilizing all of the limited number of host species in its area. In other insects, specialized forms are not the most recently derived, and in some cases specialist

lineages have reverted back to generalist feeding (e.g. Thompson 1994). More data sets are required to address this hypothesis further.

There are clearly a number of hypotheses, each with some empirical support, for how a specialized feeding habit might have evolved (Schluter 2000). Because specialization is a complex trait, we don't necessarily expect a single hypothesis to explain the phenomenon. The fact that specializing on one plant may cause loss of use of other plant species suggests that use of plants can require specific adaptations that may entail costs to maintain them. Part of the specialization process may have to do with information processing in a complex environment. In addition, it appears that the ability of insects to detoxify, use or excrete major classes of allelochemicals precludes the use of other major classes. We describe below some of the many ways in which plants have evolved to defend themselves from herbivores and how insects have evolved to circumvent these defences.

3.3 What limits the growth, reproduction and population size of herbivorous insects?

Insect populations frequently fluctuate in size, and this fact has prompted a good deal of speculation as to what factors limit the size of herbivore populations. Hairston, Smith and Slobodkin (1960) reasoned that, since herbivores rarely consume all of their plant resources (the world is green), herbivore populations are likely to be limited by parasites and predators, but not by resource abundance (Hairston et al. 1960). However, whether herbivorous insect populations are limited by food (bottom-up forces in a food web) or by predators (top-down forces) remains a hotly debated topic (Schmitz 1993; Hartvigsen et al. 1995; Floyd 1996; Stiling & Rossi 1997), and it is unlikely that either force dominates all insect populations (Roininen et al. 1996; Hunter et al. 1997; Benrey & Denno 1997; Turlings & Benrey 1998; Turchin et al. 1999). In their arguments, Hairston, Smith and Slobodkin neglected the consideration of several important aspects of the insect–plant interaction. The fact that there is a large amount of plant biomass in the environment does not mean that all of it is accessible as food to insect herbivores — not all plant material is easily converted to meet the energy and nutritional needs of insects. Insect chemical composition is often different from that of plants, and plants defend their tissues from herbivores both chemically and physically.

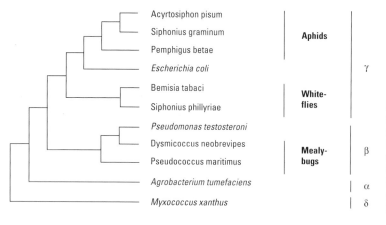

Figure 3.4 Endosymbiotic bacteria have been associated with true bugs since the origin of the animal group. Phylogeny of the subdivisions of Proteobacteria that constitute endosymbionts of different major groups of homopteran true bugs. Greek letters indicate subdivisions, and italicized binomial names are non-endosymbiotic bacteria given for reference. (Modified from Baumann et al. 1993.)

3.3.1 Nutritional requirements of insects

Insects have very much the same nutritional requirements as other animals (Southwood 1973). All animals require fats, protein, carbohydrates, vitamins and minerals. Among the complex nutrients that insects cannot synthesize, and therefore must obtain from their hosts, are the essential amino acids, sterols, linolenic acid and vitamins. The method of food processing by herbivorous insects is fairly stereotypical. Saliva, containing digestive enzymes such as amylase, is secreted into the food, usually by the labial glands. The food then is passed to the foregut, which often acts as a temporary storage area. The bulk of food digestion occurs in the midgut, where the digestive enzymes, proteases, lipases and carbohydrases convert complex nutrients to simpler compounds. Most of the nutrients are absorbed while the food resides in the midgut, the location where much of the detoxification of harmful chemicals also takes place. The remaining plant material passes to the hindgut, where water and more nutrients are absorbed. The waste, called 'frass', is eliminated through the rectum.

3.3.1.1 Energy and macronutrients

The first obstacle that an insect faces is the fact that, on average, only about 10% of the energy available to one trophic level makes it to the next trophic level. Sources of energy loss include the fact that not everything ingested can be assimilated (e.g. lignin, cellulose). Microbes and fungi are the few organisms that can convert lignin and cellulose into energy. Numerous insects cope with this problem by hosting symbiotic bacteria that allow them to digest cellulose, so much of plant material is unavailable to insects as a source of energy. For example, sap-feeding true bugs (Hemiptera) have mutualistic gut endosymbionts that enable them to utilize plant resources more fully (Fig. 3.4). Not only is more than 90% of the energy lost when herbivores consume plant material, the chemistry of plant and insect tissues is very dissimilar.

Liebig's law of the minimum states that growth is possible to the extent determined by the nutrient that is in shortest supply. For herbivores, one such nutrient is protein. Because nitrogen is relatively easy to measure and protein is not, protein content is often estimated by assaying organic nitrogen, which comprises from 15 to 18% of plant proteins (Kirk & Sawyer 1991); thus, 'protein limitation' is often referred to as 'nitrogen limitation'. Southwood (1973) made the point that herbivores should be nitrogen-limited because the amount of nitrogen in leaf tissue is typically about 2%, while the amount of nitrogen in insect tissue is of the order of 30–40%. In fact, the consumption rate of foliage often increases with decreasing nitrogen content (e.g. Griswold & Trumble 1985; Karowe & Martin 1989). Compensatory consumption is most noticeable among xylem feeders. These sap-feeding insects, like cicadas and other homopterans, often eat 100 to 1000 times their body weight per day because amino acids make up only a tiny proportion of the sap (Horsfield 1977). Nitrogen availability may also explain patterns in the use of hosts. Seasonal movements of phloem-feeding aphids from tree hosts in the spring to herbaceous hosts in summer are associated with a decrease in the soluble nitrogen content of trees. Aphids return to trees in the autumn when soluble nitrogen again increases in leaves (Mackenzie & Dixon 1991). In general, aphids and other sap-feeding insects are the most sensitive of insects to changes in nitrogen levels of their host plants (Kyto et al. 1996).

Total nitrogen concentration in leaf tissue can be misleading as an indication of protein availability to herbivores. The balance of amino acids is also important to the growth and development of herbivores (Karowe & Martin 1989). Up to 20% of an insect's exoskeleton consists of aromatic compounds, thus aromatic amino acids such as phenylalanine are in particular demand. Bernays (1982) has shown that grasshoppers show significant weight gain in response to phenylalanine as a supplement to a lettuce diet which is poor in nitrogen; no weight gain was associated with any of 15 other amino acids. Thus, as a defence, plants may store nitrogen in amino acids that are of relatively little use to insect herbivores (Bernays 1982).

Plants may also protect nitrogen in the form of non-protein amino acids that are toxic to most herbivores. Most non-protein amino acids are in the 'L' enantiomer conformation, as opposed to the 'D' conformation (Rosenthal & Berenbaum 1991). The non-protein amino acids act as neurotoxins or as substrate analogues of protein amino acids. Toxic, non-protein amino acids are often found in the seeds of plants and can be the principal N-storing metabolite. L-canavanine is famous for its highly toxic properties, as well as for a bruchid beetle that consumes it. If left in its original state, L-canavanine alters RNA metabolism and canavanyl proteins are formed (Rosenthal & Berenbaum 1991). These proteins in turn affect RNA synthesis, and DNA translation is inhibited. The beetle *Caryedes brasiliensis* is the only known predator of seeds of *Dioclea megacarpa*, a tropical tree in the pea family that stores as much as 13% of its dry weight as L-canavanine and 94% of the amino acid N in that form. The beetle has a special enzyme, arginyl-tRNA-synthetase, that detoxifies and converts L-canavanine to dietary N. Despite the fact that this beetle has the means to access this rich store of nitrogen, no other species can do this, so this form of seed defence provides protection from most enemies.

3.3.1.2 *Micronutrients*

Aside from needing macronutrients such as nitrogen, acquiring micronutrients also presents problems to herbivores. Some salts, and especially sodium (Na), are in short supply in leaf tissues. Comparison of leaf and insect tissue have shown Na concentrations >2.5 times higher in the animal than in the plant tissue, and Na can thus become a limiting factor. Male butterflies are often found 'puddling' in mud puddles, animal urine or bird droppings, a behaviour apparently induced by the presence of sodium (Arms et al. 1974). Male moths incorporate large amounts of sodium in their sperm packets, or spermatophores, as a nuptial gift for use by the female (Smedley & Eisner 1996). Nuptial gifts may increase the fitness of both the offspring produced by the female and the adult female herself (Wisner et al. 1996). In general, both micro- and macronutrients can limit the growth rate of insect herbivores.

3.3.1.3 *Distribution of nutrients seasonally and among plant parts*

Because micro- and macronutrients are unevenly distributed among plant parts, herbivores are often differentially adapted to using only a subset of resources in a plant. Different types of feeding apparatus also limit the kinds of plant resources a herbivore can use. Just as an aphid cannot chew on a leaf, neither can a beetle siphon plant juices from xylem and phloem. The combination of feeding mode and nutritional differences among plant parts means that there is never a case in which a herbivore feeds and grows equally well on any plant part at any stage. Many insects prefer young plants or tissues to old ones (Price et al.1990; Hunter et al. 1992). Some insects, like pine bark beetles, require stressed and dying trees, others require shoots (*Euura* willow gallwasps) and still others require young leaves (many chrysomelid beetles). Because many plant parts are present for only limited amounts of time during the year (like buds or young leaves) herbivores may be limited by the availability of the plant parts to which they are best adapted.

3.4 Plant defences against insect herbivores

Aside from nutritional hurdles and the limited availability of some plant parts, herbivores may also be prevented from feeding as a result of plant defences. Plants that can deprive insect enemies of their nutrients have thrived in the past, giving rise to the rich diversity of defensive adaptations evident in plants today. Adaptations include physical barriers, toxins, anti-feedants, decoys and even

Figure 3.5 Hairs on tomato stems may prevent insects from penetrating to the surface of the plant.

other organisms. Some defences are always present on the plant; we call these constitutive defences. Many others, including thorns and spines, are inducible, that is, they are augmented only after the plant is attacked (Karban & Baldwin 1997). The diverse array of characters used by plants to defend themselves is outlined below.

3.4.1 Physical barriers

Such familiar structures as thorns and trichomes come in a variety of sizes and shapes, not all of which are effective barriers to insects. Undoubtedly effective deterrents to mammals, the large branching thorns of the honey locust, which measure 20 cm or more in length, are probably of little utility against an insect attack. On the other hand, densely packed fine spines or hairs may deter insects by physically barring them from underlying vulnerable tissues (Fig. 3.5). In some cases hairs, or trichomes, also have noxious chemicals associated with them. Southern California populations of the species *Datura wrightii* are polymorphic for the presence of glandular trichomes—some plants have them while others have only non-glandular trichomes (Van Dam & Hare 1998). Whitefly females that attempt to land and lay eggs on plants with glandular trichomes become entrapped and die in the sticky exudate secreted by the glands. However, white-flies can freely land on and depart from plants lacking glandular trichomes. The polymorphism is apparently maintained in the population because another species, *Tupiocoris notatus*, uses the glandular exudates as a cue to

identify its host and is more likely to attack the plants with glandular trichomes.

Another physical defence is leaf toughness. Lignin in leaves makes them difficult to chew for many herbivores. Toughness probably serves as a defence particularly against early season herbivores that feed on young, tender leaves (Matsuki & Maclean 1994).

3.4.2 Toxins and deterrents

The list of chemicals that owe their defensive value to their ability to interfere with insect physiology or behaviour is a very long one. While the elaboration of thorns, spines and hairs is restricted largely to their size and shape, the number of possible combinations, principally of carbon, oxygen, hydrogen, nitrogen and sulfur, is enormous. When incorporation of the more exotic elements, such as bromine, fluorine and chlorine, occurs, as it does in marine plants, the possibilities are greater still. These plant constituents are commonly referred to as 'secondary' compounds. Their designation as secondary products reflects the fact that these compounds are not strictly essential to the daily functions of plants. For example, all plants contain amino acids, primary metabolites essential for the assembly of proteins, but relatively few species contain the secondary chemicals called cardenolides. Humans use secondary compounds for many different purposes—alkaloids provide the hot burning sensation that we like in spicy food, and other compounds, the terpenes, impart the minty and other herbal flavourings. Nicotine and caffeine are secondary compounds used as stimulants. Many secondary compounds have medicinal value; for example, taxol, which was originally extracted from the bark of the Pacific yew (*Taxus brevifolia*), is used to control the growth of cancerous breast tumours (Lau et al. 1999).

In the early part of the twentieth century, many natural-product chemists subscribed to the notion that these secondary compounds were nothing more than plant waste products. However, there were those who had different perspectives. The history of these ideas was summarized in a seminal paper by Fraenkel (1959) in which he discussed the possibility that secondary compounds served as defences against herbivorous insects. In 1804, Augustin de Candolle recognized that many plant genera and families were characterized by particular classes of chemicals and that these were correlated with patterns of attack by particular herbivores. In 1888, Stahl first proposed that chemical properties of plants evolved for

Plate 3.1 Floral traits of significance in pollination and defence may evolve as preadaptation from traits of other function. In many *Dalechampia* species, (A) defensive resins have been co-opted as a pollinator reward and (B) floral bracts that serve in pollinator attraction during the days close and exclude specific night-breeding predators. (A, *Dalechampia tiliifolia* [Oaxaca, Mexico], B, *Dalechampia scandens* [Belize]; photographs by courtesy of Scott Armbruster.)

Plate 4.1 (a) Male (foreground) and female (background) moose (*Alces alces*), a large browsing mammalian herbivore characteristic of subarctic and boreal habitats of the Northern Hemisphere. (Photograph by Kjell Danell.) (b) Heavy mortality of plants in Chobe National Park, northern Botswana, due to debarking by elephants (*Loxodonta africana*). Tree regeneration is also hindered there by browsing antelopes. (Photograph by Roger Bergström.)

Plate 5.1 A male red crossbill (*Loxia curvirostra*) perched next to an open ponderosa pine (*Pinus ponderosa*) cone. Although spines on the cones slow crossbills, they still extract seeds quickly by laterally abducting their lower mandibles to spread the scales apart and then protruding their tongues to the base of the scales to extract the seeds. (Photograph by William Ervin.)

facing p. 84

Plate 6.1 Floral traits are often controlled by at least one major gene, responsible for more than 25% of the variation, thus facilitating rapid diversification. Bee-pollinated *Mimulus lewisii* (a) and hummingbird-pollinated *M. cardinalis* (b) are interfertile, and produce F1 hybrids with flowers of intermediate morphology (c). QTL analysis of selfed F2 progeny (five examples in d–h) showed that 9 of 12 floral traits were controlled by at least one gene responsible for >25% of the variation. (Photographs by courtesy of Douglas W. Schemske.)

Plate 7.1 The superficial resemblance of the fruits of vertebrate-dispersed plants reflects evolutionary convergence for functional homogeneity rather than homologous origins. The portion of fruits acting as a reward to dispersers ('flesh' or 'pulp') has diverse anatomical origins, as illustrated here for a sample of bird-dispersed fruits from tropical and temperate habitats. Fruit pulp most often develops from ovary walls, as in berries ((a) *Lonicera splendida*, Caprifoliaceae) and drupes ((b) *Osyris alba*, Santalaceae). In other cases, the walls of the ovary are dry and dehiscent at maturity, and the pulp originates from the most external integument of seeds ('arillate seeds'; (d) *Euonymus sachalinensis*, Celastraceae; (e) *Dieffenbachia seguine*, Araceae; (f) *Clusia*, Clusiaceae; (h) *Stemmadenia*, Apocynaceae). Other origins include enlarged floral receptacles ((c) *Rosa sicula*, Rosaceae) and floral bracts ((g) *Taxus baccata*, Taxaceae). (All photographs by Carlos M. Herrera.)

Plate 6.2 (opposite) Models in plant–pollinator studies. Clockwise from top left: *Lavandula latifolia* with visiting *Apis mellifera*, Sierra Cazorla, Spain (photograph by C. Herrera); *Ipomopsis aggregata*, Arizona, USA; *Yucca intermedia*, insert, *Tegeticula yuccasella* pollinating yucca flower, New Mexico and Ohio, USA; split receptive fig of *Ficus congesta* (ovules lining inside), insert, emerging female of its *Ceratosolen* pollinator, Cape Tribulation, Australia (photograph by G. Weiblen); *Aconitum senanense*, Central Alps of Honshu, Japan; *Impatiens pallida*, Tennessee, USA. (All photographs by the author except where noted.)

Plate 8.1 The plant products and structures that attract and retain the services of ants are extremely varied. (a) Ants of the genus *Myrmicaria* harvest extra-floral nectar on the young leaves of a *Pometia pinnata* (Sapindaceae) sapling in the Malay peninsula. (b) The leaf-pouch domatia on *Callicarpa saccata* (Verbenaceae) which are usually inhabited by species of *Technomyrmex*, Borneo. (c) A founding queen of *Cladomyrma petalae* chewing an entrance hole to establish a domatium in *Saraca thaipingensis* (Fabaceae), Malaysia. (d) The tendril-domatium of the pitcher plant, *Nepenthes bicalcarata*, inhabited by a colony of *Camponotus schmitzi*. The ants take the copious extra-floral nectar from very large glands situated above the pitcher and attack a herbivorous weevil (Borneo). (e) A slice taken from the side of the ant-plant *Myrmecodia* sp. in NE Australia, showing the tunnels in which *Iridomyrmex* sp. live. Plants absorb nutrients from colony wastes in some of the tunnels and are generally found on nitrogen-poor soils. (f) A species of *Rhytidoponera* harvesting a diaspore of *Sclerolaena diacantha* (Chenopodiaceae) in the arid outback of Australia. The ants are attracted to the tough but malleable elaiosome on one end of an otherwise woody structure. (g) The large, soft elaisome of *Viola odorata* (Violaceae) that attracts ants of the genus *Formica* in the woods of northern Europe. (Photographs: a and c, Joachim Moog; b, Ulrich Maschwitz; d, Dennis Merbach; e, Christine Turnbull; f, Rod Peakall; g, David Culver.)

protection against attack by herbivorous animals. Verschaffelt showed in 1910 that cabbage butterfly larvae in the genus *Pieris* would feed on the leaves of non-hosts if they were treated with solutions of sinigrin, a type of glucosinolate compound associated with plants in the mustard family. He concluded that larvae are attracted by these substances. Our current understanding is that the presence of secondary compounds can deter many herbivores from using plants, but that almost every plant species has a suite of specialized herbivores that are adapted to use these compounds as attractants, as feeding stimulants or as a source of toxins for use in defence against their enemies.

In order for a plant trait to be considered truly defensive two criteria must be met: (1) the compound or other attribute must affect the degree to which a plant is attacked and (2) plants with the attribute must have greater fitness than those without it or than those with lesser amounts (Karban & Baldwin 1997). When the role of a secondary compound is defensive, it is commonly referred to as an 'allelochemical'. Many compounds owe their defensive value to their ability to interfere with primary processes in consumers — these allelochemicals are often toxins. Other allelochemicals, for reasons not clearly understood, deter herbivores from eating the plant, irrespective of whether they are toxic (Bernays & Cornelius 1992) — these compounds are called 'deterrents'. Cucurbitacins are excellent examples of compounds with both toxic and deterrent qualities, the latter of which derives from their extreme bitter taste (Metcalf et al. 1980).

Secondary compounds can be unique to only one or two plant families (e.g. the coumarins) or virtually ubiquitous (e.g. the terpenoids) (Fig. 3.6). Ultimately all secondary compounds derive from primary biochemical pathways; for example, the phenylpropanoid pathway derives from the amino acid pathway, and the terpenoid pathway derives from the mevalonic acid pathway, which is responsible for production of fatty acids and steroids. The mode of action of secondary compounds varies both within and between chemical classes; cyanogenic compounds produce hydrogen cyanide, which blocks respiratory electron transport, tannins bind with proteins rendering them indigestible; and furanocoumarins bind to DNA (Rosenthal & Berenbaum 1991). These are but a few of the great many modes of action documented for secondary compounds (Table 3.1).

Not all chemical defences are based on organic compounds, and some plant defences come relatively cheap.

Nickel is one of several metals that is abundant in serpentine soils and is toxic to most plants and insects. At least one species of plant, *Thlaspi montanum* var *montanum*, has not only adapted to tolerate nickel, it hyperaccumulates the nickel, rendering itself unsuitable for insect herbivores (Boyd & Martens 1998).

3.4.2.1 Synergism among compounds

Most plants contain a variety of secondary compounds, and while some of them are clearly toxic, others seem to have little or no effect on consumers. Synergists are chemicals that enhance the toxicity of chemicals with which they are mixed. This phenomenon has been successfully exploited by pesticide formulators to enhance the toxicity of their products while at the same time reducing the amount of the active ingredient. The most commonly used commercial synergist is piperonyl butoxide, a synthetic compound that inhibits cytochrome P450s, detoxification enzymes of widespread occurrence. Synergism has also been demonstrated among naturally co-occurring plant secondary chemicals. The common compound myristicin by itself has no effect on the mortality of larvae of the omnivorous corn earworm *Heliothis zea*, but it increased the efficacy up to fivefold of a naturally co-occurring toxin. While the degree to which synergism plays a role in plant defence is largely unknown, a growing body of data suggests that the widespread occurrence of multiple variations on a common chemical theme may be due in part to synergistic effects (Berenbaum et al. 1991).

3.4.2.2 Delivery

For the most part, delivery of allelochemicals to herbivores is passive — the herbivore simply ingests the toxins as part of its food intake. Other defences are rather more dynamic. Sticky resins secreted by wounded trees can trap and engulf an insect. These events have provided palaeoentomologists with an abundance of exquisitely preserved fossils in amber. By far the most dynamic delivery system discovered in plants is the 'squirt gun' defence employed by the shrub *Bursera schlechtendalii*, a native of southern Mexico and Guatemala. The shrub possesses a network of pressurized terpene canals, which, when severed by a herbivore, shoot a stream of terpenes that may travel distances up to 150 cm (Fig. 3.7)! This defence is effective against many but not all herbivores; a chrysomelid beetle of the genus *Blepharida* feeds on the shrub and

Chemical Defences in Plants

Compound class	Plant distribution	Number of chemicals	Example
Non-protein amino acids	2500 species 130 familes	600	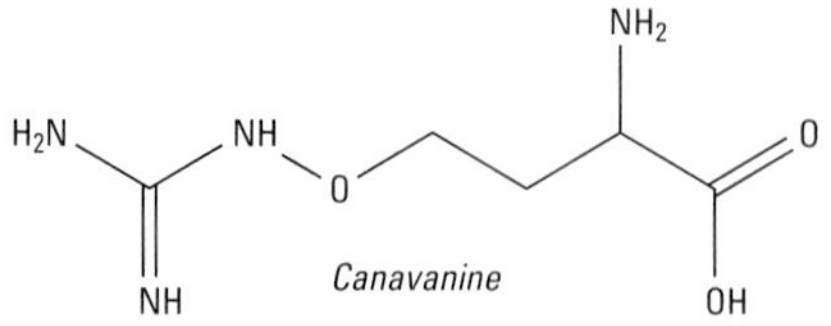 Canavanine
Alkaloids	20% of angiosperms	10 000	Nicotine 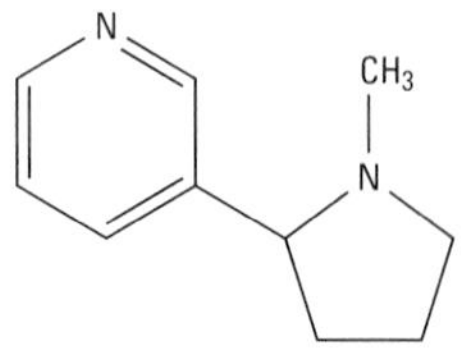
Glucosinolates	300 species 15 familes	100	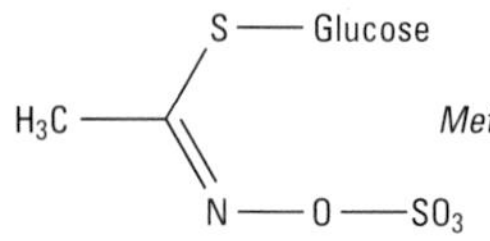 Methyl-glucosinolate
Cyanogenic compounds	2500 species 130 familes	60	Linamarin 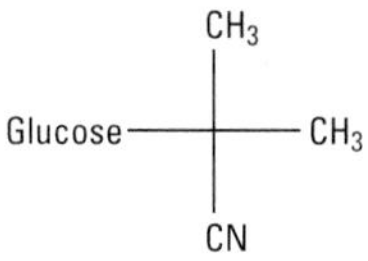
Terpenoids	all plants	15 000	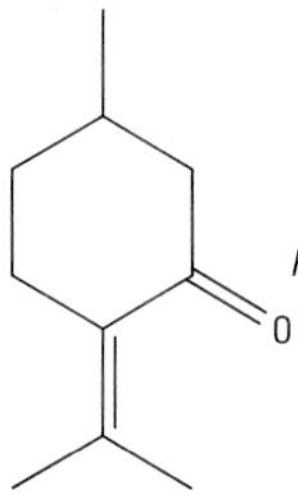 Pulegone
Coumarins	79 families	800	Coumarin 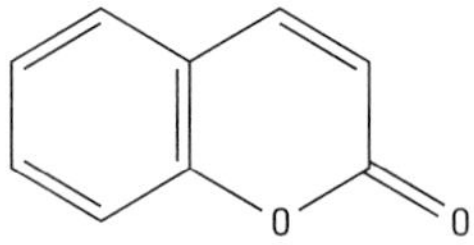
Iridoid glycosides	57 families	600	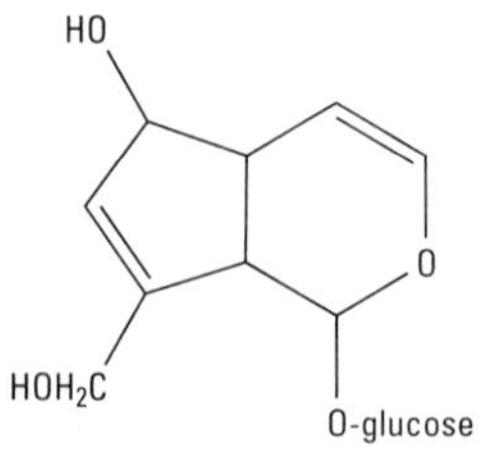 Aucubin

Figure 3.6 Examples of secondary metabolites in plants that have been linked to insect resistance.

Table 3.1 Classes of chemical compounds and their biological activity. Activities of a class of defence chemicals are often attributable to a variety of mechanisms and involve multiple targets resulting in multiple effects. This list is by no means comprehensive and the mechanisms of activity are often unknown (Rosenthal & Berenbaum 1991).

Chemical class	Sites of activity and mechanisms of effect
Non-protein amino acids	Nervous system Connective tissue malformation Interference with protein synthesis Oxime formation Impaired thyroid gland function
Cyanogenic glycosides	Electron transport
Alkaloids	Nervous system Feeding deterrence – bitterness Nucleic acids Enzyme inhibition
Glucosinolates	Antibiotic Thyroid function Organ atrophy
Terpenoids	Nervous system Feeding deterrent Growth inhibition Physical barrier Alkylation agents Protein binding Membranes Hormone analogues Development Sterol uptake
Coumarins	DNA binding Protein binding Lipid interactions Radical generation Anticoagulant Vasodilation
Cardenolides	Nervous system Feeding deterrency Emetic Toxin
Iridoid glycosides	Antifeedant Sedative Toxin
Lectins	Carbohydrate binding Cell receptor binding Intestinal mucosa Digestion inhibitor
Tannins and lignins	Protein binding Digestive inhibitor
Flavonoids	Respiration inhibitor

Figure 3.7 The squirt-gun defence of *Bursera schlechtendalii*. Members of the family Burseraceae produce high-pressure resin squirts in response to vessel cutting, thus ejecting the cutter. (Photograph by courtesy of J. Becerra.)

experiences only slightly reduced growth on only the most responsive plants (Becerra 1994).

3.4.3 Abscission defence

Yet another non-chemical defence takes the form of premature leaf abscission. Quite simply, a plant drops leaves that are infested with herbivores. This form of defence can be effective against relatively immobile herbivores, such as aphids (Williams & Whitham 1986) and leaf miners (Simberloff & Stiling 1987). However, this form of defence is costly because fully functioning leaves, complete with valuable stores of nutrients, must be jettisoned before their full photosynthetic value is realized.

3.4.4 'Third-party' defences

Plants are not above enlisting the services of an aggressive bully to protect them from herbivores. A classic example is the ant–acacia mutualism first described comprehensively by Janzen (1966). The acacia ant *Pseudomyrmex ferruginea* and closely related ant species actively defends its *Acacia* plant host from herbivores of all descriptions, both large and small. In return, the ants are supplied with housing in the form of modified thorns, carbohydrates from specialized extrafloral nectaries, and protein-rich Beltian bodies, located at the tips of the leaflets (Janzen 1983b). One potential drawback to this ant defence is that pollinators might also be repelled, causing a reduction in seed set. A recent study has determined that the ants vigorously patrol buds and older flowers but are largely absent during the time when pollinators are active. A chemical signal produced by young flowers repels the ants (Willmer & Stone 1997). Ant–plant protection associations are very widespread among land plants; ant-attracting extrafloral nectaries occur in species from groups as divergent as ferns and legumes (Chapter 8; Bronstein 1998). In subtropical Southeast Asia, species with extrafloral nectaries make up a remarkable 7.5% of the flora (Pemberton 1998). And while ants are frequently the best studied group in these interactions, it has been found that extrafloral nectaries also attract other potentially beneficial visitors such as parasitic wasps and predaceous mites.

Eating is a dangerous activity for a herbivorous insect because it requires movement, and movement attracts predators (Bernays 1997). To make matters worse for herbivorous insects, feeding damage can be used by predators and parasites as cues to the presence of prey. Wasps of the genus *Cotesia* orient to volatile compounds emitted from damaged plants (Turlings et al. 1991; Thaler 1999; Van Loon et al. 2000). To the extent that a plant can actively enlist the help of predators and parasites in the effort to rid itself of its herbivores, the interaction may properly be considered a defence.

Other groups of organisms also play a role in third-party defences. Fungal endophytes, as their name implies, live inside plants but do not cause noticeable damage. In some cases, these fungal endophytes confer resistance to herbivores in their hosts by producing toxins, such as the toxic ergot alkaloids (Clay 1988). An endophyte of tall fescue (*Festuca arundinacea*) produces toxic alkaloids that annually account for more than 600 million dollars' worth of livestock losses (Hoveland 1993). Clay (1988) has characterized these plant–endophyte associations as defence mutualisms, but not all such associations confer herbivore resistance (Tibbets & Faeth 1999).

Given the wide variety of impediments that plants have in place to dissuade herbivores from consuming them, it is likely that some plants within each population are not suitable food for herbivores, so bottom-up forces are at work in all populations of insect herbivores. However, it is

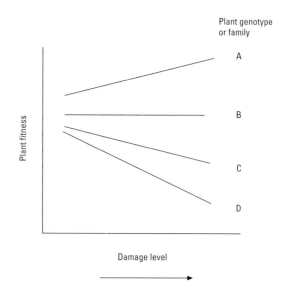

Figure 3.8 Representation of degrees of tolerance expressed by plants after damage. Plant genotype A is more tolerant than genotypes B, C or D. Genotype A overcompensates for damage (slope is positive), genotype B completely compensates, and genotypes C and D undercompensate, though C is relatively more tolerant of damage than D (it has a shallower slope).

equally clear that top-down forces are also important in many cases, particularly when plant signals emitted after damage by herbivores serve as cues alerting predators or parasitoids to the presence of a herbivore. In the latter case, both top-down and bottom-up forces are acting simultaneously in the same defence.

3.4.5 Tolerance

Another type of plant 'defence' does not rely on being deterrent or toxic, but rather on the ability to regrow after damage. Simply put, tolerance is the ability of a plant to sustain damage without consequences (Painter 1958). The mechanisms of tolerance are still poorly studied but probably involve the setting aside of a pool of resources that can be drawn upon after damage, or the ability of a plant to boost photosynthetic rates to compensate for losses (Strauss & Agrawal 1999). The tolerance of a plant genotype can be estimated by the slope of the relationship between plant fitness and the amount of damage sustained; shallow slopes are indicative of high tolerance, as they reflect the fact that changes in fitness are relatively small compared to losses of tissue (Fig. 3.8).

Unlike a true defence, tolerance may have no negative effects on herbivores, and may even have positive effects by providing additional resources for herbivores. The degree of tolerance in plants ranges from minor losses in fitness to complete compensation (no loss in fitness) to overcompensation, a phenomenon in which plants browsed by herbivores produce more seeds than their unbrowsed neighbours (see Chapter 4; Paige 1992; Gronemeyer et al. 1997; Lennartsson et al. 1998). No examples of overcompensation have as yet been documented in response to insect herbivory, although plants have been shown to compensate fully in seed production, despite high levels of leaf damage from insects (e.g. Meyer & Root 1993; Sadras 1996; Strauss et al. 1996). Tolerance is an attribute that is implicitly recognized in agricultural practice; the Economic Injury Level (EIL) of insect damage to crops indicates the amount of damage that crops can sustain before the reduction in yield has an economic impact (Higley & Pedigo 1993; Mumford & Knight 1997).

Tolerance and resistance may represent alternative strategies for coping with herbivore damage. Van der Meijden et al. (1988) surveyed a field and determined the average amount of leaf area removed by herbivores in a number of species. Subsequent experimental leaf-area removals on each species showed that the ability to regrow after damage was correlated with the mean level of damage experienced by each species in the field. Thus, they suggested that plants might either invest in defences prior to attack so that they receive less attack, or else invest in the ability to regrow after attack. The possible tradeoff between tolerance and resistance has also been examined at the intraspecific level. In *Brassica rapa*, there was a negative genetic correlation between tolerance and resistance traits in response to insect damage (Stowe 1998). However, in a related species, *Raphanus raphanistrum*, at the phenotypic level, there was no relationship between tolerance and resistance traits (Agrawal et al. 1999). There is much work being done now to determine whether resistance and tolerance really represent alternative ways in which plants cope with herbivores. Another important aspect of tolerance to keep in mind is that, like resistance, the traits involved and plant responses may depend on the identity of the herbivore.

3.5 Adaptations of insects to plant defences

Insects are not powerless to cope with plant traits that im-

pede their use of plants for food. As many means as plants have to deter insects, insects have ways of circumventing them.

3.5.1 Behaviour

One of the ways in which insects cope with plant defences is to avoid them by not consuming the defended plants or plant parts. This behaviour may be 'hard-wired' in the sense that the avoidance is genetically controlled, or it may be learned, as was the case for a pair of arctiid moths, whose caterpillars learned to avoid eating a plant that caused them to vomit (Dethier 1980). Adult female insects can avoid subjecting their offspring to unsuitable plants by not laying their eggs on those plants. These oviposition preferences have been shown to be genetically variable in butterflies (Thompson 1994). In the event that adult females are not capable of determining which plants are most suitable for its young, larvae of at least some species can and do move between plants. Another behaviour that allows herbivores to exploit hosts is defence deactivation. Faced with pressurized cardolenide-rich latex canals in the milkweed *Asclepias syriaca*, the chrysomelid beetle *Labidomera clivicollis* first bites into lateral leaf veins near the midvein. This action causes the vein to leak droplets of latex and to lose pressure. The beetle subsequently feeds at the edge of the leaf, distal to the cuts (Dussourd & Eisner 1987). Dussourd and Eisner confirmed the utility of this behaviour by comparing the acceptability to several specialist and generalist insects of *A. syriaca* intact leaves and leaves that had all of their lateral veins on one side mechanically severed. Leaves with severed canals were more readily consumed by all but one of the insect species tested. Latex and resin canals occur in many plant groups (Farrell et al. 1991), and both specialist and generalist insects have been shown to engage in trenching or cutting behaviour only when it is warranted, that is, when the plant contains canals (Fig. 3.9; Dussourd & Denno 1994). The *Blepharida* species that feed on the terpene-squirting *Bursera schlechtendalii* also cut veins. However, defusing this weapon is a very time-consuming process, taking as long as 1.5 hours to render safe an amount of tissue that is consumed in 10–20 minutes (Becerra 1994). While behavioural avoidance is an effective way for an insect to remove itself from harm's way, it has the disadvantage that the insect may spend much of its time searching for suitable food or, as in the case of *Blepharida*, disabling the defence. With chemical defences

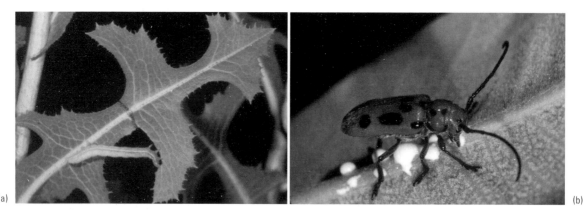

(a)   (b)

Figure 3.9 In instances where defensive compounds are distributed in veins, herbivores may be able to bypass the defence. (a) The larva of *Anagrapha falcifera*, a noctuid moth, cuts latex-transporting veins in the prickly lettuce, *Lactuca serriola*, before eating drained tissue. (b) The cerambycid milkweed beetle, *Tetraopes tetrophthalmus*, punctures the midvein of *Asclepias* leaves before feeding on distal parts. (Photographs by courtesy of David Dussourd.)

that don't rely on physical properties such as pressurization, many insects have evolved to cope with these defences after they have ingested them.

3.5.2 Detoxification

A common form of defence against host-plant toxins involves enzymatic conversion of allelochemicals to less toxic forms. Although detoxification enzymes are often produced by the insects themselves, there are numerous cases in which microbial symbionts are responsible for production of the enzymes that detoxify allelochemicals (Dowd 1991). The genetic basis for detoxification is not well studied in most natural systems, but there is evidence for genetic variation in esterases and cytochrome P450s, two enzymes that figure prominently in detoxification pathways (Lindroth & Weisbrod 1991; Berenbaum & Zangerl 1992).

Many plant toxins ingested by insects are hydrophobic, that is, they tend not to dissolve in water. Detoxification of these compounds involves their conversion to less toxic, more water-soluble structures that are more readily eliminated. This detoxification is generally accomplished via a two-step process (Williams 1974). The first step, termed Phase I, involves conversion of the toxin to a less toxic form. These products are then further modified in Phase II by their conjugation with sugars, amino acids, or other water-soluble compounds that facilitate excretion.

Prominent among the Phase I enzymes are the cytochrome P450s. These iron-containing proteins are oxidative enzymes present in virtually all eukaryotic taxa. The list of compounds metabolized by P450s is large and includes alkaloids, terpenes, terpenoids and furanocoumarins, to name just a few (Schuler 1996). The reactions catalyzed by cytochrome P450s follow the general scheme:

$$\text{substrate} + \text{NADPH} + \text{H}^+ + \text{O}_2 \rightarrow$$
$$\text{substrate-OH} + \text{NAPD} + \text{H}_2\text{O}$$

Cytochrome P450s have also been referred to as MFOs (mixed-function or multifunction oxidases) and PSMOs (polysubstrate monooxygenases). As these names imply, cytochrome P450s may take more than one substrate, a feature that makes them particularly useful to insects and mammalian herbivores that feed on a wide range of foods containing diverse secondary compounds. For example, the human cytochrome P450, CYP3A4, metabolized 50% of the drugs tested (Karanam et al. 1994), despite the fact that humans have had little or no evolutionary history with these drugs. However, one should not assume that all P450s have a broad substrate activity and are therefore of little use to specialist herbivores. For example, the cytochrome P450s of black swallowtail butterflies exhibit a high degree of specificity for ecologically relevant toxins, namely those that are contained in the insect's host plants (Ma et al. 1994). The widespread dependence of organisms on P450s as their primary means of detoxification might be related to the extraordinary shifts in substrate specificity that can accompany single amino-acid substi-

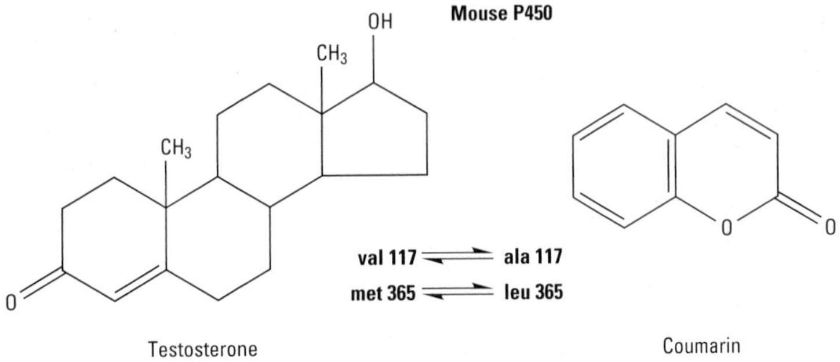

Figure 3.10 Minor change in secondary compounds can dramatically alter their specificity, thus facilitating the evolution of novel defences and detoxification capability. In vitro alteration of a mouse P450 at either of two amino acid positions causes a shift in substrate specificity from testosterone to coumarin. (From Lindberg & Negishi 1989.)

tutions in the enzyme. Substitution of amino acids at either of two positions in a mouse P450 shifted the substrate affinity from testosterone, a steroidal hormone, to coumarin, a plant allelochemical (Lindberg & Negishi 1989; Fig. 3.10). This example conveys the ease with which an enzyme responsible for metabolism of a primary metabolite, testosterone, might have subsequently evolved to detoxify secondary compounds.

The Phase II enzymes consist of a series of transferases. These enzymes are named according to the chemical groups that they transfer to their substrates. Some examples are glycosyl transferases, sulfo transferases, glutathione transferases and amino acyl transferases. These conjugating enzymes may have as substrates Phase I products which they render still more polar and thus more amenable to elimination.

In addition to the cytochrome P450s and transferases, there are enzymes effective for specific challenges. Feeding by insects often elicits an 'oxidative burst' in damaged plant tissues (Bi & Felton 1995). The oxidative burst results when oxygen radicals or active oxygen species are released. These oxygen species are very destructive if they find their way into the reduced environment of living cells. A series of enzymes, as well as non-enzymatic antioxidants, such as ascorbic acid (vitamin C), are involved in transforming these oxygen species to harmless products (Summers & Felton 1996). Another class of enzymes, the β-glycosidases, effects detoxification via reducing reactions.

The potential of detoxification as an adaptation to host defence chemistry is perhaps best illustrated by the diversity of cytochrome P450 genes in a single species. The recent completion of the sequencing of the fruit fly genome has revealed some 80 cytochrome P450 genes, the functions of most of which are unknown. While some of them are undoubtedly involved in primary metabolism, there

are presumably a fair number involved in other activities. With such diversity in only one family of enzymes potentially involved in detoxification, it is tempting to infer that detoxification is likely to figure prominently in almost all herbivore–plant interactions. Just how costly detoxification is to insects is unclear. Detoxification requires energy and synthesis of enzymes, and sometimes cofactors and conjugates, but evidence of these costs is scant (Berenbaum & Zangerl 1998). A decade ago, there was a similar lack of evidence for the cost of defence in plants, but increased study and refined methods have recently provided numerous examples (Bergelson & Purrington 1996). It seems reasonable to expect that examples in insects will emerge in the near future.

3.5.3 Conjugation

Insects employ other means of coping with toxins that can be as effective or more effective than detoxification. Conjugation is a counter-measure that relies on the ability of the insect to produce a compound that binds with harmful compounds in such a way as to render them harmless. A prime example involves a counter-measure to tannins, defensive compounds in plants that bind to proteins and ostensibly prevent them from being digested. While the ability of tannins to bind with proteins is well established, it is far from clear that insects feeding on tannin-rich diets suffer from reduced digestibility (Schultz 1989), suggesting that herbivores possess counter-adaptations. It is now known that mammals produce proline-rich salivary proteins that conjugate with dietary tannins, thereby reducing the ability of the tannins to interfere with digestion of other proteins (Ann et al. 1997). Recently, it has been discovered that lepidopteran digestive juice is rich in free glycine (Konno et al. 1996). Gut glycine concentrations are as much as 20-fold higher than any other amino acid, and glycine occurs most abundantly in the anterior portion of the midgut, where much of the digestion of food takes place. The addition of glycine, in concentrations comparable to those observed in larvae, prevented protein denaturation in *Ligustrum obtusifolium*, the host of larvae of several of the species displaying glycine-rich gut contents (Konno et al. 1998).

3.5.4 Target site insensitivity

All toxins exert their effects by interacting with chemicals in the herbivore. In some cases the 'targeted' compounds include a variety of biomolecules, but for other toxins the targets are highly specific. Many of the neurotoxins, for example, bind to specific receptors, for example, pyrethrins, which are terpenoids first extracted from *Chrysanthemum*, target sodium channel receptor proteins. Exposure of an insect to pyrethroids results in an immediate 'knock-down', i.e. the insect falls to the ground because its nervous system has been disabled. Resistance to pyrethroids appears to be due to *kdr* ('knock-down resistance') mutations, which, in several cases, cause alterations in sodium channel proteins. These mutations, shared by at least three phylogenetically distant species — diamondback moth, house fly and German cockroach — prevent binding of pyrethroids, and thus confer target-site insensitivity (Schuler et al. 1998).

Target-site insensitivity may be the principal trait that allows insects to store or sequester toxins for use in their own defence against predators. The monarch butterfly is famous for its ability to sequester cardiac glycosides from its host plants. The target sites of ouabain, one of these cardiac glycosides, are the sodium and potassium ATPases responsible for maintaining ion gradients across cell membranes. Monarch ATPase differs from ouabain-sensitive ATPases of other insect species in that it contains a histidine instead of an asparagine at a key position in the protein. By genetically engineering this amino substitution in a previously ouabain-sensitive *Drosophila* strain, Holzinger and Wink (1996) were able to produce a strain that was ouabain-resistant, confirming that this single amino acid substitution is responsible for resistance.

In the case of protease inhibitors, target-site insensitivity is an inducible condition. Proteases are enzymes that are essential for the digestion of proteins, and protease inhibitors are substances produced by plants that inhibit this process. When food rich in protease inhibitors is ingested by insects resistant to protease inhibitors, alternative proteolytic enzymes synthesized by the insect are immune to the inhibitor (Broadway 1996).

Although the frequency with which target-site insensitivity is employed by insects in defence against host-plant allelochemicals is not yet clear, this mode of resistance may have distinct advantages beyond those associated with sequestration. Once target-site insensitivity is acquired, and assuming the insensitivity does not interfere with normal insect functions, the mechanism is essentially cost-free because the insect merely produces an altered form of the target instead of the susceptible form. The cost of producing inhibitor-resistant proteases may be substantially less than the likely alternative of producing proteases in excess of the amount inhibited.

3.5.5 Excretion

Excretion of unaltered allelochemicals is yet another mechanism by which insects may avoid toxic effects. Excretion was thought to be the principal means by which the tobacco hornworm *Manduca sexta* dealt with nicotine. The Malpighian tubules, which are excretory structures located in the anterior portion of the hindgut, were shown to actively absorb nicotine from the haemolymph and transfer it to the outgoing frass (Maddrell & Gardiner 1976). On closer examination, however, it has been shown that the majority of nicotine ingested by this insect is not excreted but is metabolized by cytochrome P450s (Snyder et al. 1994). Whether excretion is an important and widespread phenomenon remains unknown.

3.6 Plant defences (generally) reduce the fitness of herbivores

The effects of plant defences on insect fitness range from subtle reductions in insect growth rate, resulting ultimately in reduced numbers of offspring produced, to outright death. There are countless examples where extracts of plants have been shown to affect insects adversely but relatively few that show a relationship between insect fitness and variation in plant defence traits. Responses often differ between species, called generalists, that feed on a wide variety of host species, and specialists, species that feed on only one or a handful of species, and both groups differ from non-adapted species that ordinarily do not encounter the defence. Non-adapted species are often intolerant of allelochemicals, even at low doses. Generalists, the minority of insect herbivores that consume a large variety of plants, tend to experience ill effects from toxins in proportion to the amounts that they consume, while specialists are generally less sensitive to allelochemicals. For example, caterpillars of the generalist gypsy moth (*Lymantria dispar*) grew poorly when given a diet containing iridoid glycosides, a class of allelochemicals, while the specialist buckeye butterfly *Junonia coenia* fared better with the allelochemicals than without them (Bowers & Puttick 1988). Many specialists, however, also suffer when allelochemicals are present in increasing concentrations. Survivorship of the specialist insect on long-leaf plantain, *Plantago lanceolata*, was negatively correlated with the concentration of aucubin, a specific iridoid glycoside (Adler et al. 1995).

3.7 Herbivore feeding (generally) reduces plant fitness

Despite the fact that some plants can compensate or over-compensate for sizeable amounts of damage, in general, tissue removal by herbivores decreases plant fitness (reviewed in Bigger & Marvier 1998). As a group, insect herbivores tend to have larger effects than mammalian herbivores on plant growth and reproduction, and this pattern is especially true for forbs as opposed to grasses (Bigger & Marvier 1998). Herbivores remove on average 18% of terrestrial plant biomass and 51% of aquatic biomass (Cyr & Pace 1993). To examine the impact of insect herbivores on tree growth in a eucalyptus forest, Morrow and LaMarche (1978) sprayed eucalyptus trees with insecticide for several years. They then used tree-ring analysis to examine the impact of insect herbivores on tree growth. There was a remarkable increase in growth in the years with reduced insect loads compared to unsprayed control trees; after three years, sprayed trees were 100% taller than unsprayed trees in a nearby site (Fox & Morrow 1992). For the tropical understorey pepper plant *Piper arieianum*, even 10% leaf-area loss reduced growth and reproduction of the plant (Marquis 1984). Sprayed oak trees in Britain produced 2.5–4.5 times as many acorns as their unsprayed counterparts (Crawley 1985). Insect herbivores with different feeding modes can differentially affect plant fitness, and it is not always easy to predict which type of herbivore is most injurious. In goldenrods, one or two spittlebugs had a greater impact on plant fitness than did either hundreds of aphids or several leaf-beetle larvae (Meyer 1993).

3.7.1 Attack from multiple herbivores may have synergistic impacts on plant fitness

Because plants are a dynamic resource and respond to damage, when a plant is attacked by one herbivore it may become more or less vulnerable to attack by others (it may experience induced susceptibility or induced resistance). Effects of multiple species of herbivores that attack plants simultaneously or in succession can have either additive or multiplicative effects on plant fitness. Sumach stems attacked by spring-feeding leaf-feeding chrysomelids were more likely to be subsequently attacked by summer stem-boring beetles and less likely to be winter-browsed by deer (Strauss 1991). Beetle-attacked stems had lower growth rates and reproduction than did browsed stems, and stems attacked by both beetles suffered the greatest mortality (Fig. 3.11). Similarly, feeding on apical meristems of goldenrods caused increased branching, and then greater attack rates by aphids (Pilson 1992). Through both architectural and chemical changes, early-feeding herbivores can alter the suitability of host plants for later-feeding species. The extensive literature on induced defences has shown that damage by one herbivore can alter host-plant suitability for other herbivores (see the section on 'The evolution of plant defence' below; Karban & Baldwin 1997). Such interactions can be used to our advantage: human-imposed damage by one species of spider mite early in the season on grapevines reduced the performance of another, more harmful mite species later in the season, and increased the quality, though not the quantity, of the grape harvest (Karban et al. 1997).

There are also cases in which feeding by different types of herbivore affects plant fitness more or less independently. In the bush lupin, *Lupinus arboreus*, there was no statistical interaction between herbivory above and below ground. Both types of herbivory had significant cumulative effects on lupin fitness. Protection from chronic above-ground herbivory increased mean cumulative seed output over three years by 78%, while suppression of below-ground herbivores increased mean cumulative seed production by 31%. Root herbivory was also associated with a higher risk of mortality (Maron 1998).

3.7.2 Herbivores may reduce plant fitness by acting as vectors for disease

Sometimes, the actual damage an insect does is not as

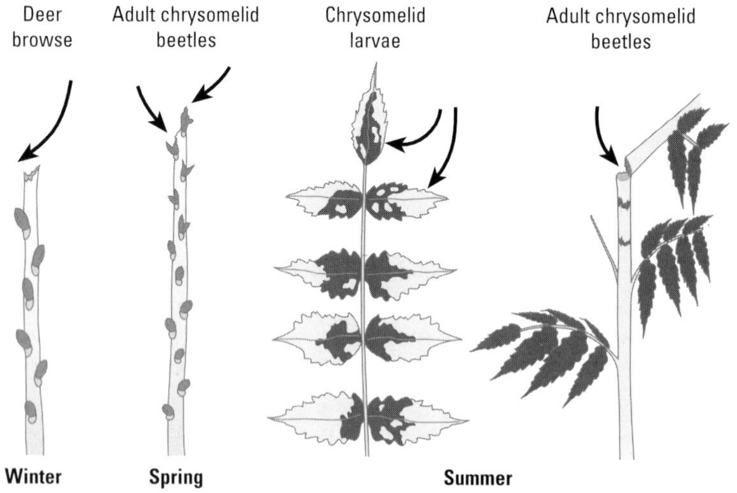

Deer browse **Adult chrysomelid beetles** **Chrysomelid larvae** **Adult chrysomelid beetles**

Winter   **Spring**   **Summer**

Figure 3.11 Multiple herbivores can have synergistic effects on total patterns of herbivory. Feeding patterns, interactions and effects of three different herbivores on sumach growth and reproduction. (From Strauss 1991.)

harmful as the opportunities the insect provides for infection by fungi and other plant diseases. Some herbivores actually enlist pathogens to enable them to feed on plants. Pits on the elytra of newly emerging *Ips pini* carry spores like those of the tree-pathogenic blue-stain fungus *Ophiostoma ips* and other fungi (Furniss et al. 1995). *Ophiostoma ips* ascospores develop on the walls of pupal chambers in the phloem of infested pines. The fungus-infected wood serves as the breeding grounds for larvae, which cannot survive in healthy trees. Fungal spores adhere to newly metamorphosed *I. pini* that then introduce the blue-stain fungus into new trees. Dutch elm disease, which struck North American forests in the 1930s and removed the elm as a major canopy-tree species, was vectored by the introduced elm-bark beetle that carries the fungus *Ceratocystis ulmi* from tree to tree. At least 380 plant viruses are known to be vectored by insects, primarily hemipterans, but also thrips, beetles and mites (Nault 1997). Some insects are such effective vectors that they have been evaluated for use in the biological control of weeds and as means of inoculation in pathogen-resistance screening programmes (Vega et al. 1995; Narayana & Muniyappa 1996; Pico et al. 1998).

3.7.3 Herbivores reduce plant fitness by impeding relationships with mutualists

Leaf and floral damage to plants affects floral traits and floral display and thus can act indirectly through pollinators to reduce plant fitness. Caterpillar damage to leaves caused plants to have reduced flower size and attractiveness to pollinators (Strauss et al. 1996; Lehtila & Strauss 1997). Because of effects on floral traits, damaged plants may be more pollinator-limited than their undamaged counterparts (Juenger & Bergelson 1997; Mothershead & Marquis 2000). Leaf damage may also decrease the ability of pollen to sire seed on other plants, and thus may decrease male components of plant fitness (Mutikainen & Delph 1997; Strauss 1997). Finally, the presence of secondary compounds in floral parts may make floral rewards less palatable or abundant to pollinators and may alter pollinators' foraging behaviour (Strauss et al. 1996). The latter point raises the interesting issue of whether the amounts of secondary compounds incorporated into plant tissues, that is, the degree to which plants can evolve to become better defended, might be constrained by the preferences of beneficial pollinators.

A fascinating example of the long-term consequences of joint selection from both herbivores and pollinators on plant traits is provided by Armbruster and colleagues (Armbruster 1997; Armbruster et al. 1997). Using present-day observations of interactions between tropical *Dalechampia* vines and their herbivores and pollinators, as well as a detailed phylogenetic analysis of the relatedness among *Dalechampia* species and related genera, this group has reconstructed the history of traits and their joint use in pollination and defence (see Plate 3.1, facing p. 84). *Dalechampia* flowers originally produced resins to protect themselves from microbial and herbivore attack, as well as nectar to attract pollinators. The resin, which was originally a defence, became a reward for pollinators that used the resin to build their nests. Apparently, the antimicrobial properties of the resin also protect bees from fungal pathogens and disease. An adaptive radiation in *Dalechampia* is associated with the production of resin as the sole reward for specialized bee pollinators; little or no nectar is produced by these plants. Thus, a trait originally used in plant defence has assumed paramount importance in the plant–pollinator relationship. Similarly, one section of the same genus has flowers that are surrounded by showy bracts that close at night. Other members of the genus have these bracts, but they do not close. Experiments show that bract closure protects flowers from night-feeding herbivores. In this case, a trait that evolved initially to attract pollinators has been enlisted as part of the plant's defence.

Other mutualists are also important to plant fitness and may interact with herbivores indirectly via the plant. Mycorrhizae are root-associated fungi that enhance a plant's ability to gather nutrients, especially phosphorus and nitrogen (Gehring & Whitham 1994). In return, the plant supplies the fungus with carbohydrates for its growth and reproduction. In pinyon pines, above-ground herbivory decreases the colonization and biomass of beneficial ectomycorrhizal associates (Del Vecchio et al. 1993). Similarly legumes, plants in the pea family, have associated bacteria in the genus *Rhizobium* that fix soil nitrogen for use by the plant. These bacteria live in nodules made by the plant on its roots. Attack above-ground by aphids decreased nodulation and N fixation by *Rhizobium* symbionts of beans, as well as plant growth (Sirur & Barlow 1984).

In summary, through direct effects of damage, as well as through indirect effects that alter plant relationships with other herbivores, mutualists and/or pathogens, herbivores usually have detrimental effects on plant fitness. As such, they act as selective agents on plants and may drive the evolution of defensive traits.

3.8   The evolution of plant defence

In previous sections, we have described the many different kinds of physical and chemical traits that plants can deploy as defences against herbivorous insects. How much of a plant's resources are dedicated to defence, and when and where such defences are to be deployed, form the subject of an evolutionary theory of optimal defence addressed here.

3.8.1   Optimal defence theory

Close examination of the distribution and abundance of defences within plants reveals considerable variation among species. McKey (1974) was the first to attempt to make sense of this variation, using his knowledge of the location and concentrations of alkaloids in plants. He reasoned that these nitrogen-containing compounds were costly to the plant and, consequently, their distribution was governed principally by two factors—a tissue's vulnerability to herbivores and its fitness value to the plant. This notion forms the basis for what is known as 'optimal defence theory'. Stated in its simplest form, the theory predicts that selection will favour defence when the benefit of that defence exceeds its cost. Although McKey's ideas have been extended and refined during nearly three decades since their publication, the basic premise remains intact.

One aspect of a plant's vulnerability to insects is its conspicuousness. Feeny (1976) advanced the concept of apparency as a factor that influences the evolution of defences. Plant phenotypes that are less easily detected by a herbivore are less likely to suffer damage, and hence are less likely to possess defences. Factors intrinsic to a plant that may influence its apparency include visual appearance, chemical composition and the life-span of the plant or of a particular plant part. Any traits that make a plant more cryptic, visually or chemically, will lower the plant's conspicuousness. Plants or plant parts that are short-lived may avoid damage because the herbivores have not had sufficient time to discover them. Apparency is also affected by factors external to the plant, such as the density of herbivores and the species and densities of neighbouring plants. If herbivores are abundant, the chances that a plant will escape detection diminish. Non-host plants, in close proximity to host plants, may emit volatiles that mask the cues used by herbivores to detect their hosts. This phenomenon is called 'associational resistance' and is the basis of sustainable farming practices like intercropping, which reduce damage from herbivores by having chemically dissimilar crops like beans, corn and squash in alternating rows.

The value of a plant tissue, as envisioned by McKey (1974), is determined by the reduction in fitness of the plant that results from the loss of that tissue. Such costs are not fixed; if resources, such as light or nutrients, are in low supply in a plant's environment, plant growth will be constrained, and the ability of the plant to replace lost resources will be diminished. The impact of herbivory on the fitness of slow-growing plants is consequently high. By contrast, plants in resource-abundant environments grow fast and easily replace lost tissues. Based on these considerations, the Resource Availability Hypothesis (Coley et al. 1985) maintains that fast-growing plants have less need to defend themselves than resource-limited plants. In a study of 41 tropical tree species in Panama, Coley (1988) found a clear inverse relationship between growth rate and defence investment in the form of tannins and other constituents that reduce digestibility. An obvious question arises as to why fast-growing plants need not defend themselves as vigorously as slow-growing plants. A possible reason is that the fittest plants in resource-rich environments are those that grow the fastest; under such circumstances, plants that commit resources to defence have reduced growth potential and are eliminated by natural selection (Herms & Mattson 1992).

3.8.2 Inducible defences

A plant's defences need not be raised at all times—inducible defences are bolstered only after a plant has been attacked. The level of defence in a plant that has not been attacked is commonly referred to as 'constitutive'. Depending on the species of plant, the constitutive level of defence may be zero, i.e. the plant is undefended. In other plants, there is always present a non-zero level of defence. Regardless of the constitutive level of defence, many plants are damage-inducible and will augment constitutive levels (Karban & Baldwin 1997). The classic example of a defence induced by an insect herbivore involves the production of a protease inhibitor, which interferes with digestion of proteins. When Green and Ryan (1972) exposed potato and tomato plants to feeding by the Colorado potato beetle, protease inhibitors increased 3-fold in the plants. To make matters more interesting, the spatial pattern of induction varies considerably among species. The protease inhibitor response in tomatoes includes intact plant parts (systemic responses) (Stout et al. 1996) while other induced responses in tomato, involving oxidative enzymes, are restricted, more or less, to the plant part that is damaged (localized responses). Precisely which evolutionary factors dictate reliance on inducible defences is not clear; conventional wisdom holds that

defences are costly and plants that put off defence until it is required will be more fit (Karban & Baldwin 1997; Agrawal et al. 1999; Tollrian & Harvell 1999). Induced defences will be especially favoured if early attack in the season is predictive of later attack. Induced defences can also persist across seasons and may affect herbivore dynamics. Birch leaves produced by trees damaged the previous year in a moth outbreak supported significantly reduced larval growth in the next growing season than leaves from sprayed trees that were protected from moths during the same outbreak (Kaitaniemi 1999). Long-term effects of induced defences may explain some of the cyclical nature of insect outbreaks.

If induced changes in plants are really an adaptive response to attack, then, in the presence of herbivores, induced plants should have higher fitness than plants that were not previously induced. Two field studies, one on wild radish (*Raphanus raphanistrum*) and one with wild tobacco (*Nicotiana attenuata*), support this notion (Agrawal 1998a; Baldwin 1998). Plants experimentally induced early in the season with methyl jasmonate (a form of a plant-signalling hormone) experienced less damage by insects later in the season, and had greater fitness, than uninduced plants. Thus, induced responses to damage appear to be a good example of adaptive phenotypic plasticity.

3.9 The evolution of resistance in insects

Unlike plants, most insects have the ability to move about at will. Thus, many options are available to herbivores that are not available to plants. For example, insects may avoid particularly unpalatable plants or defended plant parts. Nevertheless, a plant can survive without its insect herbivores, but an insect cannot survive without its hosts. Consequently, insects must continually adapt to the barriers presented by their hosts, as evidenced by the variety of adaptations previously mentioned. The evolution of these adaptations is probably governed by many of the same selective forces as those that shape plant defensive traits. For example, the benefit of a resistance mechanism, such as detoxification, must exceed its cost. Similarly, inducible defences in plants probably arose as means of minimizing costs. In fact, inducibility is probably more prevalent among herbivores than among plants. Many of the enzyme systems involved in detoxification are inducible. Presumably, this is particularly true for generalist herbivores, which must tailor their detoxification repertoire to the species of host on which it is currently feeding. An aspect of the plant–insect interaction that is unique from the perspective of the insect is the potential use of plant defences for its own defence against predators.

3.9.1 Turning adversity to advantage

Adaptations in insects that alleviate or avoid the toxic effects of chemicals, without altering their toxic properties, afford the insect an opportunity to employ these compounds in their own defence. The most familiar example is that of the monarch butterfly, *Danaus plexippus*, whose larvae sequester cardenolides from their milkweed hosts (Brower 1958). Seiber et al. (1980) observed that more polar cardenolides are far more concentrated in the insect's tissues than they are in those of the host plant. Aldehyde and ketone reductases in the insect's gut are responsible for conversion of less polar plant cardenolides to more polar forms that can be sequestered (Marty & Krieger 1984). In addition to these enzymatic adaptations and target-site insensitivity, monarchs may also contain specific carrier proteins that convey the cardenolides from the gut into the body of the insect (Frick & Wink 1995).

The value of these 'second-hand' defences is perhaps best exemplified by the role they play in sexual selection in the arctiid moth *Utetheisa ornatrix*. Larvae of this species sequester toxic pyrrolizidine alkaloids from their host plants, and these toxins are retained by the adults and passed on by the females to their eggs. Thus, all stages of development are protected in this species. There is one other transfer that occurs during mating; the unusually large spermatophore that is transmitted from the male to the female contains nutrients and the alkaloids as a nuptial gift, the size of which is correlated with male size. Females tend to select larger males as their mates because doing so lessens the risk of predation of their offspring (Iyengar & Eisner 1999).

Sequestration of plant toxins is a widespread phenomenon in insects (Rowell-Rahier & Pasteels 1991), with well-documented cases in the orders Diptera, Coleoptera, Orthoptera, Hemiptera, Hymenoptera and Lepidoptera. Like the monarch, insects that have successfully adopted this mechanism of defence typically advertise the fact by being aposematically coloured in hues of red, black, yellow and/or orange.

Another consequence of adaptation to host-plant allelochemicals is the use of these compounds by insects as cues to identify suitable hosts (Feeny 1992). Remember

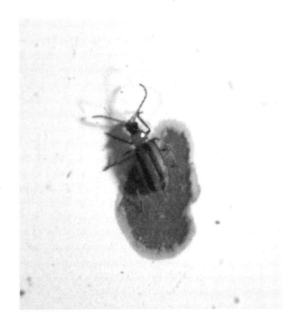

Figure 3.12 Chemical cues can be sufficient to trigger feeding in some herbivores. The striped cucumber beetle *Acalymma vittata* shown on a thin-layer silica chromatography plate where it has consumed all of the silica to which a feeding stimulant, cucurbitacin B, had been applied.

that secondary chemistry is extraordinarily diverse and that distributions of these chemicals are highly idiosyncratic. These properties have made secondary chemistry valuable traits to plant taxonomists who wish to infer phylogenetic relationships. Herbivorous insects make similar use of these chemicals to distinguish host plants from non-host plants. Volatile compounds may be used by adult females as olfactory cues to locate host plants. Nonvolatile chemicals may act as stimulants to oviposition and as feeding stimulants. Plants belonging to the family Cucurbitaceae contain cucurbitacins, chemicals that, despite being the bitterest compounds known to humans, stimulate feeding by cucurbit beetles. Confronted with a substance that is not remotely food-like but is impregnated with cucurbitacins, the beetles will feed voraciously. So compelling are these compounds as feeding stimulants that early researchers 'developed' their thin-layer chromatographic separations of cucurbitacins by allowing cucurbit beetles to eat away the silica where compounds were located (Metcalf et al. 1980; Fig. 3.12).

Not only have insects evolved a variety of counter-adaptations to plant defences, they have in some cases put these 'defences' to good use and increased their fitness.

The beneficial use of plant defences by primarily specialized species of herbivores places plants in a difficult evolutionary predicament. Their defences may be effective against a wide range of generalists, including non-arthropod species, but they may be handicaps with respect to specialists.

3.10 Defence and counter-defence may underlie the diversity of plant and insect species

Inasmuch as plants exert selection on insects and insects exert selection on plants, the potential exists for 'coevolution' between the interacting populations. Janzen (1980) defined coevolution as 'an evolutionary change in a trait of the individuals of one population in response to a trait of the individuals of a second population, followed by an evolutionary response of the second population to the change in the first'. Simply put, coevolution is reciprocal adaptation. The extent to which a plant and a herbivore coevolve will be determined by three things. First, the plant and the herbivore must be able to influence each other's fitness; in other words, each organism must exert selection on the other. Second, there must be genetic variation for characteristics of both plant and insect that influence the outcome of their interaction, and lastly, there must be a response by each organism to selection from the other.

The notion that coevolution between insect herbivores and their host plants might explain patterns of species diversity in plants and insects was first proposed by Ehrlich and Raven (1964). They noticed that phylogenetically related butterflies tended to have as their hosts phylogenetically related plants. They argued that such patterns of host use arose as a result of coevolution between plant and insect, a process involving alternating adaptive radiations. They posited that, initially, a unique genetic event (a mutation or recombination) occurs in a plant species that confers upon it resistance to most or all of its insect herbivores. Released from herbivory, the plant is able to occupy new adaptive zones and radiate. Subsequent to this adaptive radiation, an insect species experiences a genetic event that confers upon it the ability to overcome the novel plant defence. Released from competitors, the insect then undergoes an adaptive radiation of its own, and eventually these derived species will utilize many, if not all, of the plant species.

As intriguing as this idea is, testing it has been all but impossible. First, it is a phenomenon that cannot be observed while it is in progress because it occurs over a long period of time. Moreover, this scenario cannot be driven by the process of coevolution as it is understood to occur at the population level (Thompson 1999c) because there is no interaction between plant and herbivore during the time of plant diversification. Thus, coevolution as reciprocal adaptations between *interacting* populations cannot be causing plant diversification.

It is possible for population coevolution to lead to increased numbers of plant and herbivore species if interacting populations are isolated from one another. In this case, reduced gene flow among populations of plants, and also among populations of their interacting herbivores, could result in each plant–insect association having a unique evolutionary trajectory. Isolation of associations from others would prevent gene flow among populations and could result in reproductive isolation and eventual speciation. Under this scenario, plant and herbivore phylogenies may exhibit similar or congruent branching patterns, a phenomenon called 'parallel cladogenesis' (Fig. 3.11). The difficulty with comparisons of phylogenies is that the same pattern may have more than one interpretation. For example, parallel cladograms may also arise when insect herbivores track hosts that have radiated as a result of other factors, such as geographical isolation, hence a similar radiation can also occur in the absence of coevolution. If reciprocal adaptation and similar radiation can both result in parallel cladogenesis, the question immediately arises of how one might be able to distinguish between the two.

Ehrlich and Raven's hypothesis was based primarily on observations about plant chemistry, and, as discussed above, there is plentiful evidence that chemical properties of plants have impacts on insect fitness. If one could map the evolution of defences and countermeasures on cladograms, it might be possible to determine whether coevolution or similar radiation is responsible for parallel phylogenies. The best evidence to date to support the hypothesis of 'classic' coevolution using a phylogenetic approach comes from a study of beetles of the genus *Tetraopes* that feed on milkweeds in the genus *Asclepias*. Insect phylogenies, involving both *Tetraopes* species and their relatives, were reconstructed from morphological data and variation in enzymes associated principally with primary metabolism. Plant phylogenies were reconstructed based on morphological and anatomical characteristics. When Farrell and Mitter (1998) compared the phylogenies, they found strikingly similar patterns of

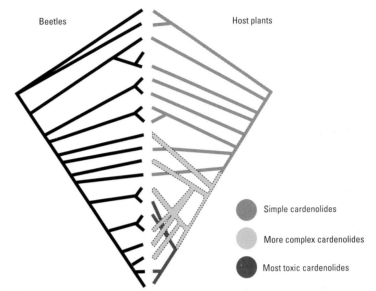

Beetles   Host plants

○ Simple cardenolides

○ More complex cardenolides

● Most toxic cardenolides

Figure 3.13 Parallel phylogenesis of *Tetraopes* species and *Asclepias* species (after Farrell & Mitter 1998). The pattern of diversification among the milkweed beetles of the genus *Tetraopes* and the milkweed genus *Asclepias* show a higher degree of parallel diversification than would be expected if the beetles had colonized the plant species at random.

insect and plant diversification (Fig. 3.13). Moreover, the most recently derived species of beetle are not only associated with the most recently derived *Asclepias*, but these younger plant groups have the most toxic cardenolides. As appealing as the *Tetraopes–Asclepias* study is, it still falls short of being able to distinguish between similar radiation and coevolution because information on the abilities of the beetle taxa to cope with cardenolides was unavailable.

Irrespective of its genesis, parallel cladogenesis is by no means the rule in plant–herbivore interactions — there are numerous examples in which there is no congruence between insect and plant lineages (Weintraub et al. 1995; Becerra 1997; Garin et al. 1999). One such study compared a phylogeny of beetles in the genus *Blepharida* to the phylogeny of its hosts, members of the genus *Bursera* (Becerra 1997). Interestingly, the insect phylogeny was matched far better by a cladogram of its hosts based on terpene chemistry than a phylogeny of its hosts based on similarity of DNA. In other words, more closely related beetles were more likely to share hosts of similar chemistry than hosts that were more closely related. In addition, lack of evidence for similar radiation does not preclude the possibility of coevolution. In the *Bursera* example, insects adapted to a particular chemical will appear as if they had not speciated with the plants when, in fact, they may have

been responding to selective forces generated by plant chemistry and were truly coevolving. An important component of using a phylogenetic approach entails specifying a mechanism that is the basis of the interaction and having *a priori* hypotheses for how these characters should map onto the phylogeny. Thus, the strength of the *Asclepias* example above is its prediction that chemicals become more toxic as the arms race escalates, and therefore we expect most toxic chemicals in the youngest groups.

3.10.1 Other hypotheses to explain patterns of diversification and host-plant use: the hybrid bridge hypothesis

Another hypothesis to explain the evolution of host-plant use by herbivores is the proposal that plant hybrid zones, through the presence of intermediate phenotypes, provide opportunities to herbivores for host-switching (Floate & Whitham, 1993). Floate and Whitham proposed that hybrids provide feeding cues from the original host-plant species mixed with attributes from the other parental plant species. This combination of traits in one phenotype might exert selection on herbivores to be able to detoxify or respond to traits belonging to the other parental plant species of the hybrids. Ultimately, they predicted that use of hybrids might allow a shift by herbivores onto the novel parental species. They examined this hypothesis in a hybrid zone of two cottonwood species, Fremont cottonwood (*Populus fremontii*) and narrowleaf cottonwood (*P. angustifolia*). In this system, the F1 hybrids (half Fremont/half narrowleaf in genotype) backcross with pure narrowleaf cottonwood parental species, but not with Fremont cottonwood. Backcrossing creates trees that grade in characters ranging from pure narrowleaf forms to F1 forms and everything in between. In contrast, because F1 hybrids do not interbreed with pure Fremont trees, there are no intermediates between pure Fremont trees and trees that are F1s. The authors predicted that there should have been more host shifts from narrowleaf to F1 cottonwoods because there were more intermediate forms in that transition. Four different insect gallers using narrowleaf cottonwood had switched onto the F1 hybrids, but not onto Fremont cottonwood. In contrast, of four different galler species that utilized Fremont cottonwood, none had switched to either the F1 hybrids or to narrowleaf cottonwood. These results are consistent with the hybrid bridge hypothesis.

The hybrid bridge hypothesis makes two unstated assumptions: (1) the factors determining host suitability (such as secondary compounds, etc.) are additively expressed in the hybrids and (2) gaps in phenotypic similarity limit the ability of insect species to switch hosts (Pilson 1999). For the former assumption, there are many cases in which we know that traits of one parental species are dominant over the other, so that hybrids have phenotypes similar to one species but different from the other (reviewed in Pilson 1999). Dominant traits would not be expressed in a gradient fashion, but rather in a qualitative, all-or-none way. Pilson surveyed species of *Helianthus* that co-occurred and did not hybridize, along with other species that co-occurred and did hybridize. She found that hybridizing pairs shared a mean of 5.6% of their fauna, whereas non-hybridizing pairs shared 6.6%. These results do not support the hybrid bridge hypothesis. Finally, in a separate study of planthoppers that feed on Hawaiian silverswords (Roderick and Metz 1997), there was no evidence that hybridization between silversword species promoted host switches. *Nesosydne* planthoppers are highly host-specific, and each species feeds on only one plant species or on closely related species that hybridize. Based on mitochondrial DNA sequences, insect diversification closely followed plant diversification. Patterns of host-plant use within plant hybrid zones did not support the hybrid bridge hypothesis.

3.10.2 Coevolution between plant and herbivore at the population level

What is lacking in all of the tree-based studies of coevolution is evidence that the insect exerts selection on its host plants and that plants exert selection on the insects (Janzen 1980). Such evidence is difficult to obtain from phylogenetic studies because the reciprocal selective forces can only be inferred from presumed adaptations that remain extant. A more tractable approach to studying coevolution and the reciprocal selective pressures imposed by plants and insect herbivores on each other is to examine microevolutionary forces affecting contemporary plant–herbivore interactions at the population level.

A case in point is the interaction of the wild parsnip and the parsnip webworm. Wild parsnip, *Pastinaca sativa*, is a member of the carrot family that was introduced from Europe to North America in the early seventeenth century by colonists who cultivated the plant for food. The parsnip webworm *Depressaria pastinacella* was introduced approximately 150 years later. The plant is now a weed in North America. The characters of interest in this interaction are the furanocoumarins, which are toxins produced by the plant, and the furanocoumarin detoxification capacity of the insect. Both plant and insect contain genetic variation for these traits and each acts as a selective agent on the other; the insect selects for increased furanocoumarin concentrations in the plants (Berenbaum et al. 1986) and the plant selects for increased detoxification capacity in the insect (Zangerl & Berenbaum 1992).

In populations where coevolutionary forces are particularly strong, there is an expectation that the frequencies of different plant phenotypes will match the frequency of insect phenotypes as a result of responses to reciprocal selection by both plant and insect. The frequencies of plant and insect phenotypes are likely to vary from population to population due, for example, to historical peculiarities and the degree of isolation, resulting in a geographical mosaic (Thompson 1994). Because the wild parsnip produces several kinds of furanocoumarins in differing amounts, and the parsnip webworm is capable of metabolizing each of them to differing degrees, the traits that influence their interaction are both quantitative and multivariate. Nevertheless, it is possible to assign individual plants and insects to one of four phenotypes characterized by the amounts of each of four furanocoumarins produced by the wild parsnip and by the rates of metabo-

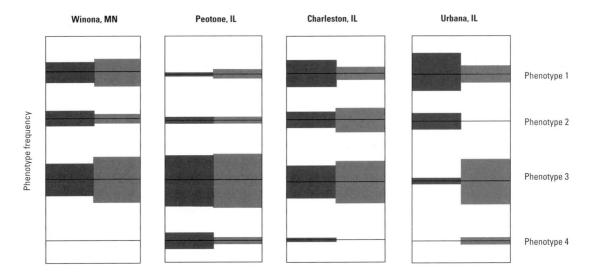

Figure 3.14 Population-level variation in plant defence and herbivore coping ability is one likely outcome of plant–herbivore interactions. Phenotype matching between wild parsnips and parsnip webworms in four midwestern populations. The degree of matching is close in all but the Urbana population. (From Berenbaum & Zangerl 1998.)

lism of those four furanocoumarins by the parsnip webworm. When frequencies of these phenotypes were compared between webworms and parsnips at four Midwestern locations, the quality of the match, with one exception, was remarkably close (Fig. 3.14). Not only are the frequency matches between plant and insect high, they also differ geographically, resulting in the mosaic envisioned by Thompson (1994).

The wild parsnip–parsnip webworm interaction illustrates the phenomenon of reciprocal selective impact of each participant on the other that is fundamental to the coevolutionary scenario described by Ehrlich and Raven. However, it also raises the question of whether a series of stepwise radiations is the only means by which coevolution results in parallel phylogenesis. Another possible path entails more or less simultaneous divergence of both insect and plant populations, leading over long periods of time to distinct interacting species. In the long run, the path travelled by coevolving herbivores and plants is of little consequence, because the result is the same—the insect exhibits adaptations that strengthen the association, while the plant exhibits adaptations to weaken it.

3.11 Plant–insect interactions and ecological communities

So far, we have addressed plant–insect interactions primarily from the perspective of single plant and insect species. We have shown that plant species often contain characteristic chemicals and structures that deter herbivores, and that some subsets of herbivores often have counter-adaptations that allow them use specific plants. We can also show many instances in which each organism has the capability to affect the other's fitness. We will now address some of the community-level consequences of insect herbivory. How do insect herbivores alter the species composition of an area? Do they affect plant distributions? Can they alter the rate at which communities change and the rate at which nutrients move through an ecosystem?

3.11.1 Effects of insect herbivores on plant distribution and abundance

The actions of mammalian herbivores can dramatically alter plant distributions (see Chapter 4); however, there have been fewer tests addressing this question for native insect herbivores. In California, seed-head herbivory by insects reduced the abundance of the shrub *Haplopappus squarrosus* in maritime areas where plants and seedlings had highest fitness. Peak densities of the shrubs were actually located further inland in hotter and drier areas, where the presence of these seed predators was greatly reduced (Louda 1982). Similarly, the mustard *Cardamine cordifolia* is restricted to shade habitats in the Rocky Mountains as a result of feeding by specialized herbivores (Louda & Rodman 1996). These herbivores prefer plants in the sun because they possess different defensive and nutritional leaf chemistry from conspecific plants located in the shade. There are a few more similar examples in the literature, but the role of native insect herbivores as agents that limit the distribution of their host plants has received relatively little attention.

We do know of many cases in which introduced insects can drastically alter the distribution and abundance of an introduced plant species. Introduced species are those that have been brought to new geographical areas either accidentally or intentionally. Some of these species can become invasive and can threaten native biodiversity as well as agricultural and range lands. One approach to stop the spread of invasive introduced plant species has been to introduce herbivores from native habitats as biological control agents. There have been a number of cases in which these biocontrol agents have proved to be the magic bullet against invasive species. Beetles from France now control the European Klamath weed (*Hypericum perforatum*) in California (Huffaker & Kennett 1959). The Brazilian weevil *Cyrtobagus salvinae* has freed Australian lakes choked with introduced floating fern, *Salvinia molesta* (Room et al. 1981). A classic example of how introduced herbivores have altered the distribution and abundance of introduced host plants is that of the cactus *Opuntia stricta*, introduced to Australia in 1839. These cacti spread rapidly and by 1925, over 240 000 km^2 were covered by *Opuntia* and inaccessible to humans and livestock. In 1925, the moth *Cactoblastis cactorum*, a herbivore of *Opuntia* in its native South America, was introduced to control *O. stricta*. By 1932 the moth had nearly eradicated *Opuntia*, and now one can find *Opuntia* and *Cactoblastis* only in small patches in Australia (Osmond & Monro 1981).

While these are the success stories, there are several caveats. Introduced herbivores can have large, unintended impacts on native plant and insect species. For example, *Cactoblastis* was also introduced to the Caribbean to control *Opuntia* there. The moth subsequently island-hopped to the Florida Keys and is threatening the last re-

maining population of the endangered cactus *Opuntia spinosissima* (Johnson & Stiling 1996). In another case, the weevil *Rhinocyllus conicus* was introduced to control exotic musk thistle (*Carduus nutans*) throughout the west. In Nebraska, the weevil has now switched to using a native thistle, Platte's thistle (*Cirsium canescens*), and could soon threaten another endangered species, Pitcher's thistle (*C. pitcheri*; Louda et al. 1997). So, in human-made situations in which plants and/or herbivores have escaped their suite of native enemies, we know that herbivores can strongly limit the distribution and abundance of plant species.

3.11.2 Effects of herbivores on plant communities

The impact of insect herbivory on plant community composition and structure is similar to that of mammals in many cases (see Chapter 4). Insect herbivory is generally greater on plants characteristic of early to intermediate successional stages (Brown et al. 1988; Brown & Gange 1992). Early colonists are typically fast-growing and rely on good colonizing ability and rapid reproduction to escape herbivores and competitors. They tend to have leaves that are more nutrient-rich. Herbivores that attack these species will tend to speed the rate of succession, as later plant invaders have an easier time out-competing these damaged species. Many preferred plant species actually fall within intermediate successional stages (Davidson 1993). Herbivory on these intermediate successional stages retards succession from an earlier stage (which normally get out-competed), and speeds succession to later stages by removing intermediate species from the community more rapidly than they would be out-competed by late successional species (Davidson 1993). In the cases studied, herbivory does not seem to alter the final outcome of succession, but does change the rate at which successional changes occur.

Insect herbivory can also affect the species diversity or composition of plant communities. Like mammals, insect herbivores that prefer competitively dominant plant species (and therefore inhibit their growth and reproduction) may serve as keystone predators that allow the persistence of other, competitively inferior plant species. For example, beetle herbivory on willows in sand dunes increases the relative abundance of herbaceous and other woody species by suppressing willow growth (Bach 1994). Similarly, herbivores that retard succession (see

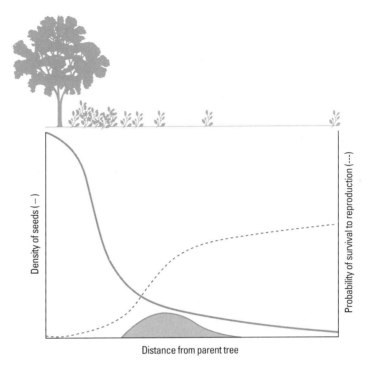

Figure 3.15 The Janzen–Connell hypothesis predicts that probability of seedling establishment is related to distance from the natal tree. The graph depicts expected relative distance from the parent tree where recruitment by a seed into the canopy is optimized. Density-dependent seed predation is high beneath the parent tree; few seeds are dispersed to distances far from the parent. Maximum probability of recruitment is at some intermediate distance from the parent tree.

above) may allow the persistence of competitively inferior plants. However, if herbivores prefer competitively inferior species, they may cause a decline in species richness (Brown & Gange 1989).

3.11.2.1 The Janzen–Connell hypothesis

Actions of specialized herbivores and pathogens are thought to be one cause of the high diversity of tree species in the tropics (proposed by Janzen (1970) and Connell (1971) independently and referred to as the Janzen–Connell hypothesis). The majority of seeds produced by a tree or plant land beneath the parent plant, regardless of whether seeds are dispersed by animals or wind, and seed densities decrease as distance from the parent plant increases. These 'seed shadows' provide varying levels of resources available to herbivores and pathogens. Since many insect herbivores and pathogens are extremely specialized with respect to the species they attack (see above), these specialists' enemies use seeds or seedlings of one species, but not necessarily those of a neighbouring, chemically disparate species. Janzen and Connell proposed that because seed and seedling resources are greatest under the

parent tree, mortality as a result of density-dependent attack by specialized herbivores and pathogens is also much greater there. Herbivores locate areas of high-resource density more readily and stay there longer (Root 1973). The likelihood that a seed will recruit into the canopy is therefore a function of its distance from the parent plant and the chance that it will be found and attacked by enemies (Fig. 3.15). Seeds immediately under the parent are consumed; on the other hand, a minute number are dispersed into habitats that are suitable and far from the parent tree. At some intermediate distance, the probability of a seed's arriving and establishing itself, and the likelihood of its escaping its enemies, is greatest. If similar recruitment processes occur for each species in the forest, then very few trees are expected to have conspecifics as neighbours. Density-dependent seed or seedling predation would thus lead to highly diverse forest tracts, as we see in many tropical forests.

What is the evidence for this scenario as it was proposed by Connell and Janzen? Numerous studies have addressed this hypothesis in several tropical forests. In Guyana, two tree species exhibited a pattern consistent with the Janzen–Connell hypothesis, but recruitment away from parents seemed more closely linked to light availability than to predation on seeds and seedlings (Zagt & Werger 1997). In an Amazonian forest, the midstorey palm *Astrocaryum murumuru* experienced density-dependent seed predation from insects, but there was no spatial relationship between distance from the parent tree and the likelihood of establishment. For the canopy-emergent legume *Dipteryx micrantha* predation and recruitment patterns were as predicted by the Janzen–Connell hypothesis in the first year, but in the second year recruitment was higher under the parent tree (Cintra 1997). On Barro Colorado Island in Panama, 67 of 83 tree species in the rainforest had population dynamics that were governed primarily by intraspecific, density-dependent processes (Wills et al. 1997). Recruitment patterns offered broad support for the mechanism proposed by Janzen–Connell in a large number of species, including the 10 most abundant species on the island. In general, it is clear that density-dependent predation on seeds and seedlings by herbivores can explain some of the patterns of plant community composition in tropical forests. The same cannot be said for temperate forests, however, where recruitment seems to be more spatially contagious, and where seed or seedling predation is not density-dependent as often (Houle 1995).

3.11.3 Effects of plant community diversity on insect herbivores

While it is clear that herbivores can affect plant community composition and species distribution, the reciprocal effect also exists: plant community composition affects insect herbivore loads. In a famous study, Root (1973) manipulated vegetational diversity to examine the impact of plant community composition on the distribution and abundance of insect herbivores. Collards, a crop in the mustard family, were grown in pure stands or in single rows interspersed among diverse meadow vegetation. Herbivores were sampled 20 times a year for three years and the numbers of individuals and biomass were recorded for each species. The diversity of herbivore species was greater on collards in the complex habitat; however, herbivore load, or the biomass of herbivores per biomass of consumable foliage, was greater in the monoculture, and this was due primarily to increased biomass of specialist insect herbivores. The diversity of predators and parasites was higher in mixed stands. Root proposed two hypotheses, which are not mutually exclusive, to explain

the observed pattern that monocultures received greater levels of attack from herbivores:

1 The 'resource concentration hypothesis': herbivores are more likely to find hosts that are concentrated, and herbivores remain longer on hosts growing in dense or pure stands.

2 The 'enemies hypothesis': increased diversity of predators and parasitoids in diverse stands may limit population densities of herbivores in these stands.

The idea that diverse plant community composition may result in reduced attack by herbivores has been called 'associational resistance'. This concept was broadened to include other ecological phenomena resulting from diverse plant assemblages (Atsatt & O'Dowd 1976). For example, 'insectary' plants provide nectar and energy for parasites and predators and can increase rates of parasitism on herbivores feeding on adjacent plants. Flowering plants surrounding fields are encouraged in organic and conventional farming for just such reasons. Insectary plants can also support alternate hosts of parasites until another herbivore host emerges. Thus, wild blackberries have herbivores that serve as hosts of the egg parasite *Anagrus* until grapevines leaf out, when the grape leafhopper becomes abundant and parasites switch to these herbivore hosts. 'Repellent' plants may repel herbivores sufficiently to provide shelter for palatable plant species. Volatile chemical deterrents may do this especially effectively, though physical protection, like growing near spiny plants, may also work. Using these principles has allowed intercropping and other planned plantings to be successful in reducing pest attack on agricultural species in many cases (Risch 1981). With intercropping, distinct unrelated crops are grown in alternating rows, and suffer less attack than equal amounts of crops planted in separate monocultures. A classic combination that effectively reduces attack is beans, squash and corn. Risch's study showed that not only did intercropped plants receive less damage, but also that this effect was a result of movement patterns and host use by the herbivores, not a result of increased predation on herbivores in polycultures.

The effectiveness of associational resistance is very dependent upon the spatial scale over which herbivores locate food. For example, buffalo and wildebeest grazed a lower proportion of *Themada triandra*, a palatable African savannah grass, from plots with higher proportions of less palatable secies. However, this associational resistance was not effective against zebra and Thompson gazelles, which are smaller and more selective feeders

(McNaughton 1978). The same kinds of phenomena have been documented for insects. Kareiva (1982), also working on collards, showed that the absolute dispersion of plants, as well as community composition, affected rates of herbivore attack. He used experimental arrays of collards surrounded by open ground or old field vegetation. He then manipulated the distance between individual plants as well as quality of plants (lush versus stunted plants), and observed the movement patterns of marked flea beetles in these arrays. Dispersal rates of herbivores increased as distances between patches of host plants decreased. In other words, patches formed stepping stones that facilitated movement. Beetles moved over open ground more readily than through diverse vegetation. Beetles also showed greater discrimination when patches of differing quality were close together. In other words, when resources were nearby, beetles were less likely to remain on stunted plants. The more mobile beetle species, *Phyllotreta cruciferae*, was better able to adjust its distribution in relation to stunted and lush patches of plants than *P. striolata* beetles, which moved less. Beetles abandoned lush patches less frequently than stunted patches, a result that supports the ideas proposed in the resource concentration hypothesis. Thus, both community composition and the dispersal abilities of herbivores in relation to the scale of community diversity will affect the degree to which plants receive damage from herbivores.

3.11.4 Effects of herbivores and plant defensive compounds on nutrient cycling through ecosystems

Herbivores can influence both plant dynamics and nutrient cycling through their feeding preferences. In several systems, legumes are preferred by herbivores over non-legume species (Hulme 1996c; Ritchie et al. 1998). For example, herbivore exclusion (both insects and mammals) in a nitrogen-limited old field in Minnesota caused increased biomass of the legume *Lathyrus venosus* and a few species of woody plants. Herbivore exclusion also modified N cycling by resulting in increased N content in above-ground plant tissue early in the growing season, and increased soil nitrate and total available N concentrations. N cycling was decelerated by herbivores by reducing the abundance of plant species with nitrogen-rich tissues (Ritchie et al. 1998). Herbivores may therefore indirectly control productivity, N cycling and succession by

consuming nitrogen-fixing and woody plants that have strong effects on plant resources.

Insect herbivory also alters nutrient cycles because insects feed on green leaves, which have greater nutrient concentrations than senescent leaves. In addition, a large amount of nutrient-rich litter falls as frass earlier in the season than typical autumnal litter fall. In southern Appalachian forests, concentrations of nitrogen in greenfall, i.e. green leaves falling as a direct result of herbivore feeding activity, occurring in May to September were always higher than concentrations of nitrogen in ageing autumn leaves (Risley & Crossley 1993). Annual inputs of greenfall nitrogen ranged from 0.08 to $0.18\,g/m^2/yr$.

Greenfall is viewed as a high-quality substrate supplying nitrogen to decomposer organisms; rapid use of nutrients in greenfall by microbes may or may not be returned efficiently to the plant community. Gypsy moth defoliation in oak forests caused a statistically significant increase in the quantities of N, P and K, and a significant decrease in the quantity of Ca contained in litter fall (Grace 1986). In undefoliated plots, 90% of the litter was deposited during the autumn, and tree leaves were the major litter component. Major litter components within the defoliated plots included insect frass, leaf fragments and tree leaves; 56% of the litter was deposited during the growing season. The nutrient content of litter comprising tree leaves on the undefoliated plots contributed 85% of all nutrients. In defoliated plots, the nutrient content of tree leaves contributed about 30% of all nutrients.

Outbreaks of the California oak moth, *Phryganidia californica*, caused increased inputs of N and P into soils during outbreak years (Hollinger 1986). The composition of the litter during the outbreak year shifted so that for *Quercus agrifolia*, almost 70% of the total N and P flow to the ground moved through frass and insect remains, while for *Quercus lobata*, approximately 60% of the N and 40% of the P moved through frass and insect remains. Short-term leaching experiments showed that nitrogen was far more rapidly lost from *Phryganidia* frass than from leaf litter of either species (Hollinger 1986). Similarly, honeydew produced by aphids (*Pterocallis alni*) infesting red alder (*Alnus rubra*) caused a reduction in available soil nitrogen, nitrogen mineralization rates, above-ground net primary production and nitrogen uptake by trees (Grier & Vogt 1990). In summary, herbivory from both leaf-feeding and sap-sucking insects can dramatically alter the timing and availability of nutrients to plants, and may impose contemporaneous as well as time-lagged responses to herbivory via altered soil nutrient regimes.

Herbivory from insects can also indirectly affect nutrient cycling by selecting for plants with particular kinds or higher concentrations of secondary compounds. Secondary compounds themselves affect nutrient cycling rates. In primary succession on river floodplains, succession goes from alder (*Alnus tenuifolia*) to balsam poplar (*Populus balsamifera*). This is a shift from an N-fixing shrub to a deciduous tree. Through this transition there are major changes in N cycling including a decrease in N-fixation, mineralization and nitrification. Secondary compounds present in balsam poplar may play an important role in soil changes. Poplar tannins inhibited both N-fixation in alder and decomposition and N mineralization in alder soils. Other poplar compounds, including low-molecular-weight phenolics, were microbial substrates and increased microbial growth and immobilization, thereby reducing net soil N availability to plants. Thus, substantial changes in soil N cycling through succession appear to be mediated by balsam poplar secondary chemicals. If selection from herbivory plays a role in concentration of these compounds (see Hwang & Lindroth 1997 for evidence in a related species), then herbivores may indirectly affect soils in these successional landscapes through selection of secondary compounds.

3.12 Conclusion

In summary, insect herbivores respond to selection exerted by plant defences and nutritional status. Plants strongly affect insect fitness so that, in general, insect herbivores are relatively specialized with respect to their diet breadth (in comparison with mammalian herbivores). Part of this specialization may arise from the fact that many insects can spend most of their larval development associated with only one or a few host plants. Plants affect insect abundance through their defences, which often entail the actions of other species, such as predaceous and parasitic enemies of herbivores.

Insects in turn affect plant fitness, and may exert selection on plant defences, both physical and chemical. There is a growing body of evidence suggesting that these defences come at some cost to the plant. On a larger ecological scale, insects affect plant distribution and abundance, as well as the species diversity of plant communities. Frass, honeydew and greenfall from insect outbreaks also alter nutrient cycling regimes in the soil and the availability of nutrients to plants.

Finally, many of the adaptations and counter-adaptations of plants and their insect herbivores support the idea that much of the biodiversity of the earth is a result of the arms race between insect herbivores and their host plants. Modern tools involving molecular genetic techniques and the building of phylogenetic trees are providing new opportunities to evaluate the long-term evolutionary relationships between groups of insect herbivores and their host plants. To date, several of these studies support the idea that reciprocal selective pressures exerted by plants and herbivores have been linked to extraordinary speciation rates in both groups.

Chapter 4 Mammalian herbivory in terrestrial environments

Kjell Danell and Roger Bergström

4.1 Introduction

Herbivory is a common trait among mammals. At least 50% of more than 1000 mammal genera in the world include plants in their diet. There is thus a potential for a tremendous number of intriguing relations between mammalian herbivores and plants in a variety of habitats. However, many of these are unknown, and we still lack good information on what about 15% of the mammalian genera really feed upon.

In this chapter, we focus on mammalian herbivores (see Plate 4.1a, facing p. 84) and their interactions with food plants. First we set the scene by presenting the plants as food, and second, we discuss the mammals as consumers, including their food-plant selection. Third, we describe the responses to herbivore feeding on individual plants, plant populations, communities and ecosystems. Finally, we will explore the feedback of these responses to the herbivores. As you will see, mammalian herbivory is more than a minor cropping of primary plant production.

4.1.1 Vertebrate versus invertebrate herbivory

Herbivorous vertebrates and invertebrates often occur in the same habitat. Do vertebrates make a more serious impact on the plants than invertebrates, and is it of a different type? The herbivores of these two groups differ in many aspects (Table 4.1). Their body masses rarely overlap, and the metabolic rate (weight for weight) of homoeotherms is about 30 times higher than for poikilotherms (McNeill & Lawton 1970). Invertebrate herbivores, on the other hand, are probably ten times more abundant than vertebrate herbivores (Peters 1983). A rough estimate is that herbivores remove around 10% of the net primary production in terrestrial environments (Crawley 1983). For the Nylsvley savanna in southern Africa, grazers and browsers remove 6%, and grasshoppers and caterpillars 5% of the above-ground primary production (Fig. 4.1). The majority of the herbivorous insects are mono- or oligophagous and feed on a few plant species, while vertebrate herbivores are more polyphagous and have a wider range of food plants. The larger body size of vertebrates also implies that many plant modules are taken per mouthful, while a typical invertebrate can grab only a part of the module. Because of their greater body size, polyphagy, individual bite size, mobility and tolerance of starvation, vertebrate herbivores should have a more immediate, and in the long term a more profound, impact on plant populations than invertebrate herbivores.

4.2 Plants as food

4.2.1 How much plant biomass is there in nature, and is it available?

The amount of food differs between biomes. The tundra has a primary production of only about $140\,g\,m^{-2}\,yr^{-1}$, while swamps and marshes reach about $3000\,g\,m^{-2}\,yr^{-1}$, i.e. a 20-fold difference between the extremes (Archibold 1995). The plant biomass, or standing crop, shows an even greater range between the least and most productive biomes, i.e. a 75-fold difference from about $600\,g\,m^{-2}$ in the tundra to 45 000 in tropical rainforests.

Estimates of food resources are vital for understanding the relations between plants and herbivores. As an example, consider the estimates of standing crop and production of grasses in two red-deer (*Cervus elaphus*) habitats shown in Table 4.2. In one habitat the production of grass was high but the standing crop low, and in the other habitat it was the opposite. This illustrates nicely the different pictures given by two commonly used methods, and the need for estimates that capture both the static and dynamic situations of the food resources.

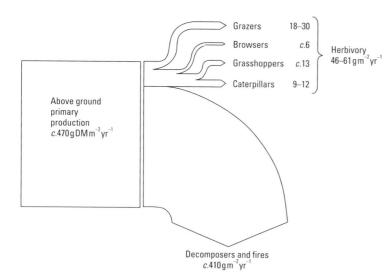

Figure 4.1 Pathways of the disappearance of plant-leaf material in the broad-leaved savanna at Nylsvley, South Africa, and their approximate magnitude. (From Scholes & Walker 1993.)

Table 4.1 A comparative summary of the main characteristics of invertebrate and vertebrate herbivores. (Based on Crawley 1989.)

Characteristic	Invertebrates	Vertebrates
Body size	Small	Large
Metabolic rate	Low	High
Population density	Large	Small
Food specificity	High	Low
Bite size	Small	Large
Mobility	Low–high	Low–high
Starvation tolerance	Low	High

Table 4.2 Standing crop and production of forage in two red deer (*Cervus elaphus*) habitats on Rhum Island, Scotland. Figures shown are mean values (± half-width of 95% confidence intervals), expressed as g dry weight per m^2. (From Clutton-Brock et al. 1982.)

	Short greens	Long *Agrostis/Festuca* and *Molinia*
Standing crop		
April	20 ± 10	324 ± 32
July	58 ± 12	402 ± 35
October	52 ± 13	461 ± 38
Production		
April–October	207 ± 48	173 ± 102

Data on production and standing crop of plants provide, however, only a vague picture of the food situation for herbivores, because not everything is actually available. The limitations are of both a physical and a 'nutritive' nature. For example, the leaves in the top of an acacia tree are physically unavailable to a black rhino (*Diceros bicornis*), but not to a giraffe (*Giraffa camelopardalis*). Mosses on the boreal forest floor are unavailable for the squirrel (*Sciurus* spp.) from a nutritive point of view, but not for the wood-lemming (*Myopus schisticolor*) which is a moss specialist. Such information needs to be incorporated into the design of studies of food resources.

4.2.2 Spatial and temporal variation in food

Mammalian herbivores are exposed to spatial and temporal variation in the abundance and quality of their food plants, and the level of variation differs between habitats. For example, the Arctic hare (*Lepus arcticus*) meets shoots of just a few plant species when feeding in the tundra in winter, while a primate feeding in the tropical rainforest has hundreds of possible food plants to chose from, and for each of these there are different parts such as leaves, shoots, flowers and fruits.

Spatial variation occurs on all possible levels, from the globe to the individual leaf. The variation in the distribution of light, temperature and nutrients creates large variations in plant quantity and quality, which we notice in gradients and patches. Most obvious is the large variation

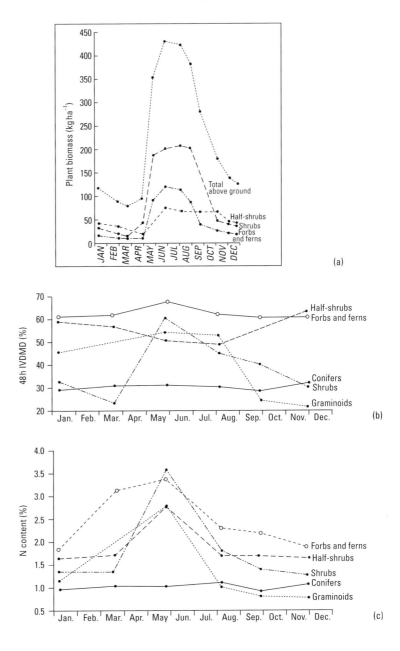

Figure 4.2 Monthly changes in (a) plant biomass (current annual growth), (b) *in vitro* dry matter digestibility (IVDMD) and (c) nitrogen content of different groups of understorey plants in south-eastern Alaska. (From Hanley et al. 1989.)

among biomes, but considerable spatial variation is also recognizable on smaller spatial scales. For example, the more vigorously growing shoots in the top of trees are often better food for herbivores than the smaller shoots in the lower part of the crown (Price 1991b), and the shoots of mature deciduous trees are more palatable than shoots from juvenile trees (e.g. Bryant 1981). The herbivores can also detect a genotypic variation in the food plants. Some

groups of lodgepole pine (*Pinus contorta*) were more palatable to microtine rodents, roe deer (*Capreolus capreolus*) and moose (*Alces alces*) than others, especially the trees of southern origin (Hansson 1985).

The temporal variation in food plants is seen as differences in food resources between years, months and weeks (Table 4.2, Fig. 4.2a). Differences in weather conditions between years are reflected in the quantity and quality of

herbivores. Mammals, like all other living organisms, have a perverse tendency to defy exact classification (Eisenberg 1981). Sixteen different categories of dietary specialization have been proposed, and seven of them refer to herbivores (Table 4.3). Of the 560 world mammalian genera classified as herbivores, the nectarivores and gumivores together constitute only 3% of the genera (Table 4.4), while the herbivores/browsers and frugivores/granivores groups each includes around 10% of total genera. More common are frugivores/herbivores and herbivores/grazers, each representing around 20% of the mammalian genera. Frugivores/omnivores make up one third of the genera, and are thus the numerically dominant group. The relative proportions of the different feeding types of mammalian herbivores remain more or less constant across continents (Table 4.4). However, nectarivores and gumivores are more common than average in Australia, as are frugivores/omnivores in the Americas, frugivores/granivores in North America, frugivores/herbivores in Australia and Europe, herbivores/browsers in Africa and herbivores/grazers in Africa and Europe. The existence of such large quite undefined group such as 'frugivore/omnivores' indicates, among other things, that we need to elaborate the system for the classification of mammalian herbivores (Langer & Chivers 1994).

4.3.2 Constraints on mammalian herbivores

In general, specialization implies greater access and efficiency in exploiting the resources, but specialization also entails restrictions for different reasons. Many factors constrain the mammalian herbivores and often these constraints are correlated with the body mass. The most obvious constraints are food and habitat structure, but predators can also be important.

Feeding on nectar is mainly restricted to small mammals ranging between 10 and 100 g. Feeding on grasses and woody plants, on the other hand, is not a trait of the smallest mammals; instead, grazers and browsers are generally larger and may exceed a ton in weight. The metabolic requirements of mammals increase with (body

Table 4.3 Major categories of dietary specialization in herbivorous mammals, according to Eisenberg (1981).

Category	Diet
Nectarivores	Nectar and pollen
Gumivores	Exudates from trees
Frugivore/omnivore	Pericarp or the fleshy outer covering of plant reproductive parts, invertebrates and small vertebrates
Frugivore/granivore	Reproductive parts of plants, including seeds
Frugivore/herbivore	Fleshy fruiting bodies and seeds of plants, storage roots and some green leafy material
Herbivore/browser	Stems, twigs, buds and leaves
Herbivore/grazer	Grasses

Table 4.4 Percentage of different feeding specialization among the 560 extant genera of mammalian herbivores. The feeding classes follow Eisenberg (1981) and are described in Table 4.3. Data on distribution of the different mammalian genera and feeding types are compiled from Nowak (1999). The information currently available on the feeding habits of many mammalian genera is still very poor, so the figures shown should be taken as reasonable approximations.

Feeding type	World	Africa	Asia	Australia	Europe	North America	South America
Nectarivore	2	2	2	7	0	0	1
Gumivore	1	1	0	5	0	1	2
Frugivore/omnivore	33	27	29	19	21	42	44
Frugivore/granivore	11	8	7	5	8	15	11
Frugivore/herbivore	22	20	26	30	30	18	20
Herbivore/browser	8	11	8	7	4	6	6
Herbivore/grazer	23	30	28	28	38	18	16

mass)$^{0.75}$ (Kleiber 1932), but the capacity of the gastrointestinal tract with (body mass)$^{1.0}$ (Parra 1978; Demment & Van Soest 1985). Smaller animals thus have higher mass-specific food requirements without any accompanying proportional increase in the gut capacity, which limits the volume of digesta retained and its passage (Iason & Van Wieren 1999). The larger animals thus have an advantage in this respect. How small can the mammalian herbivores be then? Cork (1994) gave a minimum estimate of 15 kg for ruminants and 1 kg for mammals feeding on a grass diet containing more than 50% fibre. Obviously, there are mammalian herbivores that have beaten the allometric theory by becoming smaller through various refinements of their digestive apparatus and processes.

Substrate specialization also means constraints. For example, fossorial mammals are small and have legs specialized for burrowing, but that limits mobility (Eisenberg 1981). Likewise, arboreal animals cannot be too large because of the difficulty of climbing and moving in trees; the upper limit of arboreal herbivores is 13–15 kg (Cork & Foley 1991). Most of the largest primates are mainly ground-dwelling, while the predominantly tree-dwelling ones include smaller species. Arboreal herbivores cannot be too heavy, because then the twigs break, nor so large that they can overcome the physiological difficulties in feeding on the most obvious foods in the trees, twigs and leaves. The vast majority of arboreal mammals select low-fibre, high-energy foods like fruits, flowers and invertebrates (Cork 1994). A few mammals are both arboreal and principally folivorous; all these are heavier than 700 g, and most are between 5 and 15 kg. The most intensively folivorous arboreal mammals are marsupials (Cork 1994). The koala (*Phascolarctos cinereus*) (5–13 kg) is highly folivorous and includes a large proportion of mature foliage. The greater glider (*Petauroides volans*) (0.8–1.7 kg) is also mainly folivorous, but selects young foliage. The smallest forest-dwelling marsupials avoid foliage and feed largely on sap, nectar and/or invertebrates (Cork & Foley 1991).

4.3.3 The digestive apparatus

In the mouth the food receives its first treatment, by mastication and mixing with saliva. Mastication increases the rate of digestion by facilitating microbial attachment and the release of cellular components. Its effectiveness depends on the time allocated to chewing, the rate and strength of jaw movements, and the morphology and structure of the teeth. There is a tradeoff between the rate of intake and the time allowed for chewing.

The main food values in plant materials are the digestible energy and proteins. The digestive system of the animal has to break the cell walls in order to release their contents (e.g. proteins and soluble carbohydrates) and to use the available sources of energy in the cell walls. With age the cell walls become thicker and harder (lignified) and the plant material more difficult to digest.

In animals, the structural carbohydrates cannot be digested without symbiotic gut organisms because they lack cellulases, hemicellulases or pectinases (Van Soest 1996). These enzymes are, on the other hand, found among some bacteria, protozoans and fungi. Fatty acids (acetic, propionic and butyric acids), carbon dioxide, methane, ammonia and possibly hydrogen are produced during digestion. Further, the microorganisms are digested and thereby the host animal gets amino acids, vitamins and some lipids. The plant food is thus converted into short-chain fatty acids, methane, amino acids and ammonia. Of these, methane is excreted.

Mammalian herbivores have special fermentation chambers for large volumes of food. Further, the food that is slowly digestible needs to be kept and digested during a long time (retention) and this is achieved by various structures like sieves. Another adaptation is the accelerated excretion of slowly digestible fibre and the retention of smaller particles and microorganisms in the caecum for fermentation (Björnhag 1981, 1994; Cheeke 1987; Hume et al. 1993). There are many possible solutions for incorporating the fermentation chambers into the digestive system, and the digestive sequences and associated gut morphology show considerable variation among mammals (Hofmann 1973, 1989; Stevens 1988; Hume 1989).

An excellent review of mammalian digestive systems is given by Van Soest (1996). There are two main groups: pregastric (foregut) fermentors and postgastric (hindgut) fermentors. In the first group, food is subject to fermentation before enzymatic digestion, and in the second group the situation is the reverse. In the pregastric fermentors, the microorganisms of the forestomach attack all food components coming into the fermentation chamber, i.e. also nutrients that could have been digested by the digestive enzymes of the herbivore. Microbial fermentation means 10–20% energy loss as well as loss of nitrogen (ammonia). Hindgut fermentors, on the other hand,

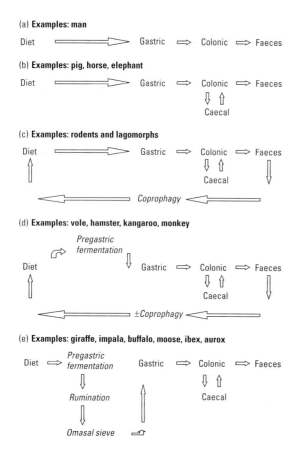

(a) **Examples: man**

Diet ⟹ Gastric ⟹ Colonic ⟹ Faeces

(b) **Examples: pig, horse, elephant**

Diet ⟹ Gastric ⟹ Colonic ⟹ Faeces

Caecal

(c) **Examples: rodents and lagomorphs**

Diet ⟹ Gastric ⟹ Colonic ⟹ Faeces

Caecal

Coprophagy

(d) **Examples: vole, hamster, kangaroo, monkey**

Pregastric fermentation

Diet ⟹ Gastric ⟹ Colonic ⟹ Faeces

Caecal

±Coprophagy

(e) **Examples: giraffe, impala, buffalo, moose, ibex, aurox**

Pregastric fermentation

Diet ⟹ Gastric ⟹ Colonic ⟹ Faeces

Rumination

Caecal

Omasal sieve

Figure 4.3 Schematic comparison of the organization of the digestive system in different mammalian herbivores and omnivores. (Slightly modified after Van Soest 1996.)

digest their food before any microbial fermentation takes place and therefore avoid these losses (Björnhag 1994).

Rumination is another adaptation evolved by herbivorous mammals to increase the efficiency of digestion; ruminants chew the cud. The ingesta is regurgitated, excess liquids are swallowed and mastication commences (Van Soest 1994). Mastication tends to extract cellular contents with saliva and enrich the fibre content of the bolus. The chewed mass, remixed with saliva, is reformed into a bolus and swallowed. The cycle is repeated for some time and may be interrupted by other activities.

The simplest digestive sequence is exemplified by man (Fig. 4.3a), who has a sacculated colon which slows down the passage of fibrous solids for a more efficient use of the fermentable energy. The pig, horse and elephant have a

developed caecum, where the fermentation starts and is continued in the thick part of colon (Fig. 4.3b). In the horse the latter parts are about four times as large as the caecum. In rodents and lagomorphs the main site of fermentation lies in the caecum (Fig. 4.3c). A 25 g vole, for example, needs about 12 times as much energy per kg body mass as a 650 kg horse (Björnhag 1994). This implies that the vole has to eat at least 12 times as much food of the same type as the horse per kg body mass. The passage rate will then be 12 times higher in the rodent, but the fermentation rate is independent of body mass. Small animals therefore have obvious difficulties in obtaining sufficient amounts of energy. The food retention time is also much shorter than what is needed for the microorganisms to reproduce. Small herbivorous animals have to combine autoenzymatic digestion in the foregut with microbial fermentation in the caecum, i.e. they have to be hindgut fermenters and they need a mechanism to retain bacteria in the fermentation chamber (Björnhag 1994). Within Marsupialia, Rodentia, Lagomorpha and Perissodactyla (Equidae) there are colonic separation mechanisms by which low-quality food can be utilized more efficiently, and microorganisms can be kept for a longer time. In all herbivorous mammals the hindgut fermention digesta always moves from the small intestine into the caecum and is mixed there; from the caecum it is moved into the colon. In rabbits (and other lagomorph animals) the small food particles and microorganisms are separated from larger, less digestible particles in the proximal part of the colon and moved back into the caecum for fermentation (Björnhag 1972). In many rodent species only microorganisms are trapped in the proximal colon and returned to the caecum to continue the fermentation (Sperber et al. 1983; Björnhag 1994). Coarse fibres are quickly excreted in faecal pellets. The surplus of microorganisms accumulated in the caecum is harvested by caecotrophy, that is, special pellets taken directly by the mouth just as they appear at the anus (Hörnicke & Björnhag 1980). The caecal matter thus eaten is rich in microbial protein and vitamins of high nutritional value.

Pregastric fermentation occurs in a wide variety of herbivorous mammals outside the large group of ruminants (Fig. 4.3d). A more complex pregastric fermentation system is found in the ruminants, e.g. giraffe, impala, moose and cattle (Fig. 4.3e). The ruminant stomach consists of four compartments. The rumen and reticulum, the first two, are the largest and take care of fermentation. The third compartment is the omasum, which acts as a filter before the fourth compartment, the abomasum, where the enzymatic processes take place.

4.3.4 Linking morphology and physiology with food choice

It is challenging to use morphological and physiological criteria to predict the feeding behaviour of herbivores and vice versa. Several interesting attempts along these lines have been made.

It is reasonable to assume that the morphology of the mouth and teeth of a herbivore should partly reflect its feeding habits. Selective feeders are supposed to have narrow muzzles and prehensile lips, while wide muzzles and high tooth-crowns (hypsodonty) should be found in generalist grazers (Janis & Erhardt 1988; Janis & Fortelius 1988). Molar complexity also varies greatly among herbivores and indicates if the animal is a browser or a grazer; grazers have more complex molar surfaces than browsers.

Based on an impressive study of the morphology of the digestive system, Hofmann (1973, 1989) proposed a system of partly overlapping feeding types: concentrate selectors, grass and roughage eaters, and intermediate, opportunistic, mixed feeders. Langer (1988) made a similar classification. Examples of concentrate selectors are roe deer, moose, giraffe and kudu (*Tragelaphus* spp.), which are unable to tolerate large amounts of fibre in the diet. Ruminants belonging to this group are characterized by large salivary glands, the rapid passage of food, a high fermentation rate, a large liver and a short intestine. Typical grass and roughage feeders are sheep, cattle, African buffalo (*Synceros caffer*) and gnu (*Connochaetes* spp.), which have small salivary glands, a slow passage of food, a low fermentation rate, a small liver and a very long intestine. Intermediate feeders are goats, red deer, impala (*Aepyceros melampus*) and Thomson gazelle.

More studies are still needed before we can fully understand and link the structural and functional components of the complex digestive system of mammalian herbivores. Gordon and Illius (1994) reported that there are no significant differences in digestive kinetics between grazers and browsers when the effect of body mass is statistically accounted for. Further, Robbins et al. (1995) found that the extent of fibre digestion varies between browsers and grazers, although fibre digestion is positively related to herbivore size. They also reported that salivary glands are generally about four times larger in browsers than in grazers, but some browsers are exceptions to this pattern

and have small glands (e.g. the greater kudu, *Tragelaphus strepsiceros*). On the other hand, Iason and Van Wieren (1999) showed that Hoffman's classes explained variation in fibre (neutral detergent fibre) digestion better than body mass. Grazers digested plant cell-wall material to a greater extent than browsers, and intermediate feeders were intermediate in their fibre digestion, as predicted by Hofmann (1989). Iason and Van Wieren (1999) recommended further studies on the differential passage of diet components, and the possible role of the hindgut in compensating for differences in retention and digestion in foregut ruminants.

There has thus been a rapid accumulation of knowledge gained by comparative physiologists and anatomists about digestive systems in mammalian herbivores. Many species have been studied and a complex picture has emerged. One way forward might be to use digestion theory and chemical reactor theory in order to reduce the complexity of form and function in the digestive tract (Hume 1989). Another challenge is to broaden the perspective and link feeding niches, behaviour, diet, morphology, digestion, intake and detoxification of plant defences, as well as plant growth and reproduction. An interesting step in the direction of linking all these aspects to ecological processes was taken by McArthur et al. (1991) and is synthesized in Fig. 4.4.

4.3.5 Food selection

To obtain adequate food, the herbivores track and adapt to spatial and temporal food variation, on both evolutionary and ecological time-scales. Within the limits set by physiological and behavioural constraints, they select their food non-randomly and show varying degrees of preference and avoidance.

Genetically determined nutritional wisdom was seen for a long time as the mechanism through which mammalian herbivores selected their food, so the choice of food was more or less fixed. Recent research, not least on domestic animals, has considerably broadened this view. For example, the food choice of young domestic rabbits was influenced by the diet of their mothers during the periods of gestation and lactation (Altbäcker et al. 1995). In an ever-changing environment, involuntary aversions and learning processes later in life may also play a role. This may take place through post-ingestive feedback or through learning from the mother or other con-specifics (Provenza & Cincotta 1993).

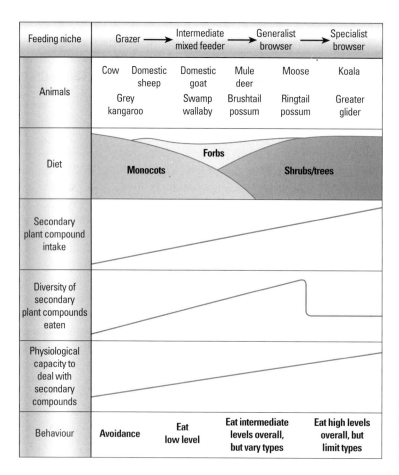

Figure 4.4 Suggested relationships between feeding niches, diet, consumption of secondary compounds and the physiological capacity to deal with these compounds. (From McArthur et al. 1991.)

4.3.5.1 Hierarchical foraging

The food selection process could be structured through the concept of hierarchical foraging, based on decision-making theory (Senft et al. 1987; Fig. 4.5). In this view, the selection process takes place at different scales, from the regional scale (expressed, for example, as seasonal migrations involving a few decisions per year) to selection among plant parts (involving thousands of decisions per day). Our division into various scales might be arbitrary from an animal's point of view, and the levels of scaling can differ depending on the aims of a study. The hierarchical foraging concept does not explain what determines selection on the different scales, but it gives a valuable structure of the process.

4.3.5.2 Selection and plant characteristics

Studies dealing with food selection have mostly focused on selection in relation to plant characteristics and the immediate surroundings of target plants. For example, morphology, chemistry, plant age, earlier herbivory, growing site conditions and the plant species mixture may all affect the feeding pattern of mammalian herbivores.

Preferences for different food types are often poorly related to proximate nutrients alone, and during the last decades, the characterization of plant tissues in relation to herbivore selection has been broadened. This is true not least for woody species, where indices based on a combination of nutrients (sometimes combined with various constraints) and anti-nutrients may give the best correlation with food selection.

Plant-centred ideas have stressed the correlation be-

Level		Examples
Region		Mountain–forest
Landscape		Valley–slope
Vegetation type		Pine forest–mire
Between species		Pine–birch–aspen
Between individuals		Slow–fast growing plants
Within individuals		Leading–lateral shoot

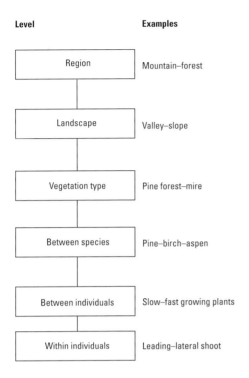

Figure 4.5 Food selection decisions by mammalian herbivores may take place at different levels, ranging from within a plant up to regional spatial scales.

tween observed preferences and allocation of resources within the plant, the plant's vigour, or the growth rate of a plant species and its modules. The resource availability and the carbon/nutrient hypotheses are both based on the assumption that the resources available to the plants determine characteristics that are important for herbivores (Bryant et al. 1983; Coley et al. 1985). The relative availability of carbon and nutrients (nitrogen) may, for example, determine the amount of carbon available for structural or chemical defences. Nutrient and light stress may result in opposite effects on palatability, through differences in the amount of carbon available for carbon-based metabolites.

Experimentally, the hypothesis of carbon-nutrient balance and its impact on herbivore feeding have been tested by fertilization, shading, root-cutting and shoot-clipping. The results have shown that the responses of different secondary compounds may or may not follow predictions from the hypothesis, and that the test of the hypotheses must involve feeding trials with relevant herbivores. To

predict food selection in relation to plant resources, further research is needed on the responses of various groups of secondary compounds and other plant characteristics, such as biomass and morphology. In intra- and interspecific comparisons, Price (1991b) draws a parallel between herbivorous insects and mammals, as some species of these two groups preferred feeding on rapidly growing plant modules. In some cases, the size *per se* can be important as it may increase the cropping rate, and in other cases the chemical composition may be more favourable in fast-growing modules. Occasionally, a simple plant vigour index can explain the risk of herbivore attack. The colour of spruce seedlings was positively correlated with large herbivore browsing in a temperate forest, and the colour was, in turn, correlated with the nitrogen concentration of seedlings (Bergquist & Örlander 1998).

Within plant populations one can find a genetically determined variation in characteristics relevant for herbivore selection. There are indications that herbivores may forage differentially on different genotypes, ecotypes or plant morphs, a pattern that may be explained by differences in morphology or chemistry (Hansson 1985; Jaramillo & Detling 1988).

A number of studies have reported on a difference in palatability between male and female plants (Ågren et al. 1999). Generally, there is a higher preference for male plants, at least by voles, hares and sheep. Sex-related differences in a plant's allocation of resources for reproduction may be one of the possible proximate causes of differential herbivory on male and female plants. The demographic and evolutionary consequences of these interactions have not yet been satisfactorily explored.

Much knowledge has been gathered on the interactions between herbivores and individual plants, but less is known about the influence of nearby vegetation or plants on the herbivory of a target plant. The risk of being eaten and the magnitude of the herbivore attack may be influenced by the neighbours. These ideas are termed plant association, plant refuge, associational resistance, associational plant defences or plant defence guilds (Hjältén et al. 1993). The mountain hare (*Lepus timidus*), for example, browsed less on birch (*Betula pubescens*), which is of medium preference, when it was mixed with a low preferred species compared to when it was mixed with a high preferred one (Fig. 4.6).

Food abundance also affects selectivity by mammalian herbivores. Based on optimal foraging theory, Weckerly and Kennedy (1992) concluded that a herbivore can be

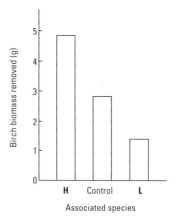

Figure 4.6 Mean biomass of birch twigs (*Betula pubescens*) removed by mountain hare (*Lepus timidus*) from experimental bundles when this species was experimentally mixed with a highly preferred plant (H, *Populus tremula*) and with a species ranking low in preferences (L, *Alnus incana*). 'Control' bundles were made up exclusively of birch twigs. (Modified from Hjältén et al. 1993.)

highly selective during high absolute food abundance, when there are more choices. The other outcome is a need for higher selectivity when there is little food, and at that time high-quality resources are the more important variable. Two large herbivore species given a supplement of high-quality food fed more selectively on natural food species compared to their feeding without any supplement (Weckerly & Kennedy 1992).

A herbivore may also affect the relative abundance of a food type to such a degree that the preference for this food type will change, but such changes may depend on the plant species composition (Edenius 1991). Brown and Doucet (1991), studying browsing white-tailed deer (*Odocoileus virginianus*) on winter yards, noted an increase in number of browse species included in the diet (equals increased generalization) during the progress of the winter and, at the same time, changed preferences.

4.3.5.3 Foraging models

There are a number of important questions to consider in relation to foraging models. Given a certain vegetation, can a herbivore optimize its intake of food (or some food component) or, considering that there are some nasty substances in plants, optimize its intake of good food from that vegetation? And on what time-scale should we see

this optimization? Is each bite a part of the best solution or do we need to see a possible optimization over a day, a year or a lifetime?

The presentation of the optimal foraging theory during the mid-1960s stimulated the development of foraging models that could be applied to mammalian herbivores. The theory of optimal foraging is based on the assumption that an animal would forage in such a way that it optimizes its fitness (survival and reproduction). To achieve that goal, the foraging animal must feed on food of above average quality and maximize its long-term net energy intake rate. Optimal foraging implies that, in a specific situation, there is one solution for the foraging herbivore. The main assumptions are of three types (Stephen & Krebs 1986): decision (which of the foragers' problems to analyse?), currency (how to evaluate various choices?) and constraint assumptions (what limits the choices?). In order to fulfil its requirements, a herbivore must obtain more than energy, and other constituents needed may also influence foraging behaviour. One major optimization problem is to get enough energy and nutrients, and at the same time minimize the intake of toxic or detrimental compounds or structures.

The early foraging models have been further developed through the inclusion of such balances and other constraints (Belovsky & Schmitz 1991). Due to intrinsic and extrinsic constraints, difficulties may arise for a herbivore trying to feed in the 'best' way. Intrinsic constraints are, for example, inherent morphological and physiological characteristics that may change mainly on the evolutionary time-scale. Extrinsic constraints could be the weather (snow, cold, heat), time or the risk of predation. Intra- and interspecific competition can exclude the herbivore from foraging on the best patch or plant. A herbivore also has to devote time to other activities, such as territorial defence and reproduction, and this may create a conflict with foraging.

Food, in terms of quantity or quality, is usually highly variable and is sometimes distributed in more or less discrete patches. Therefore, one crucial point in the optimal foraging concept will be the criteria for when to leave a feeding patch and move to another. The 'marginal value theorem' states that a herbivore should stay as long as the extraction rate is above the average for the environment as a whole. Thus, a herbivore should stay longer in high-quality patches compared to low-quality ones. Optimal foraging, however, needs an in-depth knowledge of the environment, something that the herbivore probably cannot have in every situation. But even to make an imperfect decision the herbivore must gather some information about its environment, i.e. tracking changes in food abundance, distribution and/or quality. Exploratory movements and 'sampling' (trial and error) of food plants may be a way for the herbivore to bring itself up to date on the food situation. The herbivore must also be able to remember what it has experienced through different stimuli. Self-learning, learning across generations and post-ingestive feedback may all play a role in the process of acquiring knowledge.

Understanding the decision rules used by a herbivore requires an understanding of its behavioural responses on various time-scales. It is less probable that an animal optimizes its diet at each bite, but rather that it bases future decisions on an integration over longer periods. Predictions of the optimal foraging theory have been tested and found to be in line with the theory for several herbivores, from mice to moose. In some cases it has also been possible to show that optimal foraging may result in increased fitness. Ritchie (1990), for example, reported that the ground squirrels foraging according to the theory had a considerable fitness advantage in comparison with squirrels whose feeding patterns deviated from the prediction of the theory. Considerable progress has been made in the understanding of the intake rates of mammalian herbivores (e.g. Shipley et al. 1994) and this will provide a valuable base for further modelling. By mechanistic modelling, foraging models can be scaled up from bite size and intake rate to herbivore population parameters, such as reproduction and mortality (Illius & Gordon 1999). However, the optimal foraging theory has also been questioned and Hanley (1997) argued that it is more reasonable to think of several 'good' foraging solutions instead of thinking of a single 'optimal' one.

4.4 Impact of mammalian herbivores on individual plants, populations, communities and ecosystems

Herbivores affect their environment through a number of actions, which can sometimes cascade through ecosystems and directly or indirectly control the availability of resources to other organisms by causing physical state changes in biotic or abiotic materials. This has been termed 'ecological engineering' (Jones et al. 1997). The form and magnitude of these effects are dependent on animal species via their physical capacities and behavioural characteristics. There is generally a lack of synthesis of the

Table 4.5 Major actions exerted by a mammalian herbivore on its environment with examples from a large herbivore, the moose (*Alces alces*; Plate 4.1a). The examples give rough estimates for frequencies or amounts per day. (Modified from Persson et al. 2000.) dm = dry matter.

Action	Magnitude
Feeding	10 000 bites
Trampling	25 m² trampled area
Defecation	14 pellet groups (≈2 kg dm)
Urination	10 litres

actions and the ecological engineering exerted by mammalian herbivores. The major actions and their frequencies for a large herbivore, the moose, are synthesized in Table 4.5. Other herbivores wallow, dig tunnels and burrows, build mounds or create dams, sometimes with far-reaching consequences for the abiotic environment and for other organisms. Although not necessarily engineering, one may also add the more passive roles of herbivores as food for other animals, small or large, and as carriers of plant modules and invertebrates.

4.4.1   The level of individual plants

Plants have evolved under herbivore pressure, but the latter has varied considerably in strength and duration over time. Many plant traits, such as size, modular structure, appearance in time, spinescence, leaf size and shape, compensatory mechanisms and secondary metabolites, are often interpreted as evolutionary responses to herbivory. To establish the evolutionary basis of plant traits as responses to herbivory is in itself difficult, not least because of the difficulties in separating genotypic and phenotypic variation, and the fact that several other environmental factors can act in similar ways to herbivory, i.e. causing physical damage and the removal of biomass.

Ecological impact by mammalian herbivores can be measured on various scales, but the magnitude of the impact on a given scale does not necessarily hold on another scale. For example, a decrease in the growth or fecundity of single plants after herbivory does not necessarily mean that there is an overall decrease in productivity at the community or ecosystem level. As we proceed through higher levels the complexity increases, and more interacting factors have to be accounted for. A lack of consistency may also be true across temporal scales. Again, in the shortest term, a bite is destructive for the plant, but in a longer perspective the plant may compensate for the loss, or the compensation may be overridden by other processes such as competition.

4.4.1.1 Plant modules

In contrast with the majority of animals, most plants are modular organisms. Modules are repetitive multicellular interacting units, often with one or several meristems. Shoots of trees and tillers of grasses are examples of such modules. However, each unit may have its own integrity, indicating that an impact on one place may give only localized responses and may be insignificant for the individual plant. Module size is important for the probability and intensity of herbivore attack (Price 1991), and the modular structure is significant for responses to herbivory, and may also be in itself an adaptation to herbivory. Based on studies of modular structure, Lehtilä (1996) argued that the effects of browsing on a tree depend on whether the browsing is patchily or regularly spread, and that the different outcomes depend on resource distribution systems in the plant.

4.4.1.2 Plant responses

Plants may show a number of 'strategies' in relation to herbivory, from being sensitive to it and showing no obvious defensive traits related to herbivory, to being protected by herbivore deterrent substances (avoidance) or being able to tolerate damage (tolerance). However, such strategies are not necessarily mutually exclusive.

Tolerance is a more encompassing trait than compensation, and means that the plants have various mechanisms to tolerate herbivore damage. Intense debate during the last 20 years has focused on whether such traits are evolutionary responses to herbivory (or to other ecological disturbances) and, thus, if grazing or browsing benefit the eaten plants in terms of increased fitness. For example, a coevolved mutualism between grasses and grazers has been suggested on the basis of saliva-stimulated plant growth (Owen & Wigert 1976). Increased net primary productivity of savanna grasses due to herbivory (McNaughton 1979), and increased flower production in the biennial herb *Ipomopsis aggregata* following herbivory (Paige 1992), provide further examples of the possible enhancement of plant fitness by mammalian herbivory.

Tolerance in plants is not always a matter of regrowth; it may also involve a change in the effectiveness of existing plant parts. Partial defoliation may, for example, result in a more effective use of existing leaves instead of refoliation involving production of new leaves.

The effect of herbivory on plants falls somewhere along a continuum from death to no response. In between these two extremes, there are a number of responses controlled by physiological processes such as photosynthesis, the uptake and allocation of nutrients, and hormone flow (Jaramillo & Detling 1988). The overall responses of plants subjected to herbivory may be viewed as a tradeoff between growth and defence. In comparisons between plant species, van der Meijden et al. (1988) found support for the idea that regrowth and defence were alternative strategies. However, some recent studies (Lehtilä 1996) indicate that compensatory growth and defence may sometimes act together towards an increased fitness.

Directly or indirectly, herbivory also affects reproductive tissues and the fecundity of plants, sometimes with a hypothesized increase in plant fitness. By simulating rodent branch-cutting on bilberry (*Vaccinium myrtillus*), Tolvanen et al. (1993) noted negative effects on flower production and the subsequent development of berries. Sheep grazing on a shrub in the Karoo rangeland reduced both flower production and seed size (Milton 1995). There might also be a tradeoff between seed number and individual seed mass (Bergström & Danell 1987). Simulated browsing revealed that cone production of Scots pine (*Pinus sylvestris*) was more negatively affected by clipping fast-growing pines than slow-growing ones (Edenius et al. 1995). The dynamics and long-term development of plant populations after herbivore-induced changes in plant fecundity are unclear.

Most often herbivores do not kill their food plants, because many plants have some parts with low value for herbivores and a number of compensatory mechanisms. However, mortality does occur because of direct impact, especially when herbivores feed on small plants or seedlings, possibly affecting plant recruitment. Hulme (1994b) reports on the relative impact of invertebrates and voles on grassland plant seedlings. The voles caused considerable mortality on both grasses and forbs. This impact was dependent on the overall vegetation cover, indicating an influence on the spatial distribution of grassland plants.

In ecosystems chronically exposed to fire, resprouting from stumps or roots is often the dominant regeneration mechanism. In such systems, the feeding of mammalian herbivores may have a profound impact on plant population dynamics through the reduced performance of new sprouts. For example, resprout mortality was mainly caused by herbivory in a Mediterranean ecosystem (Moreno & Oechel 1990). Similar interactions have been studied in Yellowstone National Park, where fire and herbivores in combination with weather regulate the development of aspen regeneration (Kay 1997). This indicates that periods of low herbivore pressure may contribute to 'windows of opportunity' for some plant species, which result in a strong dominance for certain age classes (cohorts).

Many mammalian herbivores may even cause mortality of older or mature trees. Such mortality is mainly caused by debarking. Elephant (*Loxodonta africana*) impact on acacias includes an increased mortality due to debarking (Plate 4.1.b). This type of feeding is a common mode of other mammalian herbivores, like deer, porcupines, squirrels, voles and hares. Even if plants are not directly killed by debarking they may be weakened and subjected to subsequent lethal damage by wind or fire.

4.4.1.3 Grazing optimization

In 1979, McNaughton presented the 'grazing optimization hypothesis' and stated that grazing increased the net primary productivity of grasses and thus their fitness. This increase was highest at intermediate levels of grazing. The concept of grazing optimization has also been applied to other plant attributes (e.g. food for herbivores, flowers, seeds). This growth stimulated by grazing has been explained by a number of mechanisms, from those intrinsic to plants, such as removal of old material blocking light, to changes in the immediate environment of grazed plants through enhanced nutrient circulation. The increased productivity of grazed plants, compared to non-grazed ones, is termed overcompensation, and the other two possible outcomes are under- or exact compensation. The capacity of plants to respond to herbivory depends on their morphological and physiological traits, such as rhizomatous growth, the meristem pool, below-ground nutrient reserves and transpiration and photosynthetic rates.

Supporting evidence for the grazing optimization hypothesis has been questioned on both conceptual and methodological grounds. Belsky et al. (1993) argued that responses to herbivory by plants are evolved general responses to many types of injury, such as fire, wind, frost, snow and trampling. In that perspective, a general response mechanism could have evolved as a way to enhance growth and compete for light. The magnitude of short-term compensatory growth and possible stimulated growth above non-grazed control plants may depend on several factors, such as light, water, nitrogen, competition, growth form and the carbon and nitrogen balances of the plants (Pastor et al. 1997).

It is increasingly clear that mammalian herbivory on a given plant species can result in a continuum of responses, depending on the characteristics of the plant, the type of herbivory and the environment. Maschinski and Whitham (1989) stressed that there is no simple typical response for a given plant species. In studying *Ipomopsis arizonica*, a biennial forb, they found that the herbivore impact depended on plant association, nutrient availability and the timing of grazing. Herbivory may in this case lead to exact, under- or overcompensation, depending on the situation. The probability of compensation decreased with plant competition, low nutrient levels and late season herbivory. Houle and Simard (1996), studying simulated above- and below-ground herbivore feeding on a willow, *Salix planifolia* ssp. *planifolia*, found that the effects of genotype, nutrient availability and type of tissue damage were additive, rather than interactive.

Intensity and timing of herbivory have been extensively studied, the former especially in order to find 'tolerable' grazing and browsing intensity levels of food species. Within plant species, herbivory during different seasons may give similar or opposite responses. Figure 4.7 shows some generalized responses of birch (*Betula* spp.) after simulated browsing during three different times of the year. The magnitude and time lag of the responses vary with the timing of the treatment and this variation can be seen as a result of herbivory differentially affecting sources and sinks of carbon.

Thus, a challenge for future research on the effects of herbivory on the biomass, productivity and fecundity of plants is to identify under which conditions one or other response occurs and to what extent various factors interact.

4.4.1.4 Plant shape, plant size

Herbivory may induce changes in the size and shape of plants, which, in turn, may have a bearing on plant competition (Huntly 1991) and subsequent effects on other

herbivory raise several questions. Do herbivores manage their own food resources in order to benefit from the responses at a later time? How can feeding loops ('cyclic grazing') be seen from the plants' point of view? The feeding loop concept is important in understanding the impact of mammalian herbivores on plant performance, but also on the development of crop and forest damage, ecosystem processes and landscape heterogeneity.

A herbivore may affect its food resources so that they become more palatable or allow a greater yield per bite. In temperate areas, winter browsing on trees makes the trees more attractive the following winter compared to earlier unbrowsed trees or seedlings (Danell et al. 1985). Such re-browsing responses may also be quantitative in the sense that heavily browsed trees are attacked at a higher rate than slightly browsed ones. Similarly, feeding loops are observed on *Acacia* trees in savannas (du Toit et al. 1990). In grazing systems like the Serengeti grasslands, grazed patches ('hot spots' or 'grazing lawns') are visited more often than previously non-grazed or only slightly grazed patches (McNaughton 1984) and so are grazed patches in the North American tall-grass prairie (Hobbs et al. 1991). The time between revisits of earlier grazed patches varies depending on the plant–herbivore system and the time of initial grazing. Rebrowsing by moose was registered one year after the initial winter browsing, while intervals of days or weeks have been observed for other herbivores.

The mechanisms underlying such feeding loops have been explained by, for instance, changes in shoot size, food density and/or improved food quality, which will benefit the herbivore. Gordon and Lindsay (1990) concluded that these changes should not be seen as food 'management' evolved as a strategy by the herbivore side, because it is unlikely that a 'managed' food resource will benefit the same animal again. There are specific situations, however, where this may occur, as in systems where the herbivore may monopolize its resources through territorial behaviour; it may also be found in colonial species. From the point of view of plants, it is not obvious how they may benefit from repeated grazing. Drent and van der Wal (1999) argued that we have to consider 'whether herbivory is in some cases the price the plants pay for persistence *in situ*'. This persistence could be enhanced by changes in mineralization rates (for example, due to herbivore impact on tissue age) and in directions or rates of succession.

In order to understand more about food dynamics and the potential impact on plant and animal communities and ecosystems, the concept of feeding loops deserves further consideration and study.

4.4.1.9 Coevolution

Plant changes caused by herbivory in evolutionary time may also cause reciprocal evolutionary changes in the herbivores, leading to coevolution. Freeland (1991) stated that 'there has been ample opportunity for coevolution between mammals and plants'. Few studies, however, currently support such coevolution in plant–mammalian herbivory interactions. The best elaborated is probably one on grasses and grazers in Serengeti (McNaughton 1984).

The prerequisites for coevolution are that a herbivore may directly influence plant characteristics, that the herbivore may discriminate between different plant morphs and that plant characteristics influence animal fitness (Freeland 1991). One obvious difficulty in identifying coevolution involving mammals is that they are not usually specialists, but feed on many plant species. Many mammalian species are also present in most environments. It is therefore reasonable to think of diffuse rather than pair-wise coevolution.

4.4.2 Community level

Two recent reviews discuss the importance of herbivory in plant communities (Huntly 1991; Augustine & McNaughton 1998). Here we highlight three aspects of community responses to herbivory: the diversity of plant species, vegetation succession and the impact on animal communities.

Studies involving herbivore effects at the community level have often been based on the use of exclosures, that is, the removal of mammalian herbivore influence (Milchunas & Lauenroth 1988). Most exclosure studies produce measurable differences between the treatment (exclosure) and the control (grazed), and they may emphasize important mechanisms. From a management viewpoint, however, studies comparing many different herbivore densities are needed, not just the two densities usually compared within or outside exclosures. Because of logistical difficulties, this is a problem that is not easily solved when it comes to studies of wild mammalian herbivores.

4.4.2.1 Herbivory and plant species diversity

Mammalian herbivory can affect plant species diversity, and herbivory is often associated with increased diversity (Olff & Ritchie 1998), but no coherent theory exists that satisfactorily explains observed variation and sometimes conflicting results as well. However, recent findings stress the importance of herbivore types, temporal and spatial scales, and the dynamics of local plant species, the latter via extinction and colonization rates.

The outcome of herbivory on plant species richness is affected by the body size and diversity of the herbivore species (Ritchie & Olff 1999). Through their capacity to roam over large areas and their ability to feed on low-quality plants, large herbivores, compared to insects or small mammals, may have stronger effects on plant diversity. A single herbivore species may have a large impact on plant diversity, and so may multi-species herbivore communities if they essentially feed on the same plants, i.e. their effects are additive. In other systems with several herbivore species, the effects can be evened out because the herbivore pressure is more widely spread among the plant species.

The local plant community may be affected by the herbivores through their potential effects on extinction and colonization. One pattern observed is that if more common or dominant species are disproportionately affected, the plant species richness may increase, and the reverse if rare or intolerant species are damaged (Huntly 1991; Olff & Ritchie 1998). In this context, palatability and competitiveness are two important characteristics of plant species that may determine the impact of herbivory on species richness (Crawley 1997). If these two properties are positively correlated, the herbivores may increase species diversity, while diversity may be reduced if the correlation is negative. A response curve frequently considered in plant community studies is the one representing the 'intermediate disturbance hypothesis' (Connell 1978), where a balance between competitiveness and grazing tolerance of the different plant species will determine the species richness in relation to herbivore pressure. This is mediated by competitive exclusion. Based on earlier conceptual work, Milchunas and Lauenroth (1993) hypothesized that the long-term changes in species diversity induced by herbivores will depend on productivity and on the evolutionary history of grazing (Fig. 4.8). The response curves determined by the grazing pressure may have different shapes depending on these two factors. If, for example, an ecosystem has been subjected to grazing during long periods and adaptation to that grazing

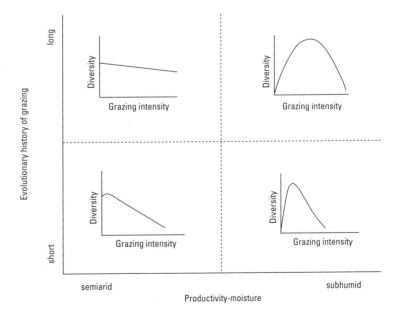

Figure 4.8 The shape of the function relating plant-species diversity to grazing intensity (inset graphs) is dependent on habitat, depending on the particular combination of the evolutionary history of grazing (vertical axis) and location along the productivity-moisture gradient (horizontal axis). (Simplified from Milchunas & Lauenroth 1993.)

pressure has taken place among component species, changed grazing intensities may give a different result from similar pressure in a system made up of non-adapted plant species.

4.4.2.2 Herbivory and vegetation succession

Can mammalian herbivores affect the rate or final outcome of primary or secondary successional vegetation changes? The impact of domestic herbivores is a plain example, but due to artificially high herbivore densities such systems are not usually comparable with those affected by wild herbivores.

Davidson (1993) reviewed the impact of small and large wild herbivores on plant succession and concluded that herbivores may either hasten or slow down the succession rate. The underlying determinants of these opposite outcomes were resource availability, plant characteristics (especially defensive chemicals) and animal feeding behaviour. If herbivores preferentially feed on early successional plants, the succession may be hastened because the late successional plant species are favoured. On the other hand, preferential feeding on late successional species may retard the succession. Davidson concluded that the various outcomes of herbivores on plant succession depended more on the attributes of the plants involved than on the different types of herbivore. However, herbivores differing in body size, ability to handle and tolerate secondary compounds, and timing of herbivory during the year may have different effects on a particular vegetation succession. One difficulty in evaluating the effects of specific herbivores on vegetation succession is that many ecosystems host a number of herbivore species with different requirements and feeding modes. The model presented by Davidson (1993) has conceptually been developed by Ritchie and Olff (1999), in the sense that, for example, they included the possible effects of several herbivore species acting concurrently on the vegetation. It must be stressed, however, that the impact of mammalian herbivores on vegetation succession cannot be considered in isolation, for environmental factors like weather and fire may interact with their effects.

4.4.2.3 Indirect effects of herbivores on other organisms

Apart from the reciprocal effects of various mammalian herbivore species on each other, they may alter the structure and chemistry of plants and plant parts, producing changes which may cascade through the system and influence a number of responses in other biota. Such alterations may have an impact on the number of available sites, the microclimate or the quality and abundance of litter, which in turn may affect the species composition or abundance of, among others, herbivorous invertebrates. The direction of these changes is dependent on the grazing pressure, the type of response organisms and the characteristics of the study area. For example, large herbivores feeding in two vegetation types in boreal forest affected ground-living invertebrate communities in a number of ways. However, the mechanism behind the observed effect was in one case (mixed forest) hypothesized to be a change in litter composition following herbivory and in the other case (pine heath) a reduction in the canopy cover (Suominen et al. 1999). The ecological importance of such interactions is not known at present, but it is an intriguing field for future research.

4.4.3 The ecosystem level

Through their actions mammalian herbivores can affect the structure of communities and interact with ecosystem processes. The trophic approach has viewed herbivores as outputs and inputs of plant communities and predators respectively. More recent approaches have focused on the impact of herbivores on other organisms and ecosystem processes. This work has made clear that herbivores may have a profound impact on ecosystem structure and processes, mainly mediated by grazing-induced changes in plants and plant communities (Hobbs 1996; Pastor et al. 1997; Mulder 1999).

Herbivore impact on vegetation has been discussed in relation to bottom-up or top-down regulations and trophic complexity in terms of, for example, food-chain lengths. Early ideas stressed that predators were regulating herbivores (top-down regulation) and not food (bottom-up), indicating a relatively low impact of herbivores on vegetation. Oksanen et al. (1981) hypothesized that the productivity of an area was determining the number of trophic levels and in turn the potential herbivore impact on vegetation. This was a simplified approach and has been questioned on the grounds that it does not take variation within trophic levels into account, and that relative consumption by herbivores is poorly correlated with ecosystem primary productivity (see Oksanen and Oksanen (2000) for further developments along these lines).

Pastor et al. (1997) listed four major mechanisms by means of which herbivores may affect boreal ecosystem properties: (i) short-term compensatory growth responses after browsing, (ii) deposition of faeces and urine, (iii) long-term successional changes in species composition, involving mainly the increased dominance of unpalatable or less preferred species and (iv) changes in the physical environment as a consequence of the three preceding effects. The importance of compensatory growth on the ecosystem level is not well understood. As discussed above, the responses of plants to herbivory will, among other things, depend on environmental characteristics. If herbivores stimulate plant regrowth beyond that of non-grazed plants, then the magnitude and persistence of this increased productivity will depend on the capacity of the environment to meet the requirements of the new level of productivity. Thus, an increase in plant productivity requires support in terms of nutrients, a situation that can prevail if, for example, the herbivores themselves bring nutrients in to the system.

Herbivores concentrate nutrients through collecting food in many places (thousands a day) and leaving organic matter and non-absorbed nutrients on localized patches (some tens a day), mainly via faeces and urine. Many herbivore species also move between resting, shading, cover and foraging areas, behaviours that indicate other ways through which herbivores may redistribute nutrients. Faeces and urine directly enrich with nitrogen and other nutrients the patches where they are deposited. The carbon–nitrogen ratio in faeces is often lower compared to plant litter, indicating that the faeces are more easily decomposed (Hobbs 1996).

Other mechanisms involved in the modification of nutrient cycling are of a more indirect nature. Holland and Detling (1990) list a number of ways by which the plant–soil system can be affected by grazing and they included changed carbon and nitrogen allocation to roots and above-ground plant parts. A browsing-induced change in the deciduous/conifer abundance ratio may affect soil attributes, such as mineralization and nitrogen content (Pastor et al. 1993). These changes are due to the different qualities (nutrients, secondary compounds) of litter from deciduous and conifer species. Vegetation changes induced by heavy reindeer (*Rangifer tarandus*) activities in Finland affect microbial activities and mineralization rates in the soil, probably mediated by soil temperature and moisture (Väre et al. 1996).

Based on the fact that many mammalian herbivores move over whole landscapes, and that earlier studies have usually been conducted on small plots, new research has tried to link herbivore activities with landscape patterns of nutrients. Within the grassland ecosystem of Yellowstone National Park, with high densities of large herbivores, nitrogen mineralization was considerably higher on grazed plots compared to non-grazed ones. The difference between the two types of plot was of the same magnitude as the range of nitrogen mineralization between landscape components (Frank & Groffman 1998). At Isle Royale, even a low browsing pressure added to the variance in soil nitrogen across the landscape. The reason was believed to be that disproportional browsing on preferred species was enough to change vegetation composition, with subsequent changes in soil characteristics via litter quality (Pastor et al. 1998).

As discussed earlier, herbivory is often very patchy. What effects this patchiness may have on the landscape or the ecosystem will depend on herbivore behaviour in terms of foraging or other activities. First, even low herbivore pressure may cause vegetation changes through a disproportional use of plant species or individual plants. Second, if herbivores revisit patches used earlier with a higher probability than patches not used before (positive feedback; the feeding loop concept discussed in Section 4.4.1.8 above), then they may have a stronger effect with a possible increased heterogeneity in space (Hobbs 1996). Conversely, if the herbivores are deterred from patches used earlier (negative feedback) or forage at random, one may expect a more even use of the landscape and decreased heterogeneity. The actual type of feedback occurring in a given situation will depend on the plants' ability to respond to herbivore foraging. A possible creation of landscape heterogeneity by herbivores may be counteracted by fire if there is an inverse relationship between grazing and fire. In reviewing the impact of ungulates on ecosystems, Hobbs (1996) stressed the potential role that herbivores may have on ecosystems via interactions with fire. Herbivores may change the load, spatial distribution and flammability of plant biomass.

Knowledge is now accumulating on the role of herbivores in different ecosystems. It is still premature to identify the collective role that herbivores may have on different types of ecosystems, but compilations like the one produced for the arctic and subarctic environment (Mulder 1999) may provide a base for ecosystem comparisons (Fig. 4.9).

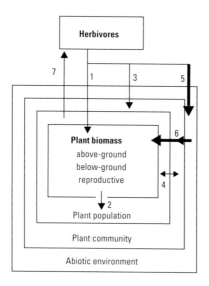

Figure 4.9 A conceptual model synthesizing the different ways in which herbivores can affect plants, showing hypothesized differences between Arctic and more temperate ecosystems. Thick arrows indicate that the effect is larger in Arctic than in temperate areas. (1) Reduction in above-ground, below-ground and reproductive biomass of plants due to consumption by herbivores; (2) changes in the plant population resulting from changes in plant size and reproductive output; (3) changes in the plant community resulting from selective removal of some species; (4) changes in the interactions between plant species due to competition and facilitation following changes in community composition; (5) changes in the abiotic environment resulting from the presence of herbivores; (6) changes in the growth rates of individual plants and in competitive abilities resulting from alterations in the abiotic environment; and (7) feedback from changes in the plant community to herbivore selectivity, movement and population numbers. (From Mulder 1999.)

4.5 Feedback of plant responses to mammalian herbivore populations

As noted previously, the abundance and quality of food plants will change after being heavily attacked by herbivores. This can in turn lead to functional responses by the herbivores, which may later be translated into numerical responses. For some of the mammalian herbivore–plant systems, like grasses and Soay sheep in the St Kilda archipelago (north-west of Scotland) and the tropical savanna grazing systems, there is now such a good understanding that functional responses can be scaled up to numerical responses through mechanistic modelling (Illius & Gordon 1999). Such approaches include the process of food intake and diet selection, from the level of the individual bite up to the daily nutrient intake, metabolism, energy balance,

reproduction and mortality. They can serve as guidelines for future studies of other systems.

4.5.1 Functional responses

If herbivores face a situation with less food per animal, more time will be allocated to the search for food and/or food of lower quality will be accepted. If there is more food per animal, the feeding rate of herbivores will increase with food availability up to a limit set by gut capacity or handling time. Food quality can either increase or decrease as a response to herbivory and the herbivores will respond to these changes. Increased food quality can result in a lowered food intake, and an increased intake of food can make up for a decrease in its quality.

4.5.2 Numerical responses

When changes in food plants have a significant impact on the mortality and/or reproduction of the herbivore population there is a numerical response. We can now raise the important question of what determines the abundance of herbivores? This is a difficult one, and Crawley (1989) stated that, in general, we cannot say whether a population of herbivores is limited by food or enemies, or even whether it is regulated at all. Six main factors could possibly determine the abundance of a herbivore: (i) predators, (ii) food supply, (iii) cover, (iv) disease, (v) parasites and (vi) climatic conditions (Krebs et al. 1999).

One of the more intriguing questions in animal ecology concerns the explanation to the population 'cycles' found in voles, lemmings, hares and their predators. The food situation is an obvious hypothesis at first glance. Many animals can reduce the food resource which then will result in fewer animals; fewer animals will allow the food resource to recover and the animal population to build up, and so on. Early work by Summerhayes (1941) found that there was plenty of food when the vole population crashed and that it seemed unlikely that food was important. However, it was not so easy to debunk such a favourite and intuitive idea, and over the decades it has persisted in different forms, e.g. Keith (1983) stressed food as an important factor for the hare cycles. The food-related hypotheses have not only included the quantitative aspects of food, but also the qualitative ones. There are three main groups of hypotheses involving the role of food plant quality in population 'cycles'. One operates on the ecosystem level (Fig. 4.10a), one on the plant

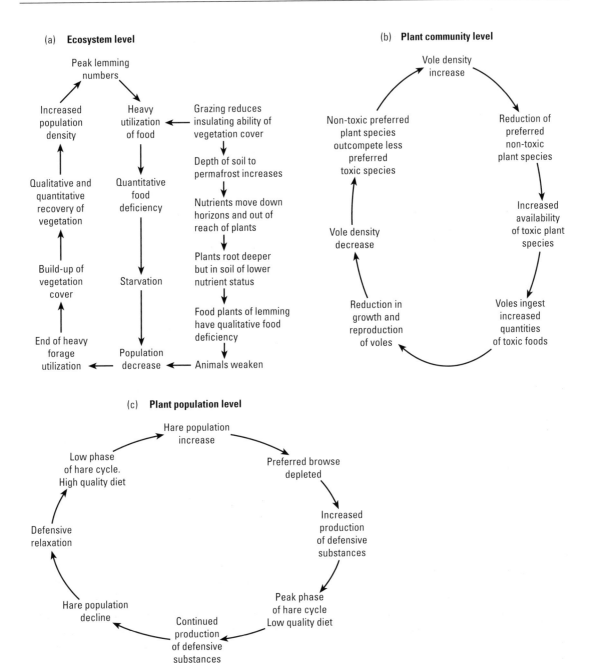

Figure 4.10 The three major 'families' of hypotheses advanced to explain the 'cyclic' population fluctuations of some mammalian herbivores (e.g. lemmings, voles and hares). These hypotheses are framed at the ecosystem level ((a), Pitelka 1964), the plant community level ((b), Freeland 1974) and the plant population level ((c), Bryant et al. 1979).

community level (Fig. 4.10b), and one on the plant individual level (Fig. 4.10c).

On the ecosystem level, the nutrient recovery hypothesis was suggested by Pitelka (1964). Following a high animal peak with strong impact on the vegetation, the plant cover and the insulating cover was reduced and this increased the permafrost. The plant nutrients moved downwards and out of reach of the plants which became nutrient poor and this negatively affected the herbivores. The herbivore populations crashed and the vegetation could recover which caused the permafrost to decrease; the nutrient level in plants increased as did the herbivores.

On the plant community level, Freeland (1974) suggested that during their peak year, voles reduced the palatable and high quality food plant species and left plant species of lower food quality; some of them even toxic. With lowered feeding pressure as a consequence of the collapsed herbivore population the more preferred plant species recovered and with that food quality improved and the herbivore population increased.

Finally, on the plant species level a major role has been ascribed to induced defences (see review by Karban & Baldwin 1997). During years with great abundance of herbivores they severely damage their food plants which, in turn, respond by increasing the levels of defense compounds which negatively affects the animal population. After the collapse of the herbivore population the plants will produce less toxic plant material and the herbivore population will build up again. Such a mechanism was proposed for the snowshoe hare (*Lepus americanus*) cycle by Bryant (1979).

4.5.3 Plants versus other factors

The three main groups of hypotheses outlined in the preceding section have stimulated much research but, unfortunately, tests are extremely difficult to conduct under realistic field conditions. They all require considerable effort, and large-scale experiments involving one or more population 'cycles' are logistically difficult. Some of the hypotheses might also be acting in concert.

A recent experiment with snowshoe hares in the Canadian boreal forest has demonstrated that no single factor can explain the hare 'cycles' (Krebs et al. 1995, 1999). Probably a combination of food shortage and predators are the main factors behind these fluctuations. The abundance of snowshoe hares and Arctic ground squirrels

(*Spermophilus parryii*) showed a mixed response to fertilization of the vegetation; their numbers were doubled by mammalian predator elimination, tripled by supplementary food and increased 9–11 times by the combined treatment of predator reduction and food addition. There is thus an interaction between food and predators. For example, when food quality declines the hares have to feed more in order to keep up with their nutrient requirements. Increased time for feeding leads to a greater exposure to predators and increased predation risk. Further, poor food can lead to a higher risk of attacks by pathogens. Not all herbivore species respond like hares and ground squirrels. Red squirrels, mice and voles were not affected by predator removal, supplementary rabbit food, or the combination treatment.

For the ungulates of the Serengeti–Mara system it was not possible to show that a single limiting factor explained the fluctuations of all grazing species (Krebs et al. 1999), but some general results were obtained: (i) large grazers are limited by food; (ii) migrant species are also limited by food even if they are small in body size; (iii) smaller resident grazers tend to be limited by predators; (iv) disease and parasites act synergistically with food; (v) the very small species may be limited by density by social behaviour as well as predators. Knowledge gained in the boreal forest and on the savanna stress that studies of herbivores and their food plants alone are not enough. Researchers who specialize in plant–animal interactions have to broaden their competence and their minds in order to make progress.

4.6 Directions for future studies

We have shown that mammal–plant systems are very complex in nature and therefore it is unlikely that narrow short-term studies, even if they are well designed and performed, will deepen understanding of the ecology of plant–mammal interactions.

There is a great need for continuation of the relatively few study systems about which we already know a lot. For these systems we can slowly build a mechanistic understanding from individual feeding behaviour and digestion, including population, community and ecosystem dynamics. The knowledge gained from such long-term ecological field studies has proved to be one of the real fundamentals of ecology and their value cannot be overestimated. We recommend that reductionist studies, which also are most welcome, should, if possible, be directed to species or systems on which there is a large amount of knowledge, then the results of more detailed studies will have an added value.

Still, we believe that the majority of future plant–mammalian herbivore studies will be relatively limited, include one or two scientists, and be performed on the landscape level. The approach of such relatively narrow studies needs to be broadened. Often they focus on the main herbivore and its dominant food plants, but we advise including other elements for as long as possible, e.g. predators, parasites and competitors, just to broaden the perspective and open it for alternative explanations. We also suggest more focus on studies which incorporate plant–herbivore processes on the ecosystem level.

Studies on a much larger scale, that is, continental and global ones, are also needed; so are syntheses in which consideration is given to ample taxonomic arrays of herbivores in order to gain an evolutionary perspective on present and past plant and animal communities for comparative studies between continents. It is time to ask the big questions. We also need to deviate from broad concepts such as 'plants', 'herbivores' and 'predators' and focus on real species and their real characteristics and activities. For example, we need to realize that a large majority of the herbivores have quite a mixed diet and also feed on animal matter. Gathering empirical data, theoretical work, modelling and experiments should remain our main methods in the future.

Chapter 5 Granivory

Philip E. Hulme and Craig W. Benkman

5.1 Herbivores as predators, seeds as prey

Granivory describes the interaction between plants and the animals (termed granivores or seed-predators) that feed mainly or exclusively on seeds. Seeds are the products of the fertilized ovules of flowering plants and consist of an embryo and food-storage organs surrounded by a protective seed coat (testa). Many animal species feed on seeds (Table 5.1) and they display a wide range of feeding habits: earthworms swallow whole seeds that are subsequently digested by gut enzymes; lygaeid bugs suck out the contents of seeds; certain lepidopteran and coleopteran larvae burrow through and feed within seeds; many birds grind up entire seeds in muscular gizzards; rodents gnaw seeds with their incisors, while ungulates crush seeds in their molar mills. A distinction is often made between pre-dispersal seed-predators that feed on seeds on the parent plant before they are dispersed (e.g. parrots, monkeys, weevils), and post-dispersal seed-predators that scavenge for seeds after they have been dispersed (e.g. pheasants, pigs, earthworms). However, many granivores act as both pre- and post-dispersal predators. Comparison of several studies that have simultaneously quantified pre- and post-dispersal seed predation (Table 5.1) reveals that: (a) a greater diversity of taxa (particularly invertebrates) feed on seeds pre- rather than post-dispersal; (b) while certain plants suffer proportionally more pre- than post-dispersal seed predation the latter is, on average, significantly more severe (47.3% vs. 61.2%); and (c) the intensities of pre- and post-dispersal seed predation are directly correlated. The lack of independence between the intensity of pre- and post-dispersal seed predation might be expected if the granivore assemblage feeding both pre- and post-dispersal were the same. This is generally not the case (Table 5.1), which suggests that there may be certain seed or crop attributes that similarly influence predation by both pre- and post-dispersal granivores. For example, seeds that are

poorly defended (either chemically or physically) may be particularly susceptible to the impacts of both pre- and post-dispersal granivores. Alternatively, large seed crops may satiate both pre- and post-dispersal granivores. The limited data available highlight the need for further studies comparing the relative impact of pre- and post-dispersal granivores.

Numerous field studies have identified granivores as having a considerable impact on seed populations (see Crawley 1992; Hulme 1998a for reviews). High rates of predation, often greater than 50%, are typical of many plant species in a number of different ecosystems (Table 5.1). Granivory is thought to play a pivotal role in the regeneration (Sarukhan 1986; Louda et al. 1990; Hulme 1994a, 1996a; Castro et al. 1999), colonization ability (Schupp et al. 1989; Myster & Pickett 1993; Picó & Retana 2000) and spatial distribution (Kollmann 1995; Hulme 1997; Forget et al. 1999) of plants. In addition, granivores have been suggested as agents of natural selection that influence seed traits (Hulme 1998b; Benkman 1999) as well as seed production strategies both within (Ruhren & Dudash 1996) and between seasons (Silvertown 1980; Jensen 1982; Curran & Leighton 2000). They may also shape the characteristics of seed dispersal syndromes involving wind (Casper 1988), ants (Ruhren & Dudash 1996), birds (Traveset 1994; Hulme 1997) and mammals (Traveset 1990; Benkman 1995).

A variety of attributes distinguish granivory from other forms of herbivory (see Chapters 3 and 4) and shape the interaction between plants and granivores:

Not all plants produce seeds. Many plant taxa produce spores rather than seeds (e.g. algae, bryophytes, lycopods, ferns). Sporivory (spore-feeding by animals) is less well documented than granivory but the available evidence suggests that it occurs infrequently. For example, of over

Table 5.1 The percentage of seed predation found in selected field studies which have separately quantified pre- and post-dispersal seed removal.

Species	Ecosystem	Pre-dispersal losses		Post-dispersal losses		Author
		%	Agent	%	Agent	
Cirsium canescens	Temperate grassland	51.8	Tephritid flies	0–99.5	Rodents	Louda et al. (1990)
Cirsium vulgare	Temperate grassland	3–17	Moth larvae	21–66	Rodent	Klinkhammer et al. (1988)
Quercus robur	Deciduous forest	25–80	Knopper galls/weevils	100	Rodents	Crawley & Long (1995)
Fagus crenata	Deciduous forest	36.9	Insects	12.3	Vertebrates	Homma et al. (1999)
Fagus sylvatica	Deciduous forest	3–17	Moth larvae	5–12	Mammals	Nilsson & Wastljung (1987)
Fraxinus excelsior	Deciduous forest	15–75	Moth larvae	25–75	Rodents	Gardner (1977)
Pinus sylvestris	Coniferous forest	80	Crossbills	67–96	Rodents	Castro et al. (1999)
Pistacia terebinthus	Mediterranean shrubland	71.7	Birds/wasps	85	Rodents/ants	Traveset (1994)
Lobularia maritima	Mediterranean shrubland	80–99	*Messor* ants	40–95	*Messor* ants	Picó & Retana (2000)
Leptospermum juniperinum	Sclerophyllous woodland	44	Beetles, wasps, moths	90	Ants	Andersen (1989)
Leptospermum myrsinoides	Heathland	64	Beetles, wasps, moths	90	Ants	Andersen (1989)
Acacia farnesiana	Deciduous tropical forest	0–37.8	Bruchid beetles	35.2–66	Rodents	Traveset (1990; 1991)
Astrocaryum mexicanum	Deciduous tropical forest	50	Squirrels	90	Mice	Sarukhan (1986)
Dryobalanops lanceolata	Malaysian rain forest	32.5	Weevils	9	Vertebrates	Itoh et al. (1995)
Tachigalia versicolor	Panamanian rain forest	20	Bruchids	43	Rodents	Forget et al. (1999)
Cecropia shreberiana	Puerto Rican rain forest	6	Vertebrates, insects	9	Ants	Myster (1997)
Cecropia polyphlebia	Costa Rican rain forest	12	Vertebrates, insects	2	Ants	Myster (1997)

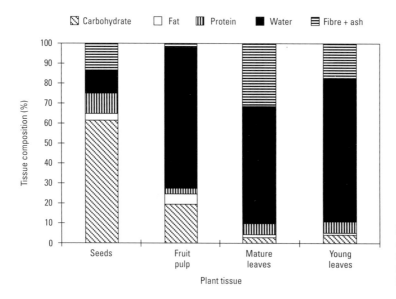

Figure 5.1 Comparison between the average composition of seeds and other plant parts. Seeds provide a concentrated nutrient source that is particularly rich in carbohydrates. (Data from Jordano 1992.)

400 species of arthropod herbivores that feed on ferns, fewer than 10 are sporivores (Balick et al. 1978). Thus, in contrast to other forms of herbivory, granivory is particular to seed plants (e.g. trees, shrubs, herbaceous plants and grasses) that nonetheless comprise the dominant elements of present-day terrestrial vegetation.

Seeds are rich in energy and nutrients. A major biological advantage of seeds over spores is that they provision the developing embryo with nutriment. The food-storage organs may be found within the embryo (e.g. cotyledons) or, in many plant species, elsewhere within the seed (e.g. endosperm). Compared to other plant tissues seeds are nutrient-rich, partly because they contain relatively little water, thus nutrients are concentrated in seed tissues (Fig. 5.1). However, even on a dry-weight basis, seeds generally have a higher energy content than roots, stems or leaves. Not surprisingly, seeds are highly sought after and this may explain why granivory is more widespread than sporivory.

Seeds vary considerably in size and shape. Plant species differ in the extent to which they provision the embryo with resources; this is reflected in the enormous range of seed sizes found in nature. The range encompasses the tiny seeds of orchids (e.g. *Goodyera repens*), weighing approximately 0.001 mg, to the seeds of the double coconut (*Loidocea maldivica*) that often weigh over 20 kg. Even

within a single plant community, the sizes of seeds of different plant species may vary across several orders of magnitude, even within a single life-form. For example, in oak–birch woodlands of north-west Europe, tree-seed size varies from 0.2 mg for the seeds of silver birch (*Betula pubescens*) to 6.44 g for the acorns of pedunculate oak (*Quercus robur*). Seeds represent a particularly diverse resource base for potential granivores.

Seeds are frequently well defended physically and/or chemically. Due to their high nutritional value, seeds often require a greater investment in anti-herbivore defence than vegetative tissue (Janzen 1971). Seeds of many plant species are contained in dry fruit that dehisce and liberate seeds when ripe (e.g. capsules, cones, follicles, legumes) or are indehiscent (e.g. nuts). These structures often form the first line of defence against granivores, and the fruit wall may be thick and woody and/or covered in spines, bristles or irritant hairs (Fig. 5.2). To crush hickory nuts (*Carya ovata*) requires a force of over 75 kg, which takes over 30 hours of processing in the gizzards of turkeys (Stiles 1989). In contrast, fleshy fruit often rely on a fibrous seed-coat to physically protect seeds (Fig. 5.3). However, plant species differ in their allocation to physical defences, which may account for as little as 5% of seed mass in spindle (*Euonymus europaeus*) to almost 90% in hawthorn (*Crataegus monogyna*). Finally, the endosperm of some seeds may be so hard that only the most deter-

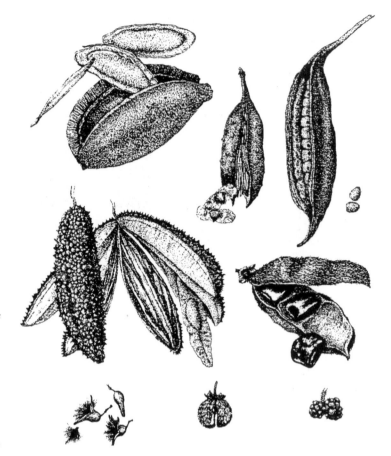

Figure 5.2 Examples of protective fruit structures from lowland tropical rainforest in Australia (redrawn from Grubb et al. 1998): top row (left to right) *Cardwellia sublimis* (Proteaceae) woody follicle, *Neosepicaea jucunda* (Bignoniaceae) woody capsule, *Brachychiton acerifolius* (Sterculiaceae) woody follicle with irritant hairs on seeds; second row *Flindersia bourjotiana* (Rutaceae) spiny woody capsule, *Mucuna gigantea* (Fabaceae) woody legume covered by irritant hairs; third row *Doryphora aromatica* (Monimiaceae) woody receptacular tissue, *Lethedon setosa* (Thymelaeaceae) woody capsule covered by irritant hairs, *Dendrocnide moroides* (Urticaceae) berries with stinging hairs.

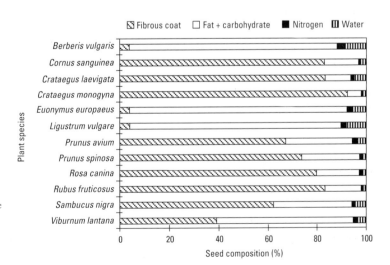

Figure 5.3 Differences in the composition of the seeds of twelve fleshy-fruited shrub species. The seeds differ considerably in their allocation to physical structures such as fibrous seed coats. (Data from Kollmann et al. 1998.)

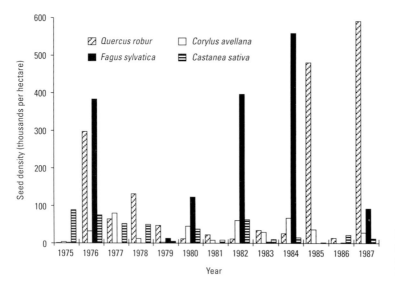

Figure 5.4 Variation over 13 years in the sizes of seed crops of four tree species within a single deciduous woodland. (Data from Gurnell 1993.)

mined granivore attempts to feed on them; such is the case of the Tagua palm nut (*Phytelephas macrocarpa*), the endosperm of which is tough enough to be used commercially as an ivory substitute. The absence of marked physical defence mechanisms in the seeds of some plant species may reflect their reliance on chemical defence. Likewise, seeds that invest in physical defence may often have low levels of toxins. Seeds are sources of some of the most toxic natural products known to humans and the secondary chemicals in seeds present formidable challenges to granivores (Bell 1978; Harborne 1993). A broad spectrum of toxins and anti-feedants occur in seeds, including non-protein amino acids (which disrupt protein synthesis), cyanogenic glycosides (which release cyanide following damage), protease and amylase inhibitors (which impede enzyme function) and phytohaemaglutinins (which reduce nutrient absorption). However, seed chemical defences can only be assessed with reference to specific target organisms since a secondary chemical may not be equally toxic to all granivores. For example, the seeds of the jojoba shrub (*Simmondsia chinensis*) contain a cyanogenic glycoside, simmondsin, that is detoxified by one species of pocket mouse (*Perognathus baileyi*) but not by a congeneric species (*P. penicillatus*) that shares the same desert habitat (Sherbrooke 1976).

Seed abundance is variable in space and time. While many seeds possess physical and chemical defences, plants may also manipulate the quantity of seeds in response to granivory. Records of the seed crops of four temperate tree species over 13 years reveal marked asynchronous variation in seed production (Fig. 5.4). In certain years (1981, 1986), few seeds were produced by any trees, whereas on five occasions (1976, 1982, 1984, 1985, 1987) one or more species produced particularly large seed crops. For the two most variable species, pedunculate oak (*Quercus robur*) and beech (*Fagus sylvatica*), seed density varied by over three orders of magnitude between successive years. Furthermore, in addition to marked differences in absolute seed densities among years, the relative abundance of the four species differed in each of the 13 years. Thus a common plant species may, at times, be rare to a granivore.

Seeds directly influence plant populations in several ways. These include: (i) the colonization of new areas at a distance from the parent population; (ii) the local increase in populations; (iii) the replacement of individuals that die in a population; and (iv) survival during unfavourable periods for plant growth. Since granivory often leads to the eradication of individuals in a plant population, whereas most forms of herbivory often result in only the partial removal of tissue from individual plants, we expect granivores to play an especially important role in plant demography and impose strong selection pressures on plants.

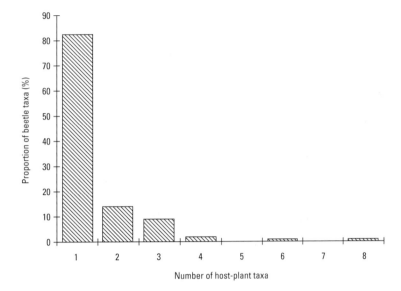

Figure 5.5 The distribution of taxa of beetle seed-predators (families Bruchidae, Curculionidae and Cerambycidae) among 975 species of dicotyledonous plants in a Costa Rican tropical deciduous forest. (From Janzen 1980b.)

5.2 The characteristics of granivory

5.2.1 The different guilds of granivores: their dependence on seeds and their impact

Most pre-dispersal seed-predators are specialists (Janzen 1971; Crawley 1992; Fig. 5.5). Seed-predators can specialize because seeds are clumped and conspicuous prior to dispersal. For the same reason selection for increased seed defences is especially strong, which leads to enhanced defences that exclude less specialized seed-predators. Plants may even respond directly to pre-dispersal seed predation by compensating for seed loss, modifying phenology to avoid predators, or by the induction of secondary chemicals to inhibit further damage (Harborne 1993). The result is the exclusion of generalists, which further favours specialists.

Pre-dispersal seed-predators not only require countermeasures to seed defences, they must also time their life cycle to the often ephemeral availability of seeds on one or a few species of plants. As a consequence, most pre-dispersal seed-predators are insects, especially in the orders Hemiptera, Lepidoptera, Coleoptera and Hymenoptera, that have life cycles synchronized with the availability of seeds from just one or a few closely related species of plants (e.g. Huignard et al. 1990). Birds and mammals lack such flexibility, so they usually consume seeds from a variety of plants and rely on seeds for only a fraction of their annual cycle (e.g. Hulme 1993). The most specialized birds (e.g. nutcrackers and crossbills) and mammals (e.g. tree squirrels and heteromyid rodents) either move between areas or store seeds in caches to ensure a more continuous supply. Such behavioural opportunism, however, may have little effect on the extent to which a granivore is morphologically specialized. For example, crossbills regularly forage on seeds of several species of conifers during a year. Yet each of at least five different 'species' of red crossbills (*Loxia curvirostra* complex; i.e. Plate 5.1, facing p. 84) in North America has a bill morphology that approximates the optimum for foraging on the conifer species each crossbill relies upon in late winter (Benkman 1993; Benkman et al. 2001).

A great variety of animals are post-dispersal seed-predators (Crawley 1992; Hulme 1998a): these include insects (especially ants and beetles), molluscs, crabs, fish, birds and mammals (especially rodents). In contrast to pre-dispersal seed-predators, post-dispersal seed-predators feed on a diverse and spatially heterogeneous resource that requires generalist feeding habits. Not only is the assemblage of post-dispersal seed-predators diverse, but its composition varies considerably among different ecosystems. In temperate woodlands, the majority of post-dispersal seed removal is attributable to one or two species of small mammals, whereas in the humid tropics,

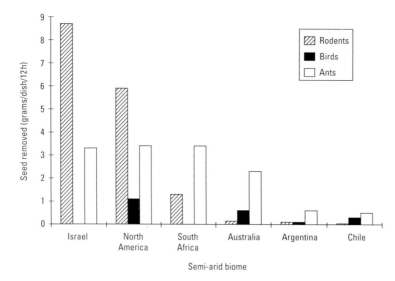

Figure 5.6 Intercontinental comparisons of post-dispersal seed removal by three guilds of post-dispersal seed-predators in semi-arid ecosystems. Data are plotted in order of overall intensity of seed removal and are adapted from sources cited by Hulme (1998b).

important seed-predators include numerous invertebrate taxa (e.g. bruchid beetles, moths and ants) as well a several species of small and large mammals (e.g. in the neotropics: agoutis, pacas, peccaries and tapirs). In arid and semi-arid ecosystems ants are significant post-dispersal seed-predators, whereas in temperate ecosystems they act mainly as seed-dispersers (see Chapter 8). However, these generalizations regarding ecosystem trends should be interpreted with caution. Experimental studies in semi-arid ecosystems reveal marked intercontinental differences in both the overall magnitude of post-dispersal seed predation and the relative importance of different guilds of seed-predator (Fig. 5.6). Rodents play a major role in Northern Hemisphere deserts whereas ants appear more important in the Southern Hemisphere, where overall rates of post-dispersal seed predation are considerably lower. Continental variation in both pre- and post-dispersal predation of *Rhizophora* propagules also reflects differences in the composition of the granivore assemblage in mangrove forests (Farnsworth & Ellison 1997). Further studies are required to assess how granivory varies across a particular plant species' geographical range.

5.2.2 The determinants of seed predation

Pre-dispersal seed-predators are initially attracted to more general features that can be easily detected by vision or olfaction. From a distance plant size and silhouette are prob-

ably important, and when closer, fruit size, structure (Mattson 1986; Brody & Waser 1995) and chemistry (e.g. Huignard et al. 1990) and abiotic factors play a role. Pre-dispersal seed-predators are also influenced by the size and maturity of the seed crop (Christensen et al. 1991). The importance of individual seed and fruit characteristics relative to other plant traits in seed and fruit choice is likely to vary depending on when choices are made. At one extreme, individual seed and fruit characteristics might have no direct impact on seed or fruit choice. This is likely when invertebrates oviposit on branches, foliage or buds before the fruits begin to develop (e.g. Brody & Waser 1995); although fruit and seed characteristics could indirectly affect oviposition if successive generations of insects remain near the host plant. In this situation, granivores discriminate between plant species but might not discriminate between individual fruits or even between plants within a species (e.g. Mattson 1986). The impact of a seed-predator is therefore influenced by how well it can deal with the plant's defences (e.g. Zangerl & Berenbaum 1997). At the other extreme, repeated use of mature seeds by many birds and mammals is based on seed characteristics and the ease with which seeds can be harvested and consumed relative to seeds on other plants (e.g. Smith 1970; Benkman 1987).

Animals are expected to forage in a manner that maximizes fitness. Consistent with this, at least some insects choose oviposition sites so as to maximize the growth and survival of their offspring (e.g. Brody & Waser 1995; Moegenburg 1996). For granivores foraging on multiple seeds, the benefits of feeding should be maximized per unit time spent foraging, while minimizing the costs. The benefits relate primarily to the nutritional quality of the seed, which is often equated with seed-energy content. Costs may be related to seed characteristics, such as the energy expended when handling seeds (e.g. penetrating a tough seed coat) and transporting seeds to nests for later consumption (e.g. heavier seeds require more energy to transport them). When energy content and handling time have been measured, at least some ants, birds and mammals have been found to harvest multiple seeds or fruits in a manner consistent with maximizing net energy gain (e.g. Benkman 1987). Moreover, energy intake rates are correlated with measures of fitness in one species of granivore (Lemon 1991).

Seed choice, however, is at times undoubtedly influenced by a variety of other factors. The value of certain seeds may also correspond to concentrations of particular minerals and/or amino acids if these are deficient in the granivore's diet, or concentrations of soluble carbohydrates if water is scarce (Hulme 1993). Costs may be related to digesting or detoxifying seed contents, which will vary among granivores. The constraint of toxins can in theory be incorporated into diet models, but these models have not been tested for granivores, although it is very clear that toxins affect their diets (e.g. Huignard et al. 1990; Hulme 1993). For example, ground-foraging finches worldwide feed mostly on grass seeds (Poaceae), which lack alkaloids, but avoid similar-sized seeds from other plant families (Leguminosae, Malvaceae, Convolvulaceae) commonly having alkaloids in their seeds (Schluter & Repasky 1991). Likewise, birds avoid grass seeds infested with fungal endophytes that produce alkaloids (Madej & Clay 1991). Other variables that will affect seed choice include the distribution (e.g. aggregated or dispersed) and detectability of seeds (e.g. Hulme 1998b), the risk of predation and the abundance of competitors (Mitchell 1977; De Steven 1981). Thus we should expect the interaction between seed predators and seeds to be shaped not only by seed traits (mass, shape, energy content, toxicity) and the characteristics of granivores (body size, susceptibility to toxins, olfactory acuity, hunger), but also by seed distribution, soil texture and habitat characteristics (Myster & Pickett 1993; Hulme 1993, 1994a). Such variation presumably helps to account for the wide differences among plant species

in rates of seed predation, even within the same habitat (Table 5.1).

Given the numerous variables influencing seed predation, we might not expect to find that large seeds, for example, are preyed on more frequently than small seeds. However, the premise that granivores prefer large rather than small seeds is widespread in the ecological literature (Crawley 1992; Hulme 1996b). While there is some support for a positive relationship between seed size and predation rate for rodents (e.g. Reader 1993) and perhaps bruchid beetles (Szentesi & Jermy 1995), many studies have found no such relationship (Janzen 1969; Myster & Pickett 1993; Hulme 1994a; Kollmann et al. 1998). Therefore while large seeds may be nutrient-rich they may also possess proportionally greater investment in physical and/or chemical defence (Grubb et al. 1998) or be too heavy for small granivores such as harvester ants to manipulate (Brown & Heske 1990).

5.2.3 Density and dependence on frequency

Granivores may respond positively, negatively or not at all to changes in seed density (Box 5.1). Many pre-dispersal seed-predators have limited abilities to increase seed predation when seed crops increase between years. Consequently, the percentage of the seed crop eaten either varies little with seed-crop size (density-independence; e.g. De Steven 1981), or more often decreases with increases in seed-crop size (inverse density-dependence; e.g. Turgeon et al. 1994). For specialist insect seed-predators especially, the percentage of predation often depends on the relative size differences between successive seed crops (e.g. De Steven 1983). For example, a greater percentage of a conifer seed crop is destroyed if the previous year's seed crop is large, so that insect populations can increase, than when the previous seed crop is small (Turgeon et al. 1994). One result is that the percentage of the seed crop damaged in one year is often positively correlated with the size of the seed crop in the preceding year. Indeed, when annual seed production is more consistent, a higher proportion of the seed crop is eaten (Mattson 1986). Inverse density-dependence can also arise because of social interactions between granivores (Pulliam & Dunning 1987). At low seed abundance sparrows consume most of the seeds and are limited by seed densities, but during years when seed densities are high only a small fraction of the seeds is consumed. Social interactions apparently limit the density of sparrows settling on their wintering grounds.

Box 5.1 Forms of response to seed density in granivory

(1) Density-independent seed predation

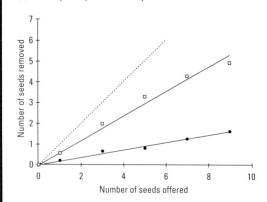

If granivores remove a constant proportion of seeds irrespective of seed density then a linear relationship between seed density and the number of seeds eaten per granivore per unit time is expected. Woodland rodents feeding on low densities of *Fraxinus excelsior* (•) and *Ulmus glabra* (□) seeds removed almost all seeds that they encountered but they encountered seeds only in a proportion of all available microhabitats. (Adapted from Hulme & Hunt 1999.)

(2) Inverse density-dependent seed predation

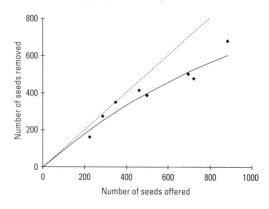

A declining proportion of available seeds are consumed as seed density increases, which produces an asymptotic curve, with a limit to the number of seeds eaten per granivore perhaps set by gut capacity, tolerance of seed toxins, handling time or intraspecific interference. Over an entire dry season, harvester ants (*Meranoplus* spp.) removed proportionally fewer *Sorghum intrans* seeds in sites where seed densities were high. (Adapted from Watkinson et al. 1989.)

(3) Direct density-dependent seed predation

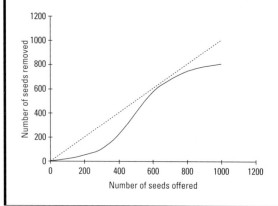

At low seed densities the proportion of seed consumed increases with increasing seed density until limits similar to those described in (2) occur, resulting in inverse density-dependence which leads to a sigmoid relationship. Most studies that have explicitly examined the effect of seed density on rates of seed removal have examined only two different seed densities and thus no published examples of sigmoid relationships are known.

The level of seed predation by generalist seed-predators or specialist seed-predators that are highly mobile or can complete several generations while feeding on a single crop is less dependent on previous seed crops. As a consequence, the percentage of the seed crop they consume often increases with increasing seed-crop sizes (direct density-dependence; Box 5.1, Fig. 3). Larger seed crops provide greater rewards and are thus more likely to attract seed predators. For example, large conifer cone crops have more seeds per cone, which results in higher feeding rates for crossbills for a greater part of the year (Benkman 1987). Because crossbills do not stay and breed when feeding rates are low, disproportionately more crossbills are attracted to and breed longer when cone crops are large than when they are small (Benkman 1987) and as a result harvest seeds from a greater percentage of the cones.

Generalist post-dispersal seed-predators may also respond to changes in seed density (Hulme 1993; Kunin 1994). For rodents, density-dependence is most commonly found for small seeds (e.g. Casper 1988; Hulme 1994a) whereas removal of relatively large seeds is often density-independent (e.g. Hulme 1996a, 1997). However, the influence of seed density on seed removal is not only a function of seed characteristics, but is also mediated by habitat characteristics, possibly related to local food abundance (Hulme 1993, 1994a).

The availability of alternative food sources may increase seed predation by sustaining granivore populations during periods of food shortage. For example, Douglas fir (*Pseudotsuga menziesii*) suffers greater seed predation by pine squirrels (*Tamiasciurus*) when they can rely on lodgepole pine (*Pinus contorta*) seeds during years that Douglas fir produces few seeds (Smith 1970). Likewise, insect seed-predators that rely on other plant parts during seed failures (e.g. conifer seed-predators that feed or develop on foliage when conifer seeds are unavailable) are less affected by fluctuations in seed crops (Turgeon et al. 1994). As for many specialist insect seed-predators, the percentage of the seed crop eaten by highly mobile specialists is probably dependent on the size differences between successive seed crops but on a larger geographical scale.

On the other hand, alternative food sources may reduce seed predation if these foods are highly preferred or simply more common. Polyphagous granivores may respond not only to the absolute abundance of seeds of a particular plant species but also to its relative abundance in relation to co-occurring seeds of other plant species (Greenwood 1985). Even if the density of seeds of a particular plant species is constant within a habitat, granivores may view the seeds as being either common or rare, depending on the relative abundance of seeds of other plant species. Frequency-dependent foraging may lead to a greater proportion of seeds being taken when the species is common and a smaller proportion when rare (pro-apostatic selection) or alternatively seeds may be preyed upon proportionally more when rare than when common (anti-apostatic selection). Although granivore foraging might be expected to be frequency-dependent (Greenwood 1985) the limited data available suggest that neither ants (Kunin 1994) nor rodents (Hulme & Hunt 1999) respond to changes in the frequency of different seeds. Evidence of frequency-dependent seed predation is limited to selection at the fruit level. Rodents, for example, are more likely to miss seeds in multi-seeded *Scheelea* palm nuts when multi-seeded nuts are rare than when they are common (Bradford & Smith 1977). Seed predation by rodents, therefore, favours a low frequency of multi-seeded nuts, even when predation by bruchids, which usually attack just one seed per nut, is high. Similarly, Mitchell (1977) suggests that frequency-dependent selection by mammals on *Cercidium* pods is responsible for the low frequency of multi-seeded pods.

5.2.4 Spatial and temporal heterogeneity

Seed predation varies across a hierarchy of spatial scales, including along topographic gradients and across a species' range, between habitats (e.g. woodland vs. grassland), among microhabitats within a single habitat (forest understorey vs. treefall gap) and at an even finer scale within a single microhabitat (Hulme 1994a, 1998a; Hulme & Borelli 1999). As might be expected, spatial variation arises because some habitats, irrespective of seed availability, are more suitable for certain granivores than others or because seed defences might vary spatially (Hulme 1998a). Frequently, fewer seeds are removed from open microhabitats (Myster & Pickett 1993; Hulme 1994a, 1996a, 1997). This appears to occur when rodents are the principal granivores since their abundance tends to be positively associated with vegetative cover which provides them with a screen from avian predators (Hulme 1993). In contrast, harvester ants appear to forage preferentially in open areas and avoid dense vegetation (Hulme 1997). Similarly, variation in granivory among habitats has also been attributed to differences in vegetation cover (Kollmann et al. 1998; Hulme & Borelli 1999). Seed-

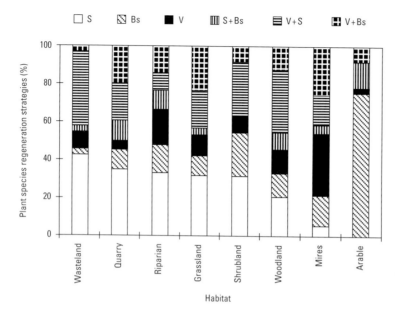

Figure 5.7 Distribution of major regenerative strategies of plant species found in the principal habitats of the Sheffield region (UK). Data and nomenclature are drawn from Hodgson et al. (1994). Key: V, regeneration by vegetative spread; Bs, regeneration via a persistent buried seed bank; S, seasonal regeneration by seed.

predators may therefore significantly modify the seed shadows of plants, both within and between habitats. We suspect that spatial variation in seed predation is widespread, and that the idiosyncrasies of the particular seed-predators and plants will play a prominent role in determining the patterns of variation.

Seed predation may vary both within and between years (Hulme 1994a, 1998a). In a study of predation on seeds of 12 species of fleshy-fruited shrubs, rates of seed removal by rodents were consistent in two different years and showed a similar seasonal trend with removal highest in summer and least in winter and spring (Kollmann et al. 1998). Temporal variation in granivory of a particular plant species may result from changes in the abundance of its seed (e.g. Gardner 1977; Nilsson & Wästljung 1987), as well as changes in other food resources (Hulme 1993) and/or in granivore densities (Hulme 1994a, 1997).

5.3 Demographic implications of seed predation

5.3.1 When is seed predation important in plant population dynamics?

The role of seed-predators in the dynamics of plant populations has received detailed attention (Andersen 1989; Crawley 1992; Hulme 1998a). Seed predation may

play only a minor role in the demography of plants if: (1) plants regenerate primarily by vegetative means; (2) seed losses to predators are buffered by the presence of a large persistent seed bank; (3) seed predators are satiated by large seed crops; (4) regeneration is microsite-limited rather than seed-limited and/or (5) granivore densities are limited by factors other than seed density (e.g. predation or parasitism) such that they cannot fully exploit seed resources.

Three major, though not mutually exclusive, regenerative strategies have been identified for flowering plants (Fig. 5.7): vegetative expansion through the formation of persistent rhizomes, stolons or suckers, and regeneration by seeds which either do or don't form a persistent bank of viable but dormant seeds in the soil. In all but two of the dominant habitats of northern England, species that reproduce exclusively by short-lived seeds are better represented than species that rely exclusively on vegetative reproduction or regeneration from a persistent soil seed bank. Even among those plant species that adopt more than one regenerative strategy, species relying to some extent on non-persistent seeds are more common than those that don't. For these species, regeneration by seed is often as or more important than vegetative reproduction or regeneration from a seed bank (Turnbull et al. 2000). It is evident that the importance of granivory will vary among

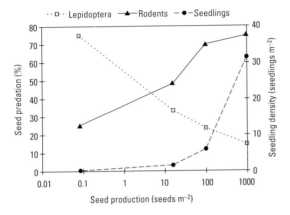

Figure 5.8 The influence of ash (*Fraxinus excelsior*) seed-crop size on seedling densities in relation to the levels of pre-dispersal seed predation by a specialist moth (*Pseudargyrotoza conwagana*) and post-dispersal seed predation by generalist rodents. (Data from Gardner 1977.)

different habitats and among species within a single habitat. Although these trends can only be taken as being representative of the flora of north-west Europe, the data suggest many plant species may potentially be influenced by seed predation.

Through the synchronous production of large seed crops at irregular time intervals (often described as masting; Fig. 5.4) plants are thought to satiate seed-predators and enhance their regenerative capacity (Silvertown 1980). The seed production of most polycarpic woody plants varies annually, with most seed crops either large or small rather than intermediate in size (Herrera et al. 1998). Large, irregular seed crops (masts) are probably more successful at limiting the impacts of specialist pre-dispersal seed-predators than generalist post-dispersal seed predators (Gardner 1977; De Steven 1983; Nilsson & Wästljung 1987; Crawley & Long 1995). Vertebrates, however, are often satiated when there is community-wide synchrony among plant species (Itoh et al. 1995; Curran & Leighton 2000) or when the plant community is dominated by one or a few species (Nilsson & Wästljung 1987; Homma et al. 1999). For example, the irregular seed crops produced by ash (*Fraxinus excelsior*) reduce the proportion of seed destroyed by invertebrates in mast years whereas, in contrast, the proportion consumed by rodents was actually higher in mast years (Fig. 5.8). This reflects the generalist feeding habit of rodents which enables them to persist in non-mast years by feed-

ing on seeds of other species (Fig. 5.4). In years of abundant food their numbers increase (Jensen 1982) and they also have the ability to exploit supra-abundant food supplies through storing food in caches (Vander Wall 1990). Perhaps the high frequency of supra-annual reproductive synchrony in New Zealand is in response to both the presence of specialist insect seed-predators and the absence of generalist seed-predators like rodents (Kelly 1994).

Densities of seedlings are often higher after mast years than non-mast years (Gardner 1977; Crawley & Long 1995; Itoh et al. 1995; Curran & Leighton 2000) since more seeds escape predation in mast years. Although the proportion of seeds destroyed by post-dispersal seed-predators remains little changed, overall more seeds escape predation in mast years. This is not the equivalent of stating that granivores have no effect during mast years, since the number of seedlings recruiting in the absence of predators is not known and could potentially be much greater. These findings for masting trees also suggest that recruitment is seed-limited for these species, since more seed production results in more seedlings (Fig. 5.8).

The extent to which plant populations are either microsite- or seed-limited is unclear because only a few field studies have simultaneously combined seed addition, disturbance and exclosure of seed predators in a factorial design (Reader 1993; Edwards & Crawley 1999). However, the few studies of plant establishment which also considered seed-predators suggest that the failure of many species to establish in dense vegetation may sometimes be due to higher rates of post-dispersal seed predation rather than to increased interference from established vegetation (Reader 1993; Edwards & Crawley 1999). These studies lend support to the view that the importance of seed limitation in communities of perennial plants may currently be underestimated (Turnbull et al. 2000). It can be argued that, to influence plant demography, seed-predators must reduce seed densities below the density of available microsites, thus reducing establishment (Crawley 1992). However, even when microsites are rare, seed predators play a pivotal role in mediating pre-emptive competition for microsites through differential mortality of seeds of various species (Brown & Heske 1990; Edwards & Crawley 1999). In microsites where seed predation is intense (e.g. beneath nurse plants), predators may markedly affect establishment probabilities even when these microsites are limiting (Hulme 1996a).

If the population size of granivores is limited by seed

abundance, then, potentially, granivores may limit seed populations and pose a strong selective force on plants. Alternatively, if predators limit the size of granivore populations to a level sufficient to restrict impacts on seed populations, then seed predation may play a minor role in plant demography. A variety of insect (Janzen 1969; De Steven 1983; Andersen 1989; Turgeon et al. 1994; Gómez & García 1997), mammalian (Gurnell 1993; Hulme 1993) and avian (Grant 1986; Schluter & Repasky 1991) granivores appear to be food- rather than predator-limited. However, in other cases granivores are predator-limited or jointly limited by predators and seeds. For example, Schluter and Repasky (1991) found evidence that ground-feeding finches in Africa and North and South America were limited jointly by predators and seed abundance; however, in the Galápagos, where predators are rare, Darwin's finches are food-limited. Janzen (1975) has argued that parasitoids on bruchids are rare in the tropics because it is difficult for a parasitoid to locate and utilize seed-predators that are so specialized. In temperate regions, however, parasitoids might often limit insect seed predators. For example, at high elevation in the Sierra Nevada (Spain), Gómez and Zamora (1994) found that predation by weevils on the shrubby crucifer *Hormathophylla spinosa* doubles when parasitoids are excluded. Yet, in another study near the same location, parasites had little impact on seed predators (a moth and two species of weevils) of a different shrub species (Gómez & García 1997). It is impossible to generalize from so few studies since it is likely that the relative importance of food and predator limitation will vary in relation to the type of granivore (e.g. pre-dispersal specialist invertebrate vs. post-dispersal generalist vertebrate), if granivores lay their eggs on the surface of the fruit or inside (e.g. Mattson 1986), whether they feed on the ground or in the canopy (e.g. Benkman 1991) and the ecosystem (e.g. tropics vs. temperate). It should nevertheless be borne in mind that whether or not granivore populations are limited by predation, predators can influence the plant–herbivore interaction through mediating where and when granivores forage (Lima & Dill 1990).

5.3.2 How does granivory influence plant demography?

By altering the size and distribution of seed populations, granivores may directly influence plant populations in several ways.

5.3.2.1 The colonization of new areas at a distance from the parent population

When certain granivores (e.g. rodents, birds and ants) encounter a seed, rather than consuming it immediately, they may move it to another location where it may be stored (often buried) for consumption at a later time. The behaviour of storing food for later consumption is termed caching, and if recovery of seed stores (caches) is less than perfect, seeds may survive to germinate. Thus the overall effect on plant populations of certain vertebrate seed-feeders may in fact be positive, resulting in a particular form of seed dispersal (see Chapter 7: Box 7.1). In these instances, seed predation is the cost of reliable seed dispersal (Janzen 1971). Two types of caching behaviour are recognized: larder-hoarding, where seeds are placed in a single large store; and scatter-hoarding, where seeds are placed in several small caches (Vander Wall 1990). Successful seed dispersal is more likely through scatter-hoarding since seeds are buried in many shallow caches, distributed among a variety of microhabitats. The large number of caches often results in less than perfect seed recovery (Vander Wall 1990). In contrast, larder-hoards are often buried more deeply (frequently within animal burrows) and the single location makes recovery of seeds highly probable (Vander Wall 1990). Even where recovery is less than perfect, the depth of burial may prevent successful germination from larder-hoards. Moreover, larder-hoards are often repeatedly used from year to year and the disturbance resulting from burying and recovering seeds often kills seedlings. Marked taxonomic differences occur in the type of caching undertaken. Certain seed-feeding birds, e.g. jays and nutcrackers, generally scatter-hoard, whereas most granivorous mammals larder-hoard, with the important exceptions of tree squirrels (*Sciurus*), chipmunks and caviomorph rodents (Vander Wall 1990).

Detailed studies of cached seeds have shown the survival and germination from naturally scatter-hoarded seeds is low: 0.02% for *Oryzopsis hymenoides* (McAdoo et al. 1983); 0–2% for *Dipteryx panamensis* (Forget 1993); 0–4% for *Fagus sylvatica* (Jensen 1982); 5–8.5% for *Purshia tridentata* (Vander Wall 1994) and 0.75–10% for *Gustavia superba* (Forget 1992). Although survivorship is low, if sufficient numbers of seeds are cached, these low percentages may translate to significant numbers of seedlings. Few studies have monitored the subsequent survival of seedlings, and those that have record high seedling mortality (Forget 1993; Vander Wall 1994) suggesting that cache locations may not necessarily be suitable for establishment. It therefore remains unclear whether regeneration is higher in the presence of scatter-hoarding seed-feeders than in their absence. Nevertheless, caution must be applied when interpreting seed removal by scatter-hoarding vertebrates as seed predation, since although most seeds removed are consumed, a small fraction may be dispersed to suitable microsites (albeit at a remarkably high cost). However, this uncertainty as regards seed fate is greatest for large-seeded species, e.g. trees, for which most evidence of scatter-hoarding exists (Vander Wall 1990).

Granivores may also significantly reduce rates and distances of seed dispersal (Sallabanks & Courtney 1992). Certain vertebrate seed-dispersal agents may preferentially disperse seeds that have not suffered pre-dispersal seed predation by invertebrate granivores. Jays (Hubbard & McPherson 1997) and mice (Crawley 1992) reject weevilled seeds, although squirrels appear not to (Steele et al. 1996). Similarly, frugivores feeding on fleshy fruit may respond to cues such as fruit colour that indicate whether or not pre-dispersal seed predation has occurred (Sallabanks & Courtney 1992). Pre-dispersal seed-predators will often reduce the size of the seed crop available for dispersal. Fewer seeds on the plant means that fewer seeds will reach any particular microsite following dispersal, thus sites a long distance away from the parent plant are less likely to receive seeds. Furthermore, if seed dispersers respond positively to seed-crop size, they may visit significantly less frequently plants that have suffered pre-dispersal seed predation (Sallabanks & Courtney 1992).

Frugivores may disperse seeds to microsites that suffer high post-dispersal seed predation, e.g. beneath shrubs (Kollmann 1995; Hulme 1996a). Often the association between shrubs and regeneration is maintained even in the face of intense seed predation. This suggests that regeneration requirements other than the escape from seed predation probably determine the spatial distribution of regeneration, e.g. requirements for shade (Hulme 1996a, 1997). However, the prevalence of high rates of seed predation in many shrub microhabitats suggests that seed predators may exert a considerable influence on the regeneration of these species (Kollmann 1995).

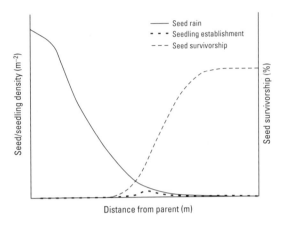

Figure 5.9 A schematic model of how distance- and/or density-responsive granivores might influence seed survival in relation to the distance from the parent plant and the consequences of granivory in terms of the distance at which peak seedling regeneration occurs. Granivores are hypothesized to prevent regeneration close to the parent plant, leading to seedling regeneration occurring farther away from the parent then if granivores were not present. (Adapted from Janzen 1970.)

5.3.2.2 The local increase in populations

Independently of the mode of seed dispersal, for many plant species most seeds fall close to the parent plant (Stiles 1989). If microsites do not vary as a function of distance from the parent plant (e.g. allelopathic and shading effects of the parent plant are negligible), then regeneration is likely to occur close to the parent. Both Janzen (1970) and Connell (1971) suggested that if granivores preferentially feed on seeds beneath the parent plant, either because they respond to the increased seed density (density-responsive granivores) or they are specialist granivores whose foraging is limited to within a certain distance of the parent (distance-responsive granivores), then maximum seedling regeneration occurs some distance from the parent (Fig. 5.9). Evidence for the Janzen–Connell hypothesis is equivocal, but suggests that invertebrate granivores are more likely to feed in a distance- and/or density-responsive manner than vertebrates (Hammond & Brown 1998). However, it is uncertain whether this spatial pattern in granivory is sufficient to limit local colonization since (a) vertebrate granivores may remove a greater proportion of seeds than invertebrates, irrespective of distance (Hulme 1998a); (b) even where a distance effect is found, it may be over such a short scale as to have negligible consequences on the spatial pattern of regeneration (Hubbell 1980); (c) microsites may also vary as a function of distance from the parent plant.

5.3.2.3 Survival during unfavourable periods for plant growth

Plants may survive unfavourable periods for growth as seeds within a soil seed bank. In almost all published studies, seed burial reduces post-dispersal seed predation (Hulme 1993). In addition, burial augments density effects by reducing losses of seeds at low density proportionally more than seeds at higher densities (Hulme 1994a). Comparisons between invertebrates and rodents show that only rodents significantly reduce buried seed populations (Hulme 1994a; Hulme & Borelli 1999). Exclusion of rodents from plant communities can therefore lead to less of a reduction of the seed bank (Kelly & Parker 1990). For buried seed, rodents locate and exploit large seeds more effectively than small seeds. It is perhaps no coincidence that the majority of plants that possess permanent seed banks (seeds remain viable but dormant in soil for >1 year) have small seeds and thus are relatively safe from predation, while most species with transient seed banks (seeds remain viable in soil for <1 year) tend to be relatively large-seeded (Hulme 1998b). Indirect support for this hypothesis is found in the arid zones of Australia where granivory by rodents, and therefore of buried seeds, is negligible (Fig. 5.10) and no relationship exists between seed-bank persistence and seed size (Leishman et al. 1995). Furthermore, and separate from any effect of seed size, rodents tend to remove a smaller proportion of buried seeds with persistent rather than transient seed banks (Fig. 5.10). Thus seeds with persistent seed banks apparently possess features than make them less easy to detect when buried. One such factor might be an impermeable seed coat that seals in any attractive odour a seed might have.

5.3.3 Seed-predators and plant species diversity

In addition to its impact on seed survival, dispersal, colonization and seed-bank persistence of seeds of particular plant species, granivory may also influence plant community structure. Granivores may prevent competitive exclusion among plant species within a particular plant community and hence enhance plant species diversity.

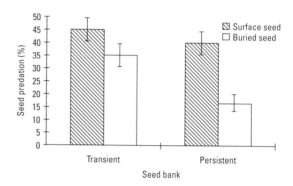

Figure 5.10 The relationship between the persistence of seeds within seed banks and seed predation by rodents on 19 species of grassland seeds either buried or placed on the soil surface. Seeds with persistent seed banks are proportionally less likely to be removed when buried than seeds with transient seed banks. (Data from Hulme 1998b.)

This may occur through a number of different mechanisms (Hulme 1996b).

Trade-offs between granivory and plant competitive ability. A large seed facilitates establishment in the face of interference competition from established plants or other seedlings (Leishman et al. 1995). If large seeds are at a greater risk from predation (Section 5.2.2) and seeds compete for microsites, then granivores may permit less competitive species (e.g. those with smaller seeds) to establish. This tradeoff would act to promote species coexistence and may enhance species diversity.

Frequency-dependence. Pro-apostatic frequency-dependent foraging by seed-predators will select against the commonest seeds within a habitat (Section 5.2.3). This may facilitate the establishment of rare species and prevent any one plant species dominating the entire habitat. Frequency-dependence can therefore strongly stabilize the dynamics of granivory and may lead to greater permissible niche overlaps between plant species. But again, there is no evidence of such frequency-dependent foraging by granivores.

Spatially heterogeneous granivory. Different species of granivores exhibit different seed preferences (Section 5.2.2) and may forage in different microhabitats (Section 5.2.4). Spatial heterogeneity in granivory may lead to different plant species regenerating more successfully in some microhabitats than others. If maintained for a relatively long period of time with respect to the lifespan of plants, consistent spatial patterns in the location where plant species regenerate may promote plant species coexistence.

The Janzen-Connell spacing model. According to this model (Fig. 5.9) granivory may lead to greater survival of seeds at a distance from the parent plant if granivores feed in a distance- or density-responsive manner. This feeding behaviour would preferentially select against regeneration of offspring within a particular distance from parent plants and therefore lead to a more uniform distribution of conspecifics. Seeds of heterospecifics could survive within this distance since they would not be fed upon by specialist distance-responsive granivores and would occur at too low a density to elicit a response from density-responsive granivores. This spacing mechanism would act to prevent any one species from dominating the plant community and thus enable more plant species to coexist.

As has been discussed above in Sections 5.2.2 and 5.3.2 of this chapter, consistent experimental evidence for any one of these mechanisms is weak. Nevertheless, granivores do appear to influence the diversity of plant communities, particularly where they preferentially feed on large seeds. This has been demonstrated in the Chihuahuan Desert, where removal of rodent granivores led to an increase in large-seeded perennials and a subsequent reduction in plant species diversity (Brown & Heske 1990). In neotropical forest gaps, high rates of predation on the relatively large seeds of primary tree species facilitate colonization by smaller-seeded pioneer species (Schupp et al. 1989). Similarly, in temperate grasslands, seed-predators may maintain species diversity by limiting tree invasion and the rate of development of woodlands (Myster & Pickett 1993).

5.4 Evolutionary implications of seed predation

5.4.1 Natural selection and seed-predators

Seed-predators potentially affect plant evolution whenever they differentially depress seed production among plants in relation to variation in some heritable plant trait. Seed-predators commonly have this potential. They have an impact on seed production (Table 5.1), and seed predation usually varies in relation to seed and fruit traits (e.g.

Smith 1970) that are often heritable (e.g. seed size (Leishman et al. 1995, but see also Silvertown 1989), seed chemistry (Zangerl & Berenbaum 1997; Berenbaum & Zangerl 1998) and fruit structure (Primack 1987)). Likewise, if certain heritable traits of a seed-predator affect its ability to exploit seeds and as a consequence affects its fitness, then seed-predators can be expected to evolve to increase their feeding efficiency. Variation between seed-predators affects both feeding abilities and fitness components (e.g. Grant & Grant 1995; Carroll et al. 1997). Moreover, traits affecting feeding ability are often heritable (e.g. detoxification capability (Berenbaum & Zangerl 1998), insect beak lengths (Carroll et al. 1997), and birds' bill size (Grant & Grant 1995)). Thus, we might expect the evolutionary interactions between seed-predators and plants to be dynamic. Well, the answer is both yes and no. In the rest of this section we would like to focus on the 'no'.

Many seemingly adaptive traits might not be heritable. For example, juniper titmice (*Baeolophus griseus*) feed on juniper (*Juniperus osteosperma*) seeds, but can distinguish empty from full seeds only after removing the surrounding pulp. As expected, titmice avoid trees with relatively high frequencies of empty seeds, which should favour trees that produce proportionately more empty seeds (Fuentes & Schupp 1998). However, the proportion of empty seeds might have very low heritability. Empty seeds are often frequent because fertilization in many gymnosperms and some angiosperms occurs well after pollination and the start of fruit development (Willson 1983). Thus, a high proportion of empty seeds might result from, for example, a high frequency of self-fertilization (e.g. Nilsson & Wästljung 1987). The benefits arising from increased numbers of empty seeds, therefore, are possibly fortuitous consequences of development and not the result of selection by seed predators (Fuentes & Schupp 1998). On the other hand, the continuing investment in the development of empty fruits so that fruits or empty seeds might act as decoys or deterrents (e.g. Fuentes & Schupp 1998) is more likely subject to selection.

Even if traits are heritable, evolutionary change can be limited when, for example, selection on seed-predators oscillates from year to year with changes in the availability of seeds (Grant & Grant 1995) or if tradeoffs exist. Tradeoffs are fundamental to life-history evolution, and tradeoffs between seed predation, dispersal and germination affect the evolution of seed and fruit characteristics (Janzen 1969; Primack 1987). For example, selection by parsnip webworms (*Depressaria pastinacella*) favours increases in two furanocoumarins, bergapten and sphondin, in seeds of wild parsnip (*Pastinaca sativa*). Apparently limiting, however, are the precursor molecules in the common biosynthetic pathways producing these two furanocoumarins. Thus, selection by webworms has led to an evolutionary stalemate because increases in bergapten result in decreases in sphondin and vice versa (Zangerl & Berenbaum 1997; Berenbaum & Zangerl 1998).

Perhaps the most evident tradeoff between seed predation and dispersal occurs for animal-dispersed seeds because protecting seeds from predators is often incompatible with enhancing the accessibility and attractiveness of seeds to dispersers. Thus, for example, fruit pulp might be less well protected because secondary chemicals deter not only seed-predators but also seed-dispersers (Janzen 1978 and Chapter 7 in this volume). This might explain why most seed defences of legumes seem to occur in the seeds rather than the pods even though most seed-predators oviposit on the pods (e.g. Johnson 1990). Likewise, a tradeoff between satiating seed predators and saturating mutualist seed-dispersers presumably causes endozoochorous woody plants to produce less variable seed crops than woody plants that are dispersed by wind or seed-predators (Herrera et al. 1998). If the tradeoffs are strong and seed predation great, selection by seed-predators might even cause plants to rely on dispersal by wind rather than by animals (Benkman 1995).

Because of tradeoffs the equilibrium level of defence depends on the strength of selection exerted by seed-predators. At least two general hypotheses might account for variation in the intensity of selection exerted by seed-predators on seed defences. First, seed-predators prefer larger seeds with higher concentrations of nutrients with a concomitant increase in the intensity of selection for seed defences (Grubb et al. 1998). (Seed preference studies (see Section 5.2.3) usually cannot address this because seed size is confounded with seed defences.) Consequently, plants with larger seeds or with, for example, higher concentrations of nitrogen in their seeds should invest more heavily in seed defences. A comparative study of 194 species of Australian rainforest plants supports this hypothesis (Grubb et al. 1998). Second, when seed crops fluctuate in size from year to year seeds may escape predation because of predator satiation (Section 5.3.1). The more seed crops vary from year to year the greater the pro-

portion of seeds that potentially escape predation. Thus, plants whose seed crops fluctuate (and hold seeds for brief periods of time) tend to invest less in seed defences than plants that produce more consistent seed crops (Janzen 1969, 1971; Smith 1970). In the following sections we discuss various seed defences and counter-defences by seed-predators, and evaluate these hypotheses further.

5.4.2 Selection on physiological seed traits and counter-adaptations of animals

Some seeds are so toxic that all seed-predators avoid them, so why aren't more seeds this toxic? One explanation is that plants are confronted with a problem of how to simultaneously maximize protection and stored reserves within a restricted space. Evidence of this tradeoff is the narrow range of concentrations of toxins within a species (Bell 1978). Presumably plants producing lower concentrations of toxins are more susceptible to predation and those producing higher levels produce seedlings that are disadvantaged in competition with seedlings having greater reserves. A potential solution to such a constraint is to have toxins that can also act as storage products, like toxic lipids and non-protein amino acids that can be metabolized and translocated in the seedling (Harborne 1993). The capacity to serve as nourishment to a seedling may favour the use of, for example, less toxic non-protein amino acids over more toxic alkaloids (Bell 1978). Another solution is to have toxins that are effective at low dosages. Most toxins are usually in small concentrations (<5%) in the seed and some alkaloids, for example, can be lethal at concentrations as low as 0.1% (Harborne 1993). An alternative explanation limiting chemical defences is autotoxicity. Chemical defences that might be autotoxic (e.g. tannins, saponins), however, are usually compartmentalized in specialized cavities, often in the seed coat or fruit (Janzen 1978). Finally, defensive chemicals can be costly to the plant. Producing more toxic chemicals deters seed-predators but at the expense of producing fewer seeds (Zangerl & Berenbaum 1997).

Like many herbivores (see Chapters 3 and 4), numerous insect seed-predators have biochemical adaptations to deal with secondary compounds, including various detoxification and sequestration mechanisms. Many bruchids, for example, have the ability to avoid incorporating toxins during biosynthesis, and at least a few species can even detoxify and degrade the toxins and then use the by-products in their metabolism (Johnson 1990). However, the difficulties of dealing with more than a few different kinds of defensive compounds has favoured the evolution of specialization in seed-predators (Johnson 1990) and helps explain why a large fraction of seed-predators are specialists (Fig. 5.5; Janzen 1980b). In contrast to insects, few if any species of bird or mammal are highly specialized on one or a few highly toxic seeds. Although some rodents can tolerate relatively toxic seeds (Sherbrooke 1976), most birds and mammals avoid toxic seeds (e.g. those with alkaloids) or use them sparingly (Harborne 1993; Hulme 1993). Many birds and mammals eat tannin-rich acorns, and several species at least can subsist on a diet consisting exclusively of acorns (Gurnell 1993). Parrots are possibly able to deal with plant toxins better than any other group of vertebrate granivores. Parrots often feed on unripe and often toxic seeds (and fruits), and can tolerate high levels of alkaloids and phenols, in part because parrots selectively feed on soil (geophagy) containing minerals with high capacities to bind plant toxins (Gilardi et al. 1999).

Two processes have potentially played a role in generating the diversity of specialist seed-predators and the diversity of chemically defended seeds. The first process is escape-and-radiate coevolution developed for herbivores and plants (see Chapters 2 and 3), which can be summarized as follows. By mutation and recombination new chemical defences arise that allow plants to escape seed-predators. These plants then radiate. Eventually seed-predators evolve counter-measures then radiate on the plants. This is roughly the scenario envisioned for bruchids and legumes by Janzen (1969). As an example, the evolution of endopeptidase inhibitors in legumes may have freed them of many seed-predators and perhaps enabled further radiation. The subsequent loss of inhibitable endopeptidases in bruchids may have enabled them to then radiate onto legumes (Janzen 1969). Given the increasing number of studies showing that seed-predators have strong impacts on plant populations, and the increasing evidence for escape-and-radiate coevolution in herbivores and plants (e.g. Chapter 3), we believe studies testing this hypothesis for seed-predators and their victims would be well worthwhile.

The second process is the result of apparent competition (*sensu* Holt 1984). This would favour divergence in chemical defences and could even provide a mechanism for escape-and-radiate coevolution. As Janzen noted, 'any pair of species with the same defence would jointly present a larger and more reliable food source to the seed predator, thereby creating strong selection favoring any mutant that caused them to diverge in secondary seed chemistry' (Janzen 1978). Such a process could contribute to the tremendous diversity of chemical defences in seeds and in turn the diversity of seed specialists. However, comparative and experimental studies designed to elucidate the importance of this mechanism for seed diversity, as has been done for other forms of competition, are lacking.

5.4.3 Selection on physical seed traits and counter-adaptations of animals

It is not surprising that granivores lack a consistent preference for larger seeds, given the tremendous variation in other seed traits between plant species (see Section 5.2.2). Such variation also appears to explain the absence of seed-size preferences by granivores within a plant species (Smith 1970). Nonetheless, several studies have found that granivores preferentially feed or oviposit on larger seeds within a plant species (Moegenburg 1996) indicating that some granivores exert selection on seed size. Seed-size selection appears most likely when seeds are outside the fruit (i.e. dispersed) and the seed-predator is specialized on one or a few species of seed. However, seed size might still not evolve much in response to seed predation for at least four reasons. First, seed-size differences between plants can often be largely the result of environmental variation rather than heritable variation (Silvertown 1989). Second, most of the variance in seed size is accounted for by variation within crops (e.g. at a level within individuals) rather than among individuals. Consequently, differential predation by seed size might not cause differential reproductive success (i.e. selection) between individual plants. Third, selection on seed size by seed-predators could be countered by selection from a variety of sources (Primack 1987). For example, the advantage of larger seed size soon after germination (Westoby et al. 1996) may counter and perhaps overwhelm selection by seed-predators. Furthermore, both studies mentioned above, showing seed-predators preferentially preying on larger seeds, also found that larger seeds had germination advantages. Moreover, in one of these studies (Moegenburg 1996) environmental variation appeared to have a large impact on seed size. Finally, a change in seed size might reduce predation by one seed-predator but increase predation from another (Willson 1983).

In contrast to the few data indicating that seed-predators influence seed-size evolution, there is considerable evidence that seed-predators have favoured the maintenance and elaboration of many structural features of the seeds and fruit. These features include background matching of dispersed seeds (Nystrand & Granström 1997), the texture, thickness and hardness of seed coats or fruits (Smith 1970; Johnson 1990; Grubb et al. 1998), pubescence or irritant hairs or spines (Grubb et al. 1998; Coffey et al. 1999) and the number of seeds per fruit (Smith 1970, but see also Casper 1988). This evidence is especially compelling for spines on pine cones, where foraging experiments and explicit phylogenetic models were combined to show that increases in spine length are related to deterring seed-predators foraging for seeds in open cones (Coffey et al. 1999). In sum, these studies support Smith's (1970) hypothesis that selection pressures from the physical environment mostly affect seed size, whereas selection pressures from seed-predators mostly affect the type and amount of tissue that protect seeds. As Smith (1970) noted, this is an oversimplification. Nevertheless, this hypothesis has been useful in guiding research.

Many studies show that variation in the physical characteristics of seeds and fruits have influenced the evolution of seed-predators. Two examples should suffice. The beak in different soapberry bug (*Jadera haematoloma*) populations has evolved to different lengths to reach the seeds of the various species of plants the bug has colonized in the past century (Carroll et al. 1997). The extensive studies by Peter and Rosemary Grant, Dolph Schluter and others (e.g. Grant 1986) show that the size and hardness of seeds available during the dry season have influenced the evolution of bill structure in Darwin's finches (*Geospiza* spp.). Moreover, bill size is highly heritable and has evolved rapidly in response to changes in seed availability (Grant & Grant 1995).

5.4.4 Selection on seed dispersal behaviour and counter-adaptations of animals

Annual variation in seed-crop size is thought to be adaptive because of the economies of scale (Norton and Kelly 1988; Kelly 1994). That is, the cost per seed and seedling declines with increasing flower or seed-crop size. The result is that for a given average amount of investment the reward from alternately producing large and small seed crops is greater than the reward of a consistent intermediate-sized seed crop (Fig. 5.11). Variable seed

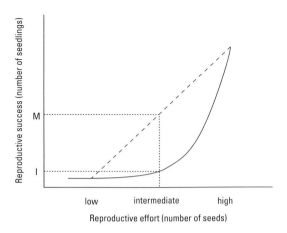

Figure 5.11 The solid curve represents a hypothetical relationship (but see Fig. 5.8) between the reproductive success of a plant and its reproductive effort. When the curve is concave-up (as shown, and as expected with the economies of scale), alternating reproductive effort between low and high seed years yields a greater average reproductive success (M) than when an intermediate (same mean) effort is made every year (I).

crops should be favoured as long as the number of seeds or seedlings surviving increases in an accelerating manner with increases in seed production (i.e. the second derivative is positive). Such an increase could result from predator satiation or, especially if unpollinated flowers are costly, improved pollination success (Lalonde & Roitberg 1992).

Selection by seed-predators has undoubtedly favored supra-annual reproductive synchrony (Kelly 1994). Evidence for selection by seed-predators is the reduction in annual variation in the absence of seed predators (Kelly 1994), a greater range of variation in seed production by plants dispersed by wind than by animals (Herrera et al. 1998), and experimental evidence that asynchronous individuals suffer higher predation rates (Crawley & Long 1995). The examples of supra-annual community-wide synchrony, including temperate conifers and hardwoods, Southeast Asian dipterocarps and bamboos, (Gurnell 1993; Kelly 1994; Herrera et al. 1998; Curran & Leighton 2000), further suggest the importance of selection by seed-predators, although in some cases community-wide synchrony might merely reflect the use of the same environmental cue to initiate large reproductive episodes (Norton & Kelly 1988).

Seed-predators in turn have adapted to variable seed

crops. At least 70 species of insect seed-predators on conifers (Diptera, Hymenoptera and Lepidoptera) extend their normal winter diapause for one to seven additional years (Turgeon et al. 1994). These insects produce progeny with a mixture of diapause lengths (risk-spreading diapause) or rely on environmental cues to emerge from diapause during a large seed crop (predictive diapause). Some insects and vertebrates move seasonally or yearly from one large seed crop to another. Birds such as crossbills (Benkman 1987) are the most effective at tracking seed fluctuations, especially considering that synchronous seed production among northern conifers can extend 500 or more kilometres (Koenig & Knops 1998). Many seed-predators eat alternative foods, but it is unclear if specialist seed-predators have evolved to become generalists in response to increasing fluctuations in seed availability. Enough is known about the feeding habits of conifer-seed-eating insects (Turgeon et al. 1994) that, given the appropriate phylogeny, one could investigate whether generalists have evolved from specialists.

Reproductive synchrony works well against specialist seed-predators, but the large pulses of seeds (which tend to lack chemical defences (Janzen 1969, 1971)) attract generalist seed predators, which may or may not be swamped (Section 5.3.2). Satiation is less likely with increasing adaptation by seed-predators. This favours even larger seed crops and possibly longer intervals between them (Silvertown 1980). Because of the costs and limitations of resource storage, most of the resources allocated to reproduction might be from current photosynthesis (Koenig & Knops 1998) limiting the size of the seed crop. Eventually asynchronous reproduction and perhaps physical or chemical defences might be favoured. However, in contrast to chemical and physical defences, the benefits of a particular temporal seed-production pattern to a plant is dependent on the temporal seed-production patterns of other plants in the population and even other species (Silvertown 1980; Curran & Leighton 2000). Once periodic and synchronous seed production evolves, selection operates against individuals that deviate (Lalonde & Roitberg 1992). It is an evolutionarily stable strategy (ESS) that cannot be invaded by *individuals* that differ from the rest of the population. It is conceivable, therefore, that with counter-adaptations of seed-predators and perhaps changes in the environment (e.g. deforestation; Curran & Leighton 2000) that periodic and synchronous seed production is no longer a good, let alone the best, strategy. But in contrast to chemical and

physical seed defences, which presumably could more easily change in response to selection, periodic and synchronous seed production might remain because it is an ESS. Change might occur only after, for example, plant population density decreases so that movement of seed-predators between plants is reduced, or physical or chemical defences increase. Thus, we should anticipate cases where predators are not satiated.

The seasonal phenology of seed production is thought to have evolved in response to selection by seed-predators, with predator satiation more likely if seeds mature quickly and are available to seed-predators only for brief periods of time (Janzen 1971; Kelly 1994). Yet some plants retain mature seeds from successive seed crops in their canopies. These seeds are held in closed seed-storing structures to be released synchronously when conditions are favourable for germination (e.g. after a fire). This is called serotiny, and is characteristic of woody perennials that occur in habitats having strongly seasonal climates and recurrent fires within the reproductive lifetime of the plant (Lamont et al. 1991). Serotiny is advantageous because it maximizes the quantity of seeds available for recruitment following fire (Lamont et al. 1991) and, like mass seeding, it satiates post-dispersal seed-predators (O'Dowd & Gill 1984). Areas where serotiny is most common are the sclerophyllous shrublands and woodlands of Australia and South Africa, and the coniferous forests of North America (Lamont et al. 1991).

The disadvantage of serotiny is that seeds are predictably and reliably held, favouring the evolution of specialist pre-dispersal seed predators (Lamont et al. 1991). This in turn has produced strong selection for increased seed defences, which has resulted in the evolution of harder or woodier structures surrounding the seeds (Smith 1970). These well-developed physical barriers also protect the seeds from damage by fire, which may explain why physical rather than chemical defences have been developed. The importance of protection from fire may also explain why monocots, which lack wood, are not serotinous (Kelly 1994). In the following section we will discuss an example of coevolution between a serotinous tree and its specialist pre-dispersal seed-predators.

5.4.5 Case studies in coevolution

Although numerous examples exist of adaptation by plants to seed predation and adaptations of animals to exploit seeds, compelling examples of coevolution between

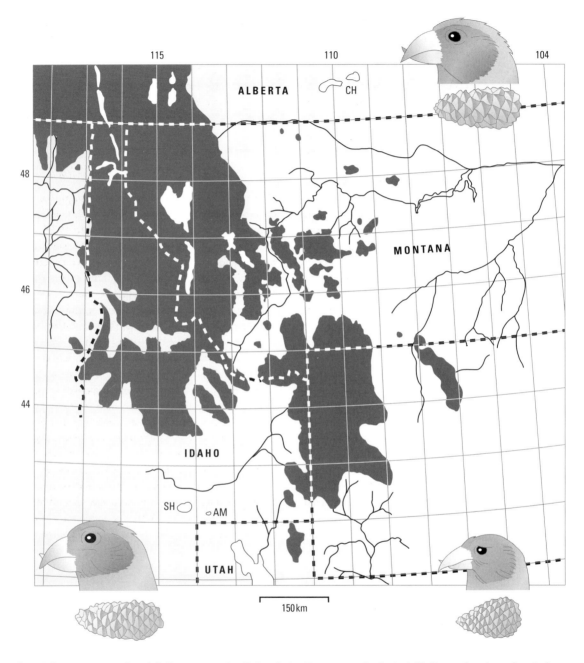

Figure 5.12 Representative red crossbills (*Loxia curvirostra*) and lodgepole pine (*Pinus contorta*; distribution in black) cones from the northern Rocky Mountains. Lodgepole pine has relatively short and wide cones (lower right) throughout most of its range in response to selection by red squirrels. The crossbill found in these forests (lower right) has an average bill size that approximates the optimum for foraging on these cones. In the South Hills (SH) and Albion Mountains (AM) in southern Idaho red squirrels are absent and crossbills are coevolving with lodgepole pine. Here, lodgepole pine cones are larger and have thicker distal scales in response to selection by crossbills, and crossbills have stouter bills to get seeds out of these cones. This was repeated in the Cypress Hills (CH) in southern Canada; however, squirrels were introduced here in 1950 apparently causing the extinction of this population of crossbills. (From Benkman 1999.)

seeds and seed-predators are rare (Johnson 1990). We discuss two examples of coevolution below. It might not be a coincidence that both represent pre-dispersal seed-predators. For at least four reasons we expect coevolution between seed-predators and seeds to be most obvious in pre-dispersal seed-predators. First, it is easier for a seed-predator to be selective of seeds prior to rather than after dispersal. For example, pre-dispersal seed-predators can more readily avoid seeds from a particularly toxic plant when the seeds are clustered on the plant than after they are scattered on the ground and mixed with seeds from other plants (Janzen 1971). Second, defences associated with dry fruits have fewer constraints than those associated solely with the seed. Third, pre-dispersal seed-predators limit the evolutionary effect of subsequent seed-predators on the subset of remaining seeds. Finally, pre-dispersal seed-predators tend to be more specialized, so that increasing seed defences are more likely to lead to counter-defences than to the predators switching to an alternative food.

5.4.5.1 Wild parsnip and parsnip webworms

The defences of wild parsnip and the counter-measures of the parsnip webworm have already been discussed in Chapter 3. Nevertheless, we would also like to lay claim to this example because seeds are a critical component of the webworm's diet, even though, like many granivores, they also eat the reproductive structure surrounding the seeds. Moreover, no other study of coevolution between insects and plants is as compelling. Rather than repeat the details of this marvellous example, some of which we have already noted, we just want to point out an intriguing

dynamic; that is, instead of simply an arms race with escalating defences and counter-defences until tradeoffs result in a stalemate (which also occurs), there are apparently cyclic chase dynamics with different populations at different points in the cycle (Berenbaum & Zangerl 1998). Although it is unclear what exactly causes the cycling, frequency-dependent selection is undoubtedly critical (Berenbaum & Zangerl 1998).

5.4.5.2 Lodgepole pine, squirrels and crossbills

One of the earliest examples of seed and seed-predator co-evolution was that between lodgepole pine and pine squirrels in western North America (Smith 1970). This study is a classic in part because Smith was able to identify fire as a variable independent of the interaction between seed and seed-predator, which determined the strength of the coevolutionary interaction. Where fire was frequent lodgepole pine evolved serotinous cones that accumulated on the tree until the next fire. This allowed for a more stable population of seed-predators, which resulted in strong selection for increased defences. The serotinous cones are so well defended (seeds represent about 1% of total cone mass) that only two seed-predators, pine squirrels and red crossbills, commonly consume seeds before the cone scales open wide and seeds are released.

Pine squirrels are the most important selective agents on the cones of lodgepole pine throughout most of its range. Squirrels preferentially harvest cones that are relatively narrow at the base and have more seeds because this maximizes both feeding rates and the mass of kernel cached per cone (Smith 1970). This results in the evolution of wider (and harder) cones (especially at the base)

Figure 5.13 The estimated contours (solid lines) for the benefit-to-cost ratio to the tree in relation to the first two principal components of seven lodgepole pine cone and seed traits. The filled circle represents the overall mean values for four sites from the Rocky Mountains, and the open circle represents the overall mean values for South Hills/Albion Mountains and the Cypress Hills, where pine squirrels are absent and crossbills are resident. (From Benkman 1993.) If pines were evolving so as to increase the benefit-to-cost ratio of their defences, they should evolve down and to the right (to high benefit-to-cost ratios). That is what was found.

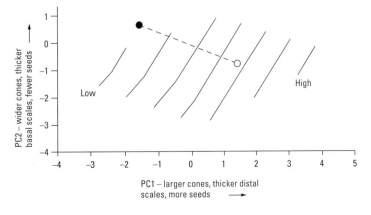

with fewer seeds (Fig. 5.11; Smith 1970; Benkman 1999). Pine squirrels, in turn, have evolved various adaptations for efficiently exploiting these cones (Smith 1970). By harvesting and caching tremendous numbers of cones soon after the seeds mature (Smith 1970), pine squirrels are effective pre-emptive competitors so that red crossbills are uncommon and apparently have little impact on cone evolution (Benkman 1999). Here crossbills have adapted to the average lodgepole pine cone (Fig. 5.12).

In two sets of mountains east and west of the Rockies crossbills have evolved in the absence of pine squirrels for the past 6000 to 10 000 years (Fig. 5.12). Here crossbills are over 20 times more abundant. Crossbills also select cones providing the highest kernel intake rates (Benkman 1987), but in contrast to squirrels, the mass of kernel per cone is not important, in part because crossbills do not remove cones from the tree. The most striking features of cones in these 'crossbill' ranges are their large size and the increased thickness of the distal scales. Cones are larger because with relaxation of selection by pine squirrels, the number of seeds per cone has increased and selection by crossbills favours larger, thicker scales (Benkman et al. unpublished). Larger, thicker distal scales make sense because the time required to extract a seed increases with increasing scale thickness (Benkman et al. 2001) and most of the seeds are located at the distal end. Not surprisingly, the composite evolution of the cones is most accurately

predicted when both the benefits of the defence, in terms of time per seed for a crossbill, and the costs of the defence, in terms of cone mass relative to seed mass, are considered. In response to selection by crossbills, lodgepole pine has increased the ratio of benefits to costs of its defences (Fig. 5.13). Crossbills in turn have adapted to these enhanced defences by evolving deeper and more strongly decurved bills (Fig. 5.12).

The interaction between wild parsnip, lodgepole pine and their seed-predators illustrates many principles common to seed and seed-predator interactions. We would like to end by emphasizing three points. First, the changes in response to selection by seed-predators can be accounted for only when both the benefits and the costs of defences are taken into consideration. Second, geographical variation in seed-predator assemblages (e.g. the presence and absence of pine squirrels), the physical environment (e.g. fire frequency) and perhaps cyclic chase dynamics (e.g. webworms and parsnip) will commonly cause divergent selection between populations and lead to a geographic mosaic of coevolution (Chapter 9). Finally, we suspect that coevolution occurs fairly commonly between plants and seed-predators. However, until we carry out careful studies testing specific models of coevolution this will remain only an opinion, which is unfortunately the basis of many conclusions concerning coevolution between plants and seed-predators.

Part 3 Mostly Mutualisms

Chapter **6**   **Pollination by animals**

Olle Pellmyr

6.1 Introduction

On finishing *The Origin of Species*, Charles Darwin turned to detailed studies of specific interactions to test the assumptions and predictions laid out in his grand conceptual framework of evolution. First he turned to the subject of pollination biology and published *The Various Contrivances by which British and Foreign Orchids are Fertilised by Insects* in 1862 (Darwin 1862). It is telling that Darwin turned first to this subject because it is rich in models, sometimes staggeringly flamboyant in their manifestations (Proctor et al. 1996), and it has been useful for exploring many fundamental aspects of evolution. The function of flower-visiting insects as pollinators had already been described in the late 1700s, notably by Sprengel (1995), but Darwin's treatise laid the foundation for an evolutionary approach to analysing mutualistic interactions in general and of plant–pollinator interactions in particular. The observational experimental approach established by Darwin and some of his contemporaries set the field on track to become a modern discipline; in fact, biologists such as Fritz Müller, Hermann Müller and Charles Riley were to use pollination systems as major examples in the early tests and corroborations of evolutionary theory.

6.1.1 The pollination process

This is the first of three chapters addressing interactions that are usually mutualisms, that is, where the interaction usually benefits both animal and plant. The lack of mobility in plants creates a physical obstacle in the dispersal of their genes. In a majority of all plants, this obstacle has been alleviated through the formation of mutualisms with animals that transport pollen grains between stigmas and also disperse seeds.

In the case of pollination, the goal for the plant is to re-ceive pollen on its stigma and to have pollen picked up and deposited on conspecific stigmas of other plants. The animal most commonly seeks a food reward. It is important to appreciate that mutualisms such as these represent reciprocal exploitation with an underlying evolutionary conflict. Selection in mutualisms favours selfish behaviour; in this, they do not differ from any other type of interaction. The optimal situation for the plant is one in which it obtains pollination service by a pollen vector without providing any reward. Similarly, a flower-visiting animal should exploit available resources as effectively as possible, and it is only rarely of any fitness consequence to a visitor whether it transports pollen or not. One manifestation of such selection for minimizing costs is the widespread phenomenon of plant species that no longer reward pollinators but instead attract visitors by deception. These are commensalistic or even antagonistic pollination systems. Likewise, many flower-visitors (if not most) do not contribute to pollination but do remove floral resources such as nectar and pollen. In a given plant–pollinator relationship, if individual plants can increase their fitness at the expense of visitors by reducing rewards, or pollinators reap more rewards while pollinating less, this conflict of interests can lead to collapse of the mutualism locally or globally. This can happen either if the plant goes extinct as seed production fails or if visitors abandon the plant in response to decreasing rewards, as less rewarding genotypes increase in frequency. But because most plant–pollinator mutualisms involve more than two species and vary across space and time, the dynamics may be more complex. A given pairwise plant–pollinator interaction may be mutualistic in one ecological context but antagonistic in another. As a consequence, evolutionary trajectories may well differ.

As a bottom line, pollination mutualisms evolve amid simultaneous antagonistic interactions; the plant is under selection to maximize the net fitness of attracting potential mutualists at the lowest net cost while minimizing the detrimental effects of non-mutualists or low-quality mutualists. This tradeoff does not exist in antagonistic interactions, such as those between a plant and a herbivore, where the plant is simply under selection for tolerance of or defence against the antagonist. Floral traits are likely to be as much the result of selection for avoidance of some animals as for attraction of others.

6.1.2 Costs and benefits of animal pollination

Animal pollination evolved from systems of abiotic pollination, involving pollen transfer mostly by wind or sometimes by water (see Chapter 2). The use of animals such as insects and birds for pollen dispersal, which now predominates among angiosperms, provides several ecological possibilities for the immobile plants. First, animals that actively seek out flowers increase the probability that a pollen grain collected on one flower will reach the stigma of a conspecific flower. This happens because animals can seek out isolated flowers and because pollen may travel farther from its source on an animal than it would on the wind. For this reason, animal-pollinated plant species can probably persist at lower population densities than wind-pollinated species, and the probability of outcrossing may increase. Second, animal-mediated cross-pollination can take place in habitats with very little wind, such as closed-canopy rain forests. Third, animal-pollinated plants can allocate a lesser proportion of their resources to pollen production than wind-pollinated plants because individual pollen grains have a higher probability of reaching and fertilizing an ovule of a different plant. Balancing these benefits, novel costs to the plants that come with animal-mediated pollination include the need to produce visual and olfactory cues that attract visitors, production of pollinator rewards such as nectar and additional pollen, and the cost imposed by non-pollinators that exploit the cues and rewards that attract and maintain the pollinators.

The benefits to the animals of visiting flowers consist of readily accessible high-quality rewards, such as nectar, pollen, oils or water, in easily identified discrete patches. In most instances there are few costs to the visiting animals. The carriage of pollen is unlikely to measurably impede the animal, except possibly when unwieldy pollen batches, termed pollinia, adhere to extremities or partially blind animals by covering their eyes. Visitation of flowers also carries the risk of predation by specialized flower-based predators, such as thomisid crab spiders, phymatid ambush bugs and some praying mantises.

6.2 The origins of animal-mediated pollination

6.2.1 Organisms involved in pollination mutualisms

6.2.1.1 Plants

Three phylogenetically separate, extant groups of plants harbour instances of pollination mutualisms: cycads, gnetalean conifers and angiosperms (Fig. 6.1). In cycads, experimental evidence exists to suggest that several of the surviving genera of this formerly large group are insect-pollinated (Tang 1985; Donaldson 1992). In the peculiar gnetalean conifers, two of the three extant genera are probably animal-pollinated. The vast majority of all extant pollination mutualisms, however, involve flowering plants, which dominate most biota on earth today. There are no modern estimates of what proportion of the approximately 250 000 species of angiosperms are animal-pollinated, in part because so few taxa have been properly studied, and even simple observations on floral biology are wanting for most tropical species. If we use the presence of either showy flower parts or floral nectar as an indication of some degree of animal-mediated pollination, it is obvious that a very large percentage fall into this category. Exceptions include wind-pollinated species (such as most grasses), water-pollinated aquatic plants and asexual or self-fertilizing species.

6.2.1.2 Animals

Given that the benefit to plants of animals as pollen vectors is transport across longer distances, it is not surprising that the three extant groups of animals that have evolved flight—insects, birds and bats—contain a very large proportion of all pollinators. Among the insects, flower-visiting species are particularly frequent within the large orders Hymenoptera (bees and wasps), Lepidoptera (moths and butterflies), Diptera (flies) and Coleoptera (beetles). Indeed, groups such as the eusocial bees, ditrysian Lepidoptera, large fly families such as the beeflies (Bombyliidae) and hoverflies (Syrphidae), and beetle families such as the longhorn beetles (Cerambycidae) and Cantharidae (soldier beetles) rely almost exclusively on floral resources during one or more of their life stages. The

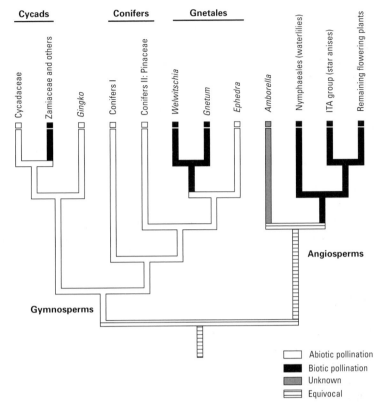

Figure 6.1 The established phylogeny of seed plants indicates three separate origins of animal pollination in major groups. A phylogeny based on DNA sequences and gene duplication data suggests three origins, within the cycads and the gnetalean conifers, and at or near the base of the angiosperms (with many secondary losses within the latter). (Phylogeny compiled from Qiu et al. 1999 and Mathews and Donoghue 1999.) An alternative topology, with Amborella + Nymphaeales as basal sister taxa (Barkman et al. 2000), would support an earlier origin of the hermaphroditic condition, but would not affect the conclusion of three separate origins of animal pollination.

Lepidoptera alone, whose coiling tongues make them flower specialists and effective consumers of nectar, constitute 11% of all described species on Earth (Wilson 1991).

Among birds, six phylogenetically independent groups have diversified as flower-visitors and often as pollinators (Sibley & Ahlquist 1990). They show convergent specialization in mouthparts for nectar feeding, including beak shape and tongue morphology, a situation that long obscured their distant relationships. Major radiations include the American Trochilidae (hummingbirds and hermits), Palaeotropical and Pacific Nectariniidae (sunbirds, flowerpeckers and sugarbirds), Australian Meliphagidae (honeyeaters and chats) and psittacid brushtongued parrots, and fringillid tanagers, New World and Hawaiian honeycreepers. Together these groups constitute over 10% of all recognized bird species. They are important pollinators in tropical and subtropical areas that can provide floral resources throughout the year

for these comparatively longlived animals; some birds utilize the flowers of over 20 species in succession throughout the year (Sazima et al. 1995). An exception to the tropical and subtropical distribution is a modest array of migratory hummingbirds in North America, which breed in temperate to alpine areas and pollinate many plant species. Macrochiropterous fruit bats are important flower-visitors in tropical regions on all continents. Other, non-flying groups in which a pollinator function has been documented in some instances include mammals such as mice, possums, lemurs and primates (Carthew & Goldingay 1997).

Because of historically lax criteria for identifying visitors as pollinators, the level of support for pollinator function in the literature must always be carefully considered (Box 6.1). For example, claims of pollinator function in slugs have been made, based on pollen observed on animals, but the actual transfer of pollen between plants was not documented.

Box 6.1   A basic protocol for determining whether a flower-visitor is a pollinator

Given the underlying conflict of interests between plants and potential pollinators, it is not surprising that visitors frequently act as parasites on the system. Thus it would be a fallacy to equate a flower-visitor with a pollinator, and the first step in studies of pollination interactions should be to quantify the role of individual participants. Ideally, a series of criteria (laid out below) should be met in experimental tests. This can be more or less of a logistical challenge, particularly for criteria, depending on the physical setting and the number of visitor species involved. For example, appropriate sample sizes for experimental tests of pollinator function may be hard to obtain for epiphytes and isolated canopy trees, or plant species that receive very infrequent visits (such as many orchids attracting by deceit). Situations with very large visitor guilds, as is often the case in plant species with easily accessible rewards, may present a challenge for selective exclusion of particular visitors. Ethical aspects of conservation must also be considered when rare and endangered plants are involved.

1 Evidence that a visitor can pollinate:

 (a) *Does the visitor pick up pollen?*
 (b) *Does the visitor deposit pollen on the stigma?*

If a flower-visitor does not pick up pollen and deposit it on conspecific stigmas, it is certain that it will not be a significant pollinator. This can be due to behaviour on a hermaphroditic flower that keeps the animal from coming into contact with either anthers and stigmas or both, selective visitation in only one phase of hermaphroditic flowers with temporally offset male and female phases (dichogamy), dioecious plant species where plants of only one sex are visited, or high levels of heterospecific pollen transfer as a result of the visitor using flowers of multiple plant species simultaneously. Analysis of

pollen on visitors' bodies can provide evidence of pollen pickup, while observations of visitors' behaviour on virgin flowers and subsequent stigma check for pollen is required to meet the second part of this criterion.

2 Evidence that a visitor pollinates:

 (a) *Does exclusion of all visitors cause lack of seed set?*
 (b) *Does visitor access cause seed set?*

Seeds can potentially be produced in the absence of visitors, through autogamous self-pollination or asexual production. Exclusion of all visitors should prevent seed production, and access by a given visitor should result in seed set. If seeds are produced in the absence of visitors, visitors may still be critical in providing cross pollination.

3 Evidence that a particular visitor contributes to pollination:

 If several species visit the flowers, does exclusion of the one you study affect seed set?

If the first two criteria are met and a visitor is considered a pollinator, we must determine whether the species contributes significantly to pollination. Within a visitor guild, species will vary in their contribution to plant male and female fitness by factors such as its relative abundance, pollination efficacy per visit and movement patterns among plants. A species that constitutes a small fraction of all visitors, causes few pollinations per visit, or is sedentary in habit is less likely to be important in pollination.

This quantification of contribution is the most context-dependent of the three criteria. Because associations between a plant and its visitors varies among populations and over time, separate tests are indicated to understand ecological dynamics and the possible coevolutionary consequences of these differences.

6.2.2   Evolutionary origins and early diversification of animal pollination

Because animal pollination is basal within the angiosperms, this type of mutualism must have existed at least since the late Jurassic. Fossil evidence reviewed by Labandeira in Chapter 2 suggests that animal-mediated pollination may have occurred earlier still in other long-extinct seed plants, such as the Palaeozoic medullosan seed ferns whose giant spores were unlikely to have travelled on the wind. The floral structure in the basal angiosperm ancestor is currently difficult to reconstruct because of a long branch between the angiosperm–gymnosperm split and the most basal surviving angiosperm family (Amborellaceae), combined with a paucity of macrofossils from that phylogenetic segment. *Amborella* flowers are small and unisexual (Endress & Igersheim 2000), whereas most of the next basal families have hermaphroditic flowers. Hermaphroditic flowers are unique to angiosperms, and in the simplest scenario would be assumed to have arisen after the origin of angiosperms. There is considerable evolutionary plasticity in sex expression among angiosperms, however, so the basal condition might have been either hermaphroditic or unisexual flowers. Sex expression is under control by the MADS box of developmental genes, a modest number of genes that diversified well before the origin of angiosperms (Lawton-Rauh et al. 2000); for this reason, the high plasticity in sex expression is not surprising. The evolution of hermaphroditic flowers at or close to the base of the angiosperms is important in considering the type of interactions that led to pollination mutualisms. Antagonistic associations between insects and plants, with the insects consuming nitrogen-rich pollen, spores or ovules, is evident in the fossil record as land plants appear from the late Silurian onward, with plant tissues being found on insect surfaces or in the gut, or in fossil faeces. Some degree of occasional pollen transport between feeding sites is a probable consequence — this is the basic mode of virtually all animal pollination — and whenever the fitness gains of such pollen grains fertilizing ovules elsewhere exceeded the cost of tissue consumed, selection would change from avoidance or deterrence to increased selective attraction of the mutualistic herbivore cum pollinator. This obviously requires that the herbivore should visit both male and female floral parts. Their physical proximity in a hermaphroditic flower could greatly facilitate the evolution of animal pollination in basal angiosperms, because a visitor seeking pollen or ovules would probably come into contact with both sets of floral parts. In contrast, in the two other origins of animal pollination, cycads and gnetalean plants are dioecious, and other mechanisms must exist to incite visitors to transport pollen from pollen-bearing male plants to non-rewarding female plants. Pollinators of cycads are attracted to female flowers by deceit (Donaldson 1997). In Gnetales, visitors lick liquid pollination droplets on ovules (Kato & Inoue 1994; Wetschnig & Depisch 1999); this droplet, which contains a low concentration of sugar and plays a role in trapping airborne pollen, is specific to gymnosperms. In contrast, carbohydrate-rich angiosperm floral nectar is not known to carry any other function than as a pollinator reward. For this reason, it is unclear why it would have existed in flowers before pollinator attraction had become established. There is no evidence to suggest that it was present as a complement to pollen in the first cases of animal-mediated pollination in angiosperms, but it appeared early in diversification as an alternative reward and is arguably the most common reward in present-day angiosperms. The production of a carbohydrate reward can reduce the cost of pollinator attraction to plants, both because gametes are not directly lost to the pollen vector, and because fewer nitrogen resources need to be allocated to pollinator rewards.

6.2.3 Pollination mutualism as a catalyst of reciprocal diversification

A longstanding issue surrounding animal pollination is whether it has been important in driving reciprocal diversification. One palaeontological and one ecological argument have been put forth to support this hypothesis. First, the angiosperms and the primary orders of insects involved in pollination experienced very rapid diversification in rough simultaneity from the latter half of the Cretaceous onward. Second, because pollinators exert immediate control on plant gene flow, the potential for assortative mating and subsequent diversification may be higher in this type of interaction.

Several lines of argument can be used to test this hypothesis. Consistent with the hypothesis, important insect pollinator groups such as Coleoptera, Lepidoptera, and Diptera emerged before the angiosperms, in the early Mesozoic or before (Chapter 2, Fig. 2.2), and after slow early diversification showed rapid rate increase once they became associated with angiosperms. A strong argument against animal pollination as a single catalyst of angiosperm success was presented by Sanderson and Donoghue (1994), who showed that diversification rate increased not at the base of the angiosperms, where animal pollination arose, but rather at a later point. They concluded that it could at best be a cofactor in later diversification, interacting with other angiosperm life habits such as herbaceous habit. A final means to test the hypothesis would be to analyze the three separate origins in an independent contrasts test, by asking whether the newly-recognized three separate origins of animal pollination led to significantly elevated diversity in the animal-pollinated group than in its sister group. While angiosperm diversity indeed is much higher than gymnosperm diversity, the animal-pollinated cycads of the Zamiaceae are not particularly more species-rich than their wind-pollinated sister cycads. Similarly, Gnetales contain fewer than 100 extant species, compared to about 262 species in the wind-pollinated sister group Pinaceae. The extant Gnetales are the last survivors within the Gnetophyta, a once-diverse group that suffered massive extinction by the end of the Cretaceous (Crane & Lidgard 1989). Unfortunately, these extinct taxa are only known from their peculiar fossil pollen and the pollination mode cannot be inferred from them. Thus it is hypothetically possible that the Gnetophyta were animal pollinated. If pollination mode can be inferred once macrofossils are recognized, it may be possible to answer this question.

Finally, it is important to recognize that the issue of early diversification is different from that of the role of specific pollinators in contributing to bursts of unilateral or reciprocal diversification. This will be addressed further in subsequent sections, particularly in the context of specificity and specialization.

6.3 Plant and animal traits relevant to pollination

The purpose of attracting pollinators is to attain outcrossing. In all hermaphroditic species, i.e., most angiosperms, this means both to receive conspecific pollen from other plants and to place pollen on visitors in such a way that those pollen grains may fertilize ovules elsewhere. In a hypothetical context, indiscriminate attraction may seem optimal, but there are fundamental factors that select for different degrees of selectivity and visitor manipulation. First, heterospecific pollen transfer is non-adaptive at best, and mechanisms that constrain visitor guilds to fewer but more efficient pollinators are expected. Second, self-fertilization through pollen transfer within or among flowers on a plant can reduce the benefits of pollinator attraction, and mechanisms that reduce the risk of such transfers are expected. Self-fertilization is disadvantageous relative to cross-fertilization for two main reasons. First, inbreeding depression will lower the fitness of selfed progeny, sometimes dramatically so. Second, pollen spent on self-fertilization may reduce the male fitness of a plant, a process known as pollen discounting (Barrett 1998). This section first addresses plant traits that can reduce the frequency of self-fertilization, and then turns to traits that affect patterns of visitor attraction.

6.3.1 Floral traits that mediate outcrossing

Angiosperms have the possibility of controlling rates of self-fertilization by mechanisms acting before the pollen has reached the stigma, once the pollen is on the stigma, or after fertilization. They are not mutually exclusive, and combinations of pre- and post-stigma mechanisms are common.

The hermaphroditic condition of most angiosperms increases the risk of intrafloral or intraplant pollination; a visitor that moves within a flower or visits numerous flowers within a plant is likely to cause some level of self-pollen

deposition on stigmas. Physiological self-incompatibility mechanisms occur in a considerable proportion of all species, acting most commonly at the stigma level or in the style, blocking pollen-tube penetration or retarding pollen-tube growth (Proctor et al. 1996; Barrett 1998). This mechanism completely blocks the possibility of selfing, and may be disadvantageous under circumstances where seed production by outcrossing is uncertain; a mixed strategy of outcrossing and facultative selfing may provide higher fitness. Pre-stigmatic mechanisms to control pollen pickup and deposition based on temporal and spatial spacing of male and female function are common, and can contribute to an elevated probability of outcrossing (Barrett 1998). Hermaphroditic flowers commonly have the male and female functional phases offset, so that a flower at any one point in time either only presents pollen or is receptive to pollen (dichogamy; Lloyd & Webb 1986). There can be a complete separation of phases, but partial overlap late in floral life may permit selfing if cross-pollination has not already occurred. Timing of the shift from one phase to the other is commonly fixed, but in some instances it is determined by the completion of the first function; the complete removal of pollen in a protandrous flower is necessary before the stigma is exposed in, for example, the disc florets of aster species (Proctor et al. 1996) and the rate of pollen removal speeds the transition to a female phase in other species (Richardson & Stephenson 1989).

Temporal offset within flowers can reduce the selfing rate, but in multiflowered plants selfing can also occur through transfer among flowers (geitonogamy) unless sexual phases are synchronized across all open flowers. When phases are not synchronized, levels of selfing tend to increase with the number of flowers as pollinators stay longer within a plant (Vrieling et al. 1999). The staggered opening of flowers is possible, with subsets of phase-synchronized flowers open at a time (Kubitzki & Kurz 1984), but such staggered patterns may involve tradeoffs in selection for a large display size at the plant level (see Section 3.2.1.2). Extreme temporal separation is present in a few cases, with synchronized phases within plants offset by several weeks (Bronstein 1992); selection for selfing avoidance is unlikely to explain such a magnitude of offsetting, and instead other factors may be examined.

An alternative to dichogamy is spatial separation of male and female floral parts (herkogamy). In hermaphroditic flowers, this is commonly indicated by the unequal exsertion of stigmas and stamens from a floral mouth; if, for example, a stigma is exserted further outside a floral tube than the stamens, an entering pollinator is more likely to deposit pollen from another source than if it were to come into contact with the anthers first (Campbell et al. 1991). There is often considerable intraspecific variation in stigma–anther distance and it is linked to variation in the selfing rate (Motten & Stone 2000). Herkogamy may be unlikely in instances where highly specific placement of pollen on the visitor is required for successful deposition on a conspecific stigma. In many tubular flowers, such as penstemons, phlox, gesneriads and lilies, stamens are first presented in a specific location, placing pollen in a specific spot on visitors. Following pollen removal, stylar elongation places the stigma in the same position, optimizing the chance of pollen capture.

Evolution of monoecy, the production of separate male and female flowers within a plant, further increases the possibility of inbreeding avoidance by independent spacing of male and female floral parts, and it offers the possibility of temporal staggering as well. But with separation of sexual functions, the risk increases that visitors will not visit flowers of both sexes, especially if only pollen is provided as a pollinator's reward. It will probably also increase the required investment in pollinator rewards, because (all else being equal) a plant of this kind needs to obtain twice as many visits as a plant with dual-function flowers to obtain the same cumulative number of pollen collections and deposits. This cost may be discounted in part by higher efficiency during each visit in unisexual flowers and lower levels of geitonogamy.

6.3.2 Attractants

Flowers offer an extraordinary range of shapes, colours and scents, reflecting high rates of evolutionary change in these traits. They affect whatever is attracted either from longer distances or at close range, and how the plant can manipulate visitors to optimize its fitness gains. Almost any flower part or even adjacent leaves are modified for the purpose of attracting pollinators. There is arguably more plasticity in these secondary reproductive traits in plants than in any other organismal group, with the possible exception of birds. Here the attractants are grouped by the senses used by visitors to perceive them: visual, olfactory or auditory cues.

6.3.2.1 *Visual cues*

6.3.2.1.1 *Shape*

Floral shape can be sorted functionally, based on the number of axes of symmetry and the presence or absence of a depth dimension. The two main categories are radial and bilateral symmetry, referred to as actinomorphic and zygomorphic symmetry respectively (Endress 1999). In some instances, very small flowers of the two forms occur side by side in the compact heads of asters, and it makes functional sense to apply these symmetry concepts at the level of resolution perceived by an approaching visitor, i.e. the flower-head. In radial symmetry an infinite number of symmetry axes can be laid through the centre of the flower or flower-head, whether it is flat or round. In contrast, a bilaterally symmetrical flower contains more information by having only a single symmetry axis running vertically through the face of the flower. Basal angiosperms are radially symmetrical, whereas zygomorphy has evolved and been lost repeatedly in the course of diversification (Ree & Donoghue 1999). The bilateral condition potentially allows the plant to manipulate the approach patterns of and selectively reward visitors, because a visitor that learns to use the symmetry axis to approach the flower in a stereotypical way can find and extract a reward more quickly. Bilateral symmetry is often combined with the concealment of nectar or other rewards in a cavity or spur behind the floral face, further increasing the difference in reward potential between naïve and experienced visitors. Invertebrate pollinators have a limited memory capacity for flowers, and temporary specialization on one zygomorphic flower form will increase their foraging rate (Lewis 1986), thus reinforcing intraspecific pollen flow. Furthermore, because visitors will prefer to approach the flower from one angle to minimize handling time, there will be fewer optimal arrangements of stamens and stigmas, potentially making for more effective use by the plant of each visitor.

The depth dimension reinforces selectivity and manipulation of visitors. The concealment of a reward at the bottom of a tube or a spur requires the ability to learn how to find the path, and also the appropriate mouthparts to reach the reward. Even though zygomorphy may select for a narrower spectrum of more effective pollinators, it is best regarded as one alternative strategy for pollination. Actinomorphy is very common, including most of the evolutionarily young and highly successful asterids,

(a)   (b)

Figure 6.2 Sterile edge flowers increase display size and hence relative attractiveness. (a) viburnum with edge flowers; (b) aster with ray flowers.

where loss of zygomorphy abounds (Ree & Donoghue 1999). The tradeoffs between the strategies remain to be studied.

Two other properties of shape commonly matter in pollinator attraction and deserve mention. First, pendant flowers have evolved in many lineages. This position is highly selective as many flower visitors lack the ability to find the floral reward, so losses to parasites may be reduced. Second, relatively subtle shape differences are used by some visitors to gauge the potential reward value of individual flowers. For example, the bowl-shaped flowers of *Anemonopsis macrophylla* contain pollen collected by bumblebees over the course of several days (Pellmyr 1988). As the flower shape changes through increased reflexing of the petaloid sepals, approaching bees assess the flower shape from a distance and selectively visit younger flowers whose pollen supplies are less likely to be depleted.

6.3.2.1.2 Size
The size of the floral display can matter in attraction both at the level of individual flowers and of the cumulative floral display of a plant. Within most plant species, pollinators generally select in favour of larger flowers, larger inflorescences and more open flowers on each plant (Bell 1985; Conner & Rush 1996; Andersson 1996). Traits that specifically increase perceived size include peripheral sterile flowers in the inflorescence of *Viburnum* that increase pollen removal (Fig. 6.2a; Englund 1994), and

sterile ray flowers in asters, whose hypertrophied petals increase the disk area (Fig. 6.2b; Andersson 1996). Display size at the plant level also varies depending on whether multiflowered plants produce flowers in sequence or in a simultaneous burst of mass flowering. The latter type is strikingly apparent in many animal-pollinated trees (Yumoto 1987), where isolated trees in bloom are readily apparent from long distances. Counteracting selection for a larger display, the risk of intraplant pollen transport increases (Vrieling et al. 1999) and greater attraction of parasitic organisms such as herbivores may increase as well (Brody 1992). One means of increasing display and attraction without increasing the risk of geitonogamy involves the retention of old, non-rewarding flowers, sometimes in conjunction with colour differences between rewarding and non-rewarding flowers. Such colour changes and retention have evolved repeatedly in many plant families (Weiss 1995).

6.3.2.1.3 Size and shape
Fluctuating asymmetry, a deviation from perfect body symmetry, is widely held to reflect the inability of an individual to maintain perfectly balanced development in response to its environment (Møller 2000). In this view, symmetrical individuals are likely to be of better genetic quality for the specific environment. While originally developed for animal models, symmetry is emerging as a potentially important trait in pollinators' selection of flowers. While applicable to both radial and bilateral sym-

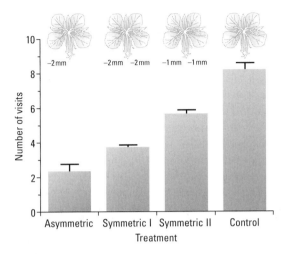

Figure 6.3 Bumblebees show a significant preference for large, symmetrical flowers. The number of bumblebee visits per time unit as a function of display size and symmetry in the fireweed, *Epilobium angustifolium*. Experimental categories had the lower petals trimmed as indicated in the figure to adjust symmetry and size. Differences were significant between all treatments. (*P* < 0.05, Scheffe F tests; from Møller 1995.)

metry, it has primarily been explored in the latter case. Møller (1995) manipulated symmetry and size of the two basal petals in *Epilobium angustifolium* (fireweed, Fig. 6.3), and found positive effects of both symmetry and floral size on the rate of visitation by bumblebees. The standing crop of nectar was negatively correlated with asymmetry, providing a direct reason why pollinators should use floral symmetry in deciding whether and when to visit a flower. The preference may be innate or learned, depending on the animals involved (Møller 2000).

6.3.2.1.4 Colour

Plants employ a range of colours that extends beyond the rainbow, having floral parts or adjacent structures that reflect from near UV (~350 nm) to deep red (~700 nm). Virtually all colours are pigment-based; major classes of pigments include anthocyanins and anthoxanthins (blues and purples, some reds), carotenoids (most yellows and oranges, some near-reds), flavones and flavonols (whites, UV, some yellows) and betaxanthines (Harborne 1993). In addition to pigment-based colours, whites can be obtained by flat reflectance from epidermal cells and glossy surfaces, providing high brightness even under low light

conditions. Such bright whites are common in species that attract visitors at dusk or in darkness; under low light conditions spectral differences are poorly perceived, whereas bright objects may remain apparent. Combinations of colours provide sharp contrast within flowers, a property that adds to visibility; for example, the disk portion of asters often contrasts sharply against the peripheral ray flowers (Fig. 6.2b). A common pattern in flowers with concealed rewards is a distinctively coloured, spectrally pure region, where sexual parts and rewards are located, surrounded by less pure areas; the entry into such patches, commonly referred to as nectar guides, triggers innate proboscis extension and probing for food in some bee species (Lunau 1996). The effect of nectar guides is apparent in a white mutant of the blue-flowered *Delphinium nelsonii*, where the nectar guide is lost through loss of the blue background pigment. Bumblebees take longer to find the probing site and discriminate against the mutants, but the nectar extraction rate and attractiveness to the bees can be restored by the addition of an artificial guide (Waser & Price 1983). Colour preferences have both innate and learned components (Gumbert 2000).

How the potential visitors perceive a given flower depends on their sensitivity and discrimination ability in different parts of the wavelength spectrum (Menzel & Backhaus 1991; Tovée 1995). Most species whose visual pigments have been analysed have trichromatic or tetrachromatic vision. Early analyses of insects' colour perception were based on analysis of spectral sensitivity in very few species, notably honeybees (*Apis mellifera*) and bumblebees (*Bombus* spp.), and they became the basis for calculations of 'bee colours' (Kevan 1983; Chittka et al. 1994). As many more insect species have been analysed, high evolutionary plasticity has been documented, with closely related species sometimes varying dramatically in spectral range and degree of resolution, even the two sexes within a species having different visual pigments. In retrospect this was to be expected, because colour vision plays a critical role in several functions, including host search and mate selection. For example, in two species of sexually dimorphic lycaenid butterflies, visual pigments facilitating host-plant recognition were exclusively present in females, and pigments of particular significance in discerning conspecific mates or rivals were over-represented in each sex (Bernard & Remington 1991). Because of such differences, perceived floral colour and pattern will often differ among potential insect pollinators, making generalizations about their powers of discrimination difficult. Strong selection on floral colour may at best be expected when the visitor guild is simple and differences among species considerable. What may be more important than identical perception is the relative distinction of flowers and their functional subunits from surrounding areas; whether a flower is perceived differently by two pollinators at the visual pigment level does not matter, as long as the ability to discern, respond and learn the visual signal is maintained.

Finally, there is the potential that floral colour may be under selection for its repellent properties on parasitic visitors, such as herbivores or poor pollinators (Schemske & Bradshaw 1999). We will return to this in the section on specialization.

6.3.2.2 Olfactory cues

6.3.2.2.1 Chemistry
Volatiles are a common feature in flowers, mostly produced in combination with visual attractants but at times as the sole attractant (Raguso 2000). Floral scents are mixtures of small compounds that vary in molecular weight and volatility; major classes of volatiles include mono- and sesquiterpenes, fatty acid derivatives and phenolics. More unusual classes, such as sulfides, appear especially in plants attracting distinctive visitor groups, such as dung- or carcass-feeders (Knudsen et al. 1993). Scent production is tissue-specific, with parts of the corolla, pollen or nectar being common sources.

6.3.2.2.2 Roles in attraction
Floral scent can carry different functions at different spatial scales. It figures prominently in long-distance attraction, particularly in species that attract pollinators such as moths and bats in situations of partial or complete darkness (Brantjes 1978; Haynes et al. 1991). While visual cues may work at short distances at best, scent plumes can travel long distances in relatively still night air. Actual distances at which a scent plume is detected and traced appear not to have been measured, but the size of the source is likely to affect the distance in which a scent plume is present to be traced back. In the noctuid moth *Hadena bicruris*, entry into a floral scent trail triggers upwind flight towards the source, and also activates a visually mediated search for white objects resembling the white flowers of its combined nectar source and host plant, *Silene latifolia* (Brantjes 1978).

Scent directly affects visitor behaviour modification at short range and serves in visitor conditioning. In some instances, it is a critical cue for manipulating an alighting visitor to approach specific flower parts (Raguso 2000). In *Cimicifuga simplex*, two scent components prolong the duration of visits by nymphalid butterflies, increasing the probability of pollination (Pellmyr 1986). Scent is also more readily learned in association with floral reward than any other type of floral trait (Menzel & Müller 1996), leading to short-term selectivity in individual insects and thus increasing the probability that pollen collected on a scented flower will reach a conspecific flower. The effect of scent learning can magnify levels of visitation, as bumblebees and other social bees learn the floral scent carried by a successful returning forager and use it in localizing flowers of the rewarding species (Fig. 6.4; Dornhaus & Chittka 1999).

Finally, in addition to its attracting or manipulative properties for mutualists, floral scent may be under selection from the effects of interactions with antagonists; repellent effects on parasitic visitors are known (Omura et al. 2000), as is the scent-based attraction of floral enemies (Raguso 2000). This is not surprising, as many floral scent compounds, such as the terpenes, figure prominently in plant chemical defences against herbivores (Pellmyr & Thien 1986).

6.3.2.3 Auditory cues

While flowers may be unlikely to produce sounds, they can serve as sound reflectors. The flower of *Mucuna holtonii* (Fabaceae), a South American liana, has an erect petal that reflects sound pulses produced by echolocating bats that visit the flowers for their copious nectar (von Helversen & von Helversen 1999). Acting as an acoustic guide within a narrow angle in front of the flower, it serves as a nectar guide by orienting the bat to the nectar and to the sexual parts of the flower. With this discovery, all senses that operate at a distance are now known to be involved in plant–pollinator interactions.

6.3.3 Rewards

Visitors seek out flowers to obtain a reward, usually a food item, but a few other types of reward are known to exist. The most common reward is **nectar**, an aqueous solution presented in glands in the flower. The primary content is water and carbohydrates of phloem origin, with the

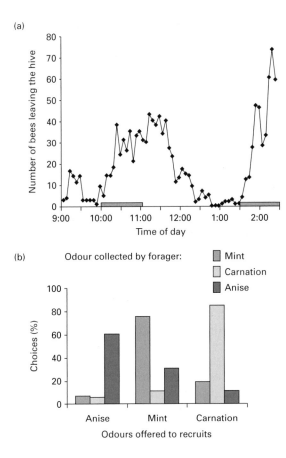

Figure 6.4 The bumblebee *Bombus terrestris* uses floral odour on a successful nest-mate to find food sources. (a) following the arrival of a successful forager (grey bars), the number of bees leaving the nest increased sharply; (b) when bees were fed artificial nectar laced with one of three common floral scent components, bees selectively sought out nectar sources with these odours. (From Dornhaus & Chittka 1999.)

monosaccharides glucose and fructose and the disaccharide sucrose constituting the sugars in the vast majority of analysed species. Relative amounts of mono- and disaccharides vary greatly among species, with one or more sugars altogether absent in some species. Total sugar concentration usually is between 15 and 45%; the higher value approaches concentrations beyond which viscosity would make drinking difficult. Modest amounts of amino acids and other compounds have been detected (Kearns & Inouye 1993), especially in flowers that attract carcass-feeders, and feeding trials show that some visitors detect and preferentially feed on nectars laced with amino acid

(Erhardt & Rusterholz 1998). The water itself is potentially a valuable reward, especially in arid conditions. Nectar production often shows a strong diurnal pattern, and standing crop is commonly between 0.1 and 10 μl, but can reach >650 μl/flower/day (Cruden et al. 1983). Data on nectar amount and sugar concentration have long been used to suggest that there is convergent evolution among plant species with similar pollinators (Baker & Baker 1982). These analyses are made problematic by frequent lack of empirical pollinator data, the assumption that all species have simple and uniform pollinator guilds across their ranges ('small bees', 'flies') and the possibility of pseudocorrelation from phylogenetic effects. Recent analyses account in part for phylogenetic effects, and indicate differentiation in sugar types and quantity for some distinctive vertebrate visitor groups (Baker et al. 1998). Whenever visitor spectra are more complex, less categorical differences are expected.

Pollen serves as a primary reward for pollen-collecting or pollen-eating flower-visitors, such as solitary bees, beetles and some flies, but more commonly functions as a secondary reward for bees that primarily track nectar sources. A rich source of nitrogen, pollen is potentially costly to the plant through both the loss of a critical nutrient and the direct loss of genetic material, and adaptations to reduce pollen losses to parasites should be expected. Poricidal anthers, which never dehisce but rather release pollen from one or more pores in response to sonic vibration by bees (Buchmann 1983), represent one mechanism limiting pollen loss to parasites. Such anthers, as well as gradual opening in dehiscent anthers, facilitate the gradual dispensation of pollen, a mechanism that can increase male reproductive success by reducing the risk of visitor grooming and the consequent opportunity for the pollen to reach a conspecific stigma (Harder & Wilson 1994).

All other types of rewards are comparatively uncommon. About 1% of all angiosperms provide fatty oils as a reward to solitary bees who use it as a provision for their larvae and possibly for nest-building (Buchmann 1987; Steiner & Whitehead 1991). Two large tropical genera, *Dalechampia* (Euphorbiaceae) and *Clusia* (Clusiaceae), contain species that produce diterpenoid **resins** as rewards for solitary bees that use them in nest-building for structural and antifungal purposes (Armbruster 1997). Neotropical members of seven very different plant families, including at least 625 orchid species, provide **fragrant terpenoids and aromatics** collected by male euglossine bees (Dressler 1982). The bees store the compounds or their derivatives in modified hindlegs. Attraction is highly specific among the 130+ bee species, depending on the exact blend of volatiles produced. Despite 40 years of study, their use has still to be documented, but it is assumed that the compounds play a role in some stage of mating success among the territorial male bees. Specialized **floral tissue**, containing high concentrations of starch, sugar, lipids or protein, is scattered among angiosperms, but particularly common in basal families such as the Annonaceae. This tissue is often located on petals or stamen tips, but is also known in other flower parts. Because consumption requires chewing, beetles are frequently involved, as are larger animals such as bats and even birds that can tear tissue. In a related vein, it has been suggested that mass-flowering trees may obtain pollination from primates that consume subsets of the flowers; there appears to be no data to support this hypothesis. Finally, in a few instances the reward is **seeds** consumed by the pollinators' progeny. Granivory is very common (see Chapter 5), often beginning while the seeds mature on the mother plant, and it is a very clear example of the evolutionary dilemma facing a plant requiring pollinators: parasitic insects can use the required attractants to find the flowers. In a few such instances, the seed parasites have also become pollinators. The most obvious examples of seeds as rewards involve yucca moths (Pellmyr & Leebens-Mack 2000) and fig wasps (Bronstein 1992), whereas seed-parasitic hadenine moths often have negative net fitness effects despite their contribution to pollination (Pettersson 1991).

6.3.4 Non-rewarding deceptive plants

The abundance of rewarding flowers and flower-visiting insects creates possibilities for parasites to steal resources or services without necessarily destabilizing mutualisms. In most instances, this happens as plants use deceit or Batesian mimicry to attract scouting flower visitors and when vagrant visitors obtain nectar or pollen without contributing to pollination.

Non-rewarding plant species constitute a substantial portion of all angiosperms, especially among orchids, but they are mostly minor components of the plant communities in which they grow. Species that deceptively advertise food rewards, such as nectar or pollen, may either coexist with a rewarding model or rely on innate or learned reward cues, often in habitats with few other flow-

ers. In an example of model–mimic coexistence, the South African orchid *Disa ferruginea* is dimorphic for flower colour in its range, matching two correspondingly coloured models in different parts of its range (Johnson 1994). A single species of satyrid butterfly draws nectar from the models and does not discriminate against the nectarless orchids. In contrast, the nectarless circumboreal orchid *Calypso bulbosa* flowers early in spring in habitats with few other flowers, and attracts overwintering bumblebee queens who explore any flowerlike objects for rewards (Boyden 1982). Similarly, *Dactylorhiza incarnata* attracts bumblebees to its showy flowers, but fruit set was reduced as the density of a rewarding, different-looking *Viola* species was increased in its proximity (Lammi & Kuitunen 1995). In the two latter examples there is no immediate model in the habitat, but rather these interactions are based on exploitation of innate search cues among potential visitors. Flower colour polymorphism may be adaptive in such models of deception as well, as long as visitors regard the different morphs as novel, potentially rewarding flowers (Nilsson 1980).

A fair number of plants mimic not flowers but rather pollinator mates or oviposition sites. Flowers of the well-studied European fly orchids (*Ophrys*) and caladeniine Australian hammer orchids provide visual, olfactory and tactile cues mistaken by naïve wasp males for conspecific females (Stowe 1988), and pollination happens as males attempt copulation with the flowers. Male insects generally emerge before females, as a result of competition for mates. Once females emerge, males quickly learn to distinguish real females from the floral mimics, thus there is selection for the plants to flower during a narrow window in time.

Oviposition site mimics most commonly attract species whose larvae feed on carcasses. Lineages in several families, notably Araceae and stapeliad Apocynaceae, have flowers or inflorescences that resemble potential oviposition sites in colour, hairy surfaces and strong putrid smells. Members of the Araceae temporarily entrap visitors inside a cavernous spathe, and cues are often sufficient to trigger oviposition in the visitors, even though their progeny will inevitably die for lack of food. Similar types of brood site mimicry include visual and olfactory resemblances to gill fungi, attracting fungus gnats, and visual mimicry of aphid clusters on flowers, attracting visitors whose larvae feed on aphids.

How can these non-mutualistic interactions persist? There are no studies measuring the fitness cost on deceived visitors in these interactions, but it is likely that most interactions are commensalistic rather than antagonistic. In food-based deception, there is a potential cost of lowered yield rate with the inclusion of non-rewarding flowers. But foraging visitors continuously encounter flowers of rewarding species that range in yield depending on recent visit history, so as long as deceptive flowers are relatively uncommon their effect is likely to be trivial. This may especially be the case in social insects, such as apid bees, where rates of deception must be high enough to affect the overall yield rate of all foragers for there to be selection against visiting such flowers. There is also little reason to assume that visits to mate-mimicking flowers carries a fitness cost, with a possible exception where sperm is expended in the interaction. In oviposition site mimicry, visitors often expend eggs during the visit, and these may be the most likely candidates for being truly antagonistic interactions, where selection can be expected on avoiding the mimics. Insects that feed on ephemeral food items such as carcasses and soft mushrooms have a comparatively low probability of finding suitable food items. In consequence, they tend to have a broader host range than, for example, herbivores. The adaptive ability to recognize a broad range of potential but rare oviposition substrates may constrain selection on such animals for specifically avoiding deceptive flowers.

We will discuss the impact of parasitic flower visitors in Section 6.6.

6.4 Convergent evolution of floral traits and the utility of pollination syndromes

The extraordinary variety of attractants and rewards among flowers of different species is evidence of considerable plasticity, and with it almost inevitably comes convergent evolution. For example, red tubular nectar-producing scentless flowers have evolved in at least 17 families. From Darwin onward, floral trait similarities among unrelated plants have been attributed to selection by shared types of pollinators. Faegri and van der Pijl (1979) formalized such trait similarities into a set of 'pollination syndromes', intended to recognize combinations of common trait states in plants attracting particular groups of visitors, such as hummingbirds, moths, carrion flies or bees (Table 6.1). They emphasized that these were to be considered as statistical constructs, with a given trait state (a given shape, colour, scent or reward) being over-represented among species relying on a given type of visi-

tor. In particular, van der Pijl, one of the most experienced pollination biologists of his day, certainly realized that their formalization could be misconstrued. They appreciated that pollinator guilds vary from a single species to large suites of very different species, that guilds may differ within species across geographical ranges and over time, and that visitor groups such as 'flies', 'beetles' and 'bees', were simplified constructs that contained much internal variation in foraging patterns. Because of these limitations, 'pollination syndromes' can serve in effect for two purposes: first, as hypotheses for testing assumptions of pollinator-mediated selection on specific traits, and second, as an educated guess in setting up suitable protocols for empirical study of plant–pollinator interactions.

Unfortunately, despite these caveats by the founders of the concept, 'pollination syndromes' are too often used in a typological sense by armchair biologists, with pollinators invoked from floral traits, and considerations of complex guilds and intraspecific variation cast aside. Quite properly, such practices have drawn the fire of empiricists (Johnson and Steiner 2000). The observations and experiments needed to document even the basic features of pollinator–plant associations are of necessity time-consuming (and increasingly so with the number of visitor species involved) but cannot be avoided if we are to understand the historical and contemporary dynamics of plant–pollinator interactions.

6.5 Specificity, selection and diversification

Specificity among visitors is a necessity for effective pollination; if animals visit flowers of different species indiscriminately, heterospecific pollen transfer will result, which reduces the probability of pollen reaching a conspecific stigma and may cause negative pollen-pollen interactions on the stigma. Specialization is in part indirectly imposed on visitors by floral traits, and in part a consequence of the innate properties of the visitors themselves.

6.5.1 Spatial and temporal scales of specialization

6.5.1.1 Specialization and specificity in the animals

The number of plant species visited varies greatly among flower-visiting species. At one extreme, most fig wasps and some pollen-collecting solitary bee species frequent the flowers of a single plant species throughout their

Table 6.1 Traits commonly associated with pollination by members of specific groups of animal pollinators. Faegri and van der Pijl developed the concept of pollination syndromes as a means to recognize traits that are over-represented in flowers that attract specific types of pollinators. This is a criterion of inclusion rather than exclusion, in the sense that other visitors may very well visit the flowers. Many plant species have flowers that will attract visitors from at least half of the categories below, and well over 100 species.

	Butterflies	Moths	Birds	Bats	Beetles (excl. carrion beetles)	Flies (excl. beeflies and carrion flies)	Carrion flies and beetles	Bees and beeflies (n.b. most are thieves)	Ants
Flowering time	Diurnal	Dusk and night, often closed during day	Diurnal	Nocturnal (mostly one night per flower)	Diurnal or crepuscular	Diurnal	Diurnal	Matinal–diurnal	Diurnal
Colours	Commonly orange, yellow, red	Mostly bright white with sheen	Bright, often scarlet red	Often drab, greenish or purple, sometimes creamy white	White, green or dull	Light	Dull, dark brown-purple-greenish	Lively; yellow and blue predominate	Greenish
Scent	Weak to moderate; pleasant	Strong, perfumy, at night	None	Strong, often stale or redolent of fermentation	Strong, fruity, aminoid	Often imperceptible mushroomy	Decaying protein, moderate	If present, sweet	Weak to none
Floral shape	With nectar in tube	With deeply concealed nectar in tube or spur	Often tubular without landing platform; hard sometimes brush type	Big, with wide mouth; sometimes a brush type	Flat- or bowl-shaped; reward exposed; no depth	Flat- or bowl-shaped; reward exposed	Often lantern type, with windows or filiform appendages	Often complex, with great depth effect; landing surfaces; reward often concealed in tube	Small, open
Floral position	Erect or horizontal	Horizontal (erect, pendent)	Often pendent, but also horizontal	Horizontal, usually at tip of branches (exposed)	Usually erect	Usually erect	Any; often near ground	Any; often pendent	Erect, near ground
Type of reward	Nectar (ample)	Nectar (more than bee, butterfly types)	Nectar (profuse)	Highest amount of nectar, pollen consumed as well	Pollen, nectar, tissue mating site	Pollen, nectar	None (deception)	Nectar, pollen, resin, scent, none	Nectar
Nectar guides	Usually not	No	Rarely	No	No	Occasionally	No	Common	No

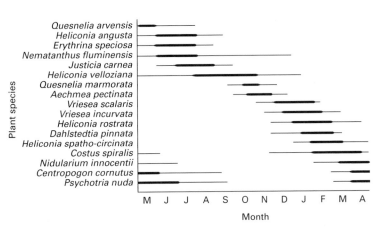

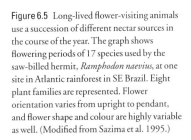

Figure 6.5 Long-lived flower-visiting animals use a succession of different nectar sources in the course of the year. The graph shows flowering periods of 17 species used by the saw-billed hermit, *Ramphodon naevius*, at one site in Atlantic rainforest in SE Brazil. Eight plant families are represented. Flower orientation varies from upright to pendant, and flower shape and colour are highly variable as well. (Modified from Sazima et al. 1995.)

ranges (Wcislo & Cane 1996). In other species, one or a few plant species may be used in any one part of the species range, whereas populations in different geographical areas use different plants. The butterfly *Pyrameis cardui* inhabits all continents except Antarctica, and specializes locally but utilizes a wide range of plant species with concealed nectar across the species range. In an analogous way, migrant flower-visitors such as some hummingbirds encounter and exploit different nectar sources as they travel. At the generalist end of the specificity spectrum, flower-visitors like many larger social bees, and opportunistic flies and beetles, may visit a very large proportion of available flowers within any habitat many hundred species within their range. In social species, such as honeybees and bumblebees, with colonies that last for months or years, and in long-lived visitors such as vertebrates, no one source may be available for the duration, but rather a succession of flowering plant species must be utilized (Fig. 6.5).

Whereas heterospecific pollen transfer will not be an issue in visitor species with limited plant spectra, it could potentially be an important factor in reducing efficacy in visitors with broad spectra (Chittka et al. 1999; Fishman & Wyatt 1999). Individual visitors often tend to specialize on a subset of potential flowers during any one foraging bout; in bees perhaps over 90% of all visits may be made to a given species, with occasional visits to other species. This short-term specialization is referred to as floral constancy. The dominant flower may vary among simultaneously foraging conspecifics, and within individual visitors on successive foraging bouts. Reasons

for such short-term selectivity have been explored in insects, and focus on the effects of foraging rate as a result of memory constraints. Insects must learn by trial and error how to effectively access a reward such as nectar in more complex flowers, as the rewards are concealed and most quickly accessed using a particular approach. Minimum handling time may be approached only after as many as 100 visits to a given zygomorphic flower, such as an *Aconitum* flower, by some bumblebees. Lewis (1986) found that lesser cabbage white butterflies (*Pieris rapae*) who first learned to effectively extract nectar from one complex flower and subsequently learned to handle another complex flower form lost their skills with the first flower form (Fig. 6.6). This suggests that such visitors may be unable to keep more than one sensorimotor protocol in active memory, thus making it a superior strategy to focus on one food source at a time rather than have lower yield rates from low extraction rates on two or more sources as a result of interference between handling protocols.

Specialization is often not in the evolutionary interest of a flower-visiting animal, as its ultimate interest is to optimize the reward harvesting rate over time. A foraging pattern that maximizes the harvesting rate of commodities such as nectar and pollen can include two or more coexisting plant species, especially if their floral structure is fairly similar so that the visitor can use a single visit behaviour protocol. Consider the flat capitula of most aster species, consisting of minute, packed-in flowers, where pollen is presented as if on a plate for any potential visitor and nectar is located in such short floral tubes that even beetles (who have mandibular mouthparts but no

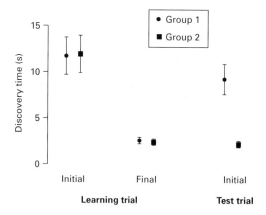

Figure 6.6 Learning a complex handling protocol can interfere with handling protocols currently in active memory in flower-visitors. *Pieris rapae* butterflies (groups 1 and 2) were trained to extract nectar from the bluebell *Campanula rotundifolia*, cutting handling time by 75%. Group 1 ($n = 17$) was then trained for 20 min on the very different flowers of *Lotus corniculatus*, whereas group 2 ($n = 20$) were kept resting. Upon return to *C. rotundifolia*, group 2 butterflies retained the same handling time, whereas Group 1 butterflies had discovery times that approached those of naïve butterflies. (From Lewis 1986.)

proboscis) can access it. There would seem to be no mechanical reason for a flower-visitor to ignore capitula of different coexisting species of this kind. Indeed, many flower-visitors such as syrphid flies and nitidulid beetles, whose mouthparts are morphologically generalized, feed on nectar and pollen on any number of species whose floral rewards are superficially located.

6.5.1.2 Specialization and specificity in the plants

In contrast to the animals, specificity is commonly in the interest of the plants visited, as their ultimate interest is to receive and deliver conspecific pollen at the lowest energy cost possible. This is a conflict of interests with fundamental implications for specialization. Ideally, a plant should obtain pollen transport without paying the pollinator any reward. Under most circumstances this would not be evolutionarily stable, because the pollinator would abandon the plant. However, selection is expected to favour traits that increase the efficacy of visitors, that is, maximizes the rate of pollen reception and successful pollen deposition on stigmas of a conspecific plant at the lowest expense possible (Box 6.2). Options for the sessile plant include selective pollinator attraction, and manipu-

lation once a visitor has landed to optimize pollen deposit and pickup. As discussed in a previous section, selective pollinator attraction can be achieved in part through visual and olfactory cues that work before an animal has alighted. Once a visitor is on the flower, the plant can make rewards less accessible by, for example, presenting nectar in a spur or tube, dispensing pollen in never-dehiscing poricidal anthers that require a specific sonic behaviour for pollen dispensing, or by placing rewards behind gate-like structures that a visitor must pry apart. These structures can evolve as a result of either or both of two sources of selection. First, they may reduce the negative effects of wholly parasitic visitors or of poor pollinators whose contributions may decrease fitness. Second, they may increase the pollination efficacy of visitors by creating a physical path to the rewards that will make pollen deposit and pickup more likely. These two effects are by no means mutually exclusive, but it is generally difficult to determine whether selection to manipulate pollinators or to avoid parasites is the main cause of selection on floral properties. The difficulty of teasing the two apart is particularly apparent when considering that the plant is under selection to minimize the cost of receiving pollination services by the pollinators. As a consequence, a trait such as spur elongation can in principle simultaneously reduce access to parasites and increase the pollination efficacy of pollinators.

The vast majority of all plants are pollinated by two or more species, an indication that specialization in plants is often limited by ecological and genetic factors. Pollinator availability may be limited by spatiotemporal variation in visitor abundance, selecting against excessive specialization. In the monkshood *Aconitum columbianum*, variation in nectary spur length reflects the highly variable composition of local pollinator bumblebee guilds throughout the plant's range (Brink 1980; Fig. 6.7). In Rocky Mountain populations of the colour-dimorphic *Aquilegia caerulea*, availability of two major pollinators — bumblebees and the hawkmoth *Hyles lineata* — with differently shaped mouthparts and innate colour preferences, vary dramatically between years as a result of population fluctuations. Seed set varies with flower colour based on the relative abundances, and as a result of these fluctuations selection for specialization on either pollinator species to the detriment of the other tends to be reversed over short timespans (Miller 1981). Similarly, flowers of many *Calochortus* lilies in California depend on different generalist visitors as pollinators in different parts of their

Box 6.2 Variation in the relative cost of pollinators

The relative cost to a plant of different pollinators within a visitor guild is determined by efficacy of pollination, which is a function of the number of pollen grains picked up and the probability that this pollen will reach a conspecific stigma, the amount of reward consumed by each visitor, and the relative genetic quality of the pollen deposited. It can be a laborious task to measure all three components simultaneously, but the genetic quality component can readily be assessed when genetic self-incompatibility is present and seed results from only interplant pollen transfer.

Differences in pollen collection and deposition efficacy were clearly demonstrated in a study by Wilson and Thomson (1991). They measured pollen removal and deposition by three distinctive visitors—bumblebees, honeybees and minute solitary bees—to the flowers of *Impatiens pallida*. The relatively large bumblebees brush against the sexual parts in the ceiling of the thimble-shaped flower as they probe for nectar in the spur. The smaller honeybees and solitary bees can enter without touching the ceiling, but they actively collect pollen for larval food. In single visits to virgin flowers, honeybees removed more pollen than bumblebees, which in turn removed more than the solitary bees. After visits to virgin flowers, stigmas of flowers visited by bumblebees had significant pollen loads, whereas honeybees (who rarely visit female-phase flowers in the first place) transferred very few grains. Solitary bees discriminated visually against female-phase flowers, and thus never deposited pollen. The resulting probability that a pollen grain will reach a stigma varies tremendously between the three types of visitor. An informal terminology of 'good, bad and ugly' visitors has emerged, in which good visitors remove and deposit much pollen, bad visitors remove and deposit little pollen and ugly visitors impose high costs by removing much pollen but depositing little or none.

Cost estimates that include all the components of pollinator efficacy and reward costs remain to be obtained.

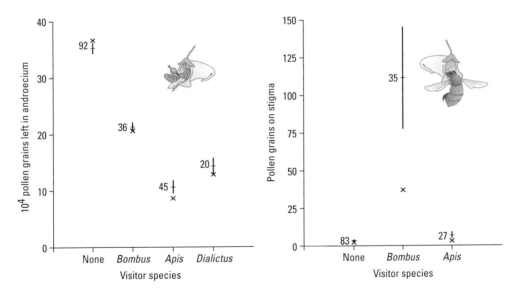

Figure 1 Effects of single visitors on pollen movement by *Bombus*, *Apis* and *Dialictus*. None indicates unvisited controls. *Bombus* collected nectar, *Apis* and *Dialictus* collected pollen on the flowers. *Dialictus* are not included in the right panel as they never visited receptive flowers. Pollen-collectors removed more pollen and deposited less than nectar-collectors. Bars are means ± 1 SE; Xs are medians; numbers are sample sizes. (From Wilson & Thomson 1991.)

ranges (Dilley et al. 2000). Herrera (1988a) documented considerable variation among and within years in the composition of a complex guild of pollinators of the southern European *Lavandula latifolia*. Under these circumstances, even if the genetic potential exists for selection on traits to further limit the visitor guild to the most effective pollinators, such selection may vary in direction over intermediate timespans.

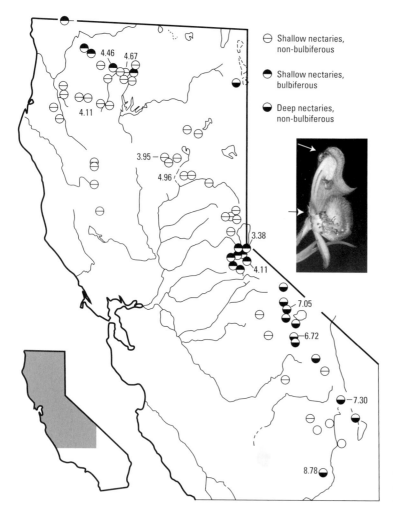

○ Shallow nectaries,
   non-bulbiferous

● Shallow nectaries,
   bulbiferous

⊖ Deep nectaries,
   non-bulbiferous

Figure 6.7 Geographical variation in floral traits reflects histories of unilateral adaptation or coevolution with local pollinator guilds. There is considerable variation in spur length and degree of asexual reproduction in the bumblebee-pollinated monkshood *Aconitum columbianum* in California. High-elevation populations have longer spurs and no asexual reproduction by aerial bulbils, whereas populations at lower elevations have shorter spurs and sometimes also produce bulbils. Spur length is on average somewhat shorter than the mean tongue length of the most long-tongued bumblebees in the area. Insert below legend shows a vertically cut *Aconitum* flower (whole flower in Plate 6.2). The light spur, with the nectary at top, is seen inside the hood; arrows indicate the region measured in the study. (Modified from Brink 1980.)

6.5.2 Selection of floral traits

6.5.2.1 Stabilizing selection and maintenance of variation within species

Selection should be detectable on floral traits that affect plant fitness. This may include traits that increase attractiveness to pollinators, increase the pollen removed and deposited on each visit, increase the degree of outcrossing, or reduce attractiveness and the availability of rewards to parasites. Quantitative measurements of selection have appeared for a handful of plant species in the last decade, including ones with quite different morphology and pollinators (Galen 1989; Stanton et al. 1989; Campbell et al.

1991; Young et al. 1994; Campbell 1996; Conner & Rush 1997). The criterion used to determine the effect of selection varies with regard to whether just female or both male and female fitness is measured, whether fitness from a single reproductive episode or lifetime fitness is measured, and whether factors extrinsic to pollination, such as edaphic factors and herbivory, are included in the analysis.

Representative species subject to extensive analyses include the phlox *Ipomopsis aggregata*, with red tubular flowers visited particularly by hummingbirds (Campbell 1996) and the radish *Raphanus raphanistrum*, whose colour-polymorphic flat flowers are visited by flies, butterflies, and small bees. Both plants flower once in their

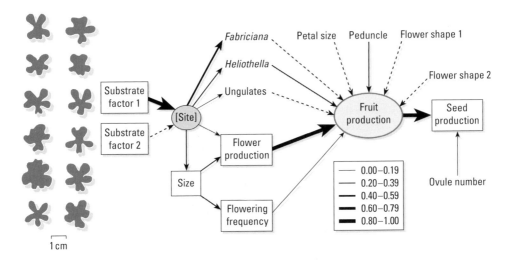

Figure 6.8 Selection on floral traits can be completely masked by selection on other traits that affect plant seed production. The spurred violet *Viola cazorlensis* shows striking variation in floral display; examples of front outline are given on the left. In a path analysis (which measures direct and direct effects) that included substrate, plant size, flower number and flowering frequency, and florivores and herbivores in addition to floral traits, significant selection on variation in floral traits only accounted for 2.1% in final seed production. *Fabriciana* is a florivore, *Heliothella* and ungulates are herbivores. Plant traits are composite variables, identified by the main factor when possible; only significant factors are included. Solid lines indicate positive effects, dashed lines negative effects, with thickness giving the relative magnitude of path coefficients (see box at lower right). Unmeasured factors are not indicated in the figure for ease of reading. (Modified from Herrera 1993.)

life, thus facilitating analysis of lifetime fitness. In *Ipomopsis aggregata*, analysis of the effects of natural variation in floral traits have revealed phenotypic selection for large corolla diameter, long tube, large nectar amount, high flower number, increased distance between stamens and stigma, and an exserted stigma (Campbell et al. 1991). Subsets of these traits have been documented as being under phenotypic selection in other model systems. Meanwhile, factors such as pathogens dispersed by pollinators may balance selection on preferred traits, such as a large corolla, in some species (Shykoff et al. 1997).

In the light of detected selection, the maintenance of often considerable levels of intraspecific variation in traits demands an explanation. Apart from the spatiotemporal fluctations of pollinator guilds discussed in the previous section, two other major explanations have been posited. First, many of the traits involved are not genetically independent, but rather covary to different degrees so selection is constrained (Mitchell et al. 1998). Second, phenotypic selection on floral traits may be completely masked by ecological factors unrelated to pollination and variation thus maintained. A well-documented example is *Viola cazorlensis*, an Iberian endemic violet pollinated

almost exclusively by the hawkmoth *Macroglossum stellatarum* (Herrera 1993; Fig. 6.8). Phenotypic selection was found on several floral traits, but when factors such as plant size, soil substrate, herbivory, and fruit production were included with floral trait variation in a path analysis, only 2.1% of variation in individual cumulative seed production was attributable to floral trait variation. While statistically significant, this source of selection becomes biologically inconsequential because of other, stronger factors. In *Paeonia broteroi*, herbivory effects were inverse of pollination effects, thus removing selection on floral variation (Herrera 2000). This highlights the need for studies of floral evolution to consider the larger context; analyses performed in artificial populations where, for example, herbivory may not be possible may yield incomplete answers that over-emphasize the role of pollinator-mediated selection in this context (Galen 1999b).

6.5.2.2 *Disruptive selection within plant species*

A general argument can be made that pollination interactions may be particularly prone to involve disruptive selection and consequent diversification, especially in plants (Thompson 1982; Schemske & Horwitz 1985). Given that the pollinators directly determine patterns of pollen-mediated gene flow, assortative behaviour among them can potentially lead to divergence and speciation (but cf. Waser 1998 for a non-canonical view).

Traditional quantitative genetic models of evolution have assumed that adaptation and reproductive isolation are typically the result of large numbers of mutations with very small individual effects (Bradshaw et al. 1995; Schemske & Bradshaw 1999). If correct, this would re- duce the potential for rapid disruptive selection in plants. In a powerful experiment, Schemske and Bradshaw used two *Mimulus* species that rely on different pollinators to test how easily pollinator-mediated selection could cause divergence. The predominantly bumblebee-pollinated *Mimulus lewisii* and the hummingbird-pollinated *M. car- dinalis* are interfertile sister species (see Plate 6.1, facing p. 84) that overlap slightly in distribution along an alti- tudinal gradient in the Sierra Nevada of California. Their flowers differ in corolla colour and shape, petal reflexion, nectar volume and concentration, and the positions of an- thers and stigmas. Quantitative trait locus (QTL) map- ping of F_2 hybrid progenies demonstrated the existence of alleles in 9 of 12 divergent traits that explained >25% of the phenotypic variance. Field trials using the hybrid plants showed assortative visitation by both bees and birds; hummingbirds favoured nectar-rich flowers high in anthocyanin, whereas bees favoured large pale flowers. The bees showed a particular aversion to carotenoids in the petals, a factor that reduced bee visitation by 80% by itself (Fig. 6.9). These effects were sufficiently strong for very few pollinator crossovers to be recorded in the zone of sympatry. Hence the presence of genes with major effects and the direct contrasting effects on visitor preferences show that diversifying selection can occur even in sympa- try in some circumstances. It is important to note that the effects can be in terms of either attraction or repulsion; the reduced attractiveness to bees exemplifies a route of selec- tion for specialization in plants against parasitic visitors in general.

Rapid diversification based on assortative pollinator behaviour may be particularly likely in situations where the attractants or rewards are ones that exploit perceptor systems with high discrimination in the pollinators. This is the case when the reproductive success of the pollinator itself is directly involved, either by the presentation of compounds used by the pollinators at some step in mate

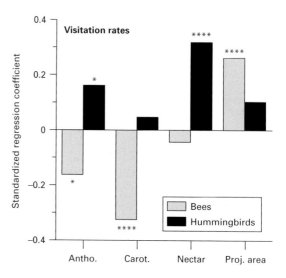

Figure 6.9 A strong aversion for floral traits can explain specificity in patterns of pollinator choice. The contribution of individual floral traits to different pollinators (bees vs. hummingbirds) were analysed in an array of F2 hybrids between *Mimulus lewisii* and *M. cardinalis* by performing separate multiple regressions on pollinator visitation rates as dependent variables and floral traits as independent vaiables. Strong aversion by bees to a petal carotenoid from *M. cardinalis* explains why effectively all visitors to this species are hummingbirds. Hummingbirds showed a strong positive response to the high amount of nectar, but no obvious aversions. Key: antho = petal anthocyanin concentration; carot = petal carotenoid concentration; nectar = nectar volume; proj. area = projected area (a composite area of surface displayed to approaching visitors). *$P < 0.05$, ***$P < 0.0001$. (From Schemske & Bradshaw 1999.)

attraction or where flowers mimic potential mates. In eu- glossine bees, minor differences in scent composition can dramatically affect the visitor guild, thus single mutations in the biochemical pathways or in genes controlling ex- pression can cause rapid isolation between plants. Fly or- chids of the genus *Ophrys* mimic receptive wasp females, and an olfactory stimulus is critical in attracting males. In *O. sphegodes*, the floral scent consists of a hydrocarbon blend derived from plant cuticle constituents (Fig. 6.10; Schiestl et al. 1999). Because plant cuticles often contain typical compounds of female wasp sex pheromones, it is easy to see how mutant plants may produce altered blends that serendipitously provide a sufficiently close resem- blance to attract inexperienced male wasps of different species. Similarly, a large number of closely related Aus- tralian caladeniine orchids are pollinated by male thyn- nine wasps that mistake the flowers for conspecific mates,

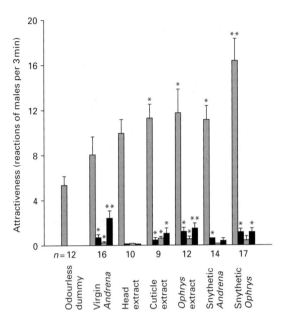

Figure 6.10 Highly specific pollinator attraction need not result from extensive trait evolution but may result from preadaptations. *Ophrys sphegodes* relies on male *Andrena* bees, who mistake the flowers for potential mates, for pollination by deceit. Attractiveness tests using dummies baited with virgin female *Andrena* scent, bee head extract, bee cuticle extract, plant cuticle extract, and synthetic mixes of bee and plant cuticular blends showed serendipitous similarities in hydrocarbon profiles that caused attraction by bees with only minor quantitative changes in the composition of volatiles. The four bars (left to right) within categories give numbers for escalating male attraction, from a close approach, to pouncing, to pseudocopulation. *$P < 0.05$, **$P < 0.01$.

giosperm flowers inherently suggests that major change is comparatively simple. Quantitative genetic analyses at the level done for the two *Mimulus* species remain to be done for other species, but emerging information about the genetic basis of floral development and specific floral traits suggest that genes or extra-genomic factors with major effects may indeed be common. For example, transposon insertions determine virtually all floral colour phenotypes in *Ipomoea*, linking highly mobile elements to rapid phenotypic changes known to affect pollinator preferences (Fry & Rauscher 1997; Clegg & Durbin 2000). Functional scent differences between the fragrant hawkmoth-pollinated *Clarkia breweri* and its bee-pollinated non-fragrant sister species *C. concinna* are determined by one gene for one compound and by epistatic biosynthetic pathway interactions for another (Raguso & Pickersky 1999); upstream mutations in the few biosynthetic pathways involved in scent production can have major phenotypic consequences in any plant species, and they are consistent with high levels of scent homoplasy among closely related species of, for example, the evening primrose family Onagraceae (R.A. Raguso, pers. comm.). Finally, single genes can control traits such as the switch between radial and bilateral symmetry (e.g. cycloidea; Coen & Nugent 1994; Cubas et al. 1999); radially symmetrical mutants are not uncommon in species with bilaterally symmetrical flowers, although such fundamental reorganization may often have a negative effect on the functional integration of the flower.

6.5.3 Selection on pollinator traits

In contrast to the documented selection of pollinators on plants, plant-mediated selection on pollinators is less well understood. Several reasons that are not mutually exclusive can account for this. First, it is logistically difficult to measure the lifetime fitness of a flower-visiting animal. For a plant, its visitors may be monitored through as little as a single reproductive episode, and pollen import/export measured to obtain an estimate of lifetime fitness. By comparison, for a flower-visitor it would entail tallying the nectar and pollen gathered from all visited flowers, and subsequently determining its relative reproductive success. This might be more tractable in a seed-parasitic pollinator, where oviposition success provides a crude estimate of pollinator fitness, but it would otherwise be quite difficult. Second, selection on animals may be more diffuse because they are mobile and thus have the

based on visual, tactile and olfactory cues (Stowe 1988). Each orchid species exploits a distinct wasp species, suggesting that diversification may have been driven by diversifying selection on floral traits.

In addition to direct selection by pollinators, plant–plant interactions at the community level may also drive diversification. In the species-rich Australian genus *Stylidium*, pollen placement on pollinators is very precise, and species with shared placement run a higher risk of heterospecific pollen transfer than those with different placement. In widely distributed species, population-level differences in floral morphology reduce the overlap of placement with other coexisting species, suggesting character displacement in response to pollen interference (Armbruster et al. 1994).

The extraordinary phenotypic diversity among an-

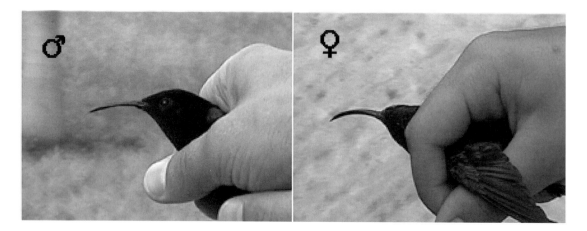

Figure 6.11 Sexual bill dimorphism can be driven by competition for floral resources. Male (left) and female (right) purple-throated caribs (*Eulampis jugularis*). Notice the significantly longer and more curved beak of the female. (Photographs courtesy of E. Temeles.)

possibility of sampling among flowers. Finally, in virtually all instances selection would be indirect, because the plant does not exert direct effects on the pollinators' gene flow but rather on food intake.

Selection on pollinators can be mediated through competitive interaction for resources. Temeles et al. (2000) showed that sexual dimorphism in beak shape in a hummingbird, the purple-throated carib (Fig. 6.11), is linked to sex-specific territories centred around two different species of nectar-producing *Heliconia*. The more aggressive males occupy the more nectar-rich patches of a species with shorter, straighter floral tubes than the other, and male beak length and shape matched the floral tubes. Females occupy patches of the other species, and have optimal beak length for its flowers. Because of this partitioning, seemingly the result of intraspecific competition among the pollinators, heterospecific pollen transfer between the two sympatric *Heliconia* species is avoided.

Community structure among flower-visiting bumblebees and birds suggests interspecific competition for floral resources. In bumblebees, the most diverse communities tend to consist of a short-, a medium-, and a long-tongued species (and sometimes a separate nectar-robber) (Pyke 1982). Similarly, in neotropical hummingbird communities, behaviour and variation in beak morphology have been interpreted as evidence of competition-driven divergence (Feinsinger 1983). A lingering difficulty in this context is the problem of inferring appropriate null models for community structure, to assess whether the distributions reflect significant differences from a random assemblage. Phylogenetic effects have not been considered either.

Intra- and interspecific competition among pollinators is integral to some of the most prominent examples of plant–pollinator coevolution, which is addressed in the next section.

6.5.4 Coevolution between plants and pollinators

Perhaps the strongest cases for plant-mediated selection on pollinators deal with coevolution, that is, reciprocal selection between plant and pollinator. The classical examples involve floral spurs or tubular corollas and insect tongues, first proposed by Darwin in 1862 (Fig. 6.12). Spurs with concealed nectar have evolved many times among angiosperms (Hodges 1997b) and can sometimes be extremely hypertrophied; some Malagasy orchids have spurs that reach 40 cm in length (Nilsson 1988; pers. comm.). Darwin suggested that long spurs or tubular corollas and equally long tongues of visitors are the result of coevolutionary contests. Consider a plant population with variation in spur length. If plant fitness is positively correlated with spur length because local pollinators can extract the nectar of short-spurred flowers without causing pollination very often, there will be directional selection for longer spurs in the population. With selection for more deeply concealed rewards, reciprocal

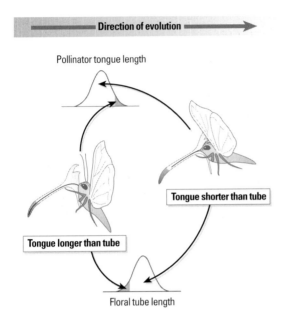

Figure 6.12 A modified version of Darwin's hypothesis for coevolution of long floral spurs and long tongues of visitors. Long-tongued visitors are not compelled to come into contact with floral sexual organs in short-tubed plant indidivudals, causing selection for longer tubes. This in turn will cause selection for long-tongued visitors as long as the degree of nectar access causes differential fitness. (From Nilsson 1988.)

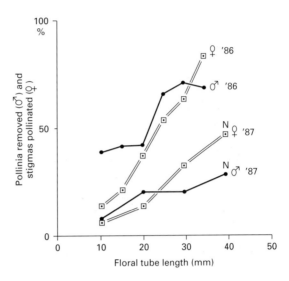

Figure 6.13 Spur length directly affects male and female fitness. Effects of experimental shortening of nectar spurs on pollinium removal and insertion during two years (1986–7) in a Swedish population of the orchid *Platanthera bifolia*. Increasing success in both male and female functions resulted from the visiting moths approaching the flower only as closely as necessary to obtain the nectar. Effective pollinium removal and deposition hinges on the placement of pollinia on the head or near it on the tongue. Effects were statistically significant (χ^2 test, $P < 0.025$) in both years for both functions. (Modified from Nilsson 1988.)

selection is expected on tongue length in the pollinator population. This process can continue as long as genetic variation is available in both parties, and will also be affected by changes in the visitor guild. This scenario was untested experimentally for over a century. Its first test used two European *Platanthera* orchids with spurs 30–50 mm long (Fig. 6.13; Nilsson 1988). Consistent with the hypothesis, pollinator moths only approached flowers with experimentally modified spur length as closely as they needed to drink the nectar, resulting in less pollination in short-spurred flowers because the orchid pollinia have a low probability of reaching a stigma unless they are attached on or near the moth's head. Reproductive success in natural plant populations was also positively correlated with spur length, showing that sufficient variation exists to cause variation in successful pollination. The final assumption, that there is differential fitness among pollinators based on tongue length, has not been directly tested in these or any other systems. Floral visitors with mouthparts that can reach concealed nectar are common in most habitats, consisting of moths, butterflies, many

bees, birds and flies, and competition among them could arguably cause differential fitness. Strong selection on mouthparts has been documented, for example, in the i'iwi, a Hawaiian honeycreeper that historically used its curved beak to access nectar in tubular lobeliads. These lobeliads are now extinct or very rare, and the bird has changed its diet to the more superficially placed nectar of *Metrosideros*. Comparison of nineteenth-century museum skins of the bird with extant individuals revealed a significant reduction in beak length since the shift to *Metrosideros* (Smith et al. 1995).

Plant communities commonly hold many species with varying degrees of spur or tubular corolla development, providing resources for long-tongued or long-beaked visitors. Because flower-visitors with very long mouthparts will act as parasites on plants with less deeply concealed nectar, their presence may cause parallel evolution among plant species with similar functional structure—in effect, plant species may become evolutionarily hijacked by the longest-tongued nectar parasites in the area. The result may be guilds of species whose flowers are almost

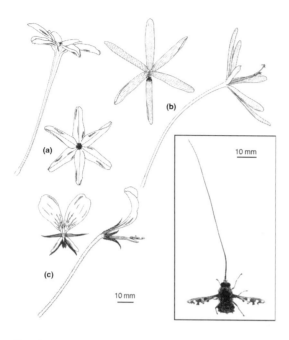

Figure 6.14 Long-tongued flower-visitors can drive convergent
selection for long spurs in large plant guilds. *Ixia paniculata* (a),
Geissorhiza exscapa (b) and *Pelargonium longicaule* (c) are some of about
200 species of plants in southern Africa that rely on three exceptionally
long-tongued flies, including the nemestrinid *Moegistorhyncis
longirostris* (insert) and two tabanids. The flowers have narrow spurs up
to 95 mm in length. (From Goldblatt & Manning 2000.)

physically damage the flower to reach nectar through a
torn spur. Honeybees routinely bite holes in corollas to
access nectar, with other insects following behind them to
steal the dregs (Roubik 1989), and birds such as flower-
piercers may puncture or tear entire flowers (Arizmendi et
al. 1996). The frequency of robbed flowers can approach
100%. High levels of nectar-robbing by bumblebees in
Mertensia paniculata had no negative effects on pollen export or seed set (Morris 1996), whereas robbing affected
both male and female fitness in *Ipomopsis aggregata* (Irwin
& Brody 2000). Among *Asclepias*, up to 185 nectar-
feeding visitor species per plant species were reported, but
only 32–81% were ever observed removing pollen
(packed in readily counted club-shaped pollinia) and the
mean probability of pollen export during a single flower
visit ranged from 0.5 to 4.5% among visitor species
groups (Fishbein & Venable 1996). The fitness consequences of these types of antagonist are rarely quantified
in pollination studies, and their evolutionary impact is little understood (Maloof & Inouye 2000).

Organisms that feed on flowers and flower parts are sufficiently common to have given rise to the term 'florivores'; numerous fly and beetle families contain many
significant flower predators (Brues (1926) quantified der-
mestid beetles eating yucca flowers by weight), and particularly in the tropics vertebrates also commonly feed on
flowers. Brody (1992) showed that seed-parasitic anthomyiid flies that oviposit on flowers counter pollinator-
driven selection for larger corollas. If this is a significant
factor in plant fitness, selection for defensive traits or tolerance of parasites and florivores may result, although
such traits can be expected to sometimes be constrained
by selection for pollinators. Perhaps the strongest evidence to date for purely aversive traits, in the sense of a
trait whose evolution had no effect on mutualists, is the
repulsion of bees by increased carotenoid pigments in
Mimulus cardinalis (Schemske & Bradshaw 1999).
Further case studies are much needed to determine the
role of selection against predators and parasites in floral
evolution.

absurdly deep; the rich communities of long-spurred orchids and long-tongued hawkmoths on Madagascar
(Nilsson 1987) are mirrored by long-tubed flowers of
species within about a dozen plant families in south-western North America (Grant 1983). Finally, one of the more
dramatic examples of convergent floral evolution is that
of long-tubed flowers of perhaps 200 species in at least 10
families in southern Africa, which rely on exceptionally
long-tongued nemestrinid and tabanid flies for their
pollination (Goldblatt & Manning 2000; Fig. 6.14).

6.6 Ecological and evolutionary impact of
parasitic or predatory flower-visitors

Like most biological processes, pollination is a sloppy procedure at best, with most individual interactions between
plant and pollinator not resulting in pollen transfer. In addition, plants may suffer loss of nectar or pollen to visitors
who cheat the plant by never transferring pollen, or even

6.7 Active pollination mutualisms as models:
the interface between pollination and herbivory

As set out in the introduction to this chapter, pollination
is almost always invariably a by-product of a visitor moving on the flower in its quest for a reward; whether the animal causes pollination or not does not affect pollinator

fitness. However, in three documented cases there is a direct link between pollination and pollinator fitness, and these models represent suitable study systems for exploring several general issues surrounding pollination mutualism.

The three model systems involve the species-rich figs and fig wasps (Herre 1999), yuccas and yucca moths (Pellmyr & Leebens-Mack 2000), and senita cacti and the senita moth (Holland & Fleming 1999). In these systems, the female insect actively pollinates flowers and her progeny consume some of the resulting seeds. With the possible exception of the senita–senita moth interaction, these are obligate mutualisms in which there are no co-pollinators. The underlying evolutionary conflicts between plant and pollinator do not differ between instances of active and passive pollination, but the reciprocal effects may be simpler to identify and quantify in systems of seed-parasitic pollinators. The effects are likely to be stronger and more immediate on seed-eating pollinators than others, and the costs of pollination to the plants are readily measured as seeds are expended for seeds. By their very nature these interactions incorporate a herbivory component of understanding the factors that determine plant fitness—as Herrera showed, a critical component to include to understand some plant–pollinator interactions. Some other seed-parasitic insects also serve as passive pollinators of their host plants, notably some *Greya* moths (Thompson 1994), hadenine moths (Pettersson 1991) and anthomyiid flies (Pellmyr 1992; Despres & Jaeger 1999), but co-pollinators are often present, and the outcomes of interactions with plants may range from facultative mutualism to antagonism.

Questions where these mutualisms have proved highly useful thus far include areas such as the identification of reciprocal limiting factors on exploitation (Pellmyr & Huth 1994; Addicott & Bao 1998), the role of pollinator dispersal patterns in host population genetic structure (Nason et al. 1998) and evolutionary transition from mutualism to antagonism (Pellmyr et al. 1996a). As phylogenies become available for the plants and animals involved, these should also provide excellent models for analysing patterns of trait coevolution in plant–pollinator interactions.

6.8 The utility of phylogenetic methods in pollination biology

Many aspects of pollination biology are best analysed in a phylogenetic context. For example, in analyses of trait evolution it is critical to have this information to separate adaptation from historical effects (Harvey & Nee 1997). The ready availability of DNA sequence data in inferring phylogenetic relationships, in combination with the rapid development of powerful tools for intra- and interspecific phylogenetic analysis (Hillis et al. 1996; Avise 2000), have opened the possibility of the rigorous testing of hypotheses about such issues as rates of diversification, patterns and frequency of shifting partners in pollination mutualisms, and patterns of trait evolution. In essence, then, phylogenetic data can provide the information needed to explore how microevolutionary processes create macroevolutionary patterns in plant–pollinator relationships. This combination of phylogeny and ecology, which has emerged in the last decade, is proving to be a powerful tool in many areas of pollination biology. Some examples of its uses are provided below.

6.8.1 Animal pollination and diversification

A longstanding hypothesis holds that animal pollination may have been critical in driving the exceptional diversification of angiosperms. Sanderson and Donoghue (1994) developed a maximum-likelihood method and applied it to then-accepted hypotheses of basal angiosperm relationships. The diversification rate increased after the origin of angiospermy and animal pollination, thus animal pollination in itself cannot explain the early diversification; the result would not differ if the most recent consensus phylogeny were used (Fig. 6.1).

Key innovations are traits that unleash rapid diversification in a lineage, and the floral spurs associated with probing pollinators (discussed in Section 6.5.4) have been hypothesized to be such a trait (Fig. 6.15). Hodges (1997b) used a comparative approach, analysing diversity between several sister groups in which members of one but not the other group have floral spurs. Comparison based on eight phylogenetically independent origins indeed supported the hypothesis that floral spurs increase rates of diversification.

6.8.2 Patterns of trait evoluton

Pollinator attractants and rewards are highly diverse within many closely related plant groups, a situation evident not only by the observed scent, shape, and floral colour variation within monophyletic genera, but also in the

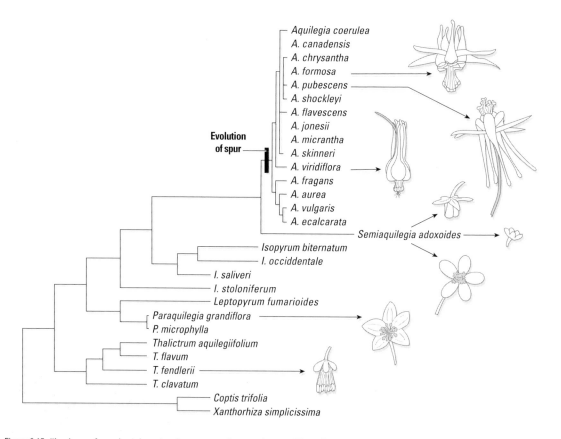

Figure 6.15 Floral spurs form a key adaptation that appears to increase the rate of diversification in angiosperms. A maximum parsimony phylogeny based on DNA sequence data for a lineage in the Ranunculaceae that includes *Aquilegia* (columbines), which have spurs. Horizontal bars indicate relative amounts of time, suggesting that the diversification of approximately 70 species of *Aquilegia* happened at a far faster rate than rates of diversification in more basal groups with spurless flowers. Comparison with other origins of spurs shows similar patterns of elevated diversification, suggesting that the spur is a key adaptation. (Modified from Hodges 1997b.)

high level of convergent evolution among unrelated plants. Under these circumstances of evolutionary lability of the traits under study, independent phylogenetic information is critical in understanding patterns of trait evolution and the role of pollinators in plant diversification. With phylogenetic and ecological data in hand, we can ask such questions as: Are there patterns of transition in relying on particular groups for pollination, or are shifts to or additions of new pollinators random? Is specialization in plants or pollinators an irreversible process, or can diet or pollinator breadth increase within a lineage? To what extent do new plant–pollinator associations build on preadaptations, and how extensive can trait coevolution be? Do sensory systems of flower-visitors constrain their choice of flowers among potential hosts?

In an early example of this type of analysis, pollination in the tropical legume genus *Erythrina* was found to have shifted at least four times from primary reliance on perching passerines to hovering hummingbirds, with floral shape and position and inflorescence position evolving convergently on each occasion (Bruneau 1997). These results suggest that this transition is evolutionarily simple in this group, and comparative analysis of other plant groups involving these and other pollinator groups may allow us to develop more general quantitative estimates of transition probabilities in pollinator shifts.

The phylogeny of the tropical genus *Dalechampia* has been used to test the hypothesis that specialization in pollination is irreversible (Fig. 6.16). Basal species of the genus are highly specialized in that each one depends on

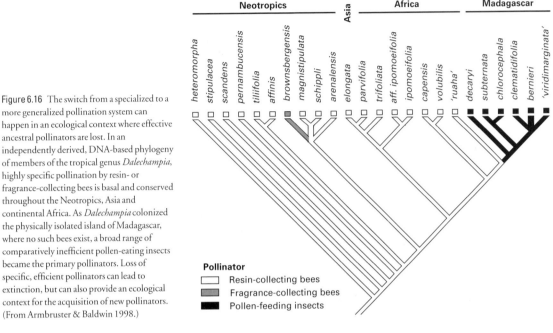

Figure 6.16 The switch from a specialized to a more generalized pollination system can happen in an ecological context where effective ancestral pollinators are lost. In an independently derived, DNA-based phylogeny of members of the tropical genus *Dalechampia*, highly specific pollination by resin- or fragrance-collecting bees is basal and conserved throughout the Neotropics, Asia and continental Africa. As *Dalechampia* colonized the physically isolated island of Madagascar, where no such bees exist, a broad range of comparatively inefficient pollen-eating insects became the primary pollinators. Loss of specific, efficient pollinators can lead to extinction, but can also provide an ecological context for the acquisition of new pollinators. (From Armbruster & Baldwin 1998.)

one or a few resin- or scent-collecting bee species for its pollination. An invasion and radiation of the genus on Madagascar coincided with a shift to reliance for pollination on a broad range of pollen-collecting bees (Armbruster & Baldwin 1998). The ancestral group of pollinators were absent in the new range, causing a situation in which comparatively inefficient visitors may become the only available mutualists. This identifies one ecological context in which evolution of the increased breadth of a pollinator guild may occur.

The obligate mutualism between yuccas and yucca moths is extremely specific. High specificity in interactions may suggest that numerous traits important to the mutualism may have resulted from coevolution once the association has been established. Alternatively, it may have resulted from colonization by organisms that already had many of the required life habits, a process Janzen (1985) called ecological fitting. In the yucca–yucca moth model, mapping of trait evolution onto an independent phylogeny supported the latter scenario, with several preadapted traits and few novel traits (Pellmyr et al. 1996b). A similar scenario applies to the *Ophrys* orchids, where the serendipitous similarity of plant cuticular waxes to some insect sex pheromones serves as a critical preadaptation in attracting pollinators (Schiestl et al. 1999). The

Figure 6.17 Loss of mutualism through evolution of obligate self-pollination often causes flower size reduction and loss of mechanisms that promote outcrossing. *Amsinckia furcata* (left) and *A. vernicosa* (right) are sister species, with *A. furcata* being obligately outcrossed and heterostylous (i.e. stigma and stamens are physically separated and take opposite positions in different morphs). The auto-fertile *A. furcata* is homostylous (i.e. one morph with stamens and stigma physically close). Autogamous species derived from outcrossing ancestors typically have inconspicuous flowers and may become extinct relatively quickly. (Photograph courtesy of D. Schoen; from Schoen et al. 1997.)

role of preadaptations in creating patterns of plant–pollinator associations in general remains to be explored (Armbruster 1997).

Finally, the evolution of self-pollination and the dissolution of mutualism can be evolutionary favourable in plants under some circumstances, and it has evolved many times in angiosperms (Schoen et al. 1997). It increases the number of gene copies of the parent in its progeny, and resources allocated to pollinator attractants and rewards can often be heavily reduced as well. Meanwhile, the dissolution of mutualism with pollinators is expected to increase the risk of extinction. If this is correct, self-pollinated plant species should be derived from animal-pollinated, outcrossed ancestors, and not survive long past their origin. Consistent with this hypothesis, Schoen et al. (1997) found multiple, recent origins of selfing from animal-pollinated outcrossing ancestors (Fig. 6.17).

6.9 Future directions

These are exciting times in the study of plant–pollinator interactions. As genetic information about plants and animals builds at an accelerating speed, it will allow the design of powerful new tools for experimental studies of interactions. QTL-based trait analyses should soon be able to be combined with direct gene manipulation in plants to test hypotheses about pollination interactions; the emerging use of mutants from studies of development is a first step in this context (Comba et al. 2000). Floral scent, pigments and shape will soon be independently manipulated to test any hypothesis. Phylogenetic information is increasing rapidly for many groups of organisms that are central in plant–pollinator interactions, and the availability of robust historical data will allow us to bridge traditional ecological and evolutionary time scales. How does history affect evolutionary potential on ecological time scales, and what part of microevolutionary variation in interactions is preserved at and above the species level? Can we detect historical patterns of trait coevolution between plants and pollinators that diversify over long periods of time?

Ecological studies are entering an important phase of considering plant–pollinator interactions in a larger context. First, multispecies interactions are the norm rather than the exception, and empirical long-term studies of them that take into account spatiotemporal variation will be crucial to understanding of their dynamics. This includes the roles of parasites and predators, which are still very poorly understood. The new integration of all pollination interaction effects into the context of other factors that affect organism fitness may eventually tell us when and how these interactions translate into evolution; if the insights from the violets at Cazorla (Herrera 1993) turn out to reflect a common situation, it will have profound implications for the field.

Last but not least, strong conservation concerns are emerging in plant–pollination interactions (Kearns et al. 1998). Issues such as the consequences of plant/pollinator population fragmentation on a landscape level, the introduction of exotic pollinators or florivores, and the extinction of either plant or pollinator in more or less exclusive associations (Schmidt-Adam et al. 2000) will require increasing attention as human effects escalate.

Chapter 7   Seed dispersal by vertebrates

Carlos M. Herrera

7.1 Introduction

Aside from the profound physiological differences derived from the ability of plants to build carbon-based organic molecules out of light, water and atmospheric carbon dioxide, the major macroscopic difference between plants and animals possibly lies in the very limited mobility of adult plants in comparison to adult animals. Adult plants remain fixed in space for their whole lives, anchored to their indispensable source of water and minerals. In the long run some adult plants may move by clonal growth but, in comparison with animals, the distances involved are negligible.

The absence of movement of adult plants entails decisive limitations at two critical stages in their reproductive cycles, namely sexual reproduction and offspring dispersal. These two processes require the movement of some reproductive structure across space. Sexual reproduction requires that pollen grains travel a variable distance from their place of origin to meet the female gametophytes, and dispersal of seeds involves their movement away from the maternal parent. Given this need for 'movement' inherently associated with the reproductive process, and provided that animals have such a quality in abundance, it should not surprise us that plants have developed a countless variety of mechanisms to use animals as the vectors of their pollen and seeds. This enormous diversity in the exploitation of animal movement for pollen and seed dispersal was once deemed one of the decisive factors responsible for the tremendous diversification and ecological success of angiosperms. More recent studies, however, are either inconsistent with that earlier view or only partly support it. Furthermore, dispersal of seeds by animals was not a functional novelty brought in by the angiosperms (Fig. 7.1). The habit is widespread among extant gymnosperms and, most likely, occurred in many extinct lineages as well, whose seeds were embedded in, or closely associated with, well-developed fleshy structures (see Chapter 2). In fact, seed dispersal by animals is proportionally much more frequent among extant gymnosperm than angiosperm lineages (64% vs. 27% of extant families respectively; Herrera 1989a).

In comparison to other aspects of plant reproduction, the study of plant–animal seed-dispersal systems from an explicitly evolutionary viewpoint is relatively recent. Darwin pioneered evolutionary studies of plant reproduction, and devoted considerable attention to studying the adaptation of plants to pollinators and the evolution of plant-breeding systems. In contrast, he paid only sporadic and cursory attention to seed dispersal in his writings. The recent emphasis on the evolutionary aspects involved in seed dispersal by animals may be traced back to a few influential studies published in the seventies (e.g. Snow 1971; McKey 1975). More recent overviews of the evolutionary ecology of plant–animal interactions for seed dispersal are provided by Janzen (1983a), Herrera (1985a, 1986) and Jordano (1992).

7.2 Seed dispersal: a summary of concepts

Dispersal is the process by which individuals move from the immediate environment of their parents to settle in a more or less distant area. Although it is not exclusive to plants, plant dispersal differs from animal dispersal in one important respect. While in animals the dispersing individual generally depends on their own locomotive powers, offspring dispersal in plants is always of a passive nature, as the seed has no control of either the dispersal process in itself or where it will eventually end up. This means that animal parents can play little or no role in determining the course of dispersal of their autonomous offspring, whereas dispersal of plant offspring will most often be determined by the traits of its maternal parent rather than by their own traits.

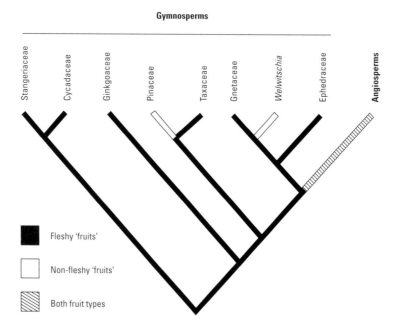

Figure 7.1 Distribution of fleshy 'fruits' (i.e. packages made up of seeds plus the accessory nutritious tissues that are used as food by animals) over the major lineages of extant seed plants. The phylogenetic arrangement shown follows the so-called 'anthophyte hypothesis' (e.g. Loconte & Stevenson 1990), whereby the Gnetales (= Gnetaceae + Ephedraceae + *Welwitschia* in the graph) are considered a sister group to angiosperms.

Seed dispersal may have different benefits to plants. These benefits may be classified in two major categories depending on whether they are related to circumstances of the 'departure' (i.e. the mere act of seeds leaving the parent) or the 'arrival' (i.e. the specific end-point of dispersal, or where the seeds eventually end up). This distinction makes sense in the light of theoretical models which have established that dispersal can provide a variety of fitness benefits even when a more favourable environment is not eventually reached by the dispersing organism (Johnson & Gaines 1990). But it can also help to explain some general evolutionary patterns exhibited by plant–vertebrate seed-dispersal systems, as shown in Section 7.8.2 later in this chapter.

7.2.1 Departure-related benefits

Merely leaving the immediate vicinity of the maternal parent may be advantageous to seeds. Escaping from the area of 'chemical influence' of the parent plant, for example, might result in increased seed germination and survival. Plants frequently produce chemicals that, after becoming incorporated to the soil, inhibit the germination not only of the seeds of other species ('allelopathy'), but of their own seeds as well ('autoallelopathy'; Solomon 1983). In most instances, however, departure-related benefits are not so straightforward, and depend in complex ways on the advantages derived from increasing the distance from the parent plant and of leaving a high-density concentration of competing siblings. As distance from the parent and seed/seedling density are usually correlated, the effects of these two factors are difficult to separate in the field without careful experimental manipulation. Seed density usually decreases away from the parent plant, hence seed dispersal may be advantageous simply because of improved survival prospects as a consequence of declining seedling competition with neighbours. In addition, some natural enemies of seeds

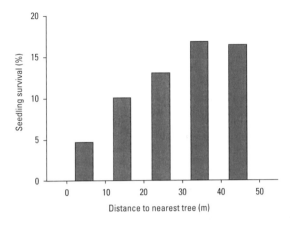

Figure 7.2 Seed/seedling survival over a 6-week period as a consequence of distance to the nearest conspecific in the tropical forest tree *Virola surinamensis*. Survival increases steadily with increasing distance to the nearest conspecific, reaching a plateau around 30–50 m. Distance-dependent herbivory and seed predation by insects and vertebrates were responsible for the observed pattern. (Drawn from data in Howe et al. 1985.)

and seedlings (e.g. pathogens, herbivores, seed-predators) respond to density and distance from the parent. In the tropical forest tree *Virola surinamensis*, for example, seed and seedling survival increase steadily with increasing distance from the nearest conspecific plant (Fig. 7.2). In this and many other investigations, the impact of fungal pathogens and animal seed-predators is greatest closest to the parent plants, which gives rise to a decline in seed and seedling mortality with increasing distance from adults (e.g. Augspurger 1984).

7.2.2 Arrival-related benefits

Seed dispersal may also be advantageous because of some particular characteristics of the point of arrival of seeds. For example, seed dispersal will be advantageous whenever it allows the quick occupation of vacant habitats or microhabitats. This is illustrated by the prevalence of species with well-developed seed dispersal mechanisms in the colonization of new volcanic islands (Whittaker & Jones 1994); by the rapid northward recolonization by forest plants of recently deglaciated territories during successive Pleistocene interglacials (Johnson & Webb 1989); and by the ability of plant species to migrate latitudinally in response to long-term climatic modifications (Huntley & Webb 1989).

Some species may have special requirements for seed germination or seedling establishment. In some mistletoes, for example, seedling establishment is optimal in a narrow range of host twig diameter (Sargent 1995). For these species with special requirements, seed dispersal may be highly advantageous if it predictably enhances the probability of seeds reaching such favourable microsites. This was termed 'directed dispersal' by Howe and Smallwood (1982), and it has been frequently suggested in relation to seed dispersal by animals. There are, however, very few well-documented examples, which probably reflects both the rarity of the phenomenon in nature and the difficulties faced by researchers when trying to objectively identify 'favourable' microsites from the viewpoint of plants. One of these few examples involves the ant-dispersed herb *Corydalis aurea*. In this species, directed dispersal of seeds to ant nests effectively increases the population growth rate because of a significant increase in the survival of seeds to reproduction (Hanzawa et al. 1988). Another well-documented instance of directed dispersal is provided by the tropical forest tree *Ocotea endresiana* and its major seed dispersers, the bellbirds (*Procnias tricarunculata*) (Fig. 7.3). The seeds dispersed by birds of this species predominantly land at microsites characterized by lower incidence of fungal pathogens and increased seedling survival rates. This directed dispersal to favourable microsites eventually results in increased seedling survival rates (Wenny & Levey 1998).

7.3 Seed dispersal by vertebrates

Seed dispersal by animals is an intrinsically heterogeneous phenomenon that involves an astounding diversity of animal and plant lineages, proximate mechanisms, and plant–animal functional, ecological and evolutionary relationships. Natural history and botanical details illustrating this enormous diversity may be found in the classical treatises of Ridley (1930) and van der Pijl (1969), and in more recent reviews focusing on the ecological and evolutionary implications of plant–animal dispersal relationships (Janzen 1983a; Sorensen 1986; Jordano 1992). With the outstanding exception of ants (see Chapter 8), invertebrates play only an anecdotal role as seed-dispersers, and this chapter will be concerned exclusively with the dispersal of seeds by vertebrates.

The diverse modalities of seed dispersal by vertebrates may be classified into one of three main *functional* categories. This will depend on whether dispersal is a casual

Figure 7.3 Seeds of the Neotropical forest tree *Ocotea endresiana* (Lauraceae) that are dispersed by frugivorous bellbirds (*Procnias tricarunculata*; Cotingidae) tend to land predominantly at forest microsites that are characterized by the lower incidence of fungal pathogens (Wenny & Levey 1998). (Drawing by Rodrigo Tavera.)

consequence of deliberate seed-harvesting by seed-predators or a result of incidental picking up of seeds. In the latter case, it will also depend on whether seed transport takes place internally or externally (Box 7.1). Rather than highlighting structural particularities, this classification emphasizes unifying functional features. As with any biological classification, however, some exceptions occur that do not fit easily. This applies, for example, to some tropical dispersal systems where vertebrates participate as primary dispersers and ants then act as secondary dispersers of vertebrate-dispersed seeds (Levey & Byrne 1993). Another exception occurs in situations where wind is the primary disperser and vertebrates then perform further dispersal of wind-dispersed seeds (Vander Wall 1992). These binary or 'two-stage' seed-dispersal systems have been little studied so far, but future studies may eventually prove that their frequency in nature is much higher than hitherto recognized.

Each of the three major categories of animal seed dispersal recognized in Box 7.1 has its own set of physical constraints, ecological requirements and life-history consequences. Harvest-based dispersal systems, for example, are associated with extensive seed mortality (see Chapter 5). Dispersal by external adhesion is constrained by seed size, as heavy seeds can hardly remain attached to animal fur or feathers for little more than a short time. Dispersal in animal interiors requires the evolution of some rewarding bait to entice animals and make them become internally 'contaminated' with inadvertently ingested seeds. This means that the three major modalities of animal seed

dispersal are ecologically and evolutionarily so disparate as to be treated separately. Animal seed dispersal based on imperfect harvesting was dealt with in Chapter 5 on granivory, because it is closer to a variant of seed-predation from both the plants' and animals' viewpoint. Dispersal by contamination of animal exteriors is not considered in this book, as it does not qualify as a plant–animal interaction. In a functional sense, it does not differ greatly from wind dispersal, in that both wind and animal agents pick up seeds incidentally and passively, with no opportunity of interaction between the plant and its animate or inanimate disperser. Seed dispersal in animal interiors will be the subject of the rest of this chapter, and the expression 'vertebrate seed dispersal' will be used hereafter to refer exclusively to this dispersal mode.

7.4 Dispersal in animal interiors

7.4.1 The mutualism: exchanging food for movement

Plants dispersed by vertebrates have evolved edible seed appendages or coverings that are ingested and digested by animals that later eject the seeds in conditions suitable for germination. The package made up of seeds plus the accessory nutritious tissues that are used as food by animals may be termed 'fruit', although it does not always originate from an ovary (the true 'fruit' in a botanical sense). The fleshy portion may originate from the seed coat (as in

Box 7.1 Main functional categories of seed dispersal by vertebrates

Seed dispersal by vertebrates comprises a mixture of hetero-geneous phenomena involving a broad variety of animal lineages, proximate mechanisms, and plant–animal functional, ecological and evolutionary relationships. All that variety, however, boils down to three major categories that differ in the essential mechanisms involved.

1 Imperfect harvesting   Animals that forage for the seeds themselves (granivores) take them away from the parent plant to be eaten (and thus killed) some time later. As a consequence of this delay from harvesting to consumption, a certain proportion of harvested seeds accidentally escape destruction, thus fortuitously converting genuine harvesters into occasional seed-dispersers. Seed dispersal based on imperfect harvesting is typically associated with situations where animals intensively harvest temporarily superabundant seed crops and store the excess food in caches for future use.

Granivorous birds and mammals are most often involved in this dispersal system (see Chapter 5). Well-known examples from temperate latitudes include dispersal of pine seeds by nutcrackers, acorn dispersal by jays and small-mammal dispersal of many forest trees. This dispersal system may be much more common in tropical forests that hitherto recognized, with rodents playing a prominent role (Forget 1993).

2 Collection of seeds   Animals do not actively seek seeds for food but perform dispersal as a consequence of becoming 'contaminated' with mature seeds. Seeds are picked up from the maternal plant (primary dispersal) or elsewhere (secondary dispersal), and then discarded some distance away in conditions suitable for germination. Both external (by adhesion following simple physical contact) and internal (by ingesting seed-'contaminated' food), picking up may occur, which leads to the following two major sub-categories.

2.1 Dispersal on animal exteriors   Mature seeds become accidentally attached to the animals' surface after fortuitous contact with the maternal plant, and are then dislodged some distance away. Characteristic examples include the dispersal of hooked or viscid seeds of terrestrial plants entangled in mammalian fur or, less often, birds' feathers (Sorensen 1986), but also the seeds of both terrestrial and aquatic plants dispersed in the muddy feet of animals.

2.2 Dispersal via animal interiors   Seeds accidentally enter the digestive system of animals, generally when they ingest plant structures closely associated with seeds. These are subsequently spat out, regurgitated or defecated in conditions suitable for germination. This category includes the dispersal of 'fleshy-fruited' plants (i.e. those producing berries, drupes or functionally analogous structures) by frugivorous vertebrates, undoubtedly the most widespread, and ecologically and evolutionarily diverse, plant–animal seed dispersal system. Birds and mammals are the organisms playing the most prominent roles by far in this mode of seed dispersal, but other vertebrates, like fishes, tortoises or lizards, may also be important dispersers for some species or in some particular habitats (Milton 1992; Souza-Stevaux et al. 1994; Valido & Nogales 1994).

---

species with arillate seeds) or from ancillary floral structures like bracts or the floral receptacle itself (see Plate 7.1 facing p. 84). This great variety of morphological and anatomical origins of the nutritious portion of fruits contrasts sharply with the homogeneity of its function. Such functional convergence suggests that (i) there have been consistent selective pressures on plants favouring the evolution of plant traits enhancing seed dispersal in animal interiors based on food rewards; and (ii) the modification of pre-existing anatomical structures to turn them into food rewards for frugivorous animals has been relatively simple to evolve, and has not involved many genetic changes. In plants, intra- and interspecific variation in important morphological and architectural traits, including many flower and fruit characteristics, are often governed by just one or two genes. This means that important phenotypic changes may be brought about with only minimal genetic reorganization (Gottlieb 1984).

The relationship between animal-dispersed plants and their dispersers is generally of a mutualistic nature, as both partners derive some benefit from their participation. The food reward provided by the plants is 'exchanged' for a service, namely the movement of seeds provided by the animals. Nevertheless, the outcome of ecological interactions is often quite context-dependent, and the mutualistic nature of the relationship between plants and the vertebrates feeding on their fruits must be corroborated in each particular instance. A given disperser may have a mutualistic relationship with some plant species but not with others. In northern temperate habitats, titmice (Paridae) predominantly behave as fruit-predators that feed on fruit pulp or the seeds themselves without performing seed dispersal, but as mutualists of a few species whose seeds they disperse successfully (Snow & Snow 1988). In some tropical frugivorous birds with lek mating systems, males probably play predominantly

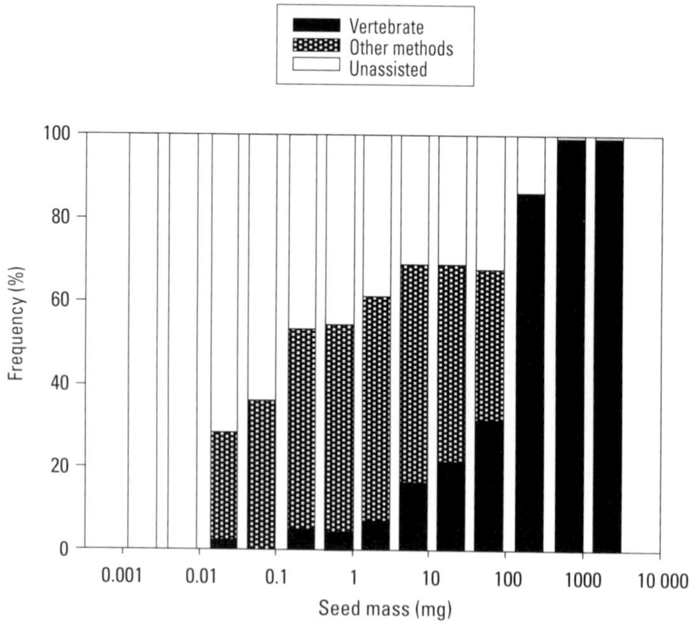

Figure 7.4 The proportion of species with different seed-dispersal methods (vertical axis) varies with seed size (horizontal axis, note the logarithmic scale). The majority of the smallest-seeded species lack special dispersal mechanisms, while vertebrate seed dispersal typically prevails among the largest-seeded species. Other seed-dispersal methods (including ant, wind, adhesive and ballistic dispersal) tend to occur most often among species with intermediate-sized seeds. (Modified from data in Hughes et al. 1994.)

non-mutualistic relationships with their food plants by generating high-density concentrations of dispersed seeds at lek sites, while females of the same species will disseminate seeds more widely and thus behave as mutualists (Krijger et al. 1997).

7.4.2 Taxonomic and ecological distribution: plants

Vertebrate dispersal occurs, and has occurred, in many and disparate seed-plant lineages, both extant and extinct (see Chapter 2 and Fig. 7.1). Starting with seed ferns (pteridosperms) in the early Carboniferous, it has evolved independently on innumerable occasions and ecological scenarios. Among extant taxa, vertebrate dispersal may characterize whole orders (e.g. Cycadales in the gymnosperms), but it occurs more often in subsets of families within orders, subsets of genera within families, or even small groups of species within large genera (e.g. species of *Hypericum* in the Hypericaceae and *Galium* in the Rubiaceae). Vertebrate dispersal also occurs sporadically in some very large families that are almost homogeneously characterized by other seed-dispersal methods, like Asteraceae (*Clibadium*, *Chrysanthemoides*) and Poaceae (*Lasiacis*, *Olmeca*). This extremely patchy distribution of vertebrate seed dispersal among and within levels of the taxonomic hierarchy, along with the variety of anatomical origins of the nutritious portion of fruits noted earlier, indicates that its evolution has been not subject to consistent morphological and ontogenetic constraints, and that it has been selectively advantageous many times and in many ecological scenarios (Herrera 1989a). There is evidence, however, that the fleshy fruit-producing habit is most frequent among basal (i.e. most primitive) angiosperm lineages (e.g. Magnoliales, Laurales), which is consistent with suggestions that the earliest angiosperms had fleshy fruits (Donoghue & Doyle 1989).

Seed dispersal by vertebrates is predictably associated with plant growth form, seed size, habitat type and geographical location. It is generally associated with a large seed size. In four British and Australian temperate floras, for example, almost all species with seeds heavier than ≈ 100 mg are dispersed by vertebrates (Fig. 7.4). The proportion of species that are dispersed by vertebrates is generally highest among trees or shrubs and lowest among herbs. In the regional flora of the north-western Iberian Peninsula, for example, 46% of woody species (shrubs and trees combined) are dispersed by vertebrates, but only 8% of herbaceous species (Buide et al. 1998). In a Brazilian tropical dry forest, proportions of woody and herbaceous species dispersed by vertebrates were 52% and 22% respectively (Gottsberger & Silberbauer-

Gottsberger 1983). As a consequence of this correlation between growth form and dispersal mode, the relative importance of vertebrate-dispersed plants differs among types of plant community, decreasing from forests to scrublands to herbaceous formations (Fig. 7.5).

Within each growth form, the frequency of vertebrate dispersal decreases with increasing latitude, altitude and aridity, and with decreasing soil fertility (Willson et al. 1990; Westoby et al. 1990). Considering woody taxa alone, for example, vertebrate-dispersed species account for ≈ 35–44% of local species in temperate forests and Mediterranean scrublands, but their importance increases to 75–90% of species in humid tropical forests (Fig. 7.6).

Availability of dispersers does not provide a general explanation of geographical differences in the frequency of vertebrate dispersal (Hughes et al. 1994). For example, frugivorous vertebrates are rather scarce in the southern hemisphere temperate rainforests of Chile, yet the frequency of vertebrate seed dispersal in these habitats is roughly comparable to that found in tropical forest communities harbouring abundant and diverse frugivorous vertebrates (Armesto & Rozzi 1989). Variation in

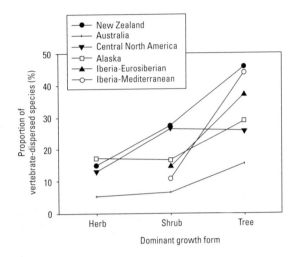

Figure 7.5 Variation among major plant-community categories from several non-tropical regions in the average proportion of plant species whose seeds are dispersed by vertebrates. In all regions considered, there is a consistent trend for the relative importance of vertebrate-dispersed species to increase from herb- through shrub- to tree-dominated plant communities. Each symbol represents the average value for a number of plant communities. (From data in Willson et al. 1990 and Guitián & Sánchez 1992.)

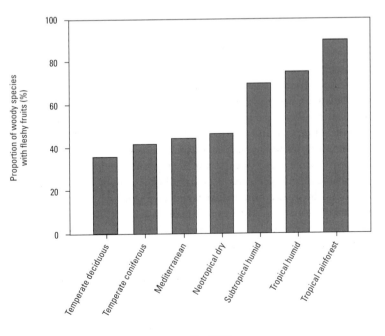

Figure 7.6 Proportion of woody species producing fleshy fruits in different forest types. (Modified from Jordano 1992.)

abiotic factors provides a parsimonious explanation for differences among plant communities in the importance of vertebrate dispersal. Variation in moisture availability and soil fertility determine differences in vegetation structure and in the relative importance of different growth forms. This will ultimately mould seed size distributions (because seed size is related to growth form) and consequently affect the proportion of different dispersal modes in plant communities (Hughes et al. 1994). The large percentage of woody plants in tropical moist forests that have vertebrate-dispersed seeds thus most likely results from strong selection for large seed size, rather than from particularities of the relationships with vertebrate dispersers (Westoby et al. 1990).

7.4.3 Taxonomic and ecological distribution: frugivores

All major lineages of vertebrates take part in fruit consumption and seed dispersal, but their importance as dispersal agents is very unequal. Birds and mammals are the only or main dispersers of the vast majority of vertebrate-dispersed plants, and an important fraction of lineages in these two major vertebrate groups maintains some links with plants related to seed dispersal. About 36% of 135 extant families of terrestrial birds, and 20% of 107 families of non-marine mammals, are partly or predominantly frugivorous (Fleming 1991). Examples of avian seed-dispersers include thrushes, waxwings and warblers in temperate habitats, and hornbills, bulbuls, toucans and manakins in tropical forests. Seed-dispersing mammals include such disparate groups as bats, lemurs, gorillas, foxes, rhinoceros and elephants. This extraordinary diversity of avian and mammalian frugivores indicates that frugivory, and associated seed dispersal, has evolved on many occasions in the vertebrate phylogeny. Fish are important seed dispersers for some tropical plants (Souza-Stevaux et al. 1994), but their quantitative importance in riparian and seasonally inundated tropical habitats is still poorly known. Tortoises and lizards are prominent seed-dispersers only in arid and insular environments (Milton 1992; Valido & Nogales 1994). There is at least one reported instance of seed dispersal by a tropical tree-frog (Da Silva et al. 1989), but dispersal by amphibians is unlikely to represent little more than a biological anecdote.

Frugivorous vertebrates are no exception to overall latitudinal trends in species diversity, and most families

of frugivorous birds and mammals occur in the tropics (Fleming et al. 1987). This tropical concentration of frugivorous taxa apparently stimulated the earlier notion that the interaction between fleshy fruit-producing plants and their dispersers was 'a quintessential tropical phenomenon' (Fleming et al. 1987), and led to the 'calumnious claim [that] the ecology of plant-animal interactions in the temperate zone is downright uninteresting' (Willson 1986). Nevertheless, a more balanced picture has emerged as recent studies have increasingly shown that vertebrate frugivory is quantitatively and qualitatively important in non-tropical regions too. Autumn bird communities of central and north-eastern North America, for example, have a greater proportion of frugivorous species than pristine Amazonian tropical forests (Terborgh et al. 1990; Willson 1991), and Mediterranean forests and scrublands are characterized by very dense populations of frugivorous birds in autumn and winter (Herrera 1995). Fruits may also be more important than hitherto recognized in the nutrition of a large proportion of the terrestrial avian community in Arctic forest-tundra habitats. Many species of Arctic birds feed heavily on fleshy fruits during the autumn migration (Guitián et al. 1994), but also during the breeding season, as fruits produced during the previous summer are preserved in good condition beneath the snow (Norment & Fuller 1997).

The earlier notion that 'specialized' frugivores relying heavily on fruit for food are a distinguishing feature of tropical habitats has also been challenged recently. On one hand, supposedly specialized tropical frugivorous birds, like some species of trogons, toucans, manakins or touracos, actually are not as extensively frugivorous as previously thought, and regularly include significant amounts of animal food in their diets (Remsen et al. 1993; Sun & Moermond 1997). And on the other hand, many non-tropical frugivores depend almost exclusively on fruits for food for extended periods. These include sylviid warblers, thrushes and waxwings among birds (Herrera 1995; Witmer 1996a), and several families of mammals in the order Carnivora (Herrera 1989b; Willson 1993). In the Mediterranean region, for example, fruits regularly contribute >90% of the diet of several species of sylviid warblers during the autumn–winter period (Herrera 1995). These and other studies have contributed to the demise of the earlier myth of plant–frugivore interactions as an essentially tropical phenomenon. It must be emphasized that the strength of the relationship of frugivores with plants can only be evaluated by detailed studies of the actual nutritional importance of fruit food. As quantitative information of this sort is still remarkably sparse for the majority of frugivores, generalizations on the taxonomic and ecological correlates of the extent of frugivory by vertebrates should be made with caution.

7.5 Plant adaptations

One central issue in the study of plant–disperser interactions from an evolutionary perspective is the degree to which selection pressures of plants on frugivores, and of frugivores on plants, have influenced morphological, physiological and behavioural traits in each group of organisms. In the case of plants, the fitness advantages derived from seed dispersal by vertebrates have selected for fruit and fruiting-related traits enhancing fruit consumption by seed dispersers. These include the timing of fruit presentation, fruit traits enhancing discovery by dispersers, and fruit size and nutritional composition, as detailed in the following sections. Plant–disperser interactions, however, do not take place in an ecological vacuum, and fruit and fruiting-related traits are also susceptible to selective forces imposed by abiotic factors and non-mutualistic organisms like seed- and fruit-predators. The possible influence of selective pressures other than those exerted by dispersal agents on dispersal-related plant traits also needs to be taken into consideration.

7.5.1 Fruiting phenology

Selection from dispersers may lead to either clumped or staggered fruiting seasons of locally coexisting plants. These two contrasting possibilities may occur in the same habitat. In central Panama, bird-dispersed species of *Psychotria* have clumped, and those of *Miconia* staggered fruiting seasons (Poulin et al. 1999). The staggering of fruiting seasons was formerly interpreted as a response to selection for decreasing competition among plants for dispersers, but subsequent studies have provided little support for this hypothesis (Wheelwright 1985a; van Schaik et al. 1993). In the case of the *Miconia* and *Psychotria* species mentioned above, their respective phenological patterns can hardly be due to selection from current dispersers, as these are roughly the same for the two plant genera.

Clumped fruiting seasons generally lead to well-defined seasonal peaks in fruit abundance. In north-

ern temperate habitats, for example, most vertebrate-dispersed plants ripen their fruits in late summer and autumn, while in Mediterranean forests and shrublands fruit ripening peaks in autumn–winter (Herrera 1995; Noma & Yumoto 1997). Although less marked than in temperate habitats, fruiting seasonality is also a salient feature of tropical and subtropical forests (van Schaik et al. 1993). Synchronous fruiting by plants sharing the same dispersers may enhance each other's dispersal by mutually attracting more dispersers than each would alone.

Local peaks of fruit availability often coincide with peaks of disperser abundance, which have frequently been interpreted as evidence of disperser selection on the time of fruit ripening (Herrera 1985b; Noma & Yumoto 1997). It is difficult, however, to distinguish this scenario from the converse, that dispersers may respond numerically to, or time their seasonal displacements to coincide with, fruiting peaks (Levey 1988; van Schaik et al. 1993). In general, there is little current support for the once-favoured notion that disperser availability has been the main selective agent shaping the time of fruit ripening in vertebrate-dispersed plants. In wet sclerophyll forests of south-eastern Australia, peak fruiting occurs during autumn, but fruit-eating birds are equally abundant from spring through autumn (French 1992). Among western European bird-dispersed plants, available evidence is contrary to the notion of phenological adjustments by individual species to the marked latitudinal and elevational variations in seasonal patterns of disperser abundance. Local fruiting peaks match disperser abundance peaks because species that fruit at times of greatest disperser abundance are locally dominant, but not because of adjustments by component species (Fuentes 1992). In the predominantly bird-dispersed *Crataegus monogyna*, for example, the more northerly European populations do not fruit significantly earlier than southern populations, as would be expected if local populations had adjusted fruiting seasons to match peaks in disperser abundance (Guitián 1998).

Abiotic factors and organisms other than dispersal agents seem to have played prominent roles in the evolution of fruiting phenologies of vertebrate-dispersed plants. Seasonality in temperature and water availability sets limits on the time of fruit development and maturation. This holds even for the weakly seasonal tropics, where the influence of insolation and water availability on the phenology of woody plants has been more pervasive than the influence of biotic factors like dispersers (van

Schaik et al. 1993). In western Europe, latitudinal shifts in the ripening season of bird-dispersed plants mainly reflect climatic constraints rather than selective pressures from dispersers (Debussche & Isenmann 1992). Microbes, invertebrates and vertebrate seed- and fruit-predators often destroy large numbers of ripe fruits and seeds before they are dispersed, thus they may also influence the evolution of ripening seasons. In central Sweden, for example, around 99% of the fruits of the bird-dispersed shrub *Viburnum opulus* are eaten by bullfinches (*Pyrrhula pyrrhula*) and bank voles (*Clethrionomys glareolus*), which are seed-predators and do not disperse seeds, while fewer than 1% of fruits are eaten by legitimate seed-dispersers (Englund 1993). Since the abundance and/or activity of fruit-damaging microbes and animals varies seasonally, then selection for escaping from these destructive agents has probably contributed decisively to drive ripening seasons closer to those times of the year when the risk is at its lowest (Herrera 1982b).

7.5.2 Advertisement

Those food resources whose fitness is reduced by vertebrate consumption (e.g. insects, seeds) have evolved adaptations that reduce detectability by harmful consumers. Fruit consumption by vertebrate dispersers, in contrast, by being advantageous to plants, has selected for fruit traits that enhance detectability by frugivores. The ripe fruits of vertebrate-dispersed plants are characterized by distinctive odours, conspicuous coloration or some combination of these.

7.5.2.1 Chemical signals

The ecological correlates of chemical signalling by wild ripe fruits remain virtually unknown, in contrast with the extensive attention received by visual signalling (see below). Evidence based on human perception suggests that wild fruits differ widely in the amount and nature of emitted volatiles, and that fruits scented to the human nose seem to be significantly associated with dispersal by nocturnal mammals (van der Pijl 1969; Herrera 1989b). Nevertheless, this may just reflect our own biased mammalian perception. Fruits that are unscented to humans may still produce volatile compounds detectable by analytical procedures and by other organisms (Scarpati et al. 1993). Fruit volatiles may mediate the relationship of fruits not only with dispersers, but also with fruit-

and seed-predators, and its study deserves more attention than it has received so far.

7.5.2.2 Visual signals

In the spectrum visible to humans, ripe fruits vary in colour from reds, blacks and blues to greens and browns. Although exceptions abound, fruits that are green or otherwise dull-coloured when ripe tend to be associated with seed dispersal by mammals, whereas fruits dispersed by birds tend to be brightly pigmented. The partial dichotomy between 'bright' and 'dull' ripe fruits has probably been selected for by the contrasting sensory capacities of birds and mammals (Janson 1983). Red and black fruits predominate among bird-dispersed plants in both tropical and temperate regions, and there is only minor geographical variation in colour spectra (Fig. 7.7). The visual conspicuousness of fruits may be further enhanced by the juxtaposition of two or more bright colours. This juxtaposition may occur in the infructescence itself, as frequently found in species where the contrasting colours of ripe and unripe fruits produce conspicuous bicoloured displays (Willson & Thompson 1982). In some species, it is the contrast between the ripe fruits and the adjacent leaves which enhances visual conspicuousness, and this kind of bicoloured fruit advertisement method was named 'foliar flag' by Stiles (1982). Ripe fruits of many bird-dispersed plants reflect ultraviolet light. This may also have evolved to enhance the visual signals of these fruits, since the colour vision of many bird species extends to the near UV (Willson & Whelan 1989).

The bright fruit colours of bird-dispersed plants represent adaptations for promoting avian frugivory. This has been proved by carefully controlled experiments showing that fruiting displays differing in coloration differ in the probability of being discovered and/or consumed by avian frugivores in the field. Frugivorous birds generally discriminate among fruits on the basis of colour and often exhibit consistent colour preferences. In the colour-polymorphic *Rubus spectabilis*, frugivorous birds select fruits on the basis of colour, and consistently favour red over orange fruits on both local and regional scales in western North America (Gervais et al. 1999). The observations that bicoloured fruit displays occur more often among species with small fruits, and among those fruiting at times of year when dispersers are relatively scarce, provide further support to the adaptive value of fruit conspicuousness (Willson & Thompson 1982; Herrera 1987).

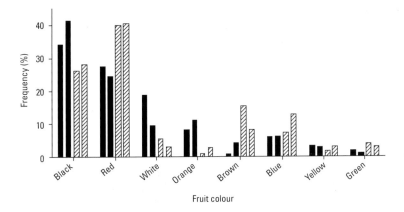

Figure 7.7 Frequencies of bird-dispersed plant species producing ripe fruits of various colours in two Neotropical (Peru and Costa Rica, $N = 134$ and 252 species respectively; filled bars) and European (central Europe and Spain, $N = 137$ and 111 species respectively; hatched bars) regions. (From data in Wheelwright & Janson 1985 and C.M. Herrera, unpublished.)

Nevertheless, factors unrelated to consumption by dispersers may also have influenced the evolution of fruit colour in bird-dispersed plants. Fruit colours may be adaptive in defending fruit against consumers that damage the fruit, either because of the deterrent properties of the pigments themselves (e.g. antifungal phenolic pigments) or because certain damaging agents cannot perceive certain colours (e.g. red fruits 'inconspicuous' to frugivorous arthropods) (Willson & Whelan 1990). Fruit colour may also be an evolutionary by-product of selection acting on some correlated character. In the Australian shrub *Rhagodia parabolica*, seed germination behaviour is correlated with fruit colour, and selection acting on seed germination, rather than selection by dispersers acting directly on fruit colour, may explain fruit colour in this species (Willson & O'Dowd 1989). Physiological factors may also sometimes be involved in the evolution of fruit colour. Plants with large or otherwise costly fruits may have green ripe fruits because photosynthesis may compensate for respiratory costs (Cipollini & Levey 1991).

7.5.3 Fruit size

Size is an important attribute of fruits, because it sets limits to ingestion by relatively small-sized dispersers that swallow them whole, like birds. Fruit size is probably less important in relation to consumption by large vertebrates with wide gapes, or by small frugivores that mandi-bulate or chew up fruits to pieces. Fruits eaten by mammals tend to be larger than those eaten by birds (Janson 1983; Herrera 1989b). Among bird-dispersed plants, interspecific differences in fruit size explain differences in the species composition of dispersers, and the mean size of ingested fruits tends to be correlated with gape width among frugivorous birds (Wheelwright 1985b; Jordano 1987a).

Geographical patterns in fruit size are related to variation in dispersers' body size. Tropical forests include considerably larger fruits than temperate-zone forests and, within the tropics, Palaeotropical fruits tend to be larger than Neotropical ones (Mack 1993). These differences are related to the greater size range spanned by tropical frugivores relative to temperate-zone counterparts, and by the relative scarcity of large frugivores in the Neotropics. In western Europe, the mean fruit diameter of bird-dispersed plants is closely correlated across habitats with the mean gape width of disperser species (Fig. 7.8).

It is difficult to assess whether all these patterns actually reflect plant adaptations to variable disperser-size distributions. The correlation across habitats between the sizes of dispersers and fruits may mean that differences in fruit size have evolved to match variations in the local disperser-size distribution, but also that local disperser assemblages differing in size distributions are built up in response to regional variations in fruit size. Like other fruit traits, fruit size is correlated with plant phylogeny (Jordano 1995), and differences in the taxonomic

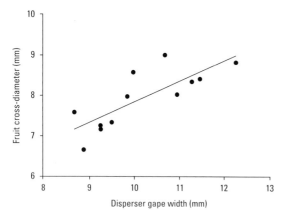

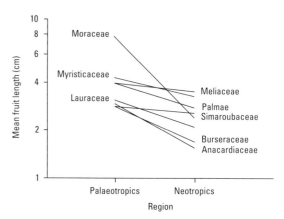

Figure 7.8 In western Europe, the average fruit cross-diameter of bird-dispersed plants is significantly related to, and slightly smaller than, the average gape width of local bird dispersers. Points in the graph represent average values for 12 localities from different habitat types. (Modified from Herrera 1985b.)

Figure 7.9 Palaeotropical fruits tend to be larger, on average, than Neotropical fruits, which is in accordance with the relative scarcity of large-sized vertebrate frugivores in the New World. As illustrated in this graph, the difference holds also within the 8 plant families represented in both the Palaeotropics and the Neotropics. The transcontinental difference thus most likely has some adaptive basis, and is not the exclusive consequence of phylogenetic correlates of fruit size and variation in the composition of fruiting plant assemblages. (From data in Mack 1993.)

composition of fruiting plants at different sites might account in part for patterns of variation in mean fruit size. An adaptive component does seem to exist, however, at least in the case of the fruit-size differences between the Neotropical and Palaeotropical fruits mentioned above, as geographical differences persist after accounting for phylogenetic correlations of fruit size (Fig. 7.9).

There is also unequivocal field and laboratory evidence indicating that fruit size may frequently be subject to selection by dispersers, at least among bird-dispersed species. Fruit choice, the handling mode and the foraging efficiency of frugivorous birds is strongly influenced by fruit size (Rey et al. 1997). As a consequence of size-based fruit preferences of frugivores, differences among individual plants in the size of fruits frequently result in differential seed-dispersal success, which provides a proximate mechanism for natural selection to operate on fruit size (Sallabanks 1993; Wheelwright 1993).

7.5.4 Pulp composition

7.5.4.1 Nutrients

Fruit pulp is the reward offered by plants to dispersers, and its nutritional value is a critical element in the plant–disperser interaction. Compared to other biological materials, fruit pulp is characterized, on average, by high water and carbohydrate content, and low protein and lipid content. There is, however, considerable interspecific variability in major nutrient composition, and this variability is in itself one distinctive feature of fruit pulp as a food resource for animals. In a set of 111 bird- and mammal-dispersed species from the Iberian Peninsula, lipid contents ranged between 0.2 and 59%, protein between 1 and 28%, and soluble carbohydrates between 26 and 93% of pulp dry mass (Fig. 7.10). Equivalent levels of variability have been found in all vertebrate-dispersed floras that have been studied so far from the viewpoint of the nutritional value of fruits (e.g. Wheelwright et al. 1984; French 1991). Fruit pulp also generally contains vitamins, carotenoids, amino acids and minerals that may play significant roles in the nutritional ecology of frugivores (Izhaki 1988; Jordano 1988). Little is known about the patterns of occurrence of these minor pulp constituents in wild fruits and their nutritional effects on consumers, but the fact that they occur at low concentrations does not necessarily mean that they are nutritionally irrelevant to frugivores. In cedar waxwings (*Bombycilla cedrorum*), for example, carotenoid pigments from ingested fruits are responsible for feather coloration (Witmer 1996b).

Earlier hypotheses on the evolution of plant–disperser interactions conferred great significance on the nutrition-

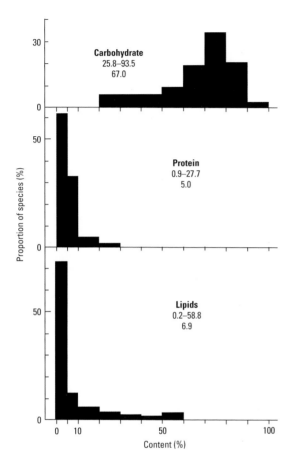

Figure 7.10 Frequency distributions of contents (as percentage of dry pulp mass) of non-structural carbohydrate, protein and lipids, in fruit pulp of vertebrate-dispersed plants from the Iberian Peninsula ($N = 111$ species). Ranges and means are shown for each constituent. (From Herrera 1987.)

al features of fruit pulp, as indicative of plants' adaptations to dispersers and the 'specialization' of the mutualism. Three main lines of evidence have subsequently de-emphasized the importance of fruit pulp quality as a trait reflecting plants' adaptations to dispersers. First, frugi-vores frequently do not discriminate on the basis of pulp nutritional quality, or their selection is inconsistent in time and/or space (Borowicz 1988; Whelan & Willson 1994). Second, there is often no relationship between the nutritional value of fruit pulp and seed-dispersal success, either within or among plant species (Herrera 1984a; Jordano 1989). And third, the nutritional composition of fruit pulp is strongly correlated with plant phylogeny, with a substantial part of interspecific variance in nutrient

concentration being attributable to taxonomic affiliation at the genus level and above (Herrera 1987; Jordano 1995). The nutritional characteristics of fruits seem relatively 'resistant' to evolutionary modification, as exemplified by the similarity in pulp characteristics of closely related species living in ecologically disparate scenarios and having their seeds dispersed by quite different agents (Fig. 7.11).

The nutritional characteristics of fruits are often related to the season of ripening, and seasonal variation sometimes matches dispersers' requirements. In Mediterranean habitats, for example, the mean water content of pulp is highest among species ripening during the dry summer, and lipid content is highest among winter-ripening ones, which may reflect plant adaptations to dispersers (Herrera 1995). Recent investigations, however, suggest that seasonality in fruit composition reflects the different fruiting phenologies of taxonomic groups that differ intrinsically in fruit composition (Eriksson & Ehrlén 1991; Herrera 1995). Similar explanations probably apply also to community-wide elevational and geographical patterns of variation in pulp composition (Herrera 1985b; French 1991). Given the close correlation between pulp composition and phylogeny, differences in the taxonomic composition of fruiting plant assemblages may explain much of the observed regional, elevational or seasonal variation in the average composition of fruits. For example, the greater average lipid content of the pulp of fruits in tropical habitats (Herrera 1981) may be explained by the greater representation there of species from plant families which produce characteristically lipid-rich fruits regardless of their geographical location. Further studies explicitly addressing the effects of phylogenetic correlations are still needed to elucidate the extent to which patterns of fruit nutritional composition actually reflect plant adaptations to dispersers (Jordano 1995; Eriksson & Ehrlén 1998).

7.5.4.2 Secondary metabolites

In addition to nutrients, fruits often contain secondary metabolites. Their concentration generally declines during ripening, but ripe fruits of many species still contain important amounts of phenolics, alkaloids, saponins, or cyanogenic glycosides in the pulp (Herrera 1982b). In some species, the concentration of secondary metabolites in ripe fruits may reach potentially lethal levels. Tropane alkaloids are so abundant in the fruits of deadly night-

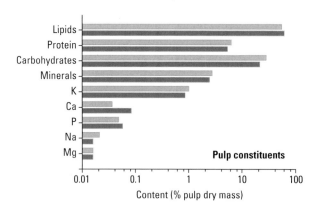 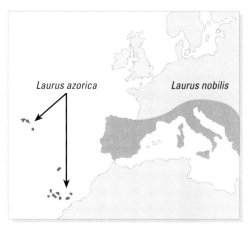

Figure 7.11 Nutritional composition of fruit pulp is closely correlated with phylogeny and may experience negligible changes over very long periods, as illustrated here by a comparison of the two extant species of *Laurus* (Lauraceae). The species derive from a common ancestor and have been geographically isolated since the Pliocene. They differ in morphological features, chromosome numbers, habitat type and seed-dispersers, but their fruits are virtually identical with respect to major pulp constituents. (Drawn from data in Herrera 1986.)

shade (*Atropa belladonna*) that their consumption often produces serious poisoning in humans and livestock. Given the mutualistic nature of the relationship between plants and their dispersers, the widespread occurrence of potentially toxic compounds in the pulp of ripe fruits is a biological paradox demanding some adaptive explanation. Secondary metabolites in the pulp of ripe fleshy fruits may inhibit seed germination, induce frugivores to leave fruiting plants early in a foraging bout, and modify the passage rate of seeds through dispersers' guts (Cipollini & Levey 1997a). Their primary function, however, is probably one of defence against microbial pathogens and invertebrate pests that may consume fruit pulp without dispersing seeds. Because of the negative incidence of such organisms in the dispersal process, the occurrence of secondary metabolites within ripe pulp presumably represents a tradeoff with respect to defence from damaging agents and palatability for dispersers. According to this hypothesis, fruiting plants that are at greater risk of attack by pests or pathogens (for example, because a low consumption rate by dispersers leads to prolonged exposure to damaging agents), should be under greater selection pressure for fruit defence than are plants with low risk of pest or pathogen attack (Herrera 1982b; Cipollini & Levey 1997a).

A number of studies provide unequivocal support for the 'palatability–defence tradeoff hypothesis'. In eastern North America, for example, the autumn-ripening, long-lasting fruits of *Vaccinium macrocarpon* are better defended against fungal fruit rot agents than the summer-ripening, quickly-removed fruits of *Vaccinium corymbosum* (Cipollini & Stiles 1993). On the other hand, fruit defence from pathogens has a measurable cost to plants in terms of reduced seed-dispersal prospects, because defensive chemical compounds in fruit pulp act to reduce the acceptability of fruits to dispersers (Cipollini & Levey 1997b; Levey & Cipollini 1998). There remains much still to be learned about the identity, ecological distribution and evolutionary significance of secondary metabolites in ripe fruits. As aptly stressed by Cipollini and Levey (1997a), secondary metabolites may eventually prove more important than other fruit attributes in understanding patterns of fruit use by dispersers.

7.5.5 Fruit 'syndromes'

Some particular combinations of fruit traits involving, for example, colour, smell, size and type of presentation, occur disproportionately more frequently in nature than other combinations. Such combinations of fruit traits have sometimes been interpeted as defining so-called 'fruit syndromes', and it often happens that suites of correlated fruit characters are related to consumption by particular groups of dispersers. In a Peruvian tropical

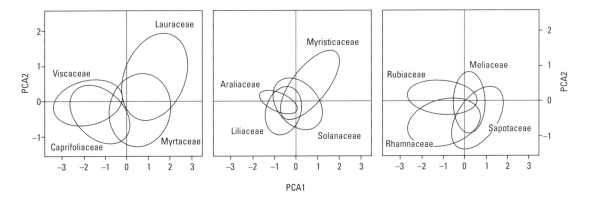

Figure 7.12 Angiosperm families inherently differ in the characteristics of their fleshy fruits. This graph depicts the relative location of a number of families on the plane defined by the first two principal components of fleshy fruit characteristics (PCA1 and PCA2). Each family is represented by its 90% equal frequency ellipse, that is, the contour line encompassing 90% of species for each family. The two principal components summarize variation in fruit mensural (e.g. size, mass), structural (e.g. number of seeds, pulp/seed ratio) and nutritional (e.g. lipids, protein) characteristics. (Modified from Jordano 1995.)

rainforest, for example, most fruits belong to one of two classes: large, dull-coloured fruits with a protective husk, or small brightly-coloured fruits without a husk. The characteristics of these two fruit classes match the size and visual ability of mammals and birds respectively, and the animals also prefer to eat one class of fruits (Janson 1983). In Mediterranean forests and scrublands, bright fruit colours tend to be associated with small size, lack of a perceptible smell and the persistence of fruits on the plants after ripening. Large fruits tend to be dull in colour, scented to the human nose, and to fall to the ground after ripening. Fruits in the former group are eaten exclusively by birds, while those in the second group are eaten by both birds and terrestrial mammals (Herrera 1989b). Similar non-random combinations of fruit traits into *statistically* distinguishable suites, and the frequent association of such suites with particular groups of fruit consumers, have been reported from other plant and frugivore assemblages from all over the world.

Do fruit 'syndromes', and their association with major groups of frugivorous vertebrates, actually reflect plant adaptations to different kinds of dispersers? This hypothesis was first critically examined by Fischer and Chapman (1993), who found that fruit character complexes are rare in nature, and that the results of analyses of covariation among fruit characters are extremely sensitive to the investigator's choice of sampling unit. In their study, no significant trait associations existed when plant genera were

used as sampling units, but they did occur when species were the units chosen. Jordano (1995), in a thorough investigation, found that dispersal syndromes are only minimally attributable to plant adaptations to dispersers, but rather they largely reflect the great influence of plant phylogeny on fruit traits (Fig. 7.12). After plant phylogeny is accounted for, there is a conspicuous lack of evolutionary correlation between the seed-dispersal agent (bird, mammal or mixed dispersal) and the vast majority of mensural, structural and nutritional fruit traits. Of all fruit traits considered by Jordano, only fruit dimensions were found to be significantly related to dispersal type after accounting for phylogenetic effects, and he concluded that 'correlated evolution [of fruit traits] with type of seed disperser is, at best, only evident for fruit diameter'.

7.6 Animal adaptations

It has been sometimes suggested that, compared with other food habits like herbivory, carnivory or nectarivory, frugivory does not require drastic morphological and physiological modifications on the part of animals (Fleming 1991). This certainly holds true for occasional frugivory, as it is a common observation that most terrestrial vertebrates are able to ingest some fruits sporadically. Nevertheless, fruits represent a substantial fraction of the diet for only a relatively small subset of terrestrial vertebrates. These 'heavy frugivores' are those most directly

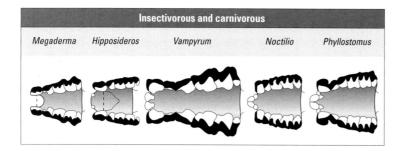

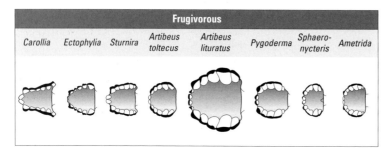

Figure 7.13 Frugivorous animals often differ from non-frugivorous relatives in the morphology of their trophic structures. Shown in this figure are ventral views of palates of insectivorous and carnivorous (upper row) and frugivorous (lower row) New World bats, drawn to the same scale. The two groups of species differ in craneal morphology, with frugivorous species having proportionally shorter and wider palates than non-frugivorous ones. (From Freeman 1988.)

relevant from the viewpoint of the evolution of plant–disperser interactions, and will be the ones considered in this section. Extensive frugivory is made possible in these species by a distinct suite of morphological, physiological and behavioural adaptations that enable them to exploit efficiently a food resource characterized by its poor food value, strong nutritional imbalance, extremely variable chemical composition and marked unpredictability in time and space.

7.6.1 External and internal morphology

Highly frugivorous passerine birds tend to have shorter, broader and flatter bills, and wider gapes, than do those that never eat fruit or do so only infrequently (Herrera 1984b). Distinct morphological trends also exist in frugivorous mammals. Among New World bats, fruit-eating species are characterized by short canines and palates that are broader than they are long, in comparison to the longer canines and palates longer than they are wide of insectivorous and carnivorous species (Fig. 7.13). The smaller tooth area of frugivorous bats is also related to their diet of juicy, soft fruit. Unlike carnivorous species, which have enlarged lower molars that occlude with the upper teeth, the premolars of frugivorous bats engage before the molars, which points to the importance of anteri-

or dentition (incisors, canines and premolars) in biting through the skin of fruits (Freeman 1988). In the case of frugivorous bats, these morphological patterns most likely reflect adaptations actually evolved to exploit fruits more efficiently. In the case of some frugivorous passerines, however, differential bill morphology probably represents a pre-adaptation enabling them to become efficient frugivores, rather than a direct adaptation to frugivory (Herrera 1984b).

The digestive system of some frugivorous birds has noticeable peculiarities that were already noted by zoologists nearly a century ago. In some species that feed heavily on fruits for most or all of the year, the gizzard tends to be smaller and less muscular, the liver larger, and the intestine shorter than in less frugivorous relatives (Pulliainen et al. 1981; Richardson & Wooller 1988). Shortening of the intestine allows for the rapid processing of fruits, while the reduction of gizzard musculature is a response to a reduced need for crushing hard food. Increased liver size may be a response to an increased demand for detoxifying capacity, derived from the frequent presence of toxins in ripe fruits noted above (Herrera 1984b). The simplification of the digestive tract reaches its extreme in the case of mistletoe birds (Dicaeidae) and other passerines feeding extensively on mistletoe fruits (Fig. 7.14). Simplification of the digestive tract, however, is only one of the array of

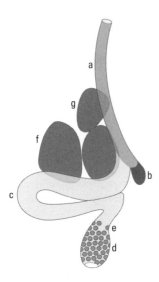

Figure 7.14 Birds feeding heavily on fruits throughout the year tend to have reduced gizzards and shorter intestines. This reduction reaches its extreme in mistletoe birds (Dicaeidae) and other passerines feeding extensively on mistletoe berries. The graph shows a ventral view of the alimentary tract of the Australian mistletoe bird (*Dicaeum hirundinaceum*) lying in its natural position (a, oesophagus; b, gizzard; c, small intestine; d, large intestine; e, caecum; f, liver; g, heart; scale line is 7.5 mm). (From Richardson & Wooller 1988.)

importance of increased intestine length for frugivory is further exemplified by the fact that, among birds, diets differing in the amount of fruit may induce changes in intestine length. In the heavily frugivorous bulbul, *Pycnonotus leucogenys*, intestine length fluctuates seasonally, being longer at times of year when the birds are most strongly frugivorous (Al-Dabbagh et al. 1987). In pine warblers (*Dendroica pinus*) experimentally fed on either insect-based, fruit-based or seed-based diets for nearly two months, intestine length and mass at the end of the experimental period were greatest in the group of experimental birds that had been feeding on the fruit-based diet (Levey et al. 1999).

Gross modifications of the digestive tract are not apparent in species that feed heavily on fruits during only part of the year, like migratory frugivorous birds from temperate latitudes (Herrera 1984b). In these cases, adaptations favouring sustained frugivory during extended periods consist of more subtle functional modifications of the digestive system, like those described in the next section.

7.6.2 Digestive physiology

As noted above, the rewarding portion of fruits is generally characterized by high-water, high-sugar and very low-protein content. In addition, the nutritional reward obtainable by frugivores is further diluted by the presence of indigestible seeds and, often, secondary metabolites in fruit pulp. Digestive adaptations to cope with nutrient dilution, nutritional imbalance and secondary compounds are thus essential for sustained frugivory.

7.6.2.1 Coping with nutrient dilution

Seeds occupy space in the digestive tract that could otherwise accomodate nutritious material. Selection has thus favoured digestive adaptations that overcome this 'gut limitation' and act to rapidly evacuate ingested seeds. In manakins, small seeds pass through the gut in only 12–15 min, while larger seeds are regurgitated within 7–9 min of ingestion (Worthington 1989). Among southern Spanish frugivorous passerines, species that rely heavily on fruits for food have shorter gut-retention times relative to body mass than occasional frugivores (Fig. 7.15), and this ability for rapidly processing fruits seems essential for their extensive and sustained frugivory (Jordano 1987a). Avian frugivores have the ability to modulate retention times in response to variations in diet composition. In

'macroscopical' digestive adaptations exhibited by frugivorous animals. In fruit pigeons of the Australasian region, for example, a distinct arrangement of connective tissue and muscle layers leads to increased elasticity of the oesophagus and glandular stomach, which is clearly advantageous in allowing larger fruits to pass through (Landolt 1987).

Shortening of the intestine, although frequently emphasized in older accounts of vertebrate frugivory, is probably not the most common adaptation evolved by frugivores for the rapid processing of many fruits. As described in Section 7.6.2 below, frugivores have evolved a variety of ways of shortening food-retention time without modification of the length of the digestive tract. In fact, increased frugivory is quite often associated with increased intestine length, as an adaptive response for increasing intestinal absorption of the water-diluted nutrients in fruit juice. Old World fruit bats (Pteropodidae) have relatively longer intestines than their insectivorous counterparts, and the intestine may be up to nine times their body length (Kunz & Ingalls 1994). The

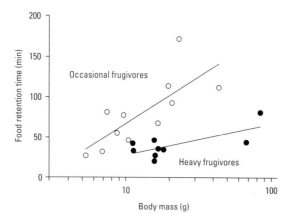

Figure 7.15 For a given body mass, food passes more rapidly through the digestive tract of 'strong' than of 'occasional' frugivores in species of southern Spanish frugivorous birds. Occasional frugivores are species feeding infrequently on fruits that do not usually perform seed dispersal, while strong frugivores are species depending heavily on fruits for food for extended periods and behaving as legitimate seed dispersers. (From Herrera 1984b.)

yellow-rumped warblers (*Dendroica coronata*) acclimatized to different diets, retention times decrease from seed-through insect- to fruit-based diets (Afik & Karasov 1995). Rapid digestive processing is not exempt from costs, however, and a tradeoff may arise between the rate and efficiency of digestive processing. The faster food is processed, the lower the nutrient extraction efficiency, and yellow-rumped warblers on diets associated with shorter retention times extracted less of the nutrients (Afik & Karasov 1995).

The ability of frugivores to adjust enzymatic activity in the gut as a function of the composition of the diet is probably one further adaptive trait allowing them to feed on relatively diluted food. Among the experimental pine warblers (*Dendroica pinus*) mentioned above that were fed on insect-, fruit- and seed-based diets, birds fed on fruit (i.e. carbohydrate-rich food) exhibited the highest intestinal activity of enzymes related to the transformation and assimilation of carbohydrates (amylase, maltase and sucrase) (Levey et al. 1999).

7.6.2.2 Coping with nutritional imbalance and secondary compounds

Most fruits are very deficient in nitrogen, which perhaps represents the most important nutritional constraint that frugivorous animals must cope with. Regular ingestion of small amounts of animal food seems to be the commonest way of complementing the poor protein intake associated with frugivory. Old and New World frugivorous bats regularly supplement their diets by hunting insects even when fruits are abundantly available (Courts 1998), and heavily frugivorous tropical birds like toucans and quetzals frequently include insects and small vertebrates in their diets (Wheelwright 1983; Remsen et al. 1993). An intrinsically low protein requirement is one further trait that facilitates the subsistence of frugivores on a diet of sugar-dominated fruits, as found by Worthington (1989) for manakins and Witmer (1998) for cedar waxwings (*Bombycilla cedrorum*). In manakins, a short-term positive nitrogen balance was achieved at very low nitrogen concentrations in a fruit diet, only about 1.3% N in dry pulp.

With very few exceptions, individuals of most species of frugivores usually ingest fruit of several species over very short periods of time. The vast majority of faecal samples of frugivorous passerines contain remains of more than one fruit species, sometimes up to 8 species (Herrera 1984a; Loiselle 1990). Given the short seed-retention times of these species, this indicates that the birds had been feeding on that many species during the short time immediately preceding capture. Varied fruit diets probably exemplify an adaptive response to overcome the nutritional imbalance of fruits and the frequent presence of secondary metabolites (Jordano 1988). By feeding simultaneously on fruits of different species with contrasting nutrient compositions, frugivores may achieve nutritional complementarity and a more balanced diet than if they relied on the fruits of a single species. Furthermore, when fruits that contain potentially harmful secondary metabolites are consumed in mixtures, the toxicity experienced by a forager due to any particular diet item can be ameliorated or diluted. This may explain the frequent observation that captive frugivorous birds fed on single-fruit diets quickly lose weight and their body condition deteriorates (Izhaki & Safriel 1989). In a set of species of frugivorous birds from southern Spanish scrublands, the extent of frugivory in the different species was positively correlated with the short-term diversity of the fruit mixtures consumed, suggesting that sustained frugivory is strongly dependent on the capacity to construct varied diets in the short term (Herrera 1984a). By means of a series of elegant field experiments, Whelan et al. (1998) have convinc-

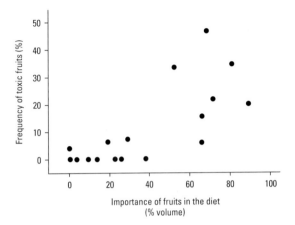

Figure 7.16 In Mediterranean scrublands, frugivorous birds that feed only occasionally on fruits tend to avoid completely species whose ripe fruits contain potentially toxic metabolites. The proportion of toxic fruit species in the diet rises disproportionately as the degree of frugivory increases. Points in the graph represent average values for 17 passerine species from three different habitats. (Plotted from data in Herrera 1985c.)

ingly demonstrated that North American migrant frugivorous birds effectively treat different fruit species as complementary resources.

Extensive frugivory is frequently associated with the capacity to exploit fruits containing secondary metabolites in the pulp, including poisonous substances or potentially harmful digestion inhibitors (Fig. 7.16). This ability of many frugivores to exploit poisonous fruits can be explained only by the greater detoxifying capacity of their guts. There is evidence suggesting that heavily frugivorous birds tolerate potentially harmful compounds found in fruits better than occasional frugivores or non-frugivores. In northern Europe, the heavily frugivorous waxwings (*Bombycilla garrulus*) metabolize ethanol, which frequently occurs in over-ripe berries, at a much faster rate ($900 \, mg \, kg^{-1} \, h^{-1}$) than either occasional frugivores like starlings (*Sturnus vulgaris*, $270 \, mg \, kg^{-1} \, h^{-1}$) or granivores like greenfinches (*Carduelis chloris*, $130 \, mg \, kg^{-1} \, h^{-1}$) (Eriksson & Nummi 1982). Frugivorous birds also seem to have better detoxification abilities than frugivorous or omnivorous mammals. The lethal dose of the alkaloid atropine, found in the berries of *Atropa belladonna*, is nearly a thousand times higher for European blackbirds (*Turdus merula*) than for humans (Seuter, 1970).

7.6.3 Behaviour

Fruits are probably more unevenly distributed in time and space than other kinds of food exploited by animals. Abundance of fruits varies markedly among years and seasons, and within as well as between habitats, which generally leads to patchy and unpredictable distributions in time and space (Levey 1988; Herrera 1998). A distinct suite of behavioural and physiological traits allow frugivores to withstand or escape from temporary situations of fruit scarcity and efficiently locate unpredictable fruit sources.

Seasonal migration and habitat shifts are the two most generalized responses of frugivores to fluctuations in fruit availability. Resplendent quetzals (*Pharomacrus moccino*) resident in Costa Rican cloud forest sequentially occupy different habitat types while closely tracking the local abundance of their highly preferred lauraceous fruits (Wheelwright 1983), and Australian fruit pigeons regularly undertake migrations across lowland rainforest in response to local variations in fruit supply (Crome 1975). The nomadic behaviour frequently exhibited by frugivorous birds, in the tropics and elsewhere, is a further behavioural trait that allows them to discover and exploit unpredictable patches of locally abundant fruits. In southern Spain, overwintering populations of blackcap warblers (*Sylvia atricapilla*) closely track fruit over large-scale mosaics of fruit abundance (Rey 1995). In African and Asian tropical forests, species of hornbills (Bucerotidae) wander over quite large areas while closely following broad-scale geographical variations in fruit availability (Kinnaird et al. 1996).

Among less mobile vertebrates, the ability to shift diet in response to occasional fruit scarcity is essential for long-term frugivory. Frugivorous primates in Amazonian tropical forest turn to feed more on nectar, leaves, insects or seeds at times when fruit become less abundant (van Schaik et al. 1993). Adjusting the time of reproduction to periods of high fruit abundance is another adaptive response of frugivores to cope with the marked temporal variability of their food supply. Birth peaks in New World frugivorous primates coincide with fruiting peaks, as do the breeding seasons of many tropical frugivorous birds and bats (Fleming 1992). In a Costa Rican cloud forest, the timing of lactation in frugivorous bats closely matches peaks in local fruit abundance (Fig. 7.17).

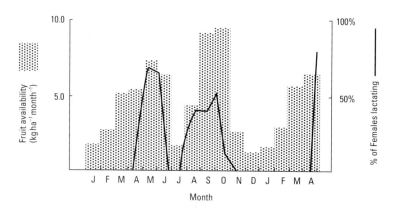

Figure 7.17 Some frugivores adjust the time of reproduction to periods of high fruit availability. In a Costa Rican tropical cloud forest, the two annual lactation periods of the frugivorous bat *Artibeus toltecus* are closely coincident with local peaks in fruit availability. (From Dinerstein 1986.)

7.7 Patterns of mutual dependence

There are no known instances of obligate plant–disperser mutualisms. For example, no animal-dispersed plant is known to depend strictly on dispersal agents for the germination of their seeds and the perpetuation of their populations. Seeds of animal-dispersed plants often germinate more easily after passing through the gut of their dispersal agents, and this observation has sometimes been construed as indicative of the dependence of plants on their dispersal agents. Nevertheless, among more than 200 species dispersed by birds, mammals and reptiles that have been tested so far, passage through the dispersers' digestive tracts had some significant effect on seed germination in only 50% of instances, and in these cases the effect could be either enhancement or inhibition (Traveset 1998). The earlier intriguing hypothesis that the endemic tree *Calvaria major* was nearly extinct in the island of Mauritius because its seeds required a passage through the digestive tract of the now extinct dodo (*Raphus cucullatus*) has been dismissed by subsequent investigations (Witmer & Cheke 1991).

Plant–vertebrate seed-dispersal systems are characterized not only by the absence of obligate partnerships, but also by weak mutual dependence between species of plants and animals, and by the prevalence of unspecific relationships. Most species of animal-dispersed plants depend on arrays of frugivorous species for dispersing their seeds, arrays often composed by species belonging to different major groups. Each species of frugivore, in turn, consumes the fruits and disperses the seeds of a number of taxonomically unrelated plant species. In a community

of eight diurnal primates in the Lopé Reserve in Gabon, individual species feed on the fruits of between 20 (black colobus, *Colobus satanas*) and 114 (chimpanzee, *Pan troglodytes*) different plant species (Tutin et al. 1997). In a Costa Rican lower montane forest, each species of frugivorous bird feeds on an average of 10.1 fruit species, while each plant species has its fruits eaten by an average of 4.5 bird species, and these figures are most likely rough underestimates (Wheelwright et al. 1984). Furthermore, the vast majority of pair-wise interactions between plant and disperser species correspond to situations in which the relative dependence of the plant on the disperser, and of the disperser on the plant, are both minimal (Fig. 7.18). A few examples are known of plant species that apparently depend on only one or perhaps a few species of animal dispersers. In tropical Africa, for example, elephants seem to be the only current dispersers of the large-seeded tree *Balanites wilsoniana*, and European robins (*Erithacus rubecula*) are the almost exclusive dispersers of *Viburnum tinus* in the western Mediterranean Basin (Chapman et al. 1992; Herrera 1995). In these and other situations of heavily dependent plants, however, the relationship is quite asymmetrical, since the single species of animal disperser of a given plant acts as the disperser of many other plants as well (Jordano 1987b).

Mutual specificity is significantly lower in plant–disperser than in plant–pollinator systems. For a habitat with *m* species of animal-dispersed plants and *n* species of frugivorous animals, the 'connectance' of the mutualistic system may be defined as the number of plant–animal pair-wise interactions actually observed divided by the maximum possible (i.e. $m \times n$). Thus defined, lower

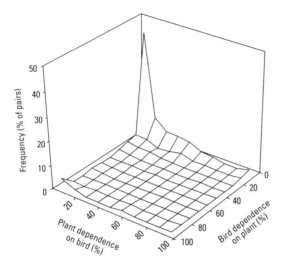

Figure 7.18 In plant–bird seed-dispersal systems, most plant–bird species pairs exemplify situations in which the dependence of the bird on the plant, and of the plant on the bird, are both negligible. For each bird–plant species pair, bird dependence on the plant was measured by the proportion of fruit food contributed by the plant, and plant dependence on the bird was assessed using the proportion of seeds dispersed. (Drawn from data in Jordano 1987b.)

connectance denotes higher average specificity, and vice versa. Jordano (1987b) found that mean specificity in both plant–pollinator and plant–disperser system increased with the increasing number of species involved, but that the increase in specificity was steeper in plant–pollinator systems. This indicates that, for any given number of plants and animals involved, the mutual specificity between plants and animals is smaller in seed dispersal than in pollination systems.

7.8 Factors hindering reciprocal specialization

The preceding sections have shown that, compared with other kinds of plant–animal interactions covered in this book, plant–vertebrate seed-dispersal systems are those for which the evidence of reciprocal specialization — or mutual adaptation — between animal and plant counterparts is the weakest. No examples of obligate partnerships exist, and the general picture is one of loose interdependence between species of plants and species of dispersers. This contrasts with earlier models and formulations of the evolution of plant–disperser systems, which were characterized by expectations of coevolution and mutual adap-

tations. Plant–disperser interactions have been thoroughly studied during the last two decades, so the notion of weak reciprocal adaptation is unlikely to stem from imperfect knowledge and is not expected to be modified substantially in future as more and more detailed studies accumulate. What, then, are the ecological and evolutionary peculiarities of animal seed dispersal that have limited the extent of mutual specialization between plants and dispersers?

7.8.1 The genetic consequences of animal seed dispersal

The main genetic consequence associated with seed dispersal by frugivorous animals appears to be preventing, or greatly reducing, the local genetic differentiation of plant populations. A small-scale, local genetic structure was traditionally hypothesized to be particularly common in plants because they are immobile as adults, and gene dispersal via pollen and seeds was often thought to be spatially restricted. Nevertheless, the distribution of dispersal distances of animal-dispersed plants frequently has very long tails. Although the vast majority of seeds are dispersed over short to medium distances, a small but ecologically and evolutionarily highly significant number of seeds are transported to very long distances (Portnoy & Willson 1993). Enhanced gene flow and increased genetic homogeneity of populations are thus expected under these circumstances, with local genetic structuring being less intense and/or frequent among animal-dispersed plants. This expectation has been confirmed by most studies that have investigated the spatial structure of vertebrate-dispersed plants using genetic markers. In a Florida population of the bird-dispersed shrub *Psychotria nervosa*, Dewey and Heywood (1988) found no evidence of spatial structuring of allele frequencies for either of the polymorphic enzyme loci studied. In the bird- and mammal-dispersed pioneer tree *Cecropia obtusifolia*, Alvarez-Buylla and Garay (1994) found high levels of gene flow and negligible population differentiation in a Mexican rainforest.

The spatial distribution of genetic variability within populations may influence significantly the evolutionary dynamics of the population, and population structure may provide the substrate for local selective forces to promote adaptive evolution, and may be a first step towards speciation. If animal seed dispersal generally enhances gene flow and persistently smoothes out spatial genetic

Table 7.1 Some of the most significant differences between seed and pollen dispersal by animals. (Based on Wheelwright & Orians 1982.)

	Pollen dispersal	Seed dispersal
Suitable site for arrival ('target')	Stigma of conspecific flower	Site appropriate for germination and establishment
Characteristics of target	Distinctive (colour, shape) and spatially predictable, often apparent at a distance	Spatially unpredictable, many subtle factors involved
Temporal availability of target	Synchronous with pollen dispersal	Unpredictable, often independent of habitat type or phenology of conspecific plants
Ability of plant to direct animal vectors to target	High, incentives (nectar, pollen) provided at target site	Low, no incentive to disperser at target site

patterns of plant populations at large spatial scales, then animal-dispersed plants will be characterized by a low probability of local or regional differentiation, including the evolution of adaptations to local dispersal agents.

7.8.2 Factors limiting plant specialization

Earlier expectations of mutual adaptation and coevolution between fleshy-fruited plants and their vertebrate dispersers were in part motivated by the superficial resemblance between pollen and seed dispersal by animals. As mutual adaptations between animal-pollinated plants and their pollinators had been known since Darwin's time, concepts from the well-established field of pollination biology were uncritically exported to the younger field of plant–disperser interactions (Wheelwright & Orians 1982). Nevertheless, pollen and seed dispersal by animals are fundamentally dissimilar (Table 7.1), and their differences have manifold evolutionary implications. The two most important distinctions are (i) that a definite target exists for dispersing pollen grains (the conspecific stigma) but not for dispersing seeds; and (ii) that the plant can control pollinators' movements by providing incentives at the target site (nectar, pollen), but there are no similar incentives for seed dispersers to drop seeds in appropriate places. These differences are best framed in terms of the departure-related versus arrival-related advantages of dispersal described in Section 7.2 earlier in this chapter.

In pollen dispersal, the existence of a definite target for pollen grains has made possible the evolution of 'payment upon delivery'. This, in turn, has promoted the evolution of complex morphological and functional floral traits

'manipulating' the behaviour of pollinators and enhancing the arrival-related component of pollen dispersal. In contrast, plants do not have a definite 'target' for their seeds, because favourable germination sites frequently do not exist when seeds are dispersed, and plants can hardly 'know' in advance where and when will they appear. Arrival-related advantages are unpredictable in time and space, and thus largely out of the control of the parent plant and hardly susceptible to natural selection. This has constrained vertebrate seed-dispersal systems to function on the basis of 'advance payment' alone. Advance payment has hampered the evolution of traits enhancing arrival-related benefits of dispersal (i.e. the quality of the germination microsites reached by dispersed seeds). But, on the other hand, the departure-related advantages of dispersal are quite predictable and straightforward, and under the control of the parent plant, which has strongly favoured the evolution of traits enhancing departure-related benefits (i.e. consumption by seed-dispersers) like fruit conspicuousness and nutritional reward. Departure-enhancing traits are easy to evolve, as shown by the wide distribution of the fleshy-fruit habit in the phylogeny of seed plants, and the morphological convergence of anatomically disparate structures into functional fleshy fruits, as noted earlier in this chapter.

There are innumerable options available to plants to restrict the range of animals visiting their flowers and to evolve adaptations to particular pollinators, as exemplified by the amazing diversity of flower colours, internal and external morphologies, and the amount and type of reward evolved by animal-pollinated angiosperms. By contrast, there are only a few rather coarse ways available to plants to restrict the range of vertebrate species that feed on their fruits, and thus few morphological options for the diversification of fleshy fruits. Fruit structure, size and colour, and the chemical composition of the pulp, all play proximate roles in restricting the assemblage of frugivores that interact with any given plant species, but only up to a certain threshold. Beyond this, the nature of the reward on which the plant–disperser interaction is based precludes further filtering based on morphological modifications and, consequently, sets limits on the possibilities of plants to specialize on the most effective or beneficial disperser(s) (Herrera 1985a; Jordano 1987b). That there are few morphological opportunities for evolutionary diversification inherently associated with animal seed dispersal is exemplified by the low frequency with which seed and fruit traits are useful to taxonomists to differentiate higher-level plant taxa. In a dichotomous key to the European flora, for example, fruit-related traits are used in only 2.9% and 21.3% of dichotomies that differentiate plant families and genera respectively. Flower-related characters, in contrast, are used in 63.4% and 54.4% of familial and generic dichotomies respectively.

7.8.3 Factors limiting frugivore specialization

Temporal inconsistency—in both the short and the long term—in the patterns of plant–disperser relationships is one important factor limiting the possibilities of frugivore specialization on plants. Although there are some notorious examples of vertebrate-dispersed plants that ripen fruit over the whole year, like some tropical fig trees or the Mediterranean hemiparasitic shrub *Osyris quadripartita* (Milton et al. 1982; Herrera 1988b), the vast majority of species have distinct fruiting seasons. Vertebrate frugivores are relatively long-lived, so the fruits of most species they feed on represent ephemeral resources that are available only during relatively short periods of the annual cycle. Specialization on the fruits of particular species is hardly possible under these circumstances.

Long-term inconstancy in the species composition of the fruit supply further contributes to limit the specialization of frugivores on particular plant species. Crop sizes of vertebrate-dispersed plants tend to fluctuate less among years than those of species with other seed-dispersal methods, but important annual fluctuations commonly occur in most species (Herrera et al. 1998). Few long-term studies have investigated in sufficient detail the consequences of this variation, but the limited evidence available suggests that plant–vertebrate seed-disperser

interactions are representative of non-equilibrial ecological systems (DeAngelis & Waterhouse 1987). A 12-yr study of the interaction between bird-dispersed plants and their main dispersers in a south-eastern Spanish Mediterranean montane scrubland has revealed loose patterns of mutual interdependence among species of plants and birds, important temporal fluctuations, the predominance of abiotic over biotic driving factors, and virtually complete decoupling of the temporal dynamics of the plant and bird species (Herrera 1998). Non-equilibrial relationships between plants and dispersal agents will lead to temporal inconsistencies in selection pressures associated with the interaction, which will generally operate against mutual adaptations of interacting species. In the 12-yr study mentioned earlier, annual variation in the species composition of the fruit diet of the two major bird dispersers (*Erithacus rubecula* and *Sylvia atricapilla*) involved drastic changes in the identity of the plant species that predominated in their diets. Similar long-term variability in the diet composition of frugivores has invariably been found whenever studies have encompassed a sufficient number of years (e.g. Loiselle & Blake 1994).

7.9 Pending issues

This chapter has summarized the main topics emerging from the numerous studies that have addressed plant–disperser interactions from an evolutionary viewpoint in the last 25 years. There are, however, a number of equally relevant topics that have not been covered because available information is scarce, inconclusive or simply non-existent. Some of these missing aspects are critical to a proper understanding of the evolution of plant–disperser interactions. It is worth highlighting some of these here, to provide both a rough guide to pending issues in the field and a list of possibilities available for the interested student.

Despite efforts to achieve a balanced treatment, the picture presented in this chapter mainly reflects current knowledge of vertebrate frugivory and seed dispersal as delineated by studies on frugivorous birds and bird-dispersed plants. Birds are doubtless prominent frugivores that deserve considerable attention, but their predominance in this chapter reflects in part the distinct 'avian bias' that has characterized recent studies of the evolutionary ecology of frugivory and seed dispersal. More studies are needed, contributing to a broader and more balanced picture of the ecology and evolution of frugivory and seed dispersal by non-avian vertebrates, particularly mammals and reptiles. Little is known, for example, on the quantitative role played by terrestrial 'carnivorous' mammals (e.g. mustelids, canids) as seed-dispersal agents in those relatively undisturbed habitats where they have not been extirpated, and on the digestive and behavioural adaptations that allow them to shift seasonally between carnivorous and frugivorous diets. Given the tremendous differences in mobility, trophic apparatus and sensory abilities among major groups of vertebrates, it seems safe to predict that a picture of plant–disperser interactions based on a more balanced knowledge of different groups of dispersal agents will be considerably richer than the one presented here.

The genetic consequences of seed dispersal by vertebrates are still very poorly understood. For example, comparative studies are needed to address the effect on the genetic structure of plant populations of seed dispersal by animals that differ in mobility and foraging patterns, like birds, bats and non-flying mammals. Dissecting the relative contribution of seed and pollen dispersal to the genetic structuring of plant populations may also help us to understand some of the evolutionary contrasts between plant–pollinator and plant–disperser systems outlined above.

Methodological problems have so far precluded attempts at connecting the departure and arrival stages of seed dispersal for maternal progenies. Much is known on the factors influencing the departure-related component of dispersal success, but virtually nothing is known on whether maternal fruit and fruiting traits translate into differential arrival-related success of vertebrate-dispersed seeds. The development of methods allowing the tracking of seed crops from the parent plant through the arrival sites has perhaps been one of the most vexing issues in the study of plant–disperser interactions. Such methods would allow, for instance, assessing the shape of the distribution of seed-dispersal distances from parent plants, particularly the reach of dispersal tails. Theoretical models predict that spatial genetic patterns will be smoothed out by gene flow at a rate which depends only on the variance of the distance between parent and offspring. The genetic consequences of seed dispersal will thus depend more closely on the variance of dispersal distances than on the modal or average distances travelled by seeds. Such variance cannot be estimated unless effective methods are developed for constructing whole distributions of dispersal distances.

Very few studies so far have examined patterns of spatial and temporal variation of the mutual relationships between species of animal-dispersed plants and their vertebrate counterparts. These investigations suggest, for example, that there is little spatial congruence between the geographical ranges of plants and the vertebrates with which they interact for seed dispersal (Jordano 1993), and that the long-term dynamics of plants and frugivores may run independently of each other (Herrera 1998). Patterns of spatial and temporal variation in plant–disperser interactions, in combination with information on the genetic structure of plant populations, are essential to evaluate mutual adaptations between plants and seed-dispersers at local or regional scales in the context of the recent 'geographic mosaic' theory of coevolution (Thompson 1994).

Part 4 Synthesis

Table 8.1 Ant species richness in different regions of the world, modified from Folgarait (1998). (Data from Hölldobler & Wilson (1990), Groombridge (1992) and Shattuck (1999).)

Region	Number of species
Africa (sub-Saharan)	2500
West Indies, Mexico, Central and South America	2233
Asia (parts of this region)	2080
Australia	1275
North America, north of Mexico	585
Europe	429
USA	400
New Guinea, New Britain and New Ireland	275
Polynesia	42
New Zealand	23

antifungal substances, allowing successful colonization of the moist, microbe-rich environment where the majority of ant species live. We will return to the theme of why ants have become plant partners *par excellence* at the end of the chapter.

In this chapter we focus mainly on research published in the last decade as prior to this there have been several extensive reviews of ant–plant interactions (Buckley 1982; Hölldobler & Wilson 1990; Huxley & Cutler 1991; Beattie 1985). We also draw readers' attention to more recent reviews of the evolutionary ecology of ant–plant symbioses (Davidson & McKey 1993) and the role of ants in ecosystem processes (Folgarait 1998). We begin with interactions that probably represent the most highly co-evolved and specialized, and proceed to those that are the most widespread and general.

8.2 Leaf-cutter ants

Processions of ants holding pieces of leaves above their heads like umbrellas are conspicuous sights in the wet forests and savannahs of the Neotropics. These ants are members of the tribe Attini (within the subfamily Myrmicinae) and they harvest leaves not directly for their own consumption, but to feed their underground fungus gardens. This ancient and highly evolved mutualism between fungus-growing ants and their fungi has become a model system for the study of symbiosis.

The Attini comprise 12 genera and approximately 200 species distributed throughout the Neotropics, where they are the dominant ant species in most habitats. The tribe has been divided into two main groups. Seven genera, including the primitive *Myrmicocrypta* and *Cyphomyrmex* comprise the 'lower attines' which are mostly inconspicuous, frequently cryptic species. These ants do not attack plants, growing their fungi instead on a diverse range of substrates including insect frass, fruits and seeds. The remaining five genera are grouped into the monophyletic 'higher attines', which comprise about half the species diversity of the tribe. Although these genera are commonly considered to be characteristic of tropical forests, they are actually more abundant and diverse in the subtropics of South America (Fig. 8.1).

It is the higher attines, especially the conspicuous genera *Atta* (15 species) and *Acromyrmex* (24 species), that have attracted the most attention. Colonies of *Atta* species may excavate nests up to 6 m below ground and contain several million ants with the collective biomass and appetite of a cow. A single nest may contain a thousand fungus gardens. Leaf-cutters are the primary consumers within New World terrestrial ecosystems, and their impact on vegetation is greater than that of any other herbivore taxon. The range of plant species collected is partly determined by the floristic diversity of the habitat. Attines that live in savannahs harvest a relatively narrow range of grasses whereas those in rainforest exploit a wide range of tree species; *Atta cephalotes*, for example, was found to exploit 30–50% of accessible plants in a Guyanese rainforest (Cherrett 1968). Selective herbivory, especially at the seedling stage, may have substantial effects on the mix of tree species in forests (Fig. 8.2). Leaf-cutters also affect patterns of plant succession and soil processes, and often prove to be the major determinants of the success or failure of a crop or plantation. The average volume of pine trees in plantations less than 10 years old, for example, can be reduced by at least 50% in a year due to defoliation where the number of *Atta laevigata* nests is greater than 30 per ha (Hernández & Jaffé 1997). The amount of vegetation cut from tropical forests by *Atta* alone has been estimated from 12 studies to be between 12 and 17% of total leaf production (Cherrett 1986).

8.2.1 Foraging

Leaf-cutter workers forage in groups and display complex cooperative behaviour. Some leaf-cutters specialize on grasses, others on dicots, and some are generalists that also

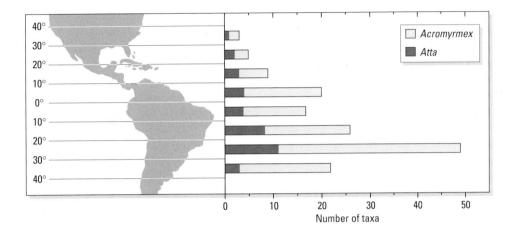

Figure 8.1 The latitudinal distribution of the taxa of leaf-cutting ants (*Atta* and *Acromyrmex*) (Fowler & Claver 1991).

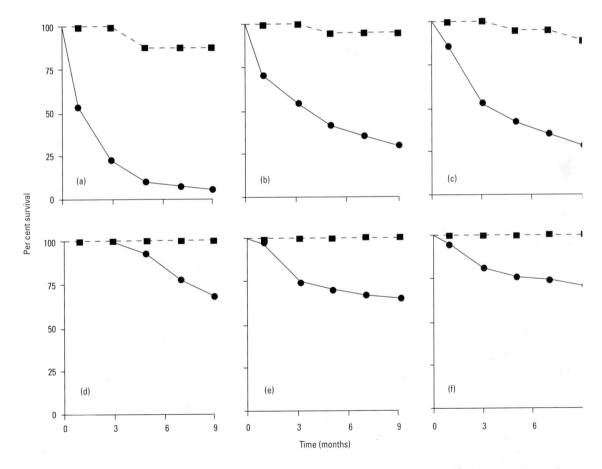

Figure 8.2 The effect of leaf-cutter ants on survivorship of tree seedlings growing on an abandoned farm near Manaus, Central Amazonia. Survivorship curves of transplanted seedlings are shown for seedlings protected from leaf-cutters (broken lines) and those unprotected (continuous lines). For each species there was an initial number of 20 protected and 80 unprotected seedlings (a) *Bellucia imperialis*, (b) *Hevea guianensis*, (c) *Trema micrantha*, (d) *Cariniana micrantha*, (e) *Vismia cayennensis*, (f) *Cecropia purpurascens* (Vasconcelos & Cherrett 1997).

collect flowers and fruits. Colonies often mount concentrated and consistent foraging efforts and a single tree may be exploited for days or weeks by very large parties of foragers. Enormous colonies in a fixed nest site make considerable demands on the resources of the immediate area and foragers may retrieve plant material from distances over 100 m from the nest. Cleared trails allow heavy traffic of workers and can lead to dense aggregations at the foraging site. Analysis of the costs and benefits of this foraging system indicates that the size of fragments cut and carried by leaf-cutters is below the size at which the individual provisioning rate or individual energetic efficiency would be maximized but that fragment size may maximize energy intake at the level of the colony by reducing idle time and congestion.

8.2.2 Ant–fungus mutualism

Cultivation of fungus by attine ants originated about 50 million years ago. The relationship between the higher attine ants and the symbiotic fungus they cultivate is obligate. Foundress queens propagate the fungus clonally by carrying a pellet of fungus in their mouths during their nuptial flight to establish new colonies. The fungi, currently identified as *Leucoagaricus gongylophorus* (North et al. 1997), produce swollen hyphae called gongylidia, bunches of which are called staphylae (see Box 8.1 for de-

finitions of terms specific to ant–plant interactions). The gongylidia are rich in glycogen in a form readily assimilated by the ants. These structures appear to have no other purpose and are derived from vegetative hyphae. It was long believed that all colony members of Attini get nutrients exclusively from eating the mycelia, but Quinlan and Cherrett (1979) found *Atta cephalotes* workers obtain 95% of their carbohydrates from plant sap and only 5% from mycophagy. In contrast, the larvae are able to subsist entirely on the staphylae.

The relationship between the attines and their fungus has been termed an 'unholy alliance' because it combines the ants' ability to circumvent plants' anti-fungal defences with the ability of the fungus to subvert plants' anti-insect defences. The ants benefit because the fungus breaks down plant tissue such as cellulose, starch and xylan, and possibly detoxifies insecticidal plant compounds. The fungus thus enables them to make use of plant material that would otherwise be unavailable and allows the ants to be truly polyphagous in the midst of diverse flora. The fungus may also provide sterols, possibly to be used for membrane synthesis or as hormone precursors and enzymes with polymer-degrading abilities that can break down casein, gelatin and starch. In the case of the 'lower attines' which do not cut leaves but instead use insect remains and excrement, the fungus breaks down products such as chitin.

Box 8.1 Glossary of terms specific to ant–plant interactions

domatia Plant structures that appear to be specific adaptations for ant occupation, often formed by the hypertrophy of internal tissue at particular locations in the plant, creating internal cavities attractive to ants.

elaiosome A lipid-rich appendage attached to seeds that is attractive to ants. Generally considered to be a specific adaptation to promote seed dispersal by ants.

extra-floral nectaries (EFNs) Secretory tissues located on leaves, twigs or the external surfaces of flowers. Nectar secreted is a mixture of nutrients, usually dominated by sugars but often containing amino acids and lipids.

food-bodies A wide variety of small epidermal structures on plants that have been interpreted as adaptations to attract ant foragers. Different types have been differentiated mainly by differences in predominant metabolites.
Beltian body: a vascularized food-body containing protein as well as lipid.
Pearl body: a food-body in which the predominant metabolite

is lipid. They are generally distinguished from other epidermal structures such as trichomes and glands by a pearl-like lustre, small size (up to 3 mm in diameter) and basal constriction.
Beccarian body: A type of pearl body, e.g. in *Macaranga*, very rich in lipid.
Mullerian body: A food-body (e.g. in *Cecropia*) containing large amounts of glycogen plus some lipid.

gongylidia Swollen hyphae of fungi tended by leaf-cutting ants, rich in glycogen and used by leaf-cutters as larval food.

honeydew Sugary liquid produced by homopterans and lepidopterans which promotes ant attendance.

metapleural gland Gland that produce antimicrobial compounds and that distinguishes ants from all other insects, but especially from bees and wasps.

myrmecochory A seed-dispersal syndrome in which seeds have an elaiosome to attract ants and encourage removal.

myrmecophyte Plant species bearing domatia (see above).

staphylae Bunches of gongylidia (see above).

The benefit to the fungus seems to derive mainly from being maintained in an environment kept virtually free of competition from other microorganisms by constant tending and application of antibiotic compounds produced by the ants (but see below). The ants, in particular the tiny 'minima' workers, actively prune the fungus garden, and their activity increases standing crops of staphylae. Workers also defecate onto substrate they have just added to the garden and their faeces supply ammonia and amino acids.

8.2.3 Research directions

The symbiosis between leaf-cutters and their fungus has attracted researchers for many decades. In the last few years, three aspects of the relationship have proved particularly exciting. First, experiments indicate that the fungus, far from being a passive partner in the relationship, may exert control over ants' foraging behaviour. Second, molecular techniques have further elucidated the taxonomic relationship of the Attini and the fungus, shedding light on the evolution of the relationship. Third, the partnership between the ants and the fungus has recently been found to be a triumvirate, with evidence that an antibiotic-producing bacterium is an important component of the symbiosis. One question that remains outstanding, however, is why this ant–fungus mutualism is confined to the Neotropics.

8.2.3.1 Fungal control of ant behaviour

Leaf-cutting ants actively avoid some plants and seem to select mainly on the basis of repellent chemicals such as terpenoids, which have strong fungicidal activity. The rejection response is sometimes delayed for 12–16 h after foraging begins, and may persist over days or weeks. Ridley et al. (1996) have shown that ants reject baits containing the fungicide cycloheximide, after an initial acceptance period of about 48 h, even though the fungicide has no effect on the ants themselves. The rejection response persists up to 30 weeks in laboratory colonies. The researchers concluded that if the forage material is toxic to the fungus, a semiochemical is produced by the fungus that acts as negative reinforcement to ants servicing that particular fungus garden. It appears that the behaviour of foraging workers is not affected directly by the release of highly volatile compounds by the fungus, but rather, close physical contact between the fungus and the foragers is required to pass on the information. Identification of this presumed semiochemical may have great practical significance, as pesticides that mimic the chemical could be developed to prevent leaf-cutting ants damaging crops.

8.2.3.2 Phylogenetic analyses

Molecular data have been decisive in identifying the evolutionary origins and phylogenetic relationships of attine fungal symbionts. Cladistic analyses of nuclear 28S ribosomal DNA and 16S-like RNA ribosome indicate that the higher atttines such as *Atta* and *Acromyrmex* cultivate a different fungus to that of the lower attines (Mueller et al. 1998). Furthermore, the higher attines appear to have clonally propagated the same fungal lineage for at least 23 million years. By contrast, the fungi cultivated by the lower attines are more genetically diverse and appear to have been repeatedly acquired from closely related free-living forms.

8.2.3.3 The third partner

To maintain the fungal monoculture free of microorganisms requires constant 'weeding' by the gardening ants. Until recently, it was believed that this was accomplished mainly by the removal of contaminants such as yeasts, bacteria or alien fungal spores by frequent licking. Antimicrobial substances produced by the metapleural and mandibular gland secretions have also been proposed as contributing to the maintenance of the gardens.

The fungus gardens are particularly prone to infection by a group of closely related, highly specialized parasites in the fungal genus *Escovopsis* (Ascomycotina). *Escovopsis* is found in gardens of virtually all species of fungus-growing ants, but not elsewhere. The parasite is usually found at low levels, but if the health of the garden is compromised it can quickly take over and destroy the fungal crop. In healthy gardens, Currie et al. (1999) have shown that the fungus is kept in check by specific antibiotics produced by *Streptomyces* bacteria living on the bodies of the ants (Fig. 8.3). The bacterium can also promote the growth of the cultivated fungi. The position of the bacterium on the ant integument is genus-specific, indicating that the association with the ants is both highly evolved and of ancient origin (Currie et al. 1999). Attine symbiosis appears to be a coevolutionary arms race between the garden parasite *Escovopsis* on one hand, and the tripartite association of the actinomycete, the ant hosts and the fungus on the

Figure 8.3 *Streptomyces* bacteria growing on the cuticle of *Acromyrmex octospinosus*. The bacteria are active against the fungus *Escovopsis*, a pathogen of attine fungus gardens (Currie et al. 1999).

Table 8.2 The frequency of ant-defended plants in temperate and tropical communities of the New World (Schupp & Feener 1991).

Site (vegetation)	% Ant-defended species	
Temperate		
Nebraska, USA (mixed)	2.3	a,e
Iowa, USA (deciduous forest)	25.9	a,f,h
Virginia, USA (deciduous forest)	25.0	a,e,h
Warm desert		
California, USA (Mojave)	1.0	b,e
California, USA (Mojave)	2.6	b,e
California, USA (Colorado)	3.2	b,e
Tropical		
Hawaii, USA (mixed)	1.2	c,e
Sao Paulo, Brazil (cerrado)	18.3	a,e,g
Barro Colorado Is., Panama (moist forest)	33.3	d,f,h

a = Woody dicotyledons.
b = All species.
c = Hawaiian native vascular species.
d = Dicotyledonous trees, shrubs, lianas, and vines.
e = Extrafloral nectaries only.
f = Includes species with large pearl bodies.
g = Mean of 5 sites, s.d. − 2.1%.
h = Ant rewards on reproductive structures not considered.

other. The relationship raises the interesting question of how the attine antibiotics have remained effective against the fungus-garden pathogens for such a long time, given that resistance to antibiotics is a well-known problem in human and other populations.

8.3 Ant-guard systems

The coevolution of ants and plants involving systems of rewards and services has resulted in a variety of elaborate and complex mutualistic interactions collectively known as ant-guard systems. Here the rewards are extra-floral nectar, specialized food bodies and nest sites, while the service is the protection of the plants from herbivory. Once they have been attracted on to the plant, ants either defend the food or nest sites against competitors or remove herbivores directly through predation. Plants may offer only one kind of reward but any combination of the three may be present, according to the plant species.

8.3.1 Extra-floral nectar

Ninety-three flowering plant and five fern families have

genera with extra-floral nectaries (Koptur 1992) and they have been found on fossil leaves aged at 35 million years (Pemberton 1992). They occur in many different plant communities around the world and there appears to be a trend of increasing frequency of species bearing extra-floral nectaries towards tropical latitudes (Table 8.2). Extra-floral nectaries are secretory tissues located on leaves, twigs or the external surfaces of flowers. They are not a component of the pollination system as, with a very few exceptions, they do not provide rewards for pollinators. The nectar is a mixture of nutrients, usually dominated by sugars but often containing amino acids and lipids.

A variety of field experiments with plant species that possess extra-floral nectaries have shown that when ants are excluded, damage by herbivores or seed-predators may increase, thus demonstrating benefits resulting from the presence of ants (Table 8.3). On the other hand, various field studies have shown no apparent benefits in a herbaceous composite, the bracken fern and tropical tree

Table 8.3 Some effects of ants on plants bearing extrafloral nectaries (Beattie 1985).

Type of damage to plant	Plant species	Plants without ants	Plants with ants
Destruction of stigmas by grasshoppers	*Ipomoea leptophylla*	74% (n=138)	48% (n=380)
Seed destruction by bruchid beetles	*Ipomoea leptophylla*	34% (n=149)	24% (n=3509)
Mean number of seeds per plant	*Ipomoea leptophylla*	45.2 (n=271)	403.2 (n=2419)
Number of sightings of parasitic flies on inflorescences (wet season)	*Costus woodsonii*	872	128
Mean number of seeds per inflorescence (wet season)	*Costus woodsonii*	183	612
Mean number of insect predators per capitulum	*Helianthella quinquenervis*	7.6	2.9
Per cent seed predation	*Helianthella quinquenervis*	43.5	27.6
Number of mature fruits per branch	*Catalpa speciosa*	0.85 ± 0.81	1.11 ± 0.81*
Average number of psyllid nymphs per shoot	*Acacia pycnantha*	351 (n=69)	145 (n=72)
Shoot tips destroyed by psyllid nymphs	*Acacia pycnantha*	34% (n=67)	7% (n=69)

*$P=0.025$. All others are significantly different at $P=0.001$.

saplings. Variation in the outcome of the ant–plant interaction has been a common finding in these studies. For example, Barton (1986) revealed very different levels of protection among three separate populations of the same plant species, *Cassia fasciculata*, in northern Florida, probably directly associated with variation in the densities of the ants and the frequency and type of herbivores at the three sites. A further source of variation may well be the variety of animals that take extra-floral nectar. Predatory and parasitoid wasps and some spiders use the nectar, and they too may benefit the plant by attacking prey species that happen to be herbivores.

8.3.2 Food bodies and nest sites

These two rewards are usually discussed separately though food bodies are mostly, but not exclusively, associated with the provision of nest sites (Table 8.4). The provision of a variety of rewards may be one solution to the variable levels of protection, such as those elicited by extra-floral nectaries, as it may lead to the evolution of specialization in the ants associated with them, generating a shift from facultative to obligate ant-guard systems. Indeed, the provision of both food and exclusive nest sites has resulted in associations between particular ant and plant species that are mutualistic symbioses.

Food bodies are small epidermal structures containing a variety of nutrients such as protein and lipids that are collected by foragers and taken back to the colony. Food bodies whose main metabolite is lipid are generally

known as pearl bodies and have been reported in 50 plant genera from 19 families. The food bodies found on *Cecropia* species are remarkable in that the principal storage product is identical to animal glycogen, a molecule that is extremely rare in plants (Rickson 1971). Food bodies are generally viewed as a specific ant attractant, although collection by ants has been seen in relatively few plant species, and their actual consumption by adults or larvae is still a rare observation (Beattie 1985). Plant species with food bodies develop them even in the absence of ants, so the trait is genetically determined. However, Risch and Rickson (1981) demonstrated a remarkable system in *Piper cenocladum*, in which food bodies are induced only by the presence of the specialist ant *Pheidole bicornis*. Plant investment in food bodies can be quite considerable. One well-studied case is *Macaranga triloba*, which invests up to 9% of above-ground tissue construction costs in food bodies located on the undersides of its stipules. Investment in terms of energy and tissue varies and the variation is attributable, in part, to the density of ant-guards (Fig. 8.4).

8.3.3 Domatia

A wide variety of plant structures are used as nest sites by ants and there is uncertainty as to precisely which ones are adaptations for ant occupancy. Opportunist nest-building occurs in all kinds of cavities such as those under bark, in galls, within the sheaths formed by the bases of leaves and in dead, hollow stems. The weaver ant,

Table 8.4 Food body–ant associations in five plant genera (Hölldobler & Wilson 1990, data modified from Beattie 1985).

Characteristics	*Acacia* sp. (Leguminosae)	*Cecropia* sp. (Moraceae)	*Macaranga* sp. (Euphorbiaceae)	*Ochroma* sp. (Bombacaceae)	*Piper* spp. (Piperaceae)
Major ant associates	*Pseudomyrmex*	*Azteca*	*Crematogaster*	*Solenopsis, Azteca*	*Pheidole*
Type of food body	Beltian, tip of pinnule and rachis	Müllerian, produced in large numbers on trichilium (pad of tissue at base of petiole)	Beccarian, on stipules or young leaves	Pearl body, on leaves and stems of sapling	On petiole margins
Principal nutrient offered	Protein, lipid	Glycogen, lipid	Lipid, starch, some protein	Lipid, perhaps some starch and protein	Lipid, protein
Anatomy	Multicellular, tissues differentiated	Multicellular	Multicellular	Multicellular	Single-celled
Extra-floral nectaries present?	Yes	No	No	Yes	Yes?
Domatia present?	Yes	Yes	Yes	No	Yes?
Plant provides complete diet for ants?	Yes	Together with coccid honeydew, probably yes	Together with coccid honeydew, probably yes	No	Yes?
Do ants protect plants?	Yes	Yes	Probably	Probably	Probably

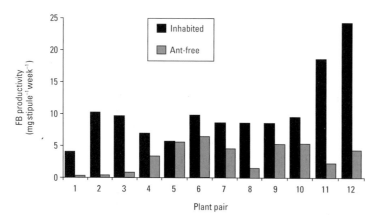

Figure 8.4 The effect of ant habitation on food-body productivity in *Macaranga triloba* ($n = 12$ plant pairs). The effect is measured in terms of total food-body (FB) production of all stipule pairs of a given plant in mg dry FBs per total plant per week. Each plant pair consisted of one inhabited and one ant-free plant of approximately the same size; they are arranged according to plant size (Heil et al. 1997).

Oecophylla, sews leaves together with larval silk to make its nests (Hölldobler & Wilson 1990). There is a large and varied group of epiphytes that are associated with ants and some tropical species form arboreal aggregations known as 'ant gardens'. Ants use the root clusters as a framework for their nests, bringing soil particles and depositing waste as the colony grows, supplying nutrients to the plants high in the canopy. Extra-floral nectar and elaiosomes are produced by some of the plants involved and there is evidence that ants defend the garden and disperse its seeds.

Plant structures known as domatia are developmentally determined and appear to be specific adaptations for

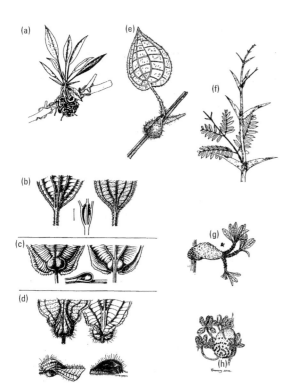

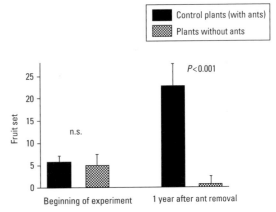

Figure 8.6 Increased fruit set following removal of adult herbivores by ant colonies inhabiting leaf-pouch domatia in *Maieta guianensis*, a myrmecophytic melastome. Data shown are the mean number of fruits (±s.e.) measured immediately before and 1 year after ant removal (*n* = 15 plants per treatment). Plants with ants had approximately 45 times the number of fruits than plants from which ants were chemically excluded. (Data from Vasconcelos 1991.)

Figure 8.5 A variety of domatia: (a) *Anthurium gracile* with an ant-inhabited root ball; (b) leaf pouch of *Clidemia* sp.; (c) leaf pouch of *Tococa coronata*; (d) leaf pouch of *Myrmedone macrosperma*; (e) stem domatia of *Clidemia tococoidea*; (f) thorn domatia of *Acacia sphaerocephala*; (g) swollen tuber of *Myremecodia brassii*; and (h) swollen tuber of *Squamellaria major*. (Sources: a, e, f, Benson 1985; b, c, d, Zizka 1990; and g, h, Huxley 1998.) See also photographs in Plate 8.1.

ant occupation. They are often formed by the hypertrophy of internal tissue at particular locations in the plant, creating internal cavities attractive to ants which, in turn, accelerate their formation by gnawing. Ant access to these cavities is critical and, in many cases, the wall surrounding the cavity is thin at a particular spot where the ants gnaw an entrance hole tailored to their particular size and behaviour. In *Cecropia* the hole is surrounded by stinging hairs (Davidson & McKey 1993). The presence of domatia and, in particular, plant tissues conducive to the creation of cavities and entrance holes, strongly suggests adaptation to ant occupation; the plant species that bear them are known as myrmecophytes.

Domatia include some of the most remarkable structures made by plants and are found primarily in stems, leaves and spines (Fig. 8.5). In central Amazon tropical rainforest, myrmecophytes can reach a density of 380 individuals per hectare, involving 16 plant and 25 ant species (Fonseca & Ganade 1996). Several studies have demonstrated the protective function of ants associated with domatia, including increased fruit set following the removal of adult herbivores by colonies inhabiting leaf pouches (Fig. 8.6). Other examples include the removal of herbivore eggs or juveniles by colonies in stem domatia (Fiala et al. 1989) and the pruning of encroaching vegetation, especially vines (Davidson et al. 1988). Letourneau (1998) showed that, in several species of *Piper*, fitness gains through ant activity result principally from their removal of stem borers and fungal spores. The ants ingest the spores, but it may be that antimicrobial secretions from the ant's metapleural glands also play some part in suppressing fungal activity. Ant recruitment to leaf damage can take as little as ten minutes in *Cecropia*, perhaps in response to the release of volatile chemicals from the affected tissues. With such short response times, even larger, more mobile herbivores may be attacked (Agrawal 1998b).

Some myrmecophytes are actually 'fed' by the ants they house. Experiments have shown that two genera in the family Rubiaceae, *Hydnophytum and Myrmecodia*, absorb nutrients from the wastes of the *Iridomyrmex* colonies

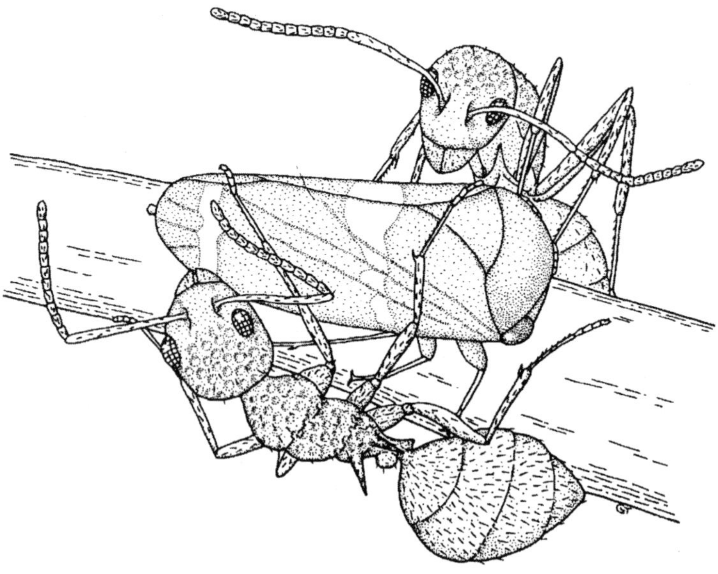

Figure 8.7 Two ants of the genus *Polyrachis* tending a cicadellid homopteran on a twig of *Eucalyptus*. The ants have been stroking the leafhopper with their antennae and the lower one is turning to the anus, anticipating a drop of honeydew. (Beattie 1985; drawing by Christine Turnbull.)

they house in tunnels inside large tubers (Rickson 1979; Fig. 8.5g). This benefit has also been demonstrated in an orchid where the domatium is a pseudobulb (Rico-Gray et al. 1989).

A variety of field studies have shown there is strong competition among ants for domatia (Davidson & McKey 1993; Fonseca & Ganade 1996). Successful occupation is the result of a variety of factors, including the physical size of individual ants, mechanisms of offence and defence, colony growth rates and the relative abundance of different-sized domatia in the habitat. While the diversity of myrmecophytes varies according to geographical location, with the greatest in South America, there is great variety in the ant species involved and there are some remarkable examples of convergent evolution. These have been succinctly described by Davidson and McKey (1993): 'In the American, African and Asian tropics, competitively dominant ants are associated with the most light-demanding and fast-growing hosts, which supply resources at the rates required to fuel rapid colony growth, interspecific aggression and other traits required for dominance. In contrast, competitively subordinate ants are restricted to plants which supply resources at rates too low to support dominant ants, or those from which dominant ants can be excluded by long, dense plant hairs, pruning of neighboring vegetation, or by other ant and plant traits which favor competitively subordinate species.'

8.3.4 Three-way systems: ants, plants and herbivores

Ant-guard systems involving extra-floral nectaries are often complicated by the presence of Homoptera or lepidopteran larvae that secrete nectar-like fluids collectively known as honeydew. In such situations, the ants have a choice of food and the outcome of these three-way interactions between plants, ants and herbivores appears to be extremely variable.

The Homoptera include herbivores such as aphids, leafhoppers, scale insects and coccids. Each animal is armed with a proboscis that penetrates plant vascular tissue, tapping into the nutrient supply. With little apparent effort, the sap enters the front end of the homopteran gut, later appearing at the back end as droplets, somewhat depleted in quality but still containing many nutrients, where it is ejected as honeydew. Many ant species harvest the honeydew and, in return, protect the homopterans from predators and parasites (Fig. 8.7). As a result, ant activity can increase levels of herbivory as well as other forms of damage, such as the transmission of viral diseases through the homopteran proboscis. Ants also tend and protect a variety of lepidopteran larvae. We include these ant–caterpillar interactions here as they also involve the production of honeydew from specialized glands; these larvae, however, are plant-chewers rather than suckers.

The main families involved are the Lycaenidae and the Riodinidae, and ant-tending can be very important in terms of reliance on ants; approximately 80% of lycaenid species in Australia, for example, are ant-tended (Eastwood & Fraser 1999).

Ant interactions with plant species that produce extra-floral nectaries, food bodies and domatia have evolved both in the presence of homopterans and lepidopteran larvae and the ant behaviour that protects them. For example, homopterans of various kinds are routinely maintained within domatia and they frequently feed on plants that bear extra-floral nectaries. This leads to the situation where plants are providing rewards for ant-guards that attack some of the plant's enemies but protect others. A solution to this apparent conflict of interest was first proposed by Janzen (1979) who suggested that the presence of homopterans was part of the cost of the ant-guard system in the same way as the provision of extra-floral nectaries and other rewards. Although they are herbivores, homopterans at least attract ants on to the plant.

Field experiments with ant–homopteran–plant and ant–lepidopteran–plant interactions have shown that this may be the case in some situations, but the net effects of these three-way interactions can be unpredictable. For example, Messina (1981) showed that the ant *Formica fusca* tended the membracid *Publilia concava* on golden-rod (*Solidago* sp.) and also attacked the plant's primary herbivore, the beetle *Trirhabda virgata*. His data showed that plants with ants were taller and yielded more seeds than those without, presumably because the positive effect on the plant of reducing beetle defoliation outweighed the negative effect of tending the membracids. Working with different three-way systems, Buckley (1983) found only adverse effects on plants and Fritz (1983) found no significant effects one way or the other. In the system that Bach (1991) studied, ants removed the primary herbivore, a caterpillar, and the experimental exclusion of ants resulted in increased deposition of honeydew on the leaves, which in turn resulted in heavy infestations of sooty mould followed by widespread leaf abscission. She concluded that there might be circumstances in which the three-way interactions result in benefits to the plant host. Horvitz and Schemske (1984) carried out a series of field exclusion experiments on a three-way interaction between *Calathea* (the plant), *Eurybia* (the ant-tended caterpillar) and an assemblage of ant attendants and found that the 33% reduction in seed set in the presence of ants was far less than the 66% reduction

Table 8.5 Seed production in *Calathea ovandensis* with and without ants, and with and without *Eurybia*, an ant-tended caterpillar (Horvitz & Schemske 1984).

Treatment	Number of inflorescences	Number of seeds produced per inflorescence	
		Mean	SD
No *Eurybia*, with ants	128	20.8	15.06
No *Eurybia*, ants excluded	105	17.5	12.10
With *Eurybia*, with ants	130	13.6	12.79
With *Eurybia*, ants excluded	131	6.2	7.94

in their absence. Given that *Eurybia* is a specialist and therefore almost certain to find *Calathea* plants, the presence of ants appears advantageous even though seed set without any of the insects was higher still (Table 8.5). To our knowledge, no studies have examined the effects of ant-tended homopterans that attack roots.

The evolution of extra-floral nectaries has itself been viewed as a defence against homopteran attack, weaning ants away from the herbivores (Becerra 1989). Homopterans are common herbivores and have been around for a very long time; thus, given their ubiquity, selection for extra-floral nectaries may have resulted in the plants exerting greater control over the ant-guards, provided ants preferred nectar to honeydew. Cushman (1991) generated a model to show how ants may behave towards their insect mutualists, switching from eating them to tending them in response to the relative quality of nectar and honeydew. If this model approximates the reality in the field, it would help to explain much of the variation in the net outcomes of these three-way interactions (Fig. 8.8).

8.4 Ant pollination

Unlike their relatives, the bees and wasps, ants are rarely pollinators. Peakall et al. (1991) listed just 10 documented cases in the literature with another 16 for which there was anecdotal evidence only. Evidence for chemical or physical barriers at or around flowers to

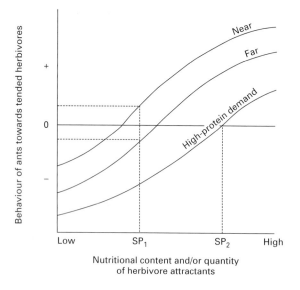

Figure 8.8 A model proposed by Cushman (1991) to explain how plant-induced variation in the nutritional content and quantity of rewards provided to ants, in association with other factors, predicts the behaviour of ants towards tended herbivores. When herbivores feed on low-quality host plants and produce rewards of low nutritional content and/or quantity, ant colonies are predicted to acquire greater fitness by acting as predators than as mutualists. Alternatively, when herbivores feed on high-quality hosts and produce rewards of high nutritional content and/or quantity, ant colonies are predicted to gain more by acting as mutualists. The first switchpoint (SP$_1$) represents a hypothetical level of reward at which ant colonies act as predators of herbivores feeding on host plants far away from their nests and mutualists on herbivores close to their nests. The second switchpoint (SP$_2$) represents the level of reward after which ant colonies with high protein demand act as mutualists.

prevent access by ants was at first thought to explain the paucity of ant pollination systems, but few asked why they were there in the first place. Galen (1999a) is one of the few researchers to perform experiments showing a negative effect of floral visitation by ants. *Formica neorufibarbus* dislodges the style in flowers of *Polemonium viscosum* and plants from which ants have been excluded set significantly more seed than those in which ants have free access. Aside from this, many explanations as to why ants would make bad pollinators have been proposed. For example, most ants do not have wings and could not effect cross-pollination, ants are not hairy and cannot carry pollen, they groom their integument too frequently or do not forage systematically like bees. An examination of the literature showed that these suggestions were either inac-

curate or insufficient to explain why there are apparently so few ant-pollination systems (Beattie 1982). Other research has focused on the metapleural glands of ants, an important characteristic that distinguishes ants from all other insects, but especially bees and wasps. These glands produce antimicrobial compounds, and a series of experiments showed that pollen grains resemble microorganisms sufficiently to be killed or disabled by the secretions when contacted. If the secretion of antimicrobials on to pollen-carrying surfaces is universal in ants, then ants are unlikely to evolve as pollen vectors (Beattie 1991).

Recent studies have shown that while the antimicrobial secretions of ants do harm pollen, there are at least two ways of circumventing the problem. The first can be seen in the Australian orchids *Leporella fimbriata* and *Microtis parviflora*, which present their pollen on stalked pollinia so that there is no contact with the ant integument. *L. fimbriata* is remarkable in that its pollinator is a winged male ant, lured to the flower along drifts of plant-originated female pheromone (Peakall 1989; Fig. 8.9). *Microtis parviflora* is also of great interest as its pollinators are wingless workers that forage systematically on the flowers (Peakall & Beattie 1989). The second strategy for dealing with ant-borne antimicrobial secretions appears to be the provision of large quantities of pollen to large numbers of ants so that at least some gets through to the stigma. Gómez and Zamora (1992) and Gómez et al. (1996), working with a variety of plant species, demonstrated reduction of pollen viability through contact with the ant species visiting the flowers. However, the large number of ants involved, together with the liberal quantities of pollen produced, still yielded reliable seed set.

In most cases of pollination by ants, floral-nectar rewards are easily accessible and ants are not the only flower-visitors, so that pollination is also effected by a variety of other insects (Gómez et al. 1996). Ramsey (1995) showed that *Blandfordia grandiflora*, an Australian herb known locally as Christmas Bells, is adapted for pollination by birds, but ant visitors can effect self-pollination, the pollen being carried chiefly on hairs on their legs, remote from the metapleural glands. Easy access to rewards is facilitated by saucer-shaped (actinomorphic) flowers and by the synchronous presentation of high densities of flowers, two traits of ant pollination anticipated by Hickman (1974). However, the orchid systems are very different as the flowers either do not offer nectar and may be highly dispersed in the habitat (*Leporella*), or the nectar is con-

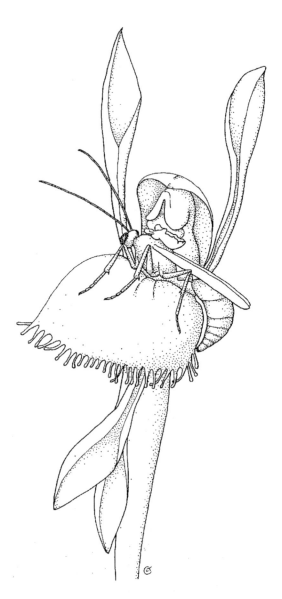

Figure 8.9 The male ant *Myrmecia urens* pollinating the orchid *Leporella fimbriata* during a visit in which it attempts copulation with the flower. (Drawing by Christine Turnbull.)

cealed in zygomorphic flowers on stems that vary greatly in density (*Microtis*).

Hickman (1974) and Beattie (1991) suggested that more ant-pollination systems would be found in environments such as high mountains and arid habitats, where ants were likely to be the most abundant and affordable potential pollen vectors. By 'affordable' we mean that

these harsh environments may constrain plant investment in pollination structures and that simple flowers with small amounts of nectar would be sufficient to attract and manipulate ants. This has proved to be the case (Gómez et al. 1996; Puterbaugh 1998).

While antimicrobials do reduce pollen viability, their presence alone is insufficient to explain the rarity of ant pollination. Nevertheless, the presence of these compounds on ant surfaces may have militated against the evolution of ant-pollination systems. Those that have evolved are of two kinds: the first is a highly specialized co-evolution of specific ant and plant species. The second is a far less specialized arrangement in which a variety of small winged or flightless insects share the rewards. When ant densities are high, ant involvement in seed production can also be high. In environments where plants are often low to the ground, flower synchronously in a short growing season and produce masses of flowers each with a very small reward, ants can be important pollinators. The habitat drives the ant–plant interaction, that is, there is convergence in characters among unrelated ant species and among unrelated plant species which enable them to take advantage of the local conditions, a theme we shall take up again at the end of this chapter.

8.5 Ants and seeds

Seeds and some of their associated dispersal structures are a conveniently packaged energy source used by many ant species. Ants act as dispersal agents for many plants, conferring a variety of benefits. Some ants are also important seed-predators and can have enormous negative impacts on plant populations. The distinction between dispersal and predation is not, however, clear-cut as many ant species both eat and disperse a proportion of the seeds they collect.

8.5.1 Myrmecochory

Over 3000 plant species (from more than 80 families) have seeds bearing a food-body known as an elaiosome. The elaiosome is highly attractive to many ant species, and is generally considered to be a specific adaptation to promote removal by ants and subsequent dispersal. Ants typically carry the intact diaspore, consisting of the seed and elaiosome, to their nests, where the elaiosome is removed and eaten (Fig. 8.10). The seed may then be discarded within the nest or on a waste midden.

Myrmecochory is a very generalized relationship between ants and seeds, with most seed species being removed by a suite of omnivorous ants with little or no specificity. Most ant-dispersed plants are found in Australia (~1500 species; Berg 1975) and southern Africa (~1300 species; Milewski & Bond 1982), and are mainly woody shrubs growing in low-nutrient, fire-prone habitats. Approximately 300 ant-dispersed species are also found in the northern hemisphere, with the majority being herbaceous species in the understorey of temperate deciduous forests (Beattie & Culver 1981). Some ant-dispersed species are also found in the tropics, North American deserts and the Mediterranean region.

The herbaceous myrmecochores of the northern hemisphere are associated with a suite of adaptations markedly different to those of the southern hemisphere sclerophyll shrubs (Table 8.6). The elaiosomes of the herbaceous species are usually soft, collapsible and short-lived, becoming desiccated and unattractive to ants within a few days. Seeds are typically presented in aggregations at ground level as the supporting stem that bears the ripe diaspores bends downwards. In contrast, the food-bodies of the southern hemisphere shrubs are firm and persistent, and usually fall to the ground singly. Elaiosomes of these seeds can retain their attractiveness to ants for several years. Some species of myrmecochores in both the northern and southern hemispheres have seeds that are ejected

Figure 8.10 The greenhead ant, *Rhytidoponera metallica*, a common seed-taking ant species in Australia, carrying a seed of *Acacia linifolia*. Note the use of the elaiosome as a handle. (Photograph by R.J. Oldfield.)

Table 8.6 Characteristics of myrmecochory in Northern (USA and Europe) and Southern Hemispheres (Australia and South Africa).

	Northern Hemisphere	Southern Hemisphere
Number of myrmecochorous plant species	~300	~1500 (Australia), plus ~1300 (South Africa)
Typical growth form	Understorey herbs	Woody shrubs
Vegetation type	Deciduous forests	Dry sclerophyll woodland and heath (Australia), fynbos (South Africa), often fire-prone, on low-nutrient soils
Elaiosome	Soft, collapsible, desiccates and becomes unattractive within a few days	Hard, long-lived, may retain attractiveness for several years
Seed presentation	Clumped on ground as peduncle bends, some species ballistic	Seeds dropped singly, often preceded by ballistic expulsion
Common plant genera	*Trillium, Viola, Sanguinaria*	*Acacia, Grevillea, Dillwynia* (Australia), *Leucodrendon, Leucospermum, Mimetes* (South Africa)
Common seed-removing ant genera	*Formica, Lasius, Myrmica*	*Rhytidoponera, Pheidole, Aphaenogaster, Iridomyrmex, Monomorium, Chelaner* (Australia), *Anoplolepis* (South Africa)
Possible advantages to plants	Protection from rodent predation, placement in nutrient-rich microsites, reduction in competition	Protection from ant and rodent predation, placement at appropriate depth for germination after fire

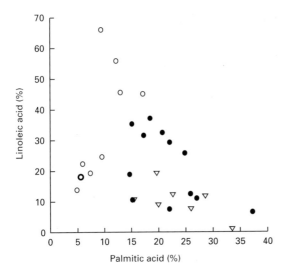

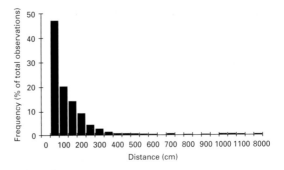

Figure 8.12 Frequency distribution of myrmecochorous dispersal distances, shown in percentages of observations within each distance category, $n = 1863$ observations (Gómez & Espadaler 1998).

Figure 8.11 Comparison of palmitic and linoleic acid composition in seeds ($n = 12$, open circles), elaiosomes ($n = 12$, solid circles) and insects ($n = 7$ orders, open triangles) (Hughes et al. 1994b).

ballistically from an explosive capsule prior to being presented to ants on the ground.

Elaiosomes have diverse morphological origins and composition. They are usually rich in lipids but may also contain amino acids, sugars and protein. Free fatty acids, especially oleic acid, have been found in high concentrations in several species and the diglyceride 1,2-diolein has been implicated as the main ant attractant in several phylogenetically diverse species. The fatty-acid composition of elaiosomes appears to be convergent with insect prey (Fig. 8.11), indicating that elaiosomes may be an insect mimic, attracting omnivorous ants that would not normally remove and eat plant material.

8.5.1.1 Benefits to the ants

The benefits to ants of removing elaiosome-bearing seeds appears to be a simple addition of energy-rich food to their diet. Specific effects of including elaiosomes in the diet have rarely been investigated, although Morales and Heithaus (1998) have found that experimental colonies of *Aphaenogaster rudis* produced significantly more gynes (reproductive females) when allowed access to elaiosomes of *Sanguinaria canadensis* than did control colonies. The production of males was not affected, resulting in the sex ratio of the colony being skewed.

8.5.1.2 Effects on plants

The fate of seeds taken by ants and the post-dispersal pattern of establishing seedlings is dependent on the behaviour and nest characteristics of the particular ant community. Seeds with elaiosomes are generally removed rapidly, within hours or days, but the removal rate may depend on diaspore size, elaiosome size, elaiosome:seed ratio and the density of other types of available seeds. The timing of seed release in some myrmecochores may have important consequences for seed fate; release of seed during the morning appears to reduce rodent predation at night. Temporal separation of release dates may also maximize exposure to seed-dispersing ants in some North American myrmecochores.

Although ants rarely carry seeds more than a few metres (Fig. 8.12), they may nonetheless have important impacts on seed fate and the subsequent establishment of seedlings. Some herbaceous myrmecochores of the northern hemisphere and tropics have been shown to benefit either by the placement of seeds in nutrient-enriched microsites, reduction of parent–offspring or sibling competition and/or reduction of rodent predation. Improved survivorship of seedlings following ant dispersal has been shown in several herbaceous species.

In areas where ant-dispersed plants are more common, such as the dry sclerophyll woodland in Australia in which myrmecochores can constitute up to 50% of the shrub species (Willson et al. 1990), the benefits of attracting ants to seeds appear to differ from the northern hemisphere system (Table 8.6). While protection from rodent predation in South African myrmecochores may be important (Bond & Breytenbach 1985), this does not seem to be the case in Australia, where the rodent fauna is relatively depauperate. Vertebrate predation on myrmecochorous legume seeds is generally low, although genera such as *Grevillea* appear to be vulnerable. Nutrient enrichment has been shown to be a potential benefit in some southern hemisphere cases, but this is not generally so. Many seed-dispersing ants do not maintain discrete nest middens to act as sites of nutrient enrichment, and frequent nest relocation often prevents significant enrichment anyway (Hughes 1991). In habitats like this, therefore, different selective pressures have probably been more important, including reduction of predation from seed-harvesting ant species, reduction of seedling aggregation, and burial at an appropriate depth for germination after fire (Berg 1975; Hughes et al. 1994a).

Removal of elaiosome-bearing seeds by ants may not, however, always benefit plants. Dispersal by ants may sometimes result in increased clumping of seeds, and, by implication, increased competition for establishing seedlings. Furthermore, at least some ant species that generally disperse seeds may occasionally eat them, although, in many Australian myrmecochores, tough seed coats confer protection.

8.5.2 Ants as secondary dispersers

Seeds with elaiosomes are not the only diaspores to be dispersed by ants. In tropical regions in particular, there is increasing interest in the role that ants play in the dissemination of seeds from fleshy fruits that are primarily adapted for vertebrate dispersal. Tropical forests produce a huge number of fleshy fruits per unit area; fallen plant diaspores may comprise up to $400\,\mathrm{kg\,ha^{-1}\,yr^{-1}}$ in the humid forests of south-east Brazil and $70\,\mathrm{kg\,ha^{-1}\,yr^{-1}}$ in cerrado vegetation (Morellato 1992, cited in Leal & Oliveira 1998). A considerable portion of these fruits reach the forest floor, either spontaneously or after being dropped by vertebrate frugivores. In these habitats ant abundance may exceed 8 million individuals per ha (Hölldobler & Wilson 1990), comprising nearly one-third of the insect biomass. These ants include a broad array of plant material in their diets, so it is not surprising

that ants are frequently observed removing both intact fruits and seeds from vertebrate faeces.

The ant species involved in seed removal from fruit are a diverse group. Seeds of the tropical tree *Miconia affinis*, for example, attracted 22 ant species along a 600 m transect in a Costa Rican rainforest (Kaspari 1993). Many of the seeds are taken by seed-predators such as *Pheidole* spp. (Levey & Byrne 1993), but seeds that escape immediate predation may gain a number of benefits. By removing the fleshy pulp from seeds, ants appear to reduce the risk of mortality due to fungal infestation and may eventually promote seed establishment. Even attine ants, traditionally regarded as plant pests, have recently been recorded facilitating seed germination in the mammal-dispersed tree *Hymenaea courbaril* by removing fleshy pulp (Oliveira et al. 1995). Pulp removal by attine ants in a study by Leal and Oliveira (1998) significantly increased the germination success of several species, even compared to control seeds where pulp was removed by hand, indicating that ant-induced mechanical and/or chemical factors facilitate seed germination. Removal of seeds by the mainly carnivorous ants *Odontomachus* and *Pachycondyla* from beneath fruiting trees may also reduce the risk of seed predation by beetles (Pizo & Oliveira 1998). Protection from fire has also been suggested as a benefit to seed removal by attines. Removal of seeds from faeces potentially also reduces both intra- and interspecific competition.

Secondary dispersal of seeds from fruit has also been found in low-nutrient, fire-prone Australian vegetation and Mediterranean scrublands.

8.5.3 Seed harvesting

Harvester ants constitute a broad assemblage representing many different evolutionary lines within the subfamilies Ponerinae, Formicinae and Myrmicinae, though the habit is disproportionately concentrated in the myrmicines (Hölldobler & Wilson 1990). Seeds form part of the diet of many ant species, and genera such as *Monomorium* and *Chelaner* eat seeds almost exclusively. Harvesting ants are dominant in deserts and drier grasslands in warm temperate and tropical regions, especially in North America, Australia, the Sahara and South Africa. Harvesting species comprise more than half of all ant colonies in some Australian sites (Briese & Macauley 1981) and in the Namib desert they make up more than 95% of the total forager biomass (Hölldobler & Wilson 1990).

There is general agreement that harvester ants can alter

the local abundance and distribution of ephemeral and annual plants, especially in deserts, grasslands and other xeric habitats where they are most abundant. Impacts on perennial plant populations, however, are poorly understood.

8.5.3.1 Effects on seed and seedling density

The abundance, size and activity level of harvester ants, such as *Pogonomyrmex* and *Messor* in North American deserts, has meant that much of the research on granivory by ants has focused on these areas. Indeed, comparative evidence suggests that granivory in North American deserts is exceptionally high, even compared to other arid zones (Lopez de Castanave et al. 1998). Ant exclusion experiments performed in some desert areas have shown that predation by ants is capable of substantially reducing seed density and the subsequent vegetative mass of the plants. For example, removal of ants from experimental plots in the Arizona desert resulted in a 50% increase in density of annual plants after two seasons compared to nearby control plots (Brown et al. 1979). Harvesting ants in these areas compete with other, such as rodents, species that rely on seeds. Brown and Davidson (1977) calculated that ant biomass in deserts is comparable to that of rodents and that together they consume most of the seeds produced. They demonstrated that rodent exclusion produced a 71% increase in ant colonies, while ant exclusion produced a 24% increase in rodent biomass. Rates of seed harvesting in arid and semi-arid areas can be extremely high. Briese (1982) estimated that ants collected nearly 4000 seeds per m^2 during an autumn harvesting period in semi-arid New South Wales. Many studies have estimated that ants can remove at least 95% of the annual seed crop.

The effects of harvesting ants on seed availability and seedling establishment are not limited to deserts. Arboreal species of *Pheidole*, for example, have been shown to limit *Ficus* reproduction in tree canopies in the tropics (Laman 1995). In eucalypt forests in Australia, predation by ants can be so heavy that many plant species are virtually absent from the soil seed bank. Experimental elimination of ants in a south-eastern Australian eucalypt unburnt woodland resulted in a 15-fold increase in seedling density of *Eucalyptus baxteri* (Andersen 1987). Predation by ants during inter-fire periods in these environments may be one of the selective influences on the phenomenon whereby fires induce a massive and synchronous release of canopy-stored seed, resulting in the temporary satiation of ant foragers.

8.5.3.2 Seed selectivity

Selective harvesting of particular seed species can affect both the relative and absolute abundance of plant species. All species of harvester ants studied so far accept a wide range of seed species under natural conditions, but most discriminate to some degree. There is a tendency to gather seeds that are most abundant and there is also a strong correlation in some species between the size of the worker caste and the size of the seeds selected. Experimental removal of ants over a 10-year period in the Chihuahuan Desert resulted in higher densities of small-seeded winter annuals (Samson et al. 1992). By contrast, rodent exclusion led to increased dominance of the winter flora by initially rare, but competitively superior, large-seeded annuals which eventually suppressed populations of small-seeded annuals. The impact of seed selection on the soil seed bank can sometimes be dramatic. The common harvester of North American desert *Pogonomyrmex occidentalis*, for example, removed 9–26% of the potentially available seed pool each year but up to 100% of preferred species (Crist & MacMahon 1992).

Differential seed removal can tip the balance in competition among some plant species and promote equilibrium in others, as illustrated by experiments using the fire ant *Solenopsis geminata* in the annual cropping system of the wet tropics of Mexico and Central America (Risch & Carroll 1986). Ants were allowed to forage on four pairwise combinations of seed species. When ants were given access to the usually dominant species they reversed the course of competition, allowing it to be excluded by the subordinate. In two combinations where the ants preferred the subordinate species these plants quickly disappeared. In the fourth combination, ants created a stable equilibrium by holding down the dominant just enough to allow the subordinate to survive.

8.5.3.3 Effects on spatial patterns

Ants can alter the spatial arrangement of seeds and establishing seedlings either by patchy foraging behaviour and preference for clumped seeds, or by seed-caching and deposition after removal. Ants can also have dramatic effects on plant communities by clearing vegetation around their nests. Species of *Pogonomyrmex* in North American deserts clear large, disk-shaped areas with diameters up to 13 m (Nowak et al. 1990), which can comprise up to 10–20% of the land surface. Defoliation by these ants can be selective, further contributing to changes in local plant community composition. Nowak et al. (1990) attributed the prevalence of the bunchgrass *Oryzopsis hymenoides* (over 70% of plants on disks) to selective defoliation of other plant species by *P. badius*, and Clark and Comanor (1975) estimated that *P. occidentalis* clip up to 226 million annual plants per ha (plus an unknown number of perennials).

8.5.3.4 Does seed predation limit recruitment?

While seed-harvesting ants have been shown to have important impacts on annual and ephemeral desert-plant populations, recruitment in stable populations of long-lived perennials often seems more limited by the availability of establishment sites. For example, in Andersen's (1987) study described above, where ant exclusion led to a 15-fold increase in seedlings, all the seedlings in the ant exclusion plots died in the first year. This suggests that at least in periods between fires, eucalypt establishment is limited by lack of suitable establishment sites and conditions rather than seed supply. Similar results for other eucalypts such as *E. salmonophloia* indicate that this may be a general phenomenon (Yates et al. 1995).

8.5.3.5 Benefits to plants

The effects of harvester ants on individual plant species are not always negative. Some harvester ants such as *Pheidole* move nests frequently, often leaving caches of viable seeds behind (Levey & Byrne 1993). Seeds that escape being eaten once they are taken to nests may not only be dispersed but reach sites advantageous for germination and establishment. For example, seeds of *Plantago insularis* and *Schismus arabicus* collected by *Pogonomyrmex rugosus* and *P. pergandei* in Arizona deserts may survive long enough to take root in refuse piles around the ant nests, where they can be at least five times more dense, on average, than in nearby sites away from nests (Rissing 1986). Several studies have found that plants growing on ant mounds can enjoy increased survivorship, growth and/or seed production. The most likely mechanism for increased seed production is that the plants benefit from nutrient enrichment at or near the nest (see Section 8.6) but reduction of competition may also be important. For example, in a glasshouse study, Harmon and Stamp (1992) showed that foraging by *Pogonomyrmex badius* lowered overall plant density and increased spatial in-

equality. Some seeds were moved into high-density patches and some were left in locations with relatively distant neighbours. Less crowded individuals grew larger and produced disproportionately more seeds, leading to an increase in overall seed production.

In summary, the omnivorous diet and sheer abundance of many ant species means that in many habitats, seeds below a certain size have a high probability of being picked up and carried to an ant nest. The identity of the seed-removing ant affects subsequent seed fate in many ways, including the distance it is carried, the probability of predation, burial depth, aggregation and establishment conditions. Many ant species play a dual role as both predators and dispersers, and to classify them as one or the other can obscure both the complexity and the generality of the interaction.

8.6 Ants and soil

While the fascinating mutualistic relationships between certain ant and plant species have received much attention, the majority of ants and plants do not interact in this way. Instead, the most ubiquitous form of interaction is probably indirect, via the physical, chemical and microbial effects of ants on soil. The relationship between ants and plants via soil is clearly not a highly coevolved mutualism, as there is no obvious mechanism by which plant growth and fecundity have a direct selective effect on ant fitness. There is, however, evidence that ants can influence plant growth and reproductive output via their effects on soil properties.

8.6.1 Physical effects

Ants rival earthworms in terms of their effects on turnover, aeration and drainage of soil. In a comparison of global rates of soil movement ants scored second ($\sim 5000\,\mathrm{g\,m^{-2}\,yr^{-1}}$) after earthworms ($15\,000\,\mathrm{g\,m^{-2}\,yr^{-1}}$, Fig. 8.13), but as ants have a wider geographical distribution they may well move more soil in total than any other group. In moist tropical and temperate systems the highest mounding rates recorded are about $10\,\mathrm{t\,ha^{-1}\,yr^{-1}}$ (Paton et al. 1995). A colony of the leaf-cutting ant *Atta sexdens* in Brazil was found to deposit on the surface an amount of soil that covered $100\,\mathrm{m^2}$, occupied $23\,\mathrm{m^3}$ and had a weight of 40 tons (Autori 1947, cited in Folgarait 1998). The activities of desert ants may also lead to high soil turnover rates, such as $420\,\mathrm{kg\,ha^{-1}\,yr^{-1}}$ in

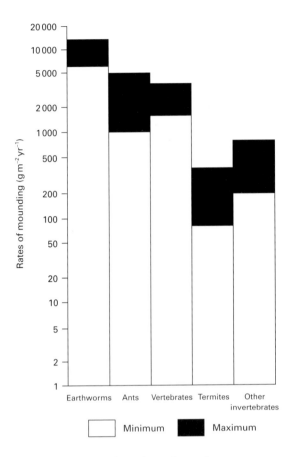

Figure 8.13 Comparison of rates of mounding or soil movement for earthworms, ants, vertebrates, termites and other invertebrates (Paton et al. 1995).

Australia (Briese 1982) to 842 kg ha^{-1} yr^{-1} in the USA (Whitford et al. 1986). As well as mixing the soil profile, ant activity alters the distribution of particle size, increases drainage and reduces bulk density. Studies in agricultural fields have shown that the effectiveness of water infiltration during irrigation depends on the spatial and temporal pattern of ant burrows.

8.6.2 Chemical effects

Ants that form large, stable nests can concentrate the organic materials they collect as forage in the vicinity of the nest. Consequently, ant-nest soils are often rich in both organic matter and mineral products of organic decomposition, especially phosphorus and nitrogen (reviewed in Beattie 1985, Folgarait 1998). These differences seem greater in poor soils and appear to depend on ant-colony size, biomass and turnover. Chemical differences between ant nests and surrounding soil can be very dramatic. McGinley et al. (1994) for example, found nests of *Pogonomyrmex* had six times the amount of ammonium and nitrate than off-mound soil.

8.6.3 Microbial and detritivore activity

While the physical and chemical effects of ant nests on soils have been recognized for a long time, the effects of ant nests on the abundance and composition of microbial and microarthropod communities has only received attention more recently. Abandoned nest mounds of *Pogonomyrmex occidentalis* in semi-arid shrub steppes of the Great Basin, USA, were found to contain a mat of densely packed roots with 2000–5000 times the vesicular arbuscular (VA) mycorrhizal fungal-spore density found in the surrounding undisturbed vegetation (Friese & Allen 1993). This level of microbial enrichment could potentially influence local plant establishment for many years after the ants move on by favouring the establishment of mycotrophic plant species.

The faunal composition of ant nests may also distinguish them from surrounding soil. Nests of *Pogonomyrmex barbatus* were found to support 30-fold higher densities of microarthropods and 5-fold higher densities of protozoa than surrounding control soils (Wagner et al. 1997). The upper surface of the nest mound of the wood ant *Formica aquilonia* harboured an abundant and functionally specialized decomposer community dominated by species concentrated at the lower trophic positions (microbes, microbivores and detritivores), some of which were ant-nest specialists (Laakso & Setälä 1998).

The microbial effects of ants on soil have important consequences for processes such as nutrient cycling and energy flux (reviewed by Folgarait, 1998). Fluxes of 13 chemicals were increased by an average of 38-fold in nests of *Atta colombica* in Panama compared to surrounding areas of forest, probably due to the greater root activity close to the ant nests. Higher rates of N mineralization, most likely due to increased microbial activity, have also been found. These changes may be particularly important in infertile environments with low levels of organic matter.

8.6.4 Effects on plants

Numerous studies have shown that plant abundance and

severely disrupt distance dispersal by ants, with flow-on effects for plant-population dynamics. A comparison of seed removal by ants on natural, disturbed and waste rock sites found that both removal rates and dispersal distances were affected (Andersen & Morrison 1998). In South Africa, a breakdown in ant–seed relationships (including a reduction in dispersal distances) due to invasion of the introduced Argentine ant *Linepithema humile* (formerly *Iridomyrmex humilis*) similarly threatens the persistence of myrmecochorous plants (Bond & Slingsby 1984).

The sensitivity of ant-species composition to changes in vegetation structure and disturbance, and the relative ease with which they can be sampled and identified, has led to their increasing use as ecological indicators, particularly in relation to mine-site rehabilitation. They are also used to monitor change due to habitat fragmentation and disturbances such as fire, grazing and logging. Whitford et al. (1999), however, found that ant communities in the south-western US desert grasslands showed no consistent patterns with grazing stress, soil disturbance, exotic plant dominance or herbicide application, suggesting that their use as indicators of ecosystem condition may be limited to particular environments.

8.8 Why are ant–plant interactions so common?

Plants are unlike animals in that their basic structure is dispersed rather than compact. A typical plant is a stem with multiple branches above ground and multiple roots below. The leaves and flowers are dispersed on the branches, and seeds and fruits may be dispersed either on the branches or on the ground. One solution to harvesting highly dispersed resources such as extra-floral nectar, food-bodies or elaiosomes is provided by the colonial organization of ants, their territoriality, their weaponry and their relatively omnivorous diets. Thus, large numbers of interdependent foragers (the worker caste) can recruit to a resource such as a tree *en masse* but then split up and harvest the small amounts of rewards scattered across the entire plant structure. The transfer of food between workers, or trophallaxis, ensures that food is shared with less successful foragers and nest-mates. The same workers may then guard the resources against competitors, armed with mandibles, stings or offensive sprays. Finally, omnivorous ants capitalize on the situation by harvesting both plant rewards and fresh meat in the form of prey, some of which will be foraging for the same resources as the ants. The presence of a soldier caste in the ants and domatia in

the plants will reinforce these processes. Behind this rationale is the assumption that wherever the plants are growing, ants will be present. Given the ubiquity, abundance and broad ecological tolerance of ants, this assumption is a fairly safe one. However, the most appropriate ants either may not be present at all or present only in low densities. This means that access to plant rewards by ant species with sub-optimal behaviours with respect to the services required by the plants can be a major source of variation in the outcome of ant–plant interactions.

8.9 Specialization and variation in the outcome of ant–plant interactions

The degree of specialization of interactions among the ants and plants in any given habitat may vary greatly. In non-symbiotic mutualisms such as those between ants and extra-floral nectary-bearing plants, assemblages of compatible plant and ant species will vary in time and space. Variations in the outcome of these interactions arise from differences in the abundance and effectiveness of visitor species and the quality of the host rewards, all of which may fluctuate markedly over short periods. On longer time-scales, variation will be produced by differences in the generation times of the participants, their relative levels of interdependence and their capacity to respond to selection relevant to the mutualism. The evolution of more highly specialized, non-symbiotic and symbiotic mutualisms may be possible and this will be discussed shortly. Increased fitness in the ants and/or plants through the acquisition of benefits reinforces selection for the mutualistic traits.

Variation in these interactions generates a continuum in the nature of the outcome that ranges from antagonism (or parasitism) to mutualism. Antagonism arises when ants take extra-floral nectar but provide no protection or even cause populations of homopterans to flourish, or when ants take floral nectar or seeds without providing pollination or dispersal. Weak mutualism may result from ant-guards attacking some herbivores but not others, or when only a small proportion of seeds is relocated to suitable places for germination. Strong mutualism is implied when plant rewards result in highly effective services with significant increases in fitness. We use the word 'implied' because there have been many studies showing increases in plant fitness resulting from ant services, but there is a paucity of studies showing increases in ant fitness resulting from the provision of plant-borne rewards. Strong interactions conferring mutual benefits may provide the basis for coevolution involving extreme specialization. An example of chemical specialization comes from the ant-guard system in *Acacia zanzibarica* and *A. drepanolobium* in East Africa. The ants attack insects on the trees thus apparently posing a conflict of interest as the young leaves most vulnerable to herbivory emerge at the same time as the flowers—which require insects for cross-pollination. Willmer and Stone (1997) showed that open flowers emit volatile chemicals that repel ants but not pollinators. They are released only while the flower is sexually active so that the ants guard both the young flower-buds and the developing fruits. The fine line between parasitism and mutualism was illustrated by Stanton et al. (1999) in a study that also involved the ants that occupy the thorn domatia of *A. drepanolobium*. Fierce competition occurs between the four species of ants that occupy these domatia, but one species in particular, *Crematogaster nigriceps*, tends to be evicted from its host tree more often than the others. This ant defends itself by pruning its host trees, thus limiting contact with adjacent ones that may harbour competitively superior ant-raiders. Unfortunately its pruning activities also remove most flowers so that, unlike the other domatia-inhabiting species that do not exhibit this behaviour and are mutualists, *C. nigriceps* is a parasite. Stanton et al. (1999) suggest that community attributes strongly affect the mix of mutualistic and parasitic interactions. Thus, the high densities of *Acacia* trees, the low density of other tree species and, of course, the presence of competitively dominant plant-ants using the same host-tree species, have been crucial in the evolution of this complex of interactions.

Another continuum can be seen in the degree of specialization of the ants and plants. In non-symbiotic mutualisms adaptations include extra-floral nectaries or elaiosomes in the plants and the appropriate foraging behaviour in the ants, and these are relatively common traits. In rare circumstances there can be extreme specialization in a non-symbiotic mutualism, as in the pollination of the orchid *Leporella fimbriata* by the ant *Myrmecia urens*. Here, the evolution of an extremely specific chemical analogue of the female ant attractant results in pollen transfer while the male ant attempts copulation with the flower. This degree of specialization among non-symbiotic mutualists is rare because the visitors generally forage on resources that are variable both in time and space, and often ephemeral. It is significant that in the case of *M. urens*, it is not nutrition that the visitor is seeking but a mate, which

converts the system from one that can be very generalized to one that must be highly specialized.

Extreme specialization is more frequent in symbiotic mutualisms, as first demonstrated by Janzen's (1966) studies of the *Acacia–Pseudomyrmex* system in which the plant provides swollen thorn domatia and food-bodies for its specialist ant-guards. In return for this complete diet and safe housing, the ants remove a variety of the plant's enemies. Parallel examples have been carefully documented for *Cecropia–Azteca* ant systems in South America, *Leonardoxa–Aphomomyrmex* systems in Africa (Davidson & McKey 1993), and *Macaranga–Crematogaster* systems in SE Asia (Fiala et al. 1991). Thompson (1994) suggested that the plants must create a 'selective sieve', that is, a mechanism that generates fidelity in a visitor species so that the host receives reliable services. In the orchid *L. fimbriata*, the sieve is a unique chemical that attracts only the relevant male ant. In many domatia-bearing plants the sieve may be the size and structure of the entrance hole to the domatium, and Davidson et al. (1988) provided evidence that the 'mesh size' of the sieve is effectively reduced by the forests of trichomes on the plant surface, which restrict the presence of ants to those species small enough to pass between them. The apparent absence of an effective 'sieve' that screens out highly specialized parasites of the system has been illustrated by a study of the ant *Allomerus demerarae*, which occupies domatia on *Cordia nodosa* in SE Peru and protects new leaves from herbivory. However, it also destroys the flowers, so that host plants produce no seed. This seems to be beneficial for the ant as deflowered plants produce extra domatia. Yu and Pierce (1998) have shown that this apparently impossible situation persists because the other coevolved plant-ants, three species of *Azteca*, are efficient ant-guards that do not damage the flowers and occupy most of the plants in any given area. The *Allomerus* is thus a highly coevolved parasite.

8.10 Convergent evolution, competition and asymmetry

Given the variation and asymmetry in ant–plant interactions, what makes them work? First, it is likely that competition among ants for the rewards is a major driving force. In many habitats there is a variety of omnivorous species competing for food and nest sites so that workers are present in large numbers, both foraging for these resources and defending them. The various services flow from this activity with varying degrees of intensity and effectiveness. Second, convergent evolution has played an important role in fostering ant–plant interactions. The evolution of similar traits in a variety of plant families assembled in particular environments is evidence for convergent evolution and the selective forces are more those determined by particular habitats rather than by particular species, although some selection may be exerted by assemblages of ant species that happen to be present. This can be seen in harsh environments such as arid zones and high mountains, which select plant forms appropriate for pollination by ants. It can also be seen in the early successional or high-light habitats, where many plants bearing extra-floral nectaries and domatia are common, and in the low-light or low-nutrient habitats where large numbers of plants produce seeds with elaiosomes. Thus, particular habitats generate similar adaptations among highly disparate plant families that inhabit them and, once they are there, ant assemblages take advantage of the adaptations, in turn generating interactions that may vary from antagonism to mutualism and, in some circumstances, from weak to strong symbiotic mutualism.

The dependence of the outcome of an interaction on local conditions may suggest limits to which partners can specialize. Numerous studies have shown that different species of ants have different levels of efficiency when it comes to servicing host plants, for example, in ant-guard systems, seed dispersal and pollination. These documented continua of effectiveness suggest that while extreme specialization among the ants may not be common, the presence of particular colony organizations and morphological and behavioural traits are clearly required. Average colony size is an important trait for domatia-dwellers (Davidson & McKey 1993), and body size can be crucial, for example, when crawling on host plants covered in dense trichomes (Davidson et al. 1988), or in response to limits to plant-borne rewards (Gaume et al. 1997). Foot morphology may also be important when negotiating host plants with slippery epicuticular wax coverings (Federle et al. 1997). Thus, although there is an impression of asymmetry in the levels of adaptation among the ants and plants, the different interactions should be considered independently. Some plants produce structures such as extra-floral nectaries or elaiosomes, which are used by a wide variety of relatively unspecialized ants. Others make domatia and food-bodies that are exploited by ant species with a more specialized range of morphologies, colony organizations and

behaviours. The obvious and sometimes dramatic adaptations in plants may divert attention from more subtle ones in ants. Also, ant adaptations may be more difficult to observe and to unambiguously designate as having evolved in response to plant rewards. Asymmetry is the focus of intriguing current research as arrays of apparently coevolved ants and plants are studied in minute detail. For example, among the subfamily Pseudomyrmecinae, which has strong associations with domatia, it appears that 37 out of 230 species are obligate to domatia and these highly specialized species inhabit 20 plant genera in 14 different families (Davidson & McKey 1993). The domatia may have evolved to be very exclusive, but even specialist ants may inhabit and service a variety of plant species.

8.11 Some questions for the future

The questions that remain to be both asked and answered are as multitudinous and diverse as the interactions themselves. Here we highlight a few areas where further research may be fruitful in increasing understanding of how these interactions evolved and how important they are in shaping community structure.

1 *The extent of plant mediation of ant behaviour by chemical signals.* Most ant species that interact with plants are omnivores, obtaining nutrients from a wide variety of sources. Plants offering rewards for specific services, such as pollination, dispersal or protection, must therefore compete for ant visits not only with each other but with other potential food sources. Plants may encourage ant visitation and fidelity in a number of ways, such as using behavioural releasers, or by providing essential dietary components that are unavailable or in short supply elsewhere (Beattie 1991). There is increasing evidence that plants harness a wide variety of chemical signals to attract not only ants in general, but possibly specific types of ants. In some systems, plants appear to have evolved 'animal-like' substances to attract omnivorous ants, such as the insect-like fatty-acid composition of elaiosomes (Hughes et al. 1994b) and the animal-like glycogen in *Cecropia* food-bodies (Rickson 1979). The finding that fungi are also apparently capable of manipulating the foraging behaviour of leaf-cutters via semiochemical production (Ridley et al. 1996) is another indication that chemical mediation of ant behaviour plays a major part in ant–plant interactions.

2 *Effects of plant rewards on ant-colony demography.* At one level, it is clear by observation that ants benefit from being provided with shelter and food by plants. However, the specific effects of plant rewards on colony fitness have rarely been measured. It is thus unclear whether ants prevented from collecting a particular type of plant reward, such as elaiosomes or nectar, would simply gain these nutrients elsewhere or, alternatively, suffer significant consequences. The finding by Morales and Heithaus (1998) that provision of elaiosomes affects colony sex ratio in *Aphaenogaster rudis* is a tantalizing indication that specific and perhaps surprising effects of plant rewards may be found with appropriate experimental manipulation.

3 *Comparative experimentation and surveys across habitats.* Most of the papers cited in this chapter involve experimental manipulations on one or a few ant and plant species in a single location. Variation in the outcomes of many of these experiments, as discussed above, suggests that generalizations about the maintenance and evolution of ant–plant interactions are more likely to be revealed if experiments are conducted on wider spatial scales, especially along environmental gradients such as latitude, aridity, soil fertility and vegetation type.

4 *What is the extent and frequency of ant specialization?* Research into myrmecophytic plants and their ant inhabitants has begun to reveal levels of specialization among ants not previously known. Aside from the chemical signals that may be involved (see above), there appear to be suites of behavioural, morphological, physiological and colony traits that both predispose ant species for coevolution with plants and determine the course of that evolution. As the literature shows that these traits can be difficult to elucidate in the field and even harder to confirm experimentally, this area requires both a deep knowledge of the natural history of the ants and a lot of imagination or, perhaps, the ability to think like an ant.

5 *How many more ant pollination systems are there?* Research to date has identified some of the constraints on the evolution of these systems. Now that we are beginning to understand them, will there be many more in those environments where they have been predicted to occur, or have both evolutionary and ecological constraints limited the number to the region of those that have been identified?

6 *The extent to which plant rewards structure ant communities.* It is possible that assemblages of plant species with particular rewards such as nectar, extra-floral nectar, honeydew, elaiosomes and domatia determine community attributes such as ant species diversity and abundance. In some habitats a large number of species bear extra-floral nectaries and an even larger number harbour honeydew-secreting homopterans. Competition between ant species and colonies, together with aggressive territorial behaviour, may mean that ant 'mosaics', first identified by Leston (1973) in tropical tree crops, may be a common feature in many other kinds of environment.

Chapter **9** **Plant–animal interactions: future directions**

John N. Thompson

9.1 Introduction

Most of the earth's diversity is in its interaction biodiversity — the tremendous variety of ways in which species are linked together into constantly evolving networks. As we have learned more about the genetic diversity of life, we have come to understand that a large fraction of those links is in the connections between photosynthetic plants and the animals that consume them. The study of plant–animal interactions has grown, diversified and matured considerably in recent years. The infusion of evolutionary biology into the ecology of plant–animal interaction has guided the development of that progress. The wealth of natural-history knowledge gathered by several generations of evolutionary ecologists and community ecologists has pointed to the big questions that we need to answer and suggested hypotheses about pattern and process. The questions have been continually refined and moved up to larger geographical scales. In this concluding chapter I concentrate on the major questions we still need to answer and the reasons why an evolutionary perspective must be part of the experimental protocols we adopt in getting to the answers.

9.2 Major general questions on the ecology and evolution of interactions

We are currently faced with nine major questions on the ecology and evolution of species interactions. The questions overlap each other, but each addresses aspects of species interactions that we need to understand if we are to conserve the earth's biodiversity among increasingly fragmented natural landscapes. The questions affect all interspecific interactions, and they provide the template for the major specific questions we need to answer on plant/animal interactions.

9.2.1 What determines the degree to which species specialize in their interactions with other species?

Much of the evolutionary history of the diversification of life is a history of specialization in interspecific interactions. All moderately large phylogenetic lineages are collections of species that differ in the degree to which they depend upon one or a few other species for survival or reproduction. Swallowtail butterflies include species restricted to feeding on only one plant species as well as other species recorded from scores of hosts (Thompson 1998b), and similar diversification in many insect lineages is linked to specialization on different hosts or combinations of hosts (Strauss & Zangerl, Chapter 3 of this volume). The same holds for plants. The plant genus *Dalechampia* includes species that depend upon resin-collecting bees for pollination and other species that rely upon a wider range of pollinator taxa (Armbruster & Baldwin 1998). All the chapters in this volume have shown clearly that the degree and pattern of specialization among species is one of the most central problems in the ecology and evolution of species interactions.

The proportion of extreme specialists and relative generalists in species interactions differs among lineages and habitats. The search for the processes producing those patterns remains one of the most important problems in biology, because it is fundamental to our understanding of how biodiversity is organized locally and globally and how species respond to environmental change. Part of the answer is in the history of lineages. We know that a large part of community structure is phylogenetically constrained, that is, species do not interact randomly with one another. Closely related species within a lineage commonly interact with closely related species in other lineages. Moreover, the traits used by species in their interactions are jury-rigged from the traits inherited from their ancestors, constraining adaptation in particular directions. As Herrera (1995, Chapter 7 of this volume) and Jordano (1995) have shown for the evolving interactions between birds and fruits, many of the major features of these interactions are shared among closely related species and sometimes genera, and cannot be interpreted as the result of direct and recent selection on each individual species.

Phylogeny, however, does not impose a straitjacket on the structure of interactions—e.g. although geometrid moths are a lineage of herbivores, some Hawaiian species have become carnivores. Rather, it imposes some historical structure on patterns of specialization and the organization of biodiversity (Futuyma & Mitter 1996). Phylogenetic and molecular tools are together providing us with increasingly powerful ways of understanding how specialization in interactions is shaped by a combination of historical constraint and current ecology. For example, Armbruster's (1997, Armbruster & Baldwin 1998) long-term pantropical studies of the evolution and ecology of pollination in the plant genus *Dalechampia* have shown that the specialization to resin-collecting bees has arisen multiple times in different ecological contexts. Moreover, those specialized adaptations have sometimes been replaced as populations have colonized geographic areas lacking resin-collecting bees, as occurred when *Dalechampia* populations from Africa colonized Madagascar (Armbruster & Baldwin 1998).

These kinds of studies are crucial if we are to understand how interactions between plants and animals have been continually moulded and remoulded over evolutionary time. They are part of a phylogenetic approach to community ecology and must become a central part of community ecology if we are truly to understand the community structure of biodiversity. Although various approaches are developing (Ricklefs & Schluter 1993; Futuyma & Mitter 1996; Losos 1996; McPeek et al. 1996; Tofts & Silvertown 2000), they are not yet part of the mainstream of community ecology.

9.2.2 What are the combined dynamics of antagonistic and mutualistic interactions in communities?

Interactions differ in both their form and their outcome. Parasitism, grazing and predation are all fundamentally different forms of antagonistic interaction, imposing different selection pressures on victims and their enemies (Price 1980; Thompson 1982, 1994). Similarly, symbiot-

ic and non-symbiotic mutualisms differ in the selection pressures and ecological dynamics they impose on interactions. In addition, many antagonistic interactions rely on mutualisms for their success, and mutualisms are always in danger of invasion by antagonists. Pellmyr's work on the evolution of non-pollinating yucca moths that both invade and rely upon the yucca/yucca-moth mutualism is one of the clearest examples (Pellmyr et al. 1996a; Pellmyr 1999). But we are only starting to understand how networks of different kinds of antagonistic and mutualistic interactions shape community dynamics and ongoing natural selection. Most analyses of community dynamics remain narrowly focused on antagonistic interactions and, more particularly, on predation and grazing.

For example, food webs and trophic cascades offer two partial views of how communities are structured. Unlike more 'stable' food webs, trophic cascades are characterized, among other things, by linear food chains, distinct trophic levels, low species diversity, positive feedback and at least occasional runaway consumption of plants (Strong 1992; Polis & Strong 1996; Jeffries 1999; Strong 1999). Although debate over the commonness of trophic cascades continues, there is some evidence that human-induced changes in ecosystems may increase the likelihood of trophic cascades and the great fluctuations in community composition that they can produce (Jeffries 1999). Consequently, these are important problems in community ecology. Nevertheless, the approaches to getting answers to these questions have so far remained focused on a small subset of trophic interactions. I do not know of a single study that has analysed in detail the relative influence of mutualistic and antagonistic interactions on trophic cascades and food webs. Individual researchers, of course, can do only so much. The future of community ecology will demand much more collaboration with researchers with expertise in parasitology, microbiology and mutualistic interactions.

9.2.3 How does the geographic structure of species shape interspecific interactions, and how do interactions shape geographic structure?

We now have thousands of studies on geographic patterns of population differentiation in species. Plants differ among populations in many traits, including their defences against herbivores and the rewards they offer to mutualists (Linhart & Grant 1996). Herbivore populations differ in their counter-defences to particular plant species (Jong & Nielsen 1999) and in their degree of specialization to one or a few local plant taxa (Singer & Thomas 1996). Extreme local specialization of insects to long-lived individual plants within populations has now been shown repeatedly, creating yet another level at which panmixis breaks down within plant–animal interactions (Mopper 1996; Mopper & Strauss 1997; Van Zandt & Mopper 1998). Local adaptation is the raw material for the processes shaping the ecology and evolution of interactions over broad geographic scales.

In addition, the species involved in particular interactions can vary tremendously over landscapes. Lodgepole pines coevolve mostly with red squirrels in the northern Rocky Mountains, but they coevolve with red crossbills in outlying mountain ranges where red squirrels are absent (Benkman 1999). The annual legume *Amphicarpaea bracteata* differs geographically in the degree to which individuals are adapted to particular genotypes of the nitrogen-fixing *Bradyrhizobium* sp. (Parker 1999). These geographic differences in species interactions create structure at levels beyond the local adaptation of one population to another.

We need to continue to refine our understanding of how the hierarchical geographic structure shapes the ecological and evolutionary dynamics of interspecific interactions across landscapes. It is part of the scaling up of ecology and evolutionary biology to assess the relevant spatial and temporal scales at which major processes shape the dynamics of biodiversity.

9.2.4 How does the coevolutionary process shape biodiversity?

As research on the evolutionary ecology and evolutionary genetics of species interactions has accelerated in recent years, it has become increasingly evident that coevolution is a pervasive process organizing biodiversity. The process sometimes produces marvellously specialized and mutually dependent interactions such as those between figs and fig wasps (Bronstein & Hossaert-McKey 1996; Herre & West 1997), yuccas and yucca moths (Pellmyr et al. 1996b; Addicott 1998), and similar coevolved interactions between plants and floral parasites (Pellmyr 1992; Fleming & Holland 1998; Pellmyr, Chapter 6 of this volume). More commonly, however, coevolution is a subtle but pervasive ecological process, continually shaping and

reshaping local populations of interacting species without always producing long-term directional selection towards extreme morphologies and behaviours.

The geographical mosaic theory of coevolution argues that coevolution is an ongoing tripartite evolutionary process that moulds evolving species interactions across landscapes (Thompson 1994). Selection acts differently on interactions in different communities (selection mosaics); selection is reciprocal on an interaction in some communities (coevolutionary hot spots) and non-reciprocal in other communities (cold spots); and the genetic landscape on which the interaction evolves constantly changes over time through gene flow among populations, random genetic drift, and extinction and recolonization of populations (trait remixing). As a result, coevolution may be a crucial organizing force on interactions across geographical landscapes without occurring in every interaction in all populations at all moments in time. Recent mathematical models have indicated that selection mosaics, coevolutionary hot spots and trait remixing are all important components of geographically structured coevolution and may result in broad regional patterns quite different from those imposed by local natural selection (Gandon et al. 1998; Hochberg & van Baalen 1998; Nuismer et al. 1999; Gomulkiewicz et al. 2000).

Appreciation of the geographic structure of coevolution has allowed for the development of specific hypotheses on the coevolutionary process, based on the population and geographic structure of species. We have moved beyond simple descriptions, such as arms races, mutualism and competition, to hypotheses based upon the genetic and ecological structure and dynamics of coevolution: geographically structured gene-for-gene coevolution, coevolutionary alternation, coevolutionary turnover, escape-and-radiate coevolution and others (Thompson 1999b). These hypotheses have pointed to coevolution as a highly dynamic, ongoing process.

Some of the best evidence for the ongoing role of coevolution in the organization of biodiversity across geographic landscapes has come from studies of interactions of plants and animals. Geographic studies of interactions including a wide range of plants and animals have now shown that coevolution is a highly dynamic process that links the diversity of life across broad geographical landscapes. These include, for examples, studies of lodgepole pines and red crossbills (Benkman 1999;

Hulme & Benkman, Chapter 5 of this volume), wild parsnip and parsnip webworm (Berenbaum & Zangerl 1998), and saxifrages and *Greya* moths (Thompson 1999d). This geographic stucturing of species interactions is part of the workaday evolution that keeps populations in the evolutionary game. It seems increasingly likely that long-term conservation of species will succeed only if these broad-scale geographical processes are also conserved.

9.2.5 How do species cope with multiple enemies and mutualists?

As evidence accumulates on the pervasive influence of coevolution as an organizing influence on biodiversity, we are being confronted with the problem of how it occurs amid the network of interactions within communities. Most eukaryotes probably interact with multiple species during their lifetimes, but the ways in which organisms cope with multiple enemies and mutualists remain mostly unresolved. Almost all the chapters in this volume confront issues related to this problem. Some field studies over the past decade have shown the inevitability of attack by multiple enemies, at least in long-lived species. During a 10-year study of a goldenrod species (*Solidago altissima*), Maddox and Root (1990) found most plants surviving for more than 5 years were attacked at least once by most of the 17 *Solidago*-feeding insect herbivores found within the local communities. During a similar 10-year study of the herbaceous perennial, *Lomatium dissectum*, all plants surviving for a decade were attacked by at least 2 of 5 local enemy species (insects and pathogens), and plants were attacked on average by 1.6 enemy species each year (Thompson 1998a).

Part of the answer to the problem of coping with multiple enemies and mutualists is through ontogenetic partitioning of interactions. Species specialize on different species at different life-history stages, turning genes on and off in the process. Genes for floral expression develop only at particular life-history stages, and plant-chewing caterpillars metamorphose into nectar-feeding butterflies. More subtle changes occur in many other taxa, producing defences only at particular life-history stages or inducing defences against particular enemies only when needed (Karban & Thaler 1999). Studies of how natural selection can shape the ontogenetic partitioning of interactions are among the major questions that remain in de-

veloping an evolutionary ecological understanding of the organization of biodiversity.

Even with ontogenetic partitioning, species must often still cope simultaneously with multiple species (e.g. Linhart et al. 1994). For plants, that can mean not only attack by multiple enemies but also linked interactions involving both herbivores and mutualists. Recent studies of the links between herbivory and pollination have demonstrated that interactions between their enemies and mutualists are linked both ecologically and evolutionarily (Pellmyr, Chapter 6 of this volume). Herbivory on leaves or flowers can influence the pattern of visitation by pollinators (Strauss 1997; Krupnick & Weis 1999; Pellmyr, Chapter 6 of this volume) and shape the pattern of selection on floral phenology (Juenger & Bergelson 1998). Moreover, plant defensive traits can be co-opted over evolutionary times as attractants for particular pollinators (Armbruster et al. 1997). These studies are beginning to dissect how natural selection shapes genomes to confront, sometimes simultaneously and sometimes sequentially, these kinds of multiple selection pressures in natural populations. In addition, controlled studies of natural selection driven by two enemies are helping to sharpen the questions by analysing the adaptive landscapes that result from such combined selection pressures (Simms & Rausher 1993; Mitchell-Olds & Bradley 1996). Recent experimental studies (Bohannan et al. 1999) using *E. coli* and two bacteriophages have indicated that such adaptive landscapes may be shaped by epistatic interactions among genes. For *E. coli*, epistasis between genes conferring resistance to the two enemies reduced the cost of multiple resistance in some environments. Bohannan et al. note that, if such a pattern is general, it may result in different coevolutionary trajectories in different environments. A major result of these conflicting selection pressures can be fluctuating natural selection (e.g. Linhart & Thompson 1999), which may be the most common form of natural selection on populations and an important force in community dynamics.

Studies of indirect interactions and tritrophic interactions are pointing to yet other ways in which multispecific interactions may be shaped within communities (Abrahamson & Weis 1997). We know that parasitoids and predatory insects can respond to plant volatiles, and there is now a group of studies suggesting that release of volatiles may be an evolved defence mechanism at least in some cases (Dicke & Vet 1999). But work on these possible plant–carnivore interactions within natural communities is only beginning. We will need larger-scale collaborative efforts to understand better how pair-wise interactions are shaped ecologically and evolutionarily within complex communities.

9.2.6 What is the role of species interactions in the diversification of life?

The role of single-celled organisms in shaping the overall diversification of life is beginning to become a major theme of research in ecology and evolution and will only grow more during the next century. The discovery of the rich diversity of bacteria, archaea and single-celled eukaryotes is just now under way (Service 1997), and the potential for lateral gene transfer among taxa adds yet another dimension to how life diversifies (Doolittle 1999).

Despite the clear evidence that most eukaryotes obligately rely upon single-celled symbionts for many basic aspects of survival and reproduction, we have yet to address it head on in our analyses of the ecology and evolution of interspecific interactions among eukaryotes. We now know that many of the major events in the history of life are not about the appearance of beneficial novel mutations but are rather about the incorporation of symbionts—whole packages of self-replicating genes (Buss 1987; Margulis & Fester 1991; Maynard Smith & Szathmáry 1995). Although one of the great ongoing events on earth has been the diversification of plant–animal interactions, we know little about the extent to which the diversification of microbial life has shaped the diversification of plant–animal interactions.

Such studies require a robust phylogenetic template for plant and animal diversification on which to infer process. Unravelling how animals and plants have shaped each other's diversification has become more feasible in the past few years as more detailed and robust molecular phylogenies have been published, using multiple genes. Recent major efforts to unravel plant phylogeny (Soltis et al. 1999) have provided the large-scale template for understanding the pattern of diversification in plants. Preliminary versions of this new template have been used to re-evaluate Ehrlich and Raven's (1964) hypothesis of escape-and-radiate coevolution (Janz & Nylin 1998). At finer phylogenetic levels, detailed phylogenies of yuccas and yucca moths are showing how patterns of geographic differentiation in interacting plant and animal populations foster further differentiation in the interacting species (Pellmyr 1999).

9.2.7 How are the genomes of organisms shaped through interspecific interactions?

The first six problems—evolution of specialization, the combined effects of antagonism and mutualism, the geographic structure of evolving interactions, the coevolutionary process, coping with multiple species and the role of interactions in diversification—are all part of the general evolutionary problem of how interactions shape the genomes of organisms and, reciprocally, how those genomes shape the ecology and evolution of interactions. Our understanding of the genetics of species interaction is changing so fast that anything we write today will soon seem like 'vague adumbrations of the truth' as Julian Huxley once wrote in regard to another major problem in biology (Gould 1977). The combined effects of nuclear and extra-nuclear genes, the effects of transposons on gene expression and the almost ubiquitous use of symbionts by eukaryotes (Price, Chapter 1 of this volume) are all becoming part of the solution of how species cope with one another.

As just one example, the rice weevil *Sitophilus oryzae* has four intracellular genomes that shape its physiology: nuclear, mitochondrial, '*S. oryzae* principal endosymbiont' (SOPE), and a *Wolbachia* species (Heddi et al. 1999). The latter two are bacteria (SOPE) or α-proteobacteria (*Wolbachia*) that have become intracellular symbionts. SOPE is highly integrated into the weevil's physiology, providing vitamins, increasing mitochondrial oxidative phosphorylation and interacting with amino-acid metabolism. In turn, gene expression in SOPE is partially controlled by its host. *Wolbachia* may be more of a parasite. Unlike SOPE, which is transmitted only vertically from parents to offspring, this *Wolbachia* is transmitted both vertically and horizontally and does not occur in all weevil populations. How these four genomes together shape the interactions with plants is only starting to be understood. Heddi et al. (1999) have argued that acquisition of SOPE allowed these weevils to colonize cereals by providing the vitamins it could not get otherwise.

9.2.8 How much of rapid ecological dynamics is caused by rapid evolutionary dynamics?

As we learn more about the population genetics of species,

we are finding that populations are often constantly changing. But the reality of ongoing, sometimes rapid evolution has not yet permeated the study of interspecific interactions. Instead, community ecology has been increasingly divorced from evolutionary ecology in recent years. Community ecology and physiological ecology, two of the mainstays of ecology during the first half of the last century, were two of the last strongholds of the Evolutionary Synthesis that took place during the middle third of the twentieth century. Three decades ago, as evolutionary ecology began to blossom as a science, it did so with close ties to both community ecology and population ecology. Communities are composed of interspecific interactions, and evolutionary ecologists began evaluating how those interactions were continually reshaped through evolutionary change to produce community patterns. But evolutionary arguments were extended beyond the data in these early studies, and answers about the links between evolution and community ecology turned out to be hard to get using short-term observational and experimental studies. The result in some quarters has been the retrenchment of community ecology into mechanisms of ecological dynamics that exclude rapid evolution of species and interactions as one of the working hypotheses.

Yet there have now been dozens of documented examples of the evolution of interspecific interactions during the last century alone (Thompson 1998c). These include a number of interactions between animals and plants. Among them are examples of the evolution of bill size in birds, which affects interactions with flowers (Freed et al. 1996) or seeds (Grant 1986), and the evolution of preference in phytophagous insects for particular plant species (Singer & Thomas 1996). Much of this selection in all interspecific interactions may be fluctuating selection, continually recycling alternative alleles in fluctuating environments. Australian wild flax and flax rust have shown great swings in resistance and virulence genes across multiple populations over the past decade (Burdon & Thrall 1999). Similarly, an 8-year study of the freshwater copepod *Diaptomus sanguineus* showed rapidly evolved changes in the timing of diapause among years, driven by natural selection imposed by predatory fish. The date in spring that females switch from the production of clutches that hatch immediately to those that remain in diapause until autumn is under genetic control. The timing and intensity of predation by sunfish, which can vary among years, is the major selective agent on diapause timing (Ellner et al. 1999). Such changes in timing of diapause surely also affect the species fed upon by the copepods. The result is a potential selective cascade of rapid evolutionary effects that may be constantly shifting within and across environments.

These accumulating examples indicate that rapid evolutionary change may be an important component of ongoing population and community dynamics. Consequently, rapid evolution needs to become one of the standard working hypotheses in the next generation of studies in community ecology. Examples of specific questions involving plant/animal interactions are these:

• To what extent do herbivores and pollinators reshape the genetic structure of local plant populations during successional change?
• To what extent does genetic change in plant population structure during local succession reshape the genetic structure of herbivore and pollinator populations?
• To what extent are the metapopulation dynamics of interacting herbivores and plants, or pollinators and plants, driven by rapid evolutionary change?

9.2.9 How do taxa differ in evolutionary rates and how do species interactions shape those rates?

Questions on rapid evolution are part of the larger group of questions we still need to confront on how phylogenetic history and rapid ongoing evolution shape species interactions at different spatial and temporal scales. Palaeobiological and neobiological approaches to questions on evolutionary rates still differ fundamentally in the questions asked. Palaeobiologists must see change in morphology to interpret change, whereas evolutionary ecologists, evolutionary geneticists and molecular evolutionists are confronted with rampant genetic changes within and among populations, most of which are never scaled up to major directional changes in morphology. The result is a range of very different views about evolutionary rates.

We need a better understanding of evolutionary rates, because our intuitions about rates shape our views of how species and their interactions evolve. Recent decades have included debates on whether parasites evolve faster than their hosts, and whether step-wise evolutionary change through simple Mendelian inheritance is more common in plants than in animals. Berenbaum and Zangerl (1999), for example, have noted the problem posed by the observation that insects have overcome synthethic organic insecticides at impressive rates, but plant defences are highly conservative. Why is it that parasites do not always win if they can evolve faster than the defences in their hosts? Part of the answer surely lies in the geographical mosaic of evolving interactions (scale effects) and fluctuating selection (temporal effects). But we still have little understanding of how geographical mosaics, fluctuating selection and rapid evolution interact to shape overall evolutionary rates in interacting species.

These questions about rates of adaptive change dovetail with questions on speciation rates, such as whether instantaneous speciation is more common in plants (primarily through autopolyploidy or allopolyploidy) than in animals? The theory of punctuated equilibrium itself has begged the question of how rates *per se* shape the organization of biodiversity through speciation and extinction events. All these questions are part of the increasing focus of research on the hierarchical structure of ecological and evolutionary processes.

9.3 The unique aspects of plant–animal interactions

In addition to the major questions that affect all species interactions, there is a group of problems that are becoming especially important to our understanding of the evolution and ecology of plant/animal interactions. These questions cut across the grain of many of the questions discussed in the earlier chapters, including the evolution of plant defence, the nutritional ecology of herbivores and the population dynamics of interacting animals and plants.

9.3.1 How does plant hybridization shape the geographical structure of plant–animal interactions and the diversification of interacting species?

Interspecific hybridization is common in many plant taxa, providing clines across landscapes and the possibility of reticulate evolution within many groups. There are now convincing data showing that hybridization among plant species shapes the intensity and geographical structure of interactions with animals, affecting not only herbivores but interactions with pollinators (Campbell et al. 1997) and even nesting birds (Fritz 1999; Whitham et al. 1999). Two major reviews, which collectively summarized 152 studies, have shown that hybridization signifi-

cantly affects antagonistic interspecific interactions in the majority of cases (Strauss 1994; Whitham et al. 1999). These studies have shifted plant hybrids from the status of freaks to that of important centres of ecological and evolutionary dynamics. Whitham and his colleagues, who have worked on a diverse range of plant hybrids in natural communities, have come to regard hybrids as important centres of biodiversity, deserving conservation in their own right (Whitham et al. 1991, 1994).

9.3.2 How has plant polyploidy shaped the evolution and ecology of plant–animal interactions and why have we ignored it for so long?

This is probably the largest gap in our understanding of plant–animal interactions. The majority of plants have arisen either directly through polyploid events or are descendants of polyploid ancestors. These polyploids have arisen either through chromosome doubling within species (autopolyploidy) or through the combination of complete chromosome sets of two species (allopolyploidy). Both forms of polyploidy can affect many plant traits and the distribution of individuals across habitats (Bretagnolle & Thompson 1996; Lumaret et al. 1997; Segraves & Thompson 1999; Segraves et al. 1999). Even though polyploidy is a pervasive force in plant evolution and diversification, and even though it is the basis of much of modern agriculture, it has been virtually ignored until recently in studies of plant–animal interactions within natural and managed communities.

Two recent studies have shown that autopolyploids can differ significantly in their interactions with herbivores (Thompson et al. 1997) and floral visitors (Segraves & Thompson 1999). These two studies, both involving the saxifrage *Heuchera grossulariifolia*, found that neighbouring diploid and polyploid individuals differ in their probability of attack by a specialist herbivore and in the suite of floral visitors they attract. Although in some populations it is easy to distinguish between diploids and polyploids in the field, in other populations the two ploidies look remarkably similar. The same is likely to be true of other plant species. Hence, even accomplished field biologists could miss these major genetic differences within and among the plant populations they are studying, if they did not directly test for plants' ploidy level.

Polyploidy is also a much more ecologically and evolutionary dynamic process than previously thought (Soltis

& Soltis 1993; Ramsay & Schemske 1998; Soltis & Soltis 1999). In *Heuchera grossulariifolia*, polyploid populations have arisen between 2 and 7 times (Wolf et al. 1990; Segraves et al. 1999). Similarly, two allopolyploid species of *Tragopogon*, which originated in the north-western US during the twentieth century from parents introduced from Eurasia, have each arisen multiple times (Novak et al. 1991; Soltis et al. 1995). These are now self-sustaining populations that have been naturalized in the Palouse region of eastern Washington State for more than half a century. They may or may not persist for millennia, but their rapid and repeated origin, coupled with their ability to sustain themselves, points to the highly dynamic nature of plant polyploidy. To the extent that polyploidy affects interactions with animals, recurrent and highly dynamic polyploid formation in autopolyploids and allopolyploids could continually reshape the geographical structure of plant–animal interactions.

9.3.3 How do symbionts shape the ecology and evolution of plant/animal interactions?

Gut symbionts and intracellular symbionts are common in herbivores, but the role they play in the evolution of specialization and speciation is mostly unknown. Most aphids obligately harbour *Buchnera* bacteria in special cells called mycetocytes. The bacteria, which are maternally inherited and are themselves obligate symbionts, provide essential amino acids that are deficient in the phloem on which the aphids feed (Douglas 1998). Molecular analyses have indicated that these highly coevolved obligate interactions originated 160–280 million years ago (Moran et al. 1993). Aphids and *Buchnera* have diversified in parallel ever since, with the *Buchnera* transmitted essentially as organelles. Only a few experiments have begun to explore the role of *Buchnera* in patterns of aphid specialization (Douglas 1998). Most other Hemiptera have bacteria-containing mycetocytes, as do many weevils and chrysomelid beetles, which are among the most diverse taxa of plant-feeding insects. Woodboring beetles commonly harbour fungal symbionts (Douglas 1998). Add to this the growing list of animal taxa that harbour *Wolbachia* bacteria, which can shape patterns of reproductive compatibility among populations (Bouchon et al. 1998; Zhou et al. 1998; Rigaud 1999) and the 'problem of symbionts' becomes one of the most pervasive problems to be solved in the study of plant–animal interactions.

9.3.4 How are plant diseases and plant–animal interactions related ecologically and evolutionarily?

Until recently the study of the evolutionary ecology of plant–insect interactions and the study of plant diseases have proceeded along independent research tracts, even though the connections are evident. Many plant diseases are transported by insect vectors, and many plant defences are shaped as much (or more) by pathogenic bacteria, fungi and viruses as by phytophagous animals. Study of the evolutionary ecology of plant–pathogen interactions has shown that major aspects of the population biology and evolutionary genetics of plants can as readily be attributed to pathogens as to interactions with animals (Burdon 1993; Parker 1993; Roy & Bierzychudek 1993; Alexander et al. 1996; Clay & Kover 1996; Antonovics et al. 1998; Burdon & Thrall 1999). Nevertheless, there are still very few studies on how plants, pathogens and animals interact within and among natural populations. Examples include studies on endophytic fungi that affect the palatability and toxicity of plants to herbivores (Clay & Kover 1996), the combined effects of fungi and insect herbivores on plant competitive ability (Clay et al. 1993), analysis of adaptive landscapes of plant defence shaped simultaneously by insects and pathogens (Simms & Rausher 1993), effects of pathogen attack on floral biology and pollination by insects (Roy 1994) and the complex relationships between insect attack and mycorrhizal associations (Gehring et al. 1997). These are the vanguard of what must be a future generation of studies on the ways in which plants, animals and microbes interact.

Work on the molecular genetics of plant defence has given further reasons to link studies of how plants respond simultaneously to different taxa of herbivores. The study of gene-for-gene interactions has been the purview of plant pathology for decades, but relatively little research has been devoted to the possibility of gene-for-gene interactions between plants and insects or nematodes. The demonstrated gene-for-gene relationships between wheat and Hessian fly have been considered the exception. But more recent work on phytophagous nematodes and insects has indicated that the genetics of the early stages of plant defence may be similar against a wide range of enemies that feed parasitically (Cook 1998; Rossi et al. 1998). The complex defences developed by plants against particular enemies are often fine-tuned and surely vary tremendously among taxa, as indicated by the chapters in this volume, but the genetics of the early stages of recognition of attack and the initiation of defence may function similarly regardless of the enemy.

9.3.5   Are there biogeographical patterns in plant defence that shape regional patterns in the ecology and evolution of plant–animal interactions?

Different plant floras have evolved with different combinations of herbivore taxa, but we have only started to explore how those differences shape overall community structure, resilience and resistance to invasion by introduced taxa. At the extreme some floras, such as those of the Hawaiian Islands, have had no evolutionary history of grazing by large mammals. In addition, there are distributed worldwide a group of mainland floras that have had few large mammals, and that history is evident in the life history, morphology and physiology of the plants. For example, although the Great Plains of North America east of the Rocky Mountains had a long evolutionary history of grazing by herds of large mammals, the grasslands and steppe communities of the northern Intermountain West to the west of the Rockies lacked these herds. Those differences in evolutionary history are stamped in the differences in the traits of grasses on either side of the Rockies (Mack & Thompson 1982). Introducing cows on either side of Rockies over the past 100 years has had completely different effects on these plant communities, altering grasslands east of the Rockies while quickly devastating grasslands of the northern Intermountain West. Unlike the rhizomatous grasses east of the Rockies, the particular caespitose forms of grasses in the northern Intermountain West died quickly under the hooves of large grazing mammals.

Similar community-wide patterns have been suggested for macroalgal communities on either side of the Pacific Ocean (Steinberg et al. 1995) and along a temperate–tropical gradient in the Atlantic Ocean (Bolser & Hay 1996). In both oceans, algal defence levels are higher in regions that may sustain higher chronic levels of herbivory. In the North Pacific, sea otters prey heavily on macroinvertebrates, thereby keeping herbivory on macroalgae to low levels. In contrast, no large keystone predator capable of reducing herbivore numbers occurs in similar habitats of temperate Australasia. The two floras differ in ways consistent with differences in levels of chronic herbivory. Australasian kelps and fucoids have phlorotannin concentrations 5–6 times that found in

similar North Pacific algae. Moreover, these plants show higher levels of tolerance to herbivory. Recent studies have shown that tolerance can evolve within natural plant populations (Strauss & Agrawal 1999). Hence, different biogeographical regions may differ in how natural selection favours the overall combination of defence levels and tolerance among plants within floras. Whether the observed major biogeographical differences in defence levels and tolerance are the result of differences in the evolutionary history of herbivory or other causes is not yet clear, but the pattern is there nonetheless.

These comparative community-wide studies of terrestrial grasslands and marine kelp forests suggest the need for more work on biogeographical patterns in plant defence, partitioning the effects of phylogenetic history, grazing history, climatic history and other indirect effects such as mutualistic interactions. We need a solid understanding of these biogeographical patterns and the processes that have produced them, because they can serve as important management tools worldwide. They can be an aid in deciding which management techniques should be tried, or at least initially avoided, in particular biogeographical areas.

9.3.6   How do plant–animal mutualisms shape biodiversity?

Much of evolution is about the co-opting of genes and gene products of other species. Nowhere is that more evident than in the evolution of pollination and seed-dispersal mutualisms. The use of other free-living species to move gametes and embryonic offspring provides the clearest examples of the importance of species interactions in the evolution and organization of biodiversity. Elimination of pollination of mutualisms would result within decades in the elimination of much of terrestrial diversity, as most plants in species-rich environments failed to reproduce. Similarly, the elimination of fleshy fruits would have devastating effects on bird and mammal species worldwide. Moreover, the global connections of taxa afforded by avian migration would be immediately disrupted, as a large proportion of migrants rely upon nectar and fleshy fruits to make transcontinental flights. Together, animal-assisted pollination and seed dispersal are so important that they deserve an ongoing central place in our studies of the ecology and evolution of biodiversity.

In fact, one of the defining characteristics of species-rich terrestrial ecosystems may be their diversity of plant–animal mutualisms. These mutualisms link species together and spread over evolutionary time as they accumulate yet more mutualists as well as exploiters of mutualism (Thompson 1982, 1994). The diversity of many of these mutualisms results both from the convergence of unrelated taxa and diversification within taxa (Beattie and Hughes, Chapter 8 of this volume; Herrera, Chapter 7; Pellmyr, Chapter 6). It is one of the major reasons why these interactions show less pair-wise reciprocal specialization than other interactions. The coevolutionary dynamics are inherently different from most other forms of interaction.

The result is a complex phylogenetic structure in mutualisms between free-living species. In most biogeographic regions of the earth, plants from different lineages have converged on similar subsets of animal mutualists that gain them nutrition, protection against enemies, pollination and seed dispersal. These convergent mutualisms are among the major organizing influences on terrestrial communities, and similar convergent mutualisms may become more evident with increasing evolutionary ecological research within aquatic communities.

Even the most highly specific mutualisms, such as those between figs and fig wasps and between yuccas and yucca moths, accumulate a whole microcommunity of associated species, many of which depend obligately on these mutualisms (Herre 1996; Pellmyr et al. 1996a). We are just now starting to realize how much these specialized mutualistic interactions act as foci for other interactions. New taxa of fig wasps and yucca moths and related 'cheater' taxa have been discovered within the past few years. One previously described yucca-moth species, *Tegeticula yuccasella*, has recently been shown to be a complex of eleven pollinator species and two non-pollinator species that exploit yucca fruits, ovipositing after the flowers have been pollinated (Pellmyr 1999).

Although it is easy to point to the potential importance of plant–animal mutualisms in terrestrial diversity, this has been translated into few major research efforts on how these interactions shape communities and ecosystems. In an analysis of 12 699 papers published in the top ten ecology and evolution journals between 1986 and 1995, Bronstein (1998) found that 33% of the articles were on species interactions but only 5% were on mutualisms. Even when included in community analyses, mutualisms are still treated somewhat as neat stories, add-ons to the fundamental food-web structure of antagonistic interactions within communities. Given what we now know,

there is no biological justification for this approach to community ecology. We need an approach that recognizes that mutualisms are an essential part of the core structure of communities.

The same problems apply to marine diversity. Marine community ecology away from the rocky intertidal is still a nascent science, as is the study of the evolutionary ecology of interactions in marine environments. Mutualisms are as ubiquitous within marine environments as they are in terrestrial environments, but we know very little about the ways in which mutualisms can act as keystone interactions within marine ecosystems. Moreover, the geographic scales at which networks of interactions evolve in marine environments is far from resolved. Many studies of rocky intertidal communities have been on small scales, focusing on interactions among near neighbours. Bertness and Leonard (1997) have argued that these studies understate positive interactions among species, because such interactions often operate over broader scales. At the other extreme, it is commonly assumed that many marine species of the open ocean show much less geographic structure than terrestrial taxa, because marine currents can move pelagic larvae great distances. Nevertheless, recent studies have been finding considerable geographic structure in some marine species (Palumbi 1985, 1994; McMillan et al. 1999; Bernardi 2000). Hence, we need careful studies not only of how mutualisms shape interactions within marine environments, both in nearshore and open ocean communities, but also of the geographical scales at which those interactions are organized ecologically and evolutionarily.

9.4 Approaches to the answers

The chapters in this volume suggest that getting answers to the major questions of plant–animal interactions demands stronger links among biological subdisciplines and increasing collaboration. The questions are too large for any single approach to provide robust answers, and the techniques are now too sophisticated for any single individual to understand all the nuances of more than a handful. Three kinds of links are providing the greatest insights in current research on plant–animal interactions and species interactions in general: studies on the hierarchical structure of species interactions, studies linking ecological and molecular approaches, and studies linking mathematical and empirical approaches.

We are in the midst of a great integration of the hierarchical structure of species interactions. Studies of the local ecological and evolutionary dynamics of interactions within communities are increasingly being placed in the context of metapopulation dynamics. These, in turn, are being placed in a context of the even larger-scale, and sometimes more stable, geographical mosaics that give structure to evolving interactions across regional and continental or oceanic landscapes. This hierarchical spatial structure is complemented by work on the phylogenetic hierarchical structure of interactions (Labandeira, Chapter 2 of this volume), placing interactions among species in the broader context of related taxa.

This overall appreciation of the hierarchical structure of ecological and evolutionary processes has become perhaps the greatest unifying theme in current ecological and evolutionary theory. It is permeating the study of species interactions, shaping our ideas on the local population dynamics of interacting plants and animals, the structure of the coevolutionary process, and the interpretation of patterns in phylogeny and the fossil record.

Linking of ecological and molecular approaches has been equally important. The ever expanding arsenal of molecular tools (e.g. DNA sequencing, RFLPs, RAPDs, AFLPs, ISSRs, microsatellies, SNPs, microarrays) will only increase in importance as we try to address questions of demographic genetics and the geographical and phylogenetic structure of species interactions. They are indispensable tools, complementing ecological studies. How best to integrate the molecular and ecological tools to probe species interactions is an ongoing process. It is tempting, for example, to use the results of molecular analyses of DNA sequences, RFLPs, microsatellites or other tools to infer the geographic structure of species interactions. But these tools provide only the overall template, primarily for neutral genes. The geographic structure of traits under selection in species interactions can be quite different, especially at finer geographic scales (e.g. Althoff & Thompson 1999, 2001)

Molecular approaches are also beginning to show how much of the genomes of organisms is devoted directly to species interactions. Comparative plant-genome research has progressed so far that over half of the 20 000 to 25 000 genes found in higher plants can now be assigned a probable function (Somerville & Somerville 1999). We now know at the molecular level that genes functioning directly to thwart enemies make up an important proportion of genomes. And many other genes, of course, have indirect effects on interactions with enemies and mutualists, adjusting life-history traits, phenology and metab-

olism. These studies will eventually be of tremendous use to ecologists and evolutionary biologists trying to understand how organisms cope with multispecific interactions.

The third major link among approaches is between mathematical and empirical studies. Recent mathematical models of coevolving antagonistic and mutualistic interactions are helping to focus empirical research by making predictions of the components of interactions most likely to drive the dynamics (e.g. Kaltz & Shykoff 1998; van Baalen 1998; Lively 1999; Nuismer et al. 1999; Parker 1999; Abrams 2000; Gomulkiewicz et al. 2000; Nuismer et al. 2000). In turn, the experiments are motivating new models with more biological realistic relationships among parameters. Such models will become increasingly important tools in guiding ecological and evolutionary experiments as we try to address questions and processes at ever larger scales. The same needs and advances apply to statistical methods, as our experiments on species interactions address increasingly multivariate questions.

9.5 Conclusions: why it matters now

This volume has focused on a strongly evolutionary approach to the ecology of interactions between plants and animals. In doing so, it has emphasized just how much we have learned about these interactions by casting an evolutionary eye on ecological questions. Species have phylogenetic histories that shape their ecological interactions and their responses to new environments. Metapopulations shape evolutionary dynamics as well as population dynamics. Specialization on other species in many plants and animals is much more evolutionarily dynamic than we expected a few decades ago. Nevertheless, we have only scratched the surface. Much of the study of the ecology of plant/animal interactions remains steadfastly non-evolutionary, and that can only constrain our ability to explain the dynamics of these relationships. We now know that interactions between plants and animals are continually evolving across complex geographical landscapes, and we know that some aspects of these interactions can evolve quickly, sometimes over the course of decades. Appreciating the role of ongoing and potentially rapid evolution is becoming increasingly important as we try to understand the dynamics of plant/animal interactions in an increasingly biologically fragmented world.

Human societies will continue to impose strong selection pressures on interactions between plants and their herbivores and mutualists in the environments that we are changing quickly worldwide. Rapid evolution of plant–animal interactions will be one of the major problems of the twenty-first century. In 1960 there were about 3 billion humans. Current population growth is about 80 million people per year. At 1.4% rate of increase, the population would double to 12 billion in about 50 years, although most demographers expect a levelling off below that number. Most of the increase will take place in developing countries, where 90% (currently 80%) of people will live by 2050. Even if the rate of increase drops linearly to zero over the next 50 years, population size will be 8 billion in 2050 (Federoff & Cohen 1999).

Understanding the evolutionary ecology of plant–animal interactions remains one of our most important imperatives if we are to sustain the biosphere while meeting the needs of human societies. Much of the world's consumption of food comes from rice, wheat and maize. Over the next 100 years, wheat is likely to surpass rice as the most important grain for consumption by the poor in developing countries. Intensive cultivation of grains and other crops to feed a human population that is 50–100% larger than today will undoubtedly magnify the human-driven surrogate coevolutionary process between plants and their herbivores that we see today. The 1970 epidemic of corn leaf blight in the US showed the vulnerability of agriculture to extensive monocultures of genetically uniform plants. Genetic resources needed to stay ahead of rapidly evolving herbivores and pathogens will become ever more valuable. The 16 CGIAR (Consultative Group on International Agricultural Research) centres devoted to preserving plant genetic resources currently hold around 600 000 samples (Hoisington et al. 1999). These gene centres are valuable resources, but they cannot be the complete answer. We need to know how to sustain genetic diversity within the natural populations that we conserve, because it is the only cost-effective way of retaining large resources of genes and gene combinations that have been tested in natural environments. Janzen (1999) has called it the gardenification of conserved wildlands.

Natural selection has already done the research and development for gene combinations that work well. It has favoured, from an almost infinite number of possible gene combinations, those that match well particular environments and problems. The job for evolutionary ecologists, evolutionary geneticists and community ecologists is to figure out how to make gardenification work in ways that retain these hard-won solutions.

The process of gardenification of wildlands therefore requires conserving the 'interaction biodiversity' of communities. The interactions are the glue of biodiversity, linking species locally and over broader geographic scales. As they weaken, communities and community networks begin to fall apart. Particularly important are the mutualistic interactions on which so many species depend but which have been ignored until recently in many conservation efforts and decisions. We cannot, however, understand how to conserve interaction biodiversity in a gardenified world unless we have some remnants of relatively pristine, geographically complex landscapes to use as touchstones for how to do it right (Thompson 1996). These few remaining relatively unexploited communities are the only way to know how different landscape schemes affect local levels of biodiversity over long periods of time.

Most of the earth, however, is now different from these unexploited landscapes. The process of gardenification and the maintenance of interaction biodiversity must therefore also require a better understanding of how to minimize the increasingly pervasive influence of invasive species. That again comes back to understanding species interactions. Introduced plants now clog the trails of even the most remote wilderness areas left in the United States outside of Alaska, and the same is increasingly true worldwide. Invasive species often outcompete native species and change the interaction structure of communities. Introduced plants attract pollinators (Parker 1997) with unknown effects on the pollination biology of native species. Biological control is one of the solutions, but avoidance of disasters in these efforts will demand a much greater refinement of our understanding of the ecology and evolution of specialization in interspecific interactions.

As in all maturing sciences, the remaining hard questions require a combination of expertise. Science has always been a social enterprise, but it is becoming even more so as the questions get larger. Few scientists are equally well trained in the biology of, say, angiosperms, insects and bacteria. Few have the combined skills in modelling, knowledge of natural history and experience in experimental design to design efficient and meaningful studies of how interactions are shaped over large geographical and temporal scales. Hence, our future of work on plant–animal interactions must necessarily involve collaborations among scientists with complementary expertise — that is, interactions to study interactions.

Appendix
Supplementary information for Chapter 2

This Appendix contains additional material for Chapter 2, 'The history of associations between plants and animals', by Conrad C. Labandeira. It is intended to contain more detailed explanations of some of the figures appearing in that chapter and supplements the captions that appear with the figures.

Fig. 2.3 A study by Wilf and Labandeira (1999) showing the response of plant–insect associations to the Early Caenozoic Thermal Interval (ECTI), an interval of elevated global warming. A greater variety of insect damage per host species and increased attack frequencies characterize Early Eocene plants, occurring under considerably warmer conditions compared to earlier Late Palaeocene plants. In addition, herbivory is elevated for both Late Palaeocene and Early Eocene members of the Betulaceae (birch family), a group of plants that are readily located by herbivores (Feeny 1976). (a) Sampling areas indicated by black polygons. (b) Damage census data for Late Palaeocene and Early Eocene leaves. From bottom to top: leaves with any insect damage, leaves externally consumed and the percentage of damaged leaves bearing more than one type of damage. These categories are each analysed separately for all leaves (**All**), Betulaceae only (**Bet**) and taxa other than Betulaceae (**NBet**). Error bars are one standard deviation of binomial sampling error; see Wilf and Labandeira (1999) for details of sample size, leaf area examined and locality information. (c) Diversity of insect damage for each plant-host species (vertical axis) plotted against the percentage of localities (49 Late Palaeocene and 31 Early Eocene) in which the species occurs (horizontal axis). Each data point is one species; many data points overlap at the lower left; survivors are plotted twice. Grey lines show divergence of 1 standard deviation (68%) confidence intervals for the two regressions. See Wilf and Labandeira (1999) for details of regressions and coefficients of determination.

Fig. 2.4 Summary of palaeobiological versus biological approaches in examinations of the temporal dimension of plant–animal associations. Palaeobiological approaches rely extensively on fossil material and extinct taxa for inferences regarding past plant–animal associations; biological approaches deduce past associations from extant taxa, often with known biologies. This distinction is occasionally unclear, but the separation proposed here approximates the differing subject material or methodologies between the two subdisciplines. The types of data and their categorization are presented at the top and are a representative sample of the literature. The geochronological placement of the data is provided as horizontal lines within vertical bars. Palaeobiological approaches typically span the terrestrial Phanerozoic record, but there are important earlier Palaeozoic and mid-Mesozoic hiatuses. By contrast, biological approaches, with the exception of modern mouthpart classes and perhaps feeding inferences in phylogenetic analyses, best describe the Cretaceous and Caenozoic, and are approximately equivalent to the duration of most lower-level extant clades. Abbreviations: Sil. = Silurian, Miss. = Mississippian, Penn. = Pennsylvanian, Neog. = Neogene.

For palaeobiological approaches, data for quantitative analyses (a) are the following, with geological stages in parentheses from oldest to youngest (see Harland et al. 1990 for nomenclature). From the Lower Permian of north-central Texas, the Coprolite Bone Bed flora (Sakmarian) (Greenfest & Labandeira 1998), the Brushy Creek flora (Artinskian) and the Taint flora (Artinskian) (Beck & Labandeira 1998); the Meeteetse flora from the Upper Cretaceous (Maastrichtian) of north-western Wyoming (Labandeira et al. 1995); several floras from the Palaeocene to Eocene transition from south-western Wyoming (Wilf & Labandeira 1999); and from the Middle Eocene (Lutetian), the Republic flora from north-eastern Washington state (Palmer et al. 1998) and the Green River flora from eastern Utah (Wilf et al.,

2001). Compendia of data for qualitative analyses (b) and dispersed coprolites (c) are provided in Labandeira (2002). Data for gut contents (d) are, from oldest to youngest: Mazon Creek, from the Middle Pennsylvanian (Moscovian) of north-central Illinois; Chekarda, from the Lower Permian (Kungurian) of the Perm Region, Russia; Karatau, from the Upper Jurassic (Kimmeridgian) of southern Kazakhstan; Baissa, from the Lower Cretaceous (Neocomian) of the Buryat Republic, Russia; Santana, Lower Cretaceous (Aptian) of Ceará, Brazil; Orapa, Upper Cretaceous (Cenomanian) of central Botswana; Messel, from the Middle Eocene (Lutetian) of Germany; Baltic amber, from the Late Eocene (Priabonian) of various localities in northern Europe; and Dominican amber, from the Lower Miocene (Aquitanian) of the northern Dominican Republic. See also Figures 2.15 and 2.16, and Labandeira (1998a, 1998b). Data for fossil mouthparts (e) is extensively documented in Labandeira (1990), abridged in Labandeira (1997). Examinations of fossil plant reproductive biology with regard to insect associations (f) are relatively few, and are provided in Labandeira (2002).

For biological approaches, data for the assignment of highly stereotyped damage in the fossil record to extant taxa (g) can be found in Labandeira (2002). The extension of modern pollination mutualisms by reference to diagnostic plant and insect morphological attributes in fossils (h) is also documented in Labandeira (2002). Inference of plant–insect associations based solely on insect fossils of modern lineages and modern ecological affiliations, biogeographical patterns or other supportive biological data (i) occur principally in major *Lagerstätten*. They are, from oldest to youngest: Issyk-Kul from the Lower Jurassic (?Hettangian) of north-eastern Kyrgyzstan; Karatau from the Upper Jurassic (Kimmeridgian) of southern Kazakhstan; from the Lower Cretaceous, Baissa (Neocomian) of eastern Russia, Jezzine amber (Barremian) of Lebanon, Yixian (Barremian) of north-eastern China, Montsec (Barremian/Aptian) of north-western Spain, and Santana (Aptian) of Brazil; from the Upper Cretaceous, Bezzonais and Durtal amber (Cenomanian) of France, Orapa (Cenomanian) of Botswana, Sayreville amber (Turonian) of New Jersey, Taimyr amber (Coniacean or Santonian) of Russia, and Cedar Lake and coeval amber (Campanian) of south-central Canada; from the Palaeogene, the Fur Formation (Thanetian) of Denmark, Hat Creek amber (Ypresian) from southern British Columbia, Green River (Lutetian) from western Colorado

and south-western Wyoming, Messel (Lutetian) from Germany, Angelsea (?Priabonian) of southern Australia, Baltic amber (Priabonian) of northern Europe, Florissant (Priabonian) of Colorado, Ruby River (Rupelian) from south-western Montana, and Aix-en-Provence and associated sites (Chattian) from central France; and from the Neogene are Rott (Aquitanian) from Germany, Chiapas amber (Aquitanian) from southern Mexico, Dominican amber (?Aquitanian) from the northern Dominican Republic; Radaboj (Burdigalian) from Croatia, Latah (Langhian) from northern Idaho, Oeningen (Tortonian) from Switzerland; Randecker Maar (Messinian) from southern Germany, and Willershausen (Piacenzian) from northern Germany.

Assessments from phylogenetic analyses of the time of origin of plant-related feeding attributes (j) are typically approximate, and midpoints have been assigned to time interval estimates from the literature. They are, from oldest to youngest: the origin of the Cicadomorpha (Hemiptera) during the Early Permian by Campbell et al. (1994); an Early Permian origin of the Auchenorrhyncha (Hemiptera) by Sorenson et al. (1995); a Late Permian origin of non-phloem-feeding homopterous Hemiptera by Campbell et al. (1994); the split during the Early Triassic between basal pollinivorous panorpoid lineages into the Hymenoptera and Lepidoptera by Kristensen (1995); basal radiation of phytophagous Sternorrhyncha (Hemiptera) during the Late Triassic by Campbell et al. (1994); Middle to Late Jurassic origin of nectarivorous Apioceridae and Mydidae (Diptera) from area cladograms by Yeates and Irwin (1996); Late Jurassic origin of Cynipoidea (Hymenoptera) by Ronquist (1995); the radiation of earlier lineages of Scarabaeidae (Coleoptera) by Scholtz and Chown (1995); the origin of the Matsuococcidae (Hemiptera) by Foldi (1997); the origin of sporophagous Keroplatidae (Diptera) by Matile (1997); the Early to Late Cretaceous origin of melponine bees by Roubik et al. (1997); the origin of *Urytalpa* (Diptera) plant feeding during the Late Cretaceous by Matile (1997); the commencement of host-plant associations within birdwing butterfly clades (Lepidoptera) during the Cretaceous to Caenozoic transition by Parsons (1996); radiations of more derived scarabaeid clades (Coleoptera) during the earlier Eocene by Scholz and Chown (1995); the origin and early diversification of *Larinus* weevils (Coleoptera) on thistles by Zwölfer and Herbst (1988); and the origin and early expansion of phytophagous Heliothinae (Lepidoptera) by Mitter et al. (1993). Data for the geochronological extension of modern mouthpart classes (k) can be gleaned from a compilation in Labandeira (1990), condensed in Labandeira (1997).

From a brief survey of analyses demonstrating congruent associations between plants and insects, a sample of 13 estimates is provided for the origin of subclades (l). These data are, from left (youngest) to right (oldest): an initial time estimate for the colonization of North American Asteraceae by *Ophraella* leaf-beetles (Futuyma & McCafferty 1990); the colonization of Hawaiian silverswords (Asteraceae) by *Nesosdyne* planthoppers (Roderick 1997; Roderick & Metz 1998); the colonization of *Eriogonum* (Polygonaceae) by *Euphilotes* butterflies (Shields & Reveal 1988); the initial occurrence of *Tetraopes* longhorn beetles on North American milkweed (Asclepiadaceae) hosts (Farrell & Mitter 1998); a subsequent estimate for the colonization of North American Asteraceae by *Ophraella* leaf beetles (Funk et al. 1995); the colonization of plant hosts by ancestral lineages of *Rhagoletis* fruit flies (Diptera) (Berlocher & Bush 1982; Berlocher 1998); the expansion of *Yponomeuta* (Lepidoptera) on Celastraceae, Rosaceae, Crassulaceae and Salicaceae (Menken 1996); the initial colonization of monocots by delphacid planthoppers (Hemiptera) (Wilson et al. 1994); the inception of associations between Lamiaceae and *Phyllobrotica* leaf beetles (Farrell & Mitter 1990); the ancestral association between rosid hosts and papilionoid butterflies (Janz & Nylin 1998); a broad estimate for the association between the Chrysomeloidea (longhorn beetles, leaf beetles and weevils) and seed plants (Farrell 1998); the initial association between *Chrysorthenches* moths (Lepidoptera) and conifers (Dugdale 1996); and the ancient colonization of the Malvales by the Carsidaridae (Hemiptera) (Hollis 1987).

Fig. 2.5 Types of evidence for plants, insects and their associations for 21 biotas, a small but representative sample selected from the fossil record. The five major categories of evidence for plant–insect associations range from those centred primarily on the plant (reproductive biology) to those focusing on insect structures that reveal herbivory (mouthparts and ovipositors). The fossil record of plant–insect associations (centre panel with links) is connected to the parallel but mostly separate fossil records of plants (left panel) and insects (right panel). Data provided from the literature illustrate biotas where one to four of the five types of evidence have been used to establish direct to indirect links between plants and insects. These biotas are not a complete inventory but represent best-case

examples, and are placed in approximate geochronological positions, with stage designations in parentheses. They are from youngest to oldest: (a) Downton Castle Formation and Ditton Group, Late Silurian (Pridoli) and Lower Devonian (Lochkovian) respectively, western United Kingdom (Edwards et al. 1995; Edwards 1996); (b) Rhynie Chert, Lower Devonian (Lochkovian) of Rhynie, Scotland (Kidston & Lange 1921; Kevan et al. 1975); (c) Battery Point Formation, Lower Devonian (Pragian) of Gaspé, Quebec (Banks 1981; Labandeira et al. 1988; Banks & Colthart 1993; Hotton et al. 1996); (d) Carbondale Formation, Middle Pennsylvanian (Moscovian) of Mazon Creek, north-western Illinois (Richardson 1980; Scott & Taylor 1983; Shear & Kukalová-Peck 1990; Labandeira & Beall 1990; Labandeira 1997); (e) Mattoon Formation, Late Pennsylvanian (Krevyakinskian) of eastern Illinois (Dilcher 1979; Millay & Taylor 1979; Retallack & Dilcher 1988; Labandeira & Phillips 1996; Labandeira et al. 1997); (f) La Magdalena coalfield, Late Pennsylvanian (Podolskian), of León, Spain (Amerom 1966; Castro 1997); (g) Waggoner Ranch Formation, Early Permian (Artinskian) of north-central Texas (Beck & Labandeira 1998); (h) Koshelevo Formation, Early Permian (Kungurian) of Chekarda, Perm Region, Russia (Becker-Migdisova 1940; Rasnitsyn 1977; Rasnitsyn & Krassilov 1996a, 1996b; Krassilov & Rasnitsyn 1997; Krassilov et al. 1997, 1999; Rasnitsyn & Novokshonov 1997; Novokshonov 1998b; Naugolnykh 1998); (i) Upper Buntsandstein and Lower Keuper Formations, Middle Triassic (Anisian and Ladinian) of western Germany and eastern France (Linck 1949; Kelber 1988; Kelber & Geyer 1989; Grauvogel-Stamm & Kelber 1996); (j) Chinle Formation, Late Triassic (Carnian) of north-western Arizona (Walker 1938; Ash 1997, 1999); (k) Karabastau Formation, Late Jurassic (Kimmeridgian) of Karatau, eastern Kazakhstan (Rohdendorf 1968; Crowson 1981; Arnol'di 1992; Labandeira 1997; Krassilov et al. 1997; Novokshonov 1997); (l) Wealden Formation, Early Cretaceous (Berriasian to Valanginian) of southern England (Jarzembowski 1990); (m) Yixian Formation, Early Cretaceous (Barremian) of western Liaoning, China (Ren 1998; Labandeira 1998b); (n) Kootenai Formation, Early Cretaceous (?Albian) of the Black Hills, South Dakota (Seward 1923; Delevoryas 1968; Crepet 1972, 1974); (o) Santana Formation, Early Cretaceous (Albian) of Ceará, Brazil (Caldas et al. 1989); (p) Dakota Formation, earliest Late Cretaceous (Cenomanian) of Nebraska and Kansas (Basinger & Dilcher

1984; Labandeira et al. 1994; Elliott & Nations 1998); (q) Hell Creek and Fort Union Formations, latest Cretaceous to earliest Palaeocene boundary (Maastrichtian to Danian) of the Williston Basin, south-western North Dakota (Labandeira et al. 2002); (r) Klondike Mountain Formation, Middle Eocene (Lutetian) of Republic, north-eastern Washington (Lewis 1994; Lewis & Carroll 1991; Wehr 1998); (s) Baltic amber, Late Eocene (Priabonian) of northern Europe (Conwentz 1890; Peyerimhoff 1909; Schedl 1947; Willemstein 1980); (t) Passamari Formation, Late Oligocene (Chattian) of Ruby River, south-western Montana (Becker 1965; Lewis, 1972, 1976); (u) the Störungszone, Pliocene (Piacenzian) of Willershausen, Germany (Straus 1967, 1977).

Fig. 2.7 The fossil history of insect external foliage-feeding. Geological stage-level resolution is provided in parentheses. (a) Transverse section of a permineralized specimen of the trimerophyte *Psilophyton dawsonii*, showing cortical collenchyma and wound response at top; inner tissues missing. This specimen is from the Lower Devonian (Pragian) of Gaspé, Quebec (Banks 1981). (b) Detail of wound response tissue in (a), indicated by arrows, suggesting surface grazing (Banks 1981). (c) A pinnule of the seed-fern *Paripteris pseudo-gigantea* displaying cuspate marginal-feeding excisions, from the Middle Pennsylvanian (Podolskian) of northern France (Amerom & Boersma 1971). (d) The seed-fern *Linopteris neuropteroides*, from the Late Pennsylvanian (Noginskian) of north-western Spain, exhibiting a bite mark (arrow) attributable to an external feeder (Castro 1997). (e) A specimen of the seed-fern leaf *Glossopteris*, showing extensive, scalloped margin feeding (arrow), from the Lower Permian (?Assellian) of north-eastern India (Srivastava 1987). (f) The cycadophyte *Taeniopteris*, from the Lower Permian (Artinskian) of north-central Texas, displaying extensive margin-feeding (arrows) (Beck & Labandeira 1998). (g) From the same deposit as (f) is the gigantopterid *Cathaysiopteris yochelsoni*, displaying hole-feeding and surrounding necrotic blotches (arrows) (Beck & Labandeira 1998). (h) A specimen of the seed-fern *Glossopteris*, showing cuspate excavations (arrow) along the leaf margin and extending almost to the midrib, from the Upper Permian of South Africa (Plumstead 1963). (i) A short leaf segment of the cycadophyte *Taeniopteris*, showing three serial, cuspate excavations that have projecting veinal stringers (Kelber & Geyer 1989). This specimen is from the Middle Triassic (Ladinian) of western Germany. (j) A leaf fragment of the filicalean fern

Cynepteris lasiophora, showing marginal- and hole-feeding traces (arrows), from the Upper Triassic (Carnian) of north-western Arizona (Ash 1997). (k) Detail of hole-feeding trace in the upper part of (j), exhibiting a reaction rim (Ash 1997). (l) Cuspate margin-feeding on the probable seed-fern *Sphenopteris arizonica*, from the same provenance as (j) (Ash 1999). (m) A cycad leaf from the undifferentiated Middle Jurassic of northern England, exhibiting marginal feeding (Scott & Paterson 1984). (n) An unidentified angiosperm leaf with hemispheric- to deltoid-shaped feeding holes between secondary veins, from the Middle Eocene (Lutetian) of western Tennessee (Stephenson & Scott 1992). (o) Damage by an adult leaf-cutter bee (Megachilidae) on a leaf of *Prunus* (Rosaceae), from the Middle Eocene (Lutetian) of north-eastern Washington (Lewis 1994). (p) Insect bud-feeding on the chestnut *Castanea atavia* (Fagaceae) from the Lower Pliocene (Zanclian) of Germany (Berger 1953). Scale bars for this and succeeding figures: crosshatched = 10 cm, solid = 1 cm, striped = 0.1 cm, dotted = 0.01 cm (100 μ), and backslashed = 0.001 cm (10 μ).

Fig. 2.8 The fossil history of piercing-and-sucking. Geological stage-level resolution is provided in parentheses. (a) An oblique, longitudinal section of an axis of the rhyniophyte, *Rhynia*, showing a lesion plugged with opaque material and extending to subjacent vascular tissue (Kevan et al. 1975). This specimen originates from the Lower Devonian (Lochkovian) Rhynie Chert of Scotland. (b) Another *Rhynia* axis from the Rhynie Chert, in transverse section, displaying hypertrophied cortical cells and associated opaque material (Kevan et al. 1975). (c) A specimen of the Early Devonian (Pragian) trimerophyte, *Psilophyton*, from Gaspé, Québec, exhibiting three sites of piercing (arrows) (Banks & Colthart 1993). Below each puncture site is a cone of lysed tissue, on a base of unaltered periderm tissue (Banks & Colthart 1993). (d) Detail of (c), showing an enlarged cone of lysed subepidermal tissue with radiate stylet tracks and damaged epidermal cells at top (Banks & Colthart 1993). (e) Three piercing wounds on an unnamed trimerophyte, eliciting the formation of light-hued response periderm (arrows) (Banks & Colthart 1993). This specimen is from the Lower Devonian (Pragian) of Québec. (f) Damage to a fern (*Etapteris*) petiole, probably by an insect with stylate mouthparts, showing disorganized tissues enveloping the puncture wound. The specimen is from the Middle Pennsylvanian of the eastern United States (Scott & Taylor 1983). (g) A stylet track with terminal feeding pool, surrounded by reaction tissue, on the seed-fern pollen organ *Bernaultia*. This specimen is from the Calhoun Coal of Late Pennsylvanian age from eastern Illinois (Schopf 1948; Millay & Taylor 1979; Retallack & Dilcher 1988; Labandeira 1998a). (h) Two stylet tracks targeting vascular tissue (*xy*lem and *ph*loem) of the marattialean fern, *Psaronius*, also from the Calhoun Coal of Illinois (Labandeira & Phillips 1996a). The right track, approximately 3 mm long, is sectioned lengthwise, and shows surrounding reaction tissue (*rt*) and a terminal feeding pool. (i) Detail of the left stylet track in (h), showing stylet track (*st*) surrounding opaque material (*om*), penetration of undifferentiated parenchyma (*pa*) and avoidance of large gum-sac cells (*gs*) (Labandeira & Phillips 1996a). (j) The head and 3.2 cm-long stylate mouthparts of the palaeodictyopterid, *Eugereon boeckingi* Dohrn, from the Lower Permian (?Assellian) of Germany (Müller 1978). (k) Reconstruction of the early hemipteran, *Permocicada integra* Becker-Migdisova from the Lower Permian (Wordian) of Russia (Becker-Migdisova 1940). (l) Two stylet probes, with surrounding rims of opaque material, on the cheirolepidiaceous conifer *Pseudofrenelopsis*, from the Lower Cretaceous of Texas (Watson 1977). See Figure 2.7 for scale-bar conventions.

Fig. 2.9 The fossil history of boring. Geological stage-level resolution is provided in parentheses. (a) Oribatid mite borings in gymnospermous *Australoxylon mondii* wood, of probable glossopterid or cordaite origin, from the Late Permian of the northern Prince Charles Mountains, Antarctica (Weaver et al. 1997). (b) Enlarged region of a boring in (a), showing ellipsoidal coprolites and associated undigested frass (Weaver et al. 1997). (c) Insect borings in cambium of *Araucarioxylon arizonicum*, from the Late Triassic (Carnian) of Petrified Forest National Monument, Arizona (Walker 1938). These borings were made by unknown beetles. (d) Cambium borings in *Araucarioxylon arizonicum* as in (c), but fabricated by a different insect species (Walker 1938). (e) Scanning electron micrograph of a probable beetle boring in *Protocupressinoxylon* wood, healed by parenchymatous tissue, from the undifferentiated Middle Jurassic of Henan, China (Zhou & Zhang 1989). (f) Beetle borings, assignable to the family Cupedidae, in wood of *Hermanophyton*, a gymnosperm of uncertain affinities (Tidwell & Ash 1990). This material is from the Late Jurassic (Tithonian) of south-western Colorado. (g) A scolytid (bark beetle) cambium boring by a beetle in an unnamed conifer from the Early Cretaceous (Berriasian to Valanginian) of south-

ern England (Jarzembowski 1990; Chaloner et al. 1991). (h) Beetle invasion of the androecium of the bennettitalean, *Cycadeoidea*, showing consumption of synangial-associated tissues (*s*, synangium), a gallery (*g*), and an exit or entry tunnel (*b*) across the microsporophyll (*ms*) which are bract-like, enveloping structures (Crepet 1972). This material is from the Lower Cretaceous (?Aptian) of South Dakota. (i) A cerambycid (longhorn beetle) boring in unknown wood, from the Late Oligocene or Early Miocene (Chattian or Aquitanian) of Germany (Linstow 1906). (j) A scolytid (bark beetle) cambium boring in unknown wood from the Middle Miocene of Shandong, China (Guo 1991). (k) The dipteran cambium miner *Palaeophytobia platani* (Agromyzidae) in sycamore wood (*Platanoxylon*), displaying tissue damage at top, from the Upper Miocene (Tortonian) of Hungary (Süss & Müller-Stoll 1980). (l) A boring of a longhorn beetle (Cerambycidae) in wood of the pinaceous conifer *Larix* (larch), from the Late Pliocene (Piacenzian) of Peary Land, northern Greenland (Böcher 1995). (m) The bark beetle *Eremotes nitidipennis* (Scolytidae) within a boring of unknown wood, from the Holocene of southern Finland (Koponen & Nuorteva 1973). See Figure 2.7 for scale-bar conventions.

Fig. 2.10 The fossil history of leaf-mining. Geological stage-level resolution is provided in parentheses. (a) A possible U-shaped blotch mine on the seed-fern leaf *Neuropteris subauriculata*, from the Middle Pennsylvanian (Moscovian) of Zwickau, Germany (Müller 1982). (b) A holometabolan serpentine leaf-mine on the corystosperm seed-fern *Pachypteris*, from the Jurassic to Cretaceous boundary of northern Queensland, Australia (Rozefelds 1988a). These leaf-mines antedate the earliest documented angiosperms by approximately 15 million years. (c) Another serpentine leaf-mine on a different specimen of *Pachypteris* (Rozefelds 1988a). (d) A serpentine leaf-mine (Lepidoptera: Gracillariidae) on the magnoliid dicot *Densinervum*, showing oviposition site, frass trail and pupation chamber, from the Late Cretaceous (Cenomanian) of Nebraska (Labandeira et al. 1994). (e) A serpentine leaf-mine of *Stigmella* (Lepidoptera: Nepticulidae) on the hamamelid dicot *Cercidiphyllum*, from the Late Cretaceous (Maastrichtian) of Wyoming (Labandeira et al. 1995). (f) A serpentine mine of ?Bucculatricidae on unknown dicot, from the Upper Palaeocene (Thanetian) of England (Crane & Jarzembowski 1980). (g) Portion of a *Cedrela* leaflet (Meliaceae) with a serpentine *Phyllocnistis* leaf-mine (Lepidoptera: Gracillariidae) bearing a frass trail (Hickey & Hodges 1975). This specimen is from the Early Eocene (Ypresian) of north-western Wyoming. (h) Circular leaf-mines of Lepidoptera (Incurvariidae) on *Macginitiea* (Platanaceae), from the Green River Formation of Bonanza, Utah, of Middle Eocene age (Lutetian) (Labandeira 1998c). (i) An undetermined dicotyledonous leaf bearing a serpentine leaf-mine assignable to the Lepidoptera (?Nepticulidae), from the Middle Eocene of England (Stephenson & Scott 1992). (j) Detail of leaf-mine in (i), showing complete developmental progression from oviposition to pupation (Stephenson & Scott 1992). (k) A leaf of Lauraceae exhibiting a serpentine lepidopteran mine (Nepticulidae), from the Late Eocene (Priabonian) of Victoria, Australia (Rozefelds 1988b). (l) Camera lucida enlargements of two mines in (k) (Rozefelds 1988b). (m) Dipteran blotch mine (Agromyzidae) on *Cinnamomum* (Lauraceae), from the Upper Miocene (Tortonian) of Bosnia (Berger 1949). See Figure 2.7 for scale-bar conventions.

Fig. 2.11 The fossil history of galling. Geological stage-level resolution is provided in parentheses. (a) Probable gall, *Acrobulbillites*, on the stem terminus of the sphenophyte *Asterophyllites longifolius*, from the Middle Pennsylvanian of northern Europe (from Weiss (1876), refigured by Amerom [1973]). (b) The 'fructification' *Paracalamostachys spadiciformis*, reinterpreted as a probable stem gall by Amerom (1973), from the Middle Pennsylvanian of northern England (Thomas 1969). (c) Gall of a holometabolous insect on the rachis of *Psaronius chasei*, a marattialean fern from the Late Pennsylvanian (Krevyakinskian) of Illinois (Labandeira & Phillips 1996b). Note the barrel-shaped coprolites and frass in the central lumen, and surrounding tufts of hyperplasic and hypertrophic parenchyma. (d) A three-dimensional reconstruction of the same gall type in (c); same provenance (Labandeira & Phillips 1996b). (e) Aborted cone of the herbaceous voltzialean conifer *Aethophyllum stipulare*, showing a basal stem expansion interpreted as a gall (Grauvogel-Stamm & Kelber 1996). This material is from the Middle Triassic (Anisian) of France. (f) Round, oval or deltoid leaf galls on the enigmatic gymnosperm, *Dechellyia gormanii*, expressed as swellings typically occurring about 1 to 1.5 cm from the base of the leaf (Ash 1997). These galls originate from the Upper Triassic Chinle flora (Carnian), from north-eastern Arizona. (g) Enlargement of a gall in (f) (arrow), showing deltoid shape and extension beyond the leaf margin (Ash 1997). (h) Abundant, bulbous *Wonnacottia* galls on the leaf of a

bennettitalean (probably *Anomozamites nilssoni*), from the Middle Jurassic (Bajocian) of northern England (Harris 1942; Alvin et al. 1967). These galls often occur in clusters and are preserved three-dimensionally. (i) Large spheroidal gall on an angiosperm leaf (Scott et al. 1992), similar in shape and size to those produced by gall wasps (Hymenoptera, Cynipidae), from the Late Cretaceous (Maastrichtian) of Tennessee. (j) Cynipid spindle galls (Hymenoptera) on a leaf of the oak *Quercus hannibali* (Fagaceae), from the Middle Miocene of north-western Nevada (Waggoner & Poteet 1996). These distinctive galls are most similar to those of the extant gall wasp *Antron clavuloides*, which parasitizes *Quercus*. (k) A basal petiolar gall, attributed to a *Pemphigus* gall aphid, on *Populus latior* (Salicaceae), from the Upper Oligocene (Chattian) of Germany (Mädler 1936). (l) Enlargement of a pemphigid gall in (k), showing a characteristic stem expansion (Mädler 1936). (m) A cone-mimicking gall of the cecidomyiid dipteran (*Thecodiplosis*), a gall midge on the conifer *Taxodium* (bald cypress), from the Middle Miocene of northern Idaho (Lewis 1985). See Figure 2.7 for scale-bar conventions.

Fig. 2.12 The fossil history of seed predation. Geological stage-level resolution is provided in parentheses. (a) A sandstone cast of the seed *Trigonocarpus*, from the Pennsylvanian of the United Kingdom, displaying a plug infilling a presumptive hole in the seed coat (Scott et al. 1992). (b) External surface of the cordaitalean seed *Samaropsis* with a hole and surrounding rim, from the Middle Pennsylvanian of Siberia, Russia (Sharov 1973). The culprit may be a palaeodictyopterid insect in the same deposit possessing a beak width approximating the hole diameter. (c) A similarly bored seed from the same deposit, in longitudinal section (Sharov 1973). (d) A boring in a lycopsid *Setosisporites* megaspore, from the Middle Pennsylvanian of Yorkshire, United Kingdom, showing the entire spore (inset) and detail of the bored margin (Scott et al. 1992). (e) A caddisfly case of seeds from the gingkophyte *Karkenia*, from the Lower Cretaceous of Shin Khuduk (stage unspecified), Mongolia (Krassilov & Sukacheva 1979). (f) Longitudinal section of a permineralized dicotyledonous seed or fruit from the probable Upper Cretaceous (stage unspecified) of southern Argentina, showing a *Carpoichnus* boring (Genise 1995), with exit hole at arrow. (g) A dicotyledonous *Rutaspermum* seed (Rutaceae) with an insect exit hole, from the Middle Eocene (Lutetian) Messel deposit in Germany (Collinson & Hooker 1991). (h) Pristine (top) and bored (bottom) dicotyledonous seeds of *Zanthoxylum* (Rutaceae) from the Early Oligocene (Rupelian) of Vermont (Tiffney 1980). (i) Bored stone fruits of *Celtis lacunosa* (Fabaceae), assigned to the ichnogenus *Lamniporichnus*, from the Early Miocene (Aquitanian) of the Czech Republic (Mikuláš et al. 1998). This damage is typical of weevils. (j) Additional damage of *Celtis*, from a Pleistocene sinter-cemented breccia of the Czech Republic (Mikuláš et al. 1998). See Figure 2.7 for scalebar conventions.

Fig. 2.13 The fossil history of surface fluid-feeding. Geological stage-level resolution is provided in parentheses. (a) *Pseudopolycentropus latipennis*, a scorpion fly with elongate mouthparts (arrow) from the Late Jurassic (Kimmeridgian) of Karatau, Kazakhstan; redrawn from a camera lucida sketch in Novokshonov (1997). (b) Head and mouthparts (arrow) of the tanglevein fly *Protnemestrius rohdendorfi* (Nemestrinidae) from the same provenance as (a), redrawn from a camera lucida sketch in Mostovski (1998). (c) The tanglevein fly *Florinemestrius pulcherrimus* (Nemestrinidae) from the Early Cretaceous (Barremian) of western Liaoning, China. Note elongate mouthparts (arrow) which are similar in form to extant nectar-feeding nemestrinids (Ren 1998). (d) Head and mouthparts of the crane fly *Helius botswanensis* (Tipulidae), extant descendants of which feed on flowers (Rayner & Waters 1991). This specimen is from the Late Cretaceous (Cenomanian) of Orapa, Botswana. (e) From the same deposit as (d) is a funnel flower consisting of fused petal bases and a relatively deep throat, indicative of insect pollination (Rayner & Waters 1991). (f) Scanning electron micrograph of a small charcoalified flower showing a nectary disk (arrow) above a region of petals from the Late Cretaceous of Skåne, Sweden (Friis 1985). (g) A showy, zygomorphic, papilionoid flower (Fabaceae) exhibiting an upper banner petal (*s*), two lateral wing petals (*w*) and bottom keel petals (*k*), associated with insect, especially bee, pollination (Crepet & Taylor 1985). This specimen comes from Palaeocene to Eocene boundary strata in western Tennessee. (h) Extra-floral nectaries (arrows) at the junction of the leaf blade and petiolar base in *Populus crassa* (Salicaceae) from the uppermost Eocene (Priabonian) of the Florissant lake beds in central Colorado (Pemberton 1992). (i) A worker of the stingless bee *Proplebia dominicana* (Apidae) bearing conspicuous resin balls (arrows) attached to hind-leg corbiculae (Poinar 1992). This specimen is from the Early Miocene (Aquitanian)

Dominican amber deposit of the Dominican Republic. See Fig. 2.7 for scale-bar conventions.

Fig. 2.15 The fossil history of spore and pollen consumption, and pollination. Geological stage-level resolution is provided in parentheses. (a) An elliptical, somewhat flattened coprolite containing plant cuticle and occasional spores, from the Late Silurian (Pridoli) of the United Kingdom (Edwards et al. 1995). (b) Contents of coprolite containing abundant *Streelispora*, *Aneurospora* and ?*Emphanisporites* spores, from the Lower Devonian (Lochkovian) of the United Kingdom (Edwards et al. 1995). (c) Sporangial fragments and isolated spores of the coprolite *Bensoniotheca*, from the Mississippian (Asbian) of England, attributed to a lyginopterid seed-fern (Rothwell & Scott 1988). (d) Two relatively intact sporangia in a *Bensoniotheca* coprolite from the same provenance as (c) (Rothwell & Scott 1988). (e) The coprolite *Thuringia*, consisting of the digested remnants of a pollen organ referable to the peltasperm seed-fern *Autunia* (Kerp 1988). This specimen is from the Pennsylvanian to Permian boundary of Germany. (f) The hypoperlid insect *Idelopsocus diradiatus* (Family Hypoperlidae), from the Lower Permian (Kungurian) of Chekarda, Russia, containing a plug of pollen (arrow) preserved in its gut (Krassilov & Rasnitsyn 1997). (g) Detail of a *Lunatisporites*-type pollen grain extracted from (f), attributable to a voltzialean conifer (Krassilov & Rasnitsyn 1997). (h) Reconstruction of the head and mouthparts of *Synomaloptila longipennis*, a hypoperlid insect (Family Synomaloptilidae) with pollinivorous habits, from the same provenance as (f) (Rasnitsyn 1977; Labandeira 1997b; Novokshonov 1998a). (i) The prophalangopsid grasshopper *Aboilus*, from the Late Jurassic (Kimmeridgian) Karatau locality of Kazakhstan, with an arrow indicating a bolus of pollen preserved in its intestine (Krassilov et al. 1997). (j) A cluster of digested *Classopollis* pollen from a cheirolepidiaceous conifer, extracted from the gut of *Aboilus* in (i) below. (k) The nemonychid weevil *Archaeorrhynchus paradoxopus*, from the same provenance as (i), showing a prolonged and decurved rostrum (Arnoldi et al. 1992). (l) The xyelid sawfly *Ceroxyela dolichocea*, from the Lower Cretaceous (Neocomian) Baissa locality of Russia, with preserved gut contents of pollen indicated by an arrow (Krassilov & Rasnitsyn 1983). (m) A pollen grain of *Pinuspollenites*, from a pinalean conifer, found in the gut of *Ceroxyela* in (l) (Krassilov & Rasnitsyn 1983). (n) A staphylinid beetle, from the Late Cretaceous (Cenomanian) of Orapa, Botswana, displaying pollen transportation (arrow) (Rayner & Waters 1991). (o) Enlargement of the terminal abdominal region of (n), showing pollen grains trapped among hairs (Rayner & Waters 1991). (p) Lateral view of a charcoalified flower belonging to the Ericales/Theales complex of dicotyledonous angiosperms, from the Upper Cretaceous (Turonian) of Sayreville, New Jersey (Crepet 1996). (q) Contents of anther from same taxon as (p), displaying the characteristic network of viscin threads, implying a pollinator mutualism (Crepet 1996). (r) A clump of two types of pollen from the external abdominal surface of the stingless bee *Proplebeia dominicana*, revealed by scanning electron microscopy (Grimaldi et al. 1994). This amber specimen is from the Lower Miocene (Aquitanian) of the Dominican Republic. See Fig. 2.7 for scale-bar conventions.

Fig. 2.19 The fossil history of oviposition. Geological stage-level resolution is provided in parentheses. (a) Reconstruction of the terminal abdominal region of a female *Uralia* (Diaphanopterodea), illustrating a vertically compressed, swath ovipositor typical of those used in endophytic oviposition. *Uralia* originates from the Lower Permian (Kungurian) of Chekarda, Perm Region, Russia; redrawn from Kukalová-Peck (1991). (b) The pollinivorous hypoperlid insect *Mycteroptila* (Family Permarhaphidae) from the same provenance as (a). Note the pronounced, flattened ovipositor, which was probably used for endophytic insertion of eggs; redrawn from a camera lucida drawing by Novokshonov (1998a). (c) Elongate-oval oviposition scars on the sphenophyte *Equisetites foveolatus*, from the Middle Triassic (Ladinian) of Germany (Kelber 1988). (d) Camera lucida drawing of a leaf sheath from the sphenophyte *Equisetites arenaceus*, from the Middle Triassic (Ladinian) of Germany, showing dense, elongate-oval oviposition scars arranged in a zigzag pattern (Grauvogel-Stamm & Kelber 1996). (e) Probable insect eggs in or on the cycadophyte, *Taeniopteris angustifolia*, from the Middle Triassic (Ladinian) of Germany (Grauvogel-Stamm & Kelber 1996). (f) Enlargement of oval oviposition scars in (e) linearly and obliquely placed on the *Taeniopteris* leaf between the midrib and margin (Grauvogel-Stamm & Kelber 1996). (g) Oviposition scars of odonatan eggs, inserted as eccentric arcs (e.g. black arrow) on *Alnus* (Betulaceae), from the Middle Eocene (Lutetian) of Republic, Washington State (Lewis & Carroll 1991). (h) Oviposition scars on an unidentified

leaf, similar to and approximately contemporaneous with those of (g), and presumably produced by an odonatan (Schaarschmidt 1992). The white arrow refers to an arcuate row of scars; this leaf is from the Middle Eocene (Lutetian) of Messel, Germany. (i) An angiosperm leaf (?Juglandaceae) showing odonatan oviposition scars typical of the Coenagrionidae, from the Upper Miocene (Messinian) of Randecker Maar, Germany (Hellmund & Hellmund 1996). (j) A leaf of the angiosperm *Carpinus grandis* (Betulaceae) exhibiting petiolar oviposition scars (arrows) typical of the odonatan family Lestidae, from the Middle Oligocene (Chattian) of Germany (Hellmund & Hellmund 1996). See Fig. 2.7 for scale-bar conventions.

Fig. 2.20 The fossil history of plant–vertebrate associations. Geological stage-level resolution is provided in parentheses. (a) Crown of a 'molariform' cheek tooth from *Diadectes*, a high-fibre herbivore from the latest Pennsylvanian (Noginskian) of ?Ohio (Hotton et al. 1997). (b) Detail of striations in (a) on the wear facet parallel to the long axis of the jaw (Hotton et al. 1997). (c) Gut contents of the pareiasaur *Protorosaurus speneri*, illustrating a food bolus (delimited by white arrows) that contains conifer ovules (dark circular or hemispherical structures) and quartzose gastroliths (*Q*) (Munk & Sues 1993). This skeletal material is from the Upper Permian of Hessen, Germany. (d) A cluster of small, loose, pellet-like coprolites attributed to an ornithopod dinosaur by Hill (1976), each containing abundant leaf cuticle from the bennettitalean *Ptilophyllum*, from the Middle Jurassic of northern England (Thulborn 1991). (e) A herbivorous dinosaur coprolite, showing constrictions and resembling the segmented faeces of extant herbivorous mammals (Matley 1941; Thulborn 1991). (f) A herbivore coprolite with two backfilled burrows (one at upper arrow) within a dark groundmass composed of comminuted xylem fragments, and the other light-coloured (lower arrow) packed with intermixed sediment and dung. Both burrows are inferred to have been made by dung beetles (Scarabaeidae), based on extant burrowing patterns; specimen from the Late Cretaceous (Campanian) of Montana (Chin & Gill 1996). (g) A serpentine dinosaur coprolite displaying conifer twigs (arrows), from the Upper Cretaceous (Maastrichtian) of southern Saskatchewan, Canada (Nambudiri & Binda 1989). (h) An upper molar of the rodent *Thalerimys headonensis*, a herbivorous browser, from the Eocene of southern England (Collinson &

Hooker 1991). Note the broad shelf for seed grinding. (i) The rodent *Treposciurus mutabilis*, from the Late Eocene (Priabonian) of England, with sub-parallel microwear scratches and pits on upper deciduous molars, suggesting the consumption of indurated food (Collinson & Hooker 1991). (j) Coprolite of the extinct folivorous moa-nalo, *Thambetochen chauliodous* (Anatidae), from the Holocene of Maui, Hawaii (James & Burney 1997) (k) A monolete fern spore from a coprolite similar to (j); ferns were a major component of the diet of *Thambetochen* (Burney 1997). (l) A heap of coprolites from the extinct goat-like bovid *Myotragus balearicus*, in a Holocene cave from the Balearic Islands, Spain (Alcover et al. 1999). (m) Pollen of *Buxus balearica* (Buxaceae) isolated from coprolites of *Myotragus balearicus* in (l), the major dietary constituent. See Fig. 2.7 for scale-bar conventions.

References

Alcover, J.A., Perez-Obiol, R., Errikarta-Imanol, Y. and Bover, P. (1999) The diet of *Myotragus balearicus* Bate 1909 (Artiodactyla: Caprinae), an extinct bovid from the Balearic Islands: evidence from coprolites. *Biological Journal of the Linnean Society*, **66**, 57–74.

Alvin, K.L., Barnard, P.D.W., Harris, T.M., Hughes, N.F., Wagner, R.H. and Wesley, A. (1967) Gymnospermophyta. In *The Fossil Record*, ed. W.B. Harland, C.H. Holland, M.R. House, N.F. Hughes, A.B. Reynolds et al. London: Geological Society of London, pp. 247–268.

Amerom, H.W.J. van (1966) *Phagophytichnus ekowskii* nov. ichnogen. & nov. ichnosp., eine Missbildung infolge von Insektenfrass, aus dem spanischen Stephanien (Provinz León). *Leidse Geologische Mededelingen*, **38**, 181–184.

Amerom, H.W.J. van (1973) Gibt es Cecidien im Karbon bei Calamiten und Asterophylliten? In: *Compte Rendu Septième Congrès International de Stratigraphie et de Géologie du Carbonifère*, ed. K.-H. Josten. Krefeld, Germany: Van Acken, pp. 63–83.

Amerom, H.W.J. van and Boersma, M. (1971) A new find of the ichnofossil *Phagophytichnus ekowskii* Van Amerom. *Geologie en Mijnbouw*, **50**, 667–670.

Arnol'di, L.V., Zherikhin, V.V., Niktritin, L.M. and Ponomarenko, A.G. (1992) *Mesozoic Coleoptera*, transl. N.J. Vandenberg. Washington, DC: Smithsonian Institution Libraries and National Science Foundation.

Ash, S. (1997) Evidence of arthropod-plant interactions in the Upper Triassic of the southwestern United States. *Lethaia*, **29**, 237–248.

Ash, S. (1999) An Upper Triassic *Sphenopteris* showing evidence of insect predation from Petrified Forest National Park, Arizona. *International Journal of Plant Science*, **160**, 208–215.

Banks, H.P. (1981) Peridermal activity (wound repair) in an Early Devonian (Emsian) trimerophyte from the Gaspé Peninsula, Canada. *Palaeobotanist*, **28/29**, 20–25.

Banks, H.P. and Colthart, B.J. (1993) Plant-animal-fungal interactions in Early Devonian trimerophytes from Gaspé, Canada. *American Journal of Botany*, **80**, 992–1001.

Basinger J.F. and Dilcher, D.L. (1984) Ancient bisexual flowers. *Science*, **224**, 511–513.

Beck, A.L. and Labandeira, C.C. (1998) Early Permian insect folivory on a gigantopterid-dominated riparian flora from north-central Texas. *Palaeogeography, Palaeoclimatology, Palaeoecology*, **142**, 139–173.

Becker, H.F. (1965) Flowers, insects, and evolution. *Natural History*, **74**, 38–45.

Becker-Migdisova, E.E. (1940) Fossil Permian cicadas of the family Prosbolidae from the Sojana River. *Transactions of the Paleontological Institute*, **11/2**, 1–98, pls. 1–6 [in Russian].

Berger, W. (1949) Lebensspuren schmarotzender Insekten an jungtertiären Laubblättern. *Sitzukngsberichte der Österreichische Akademie der Wissenschaften in Wien, Mathematisch-Naturwissenschaftliche Klasse*, **158**, 789–792.

Berger, W. (1953) Missbildungen an jungtertiären Laubblättern infolge Verletzung im Knospenzustand. *Neues Jahrbuch für Geologie und Paläontologie Monatshefte*, **1953**, 322–333.

Berlocher, S.B. (1998) Can sympatric speciation via host or habitat shift be proven from phylogenetic and biogeographic evidence? In *Endless Forms: Species and Speciation*, ed. D.J. Howard and S.H. Berlocher. New York: Oxford University Press, pp. 99–113.

Berlocher, S.B. and Bush, G.L. (1982) An electrophoretic analysis of *Rhagoletis* (Diptera: Tephritidae) phylogeny. *Systematic Zoology*, **31**, 136–155.

Böcher, J. (1995) Palaeoentomology of the Kap København Formation, a Plio-Pleistocene sequence in Peary Land, North Greenland. *Meddelelser om Grønland, Geoscience*, **33**, 1–82.

Caldas, M.B., Martins-Neto, R.G. and Lima-Filho, F.P. (1989) *Afropollis* sp. (polém) no trato intestinal de vespa (Hymenoptera: Apocrita, Xyelidae) no Cretáceo da Bacia do Araripe. *Atas II Simpósio Nacional de Estudos Tectónicos, Sociedade Brasileira de Geologia*, **11**, 195–196, abstract.

Campbell, B.C., Steffen-Campbell, J.D. and Gill, R.J. (1994) Evolutionary origin of whiteflies (Hemiptera: Sternorrhyncha: Aleyrodidae) inferred from 18S rDNA sequences). *Insect Molecular Biology*, **3**, 73–88.

Castro, M.P. (1997) Huellas de actividad biológica sobre plantas del Estefaniense Superior de la Magdalena (León, España). *Revista Española de Paleontología*, **12**, 52–66.

Chaloner, W.G., Scott, A.C. and Stephenson, J. (1991) Fossil evidence for plant-arthropod interactions in the Palaeozoic and Mesozoic. *Philosophical Transactions of the Royal Society of London B*, **333**, 177–186.

Chin, K. and Gill, B.D. (1996) Dinosaurs, dung beetles, and conifers: participants in a Cretaceous food web. *Palaios*, **11**, 280–285.

Collinson, M.E. and Hooker, J.J. (1991) Fossil evidence of interactions between plants and plant-eating mammals. *Philosophical Transactions of the Royal Society of London B*, **333**, 197–208.

Conwentz, H. (1890) *Monographie der Baltischen Bernstein-Bäume. Vergleichende Untersuchungen über die Vegetationsorgane und*

Blüten, sowie über dans Harz und die Krankheiten der Baltischen Bernsteinbäume. Gdansk: Wilhelm Engelmann.

Crane, P.R. and Jarzembowski, E.A. (1980) Insect leaf mines from the Palaeocene of southern England. *Journal of Natural History*, **14**, 620–636.

Crepet, W.L. (1972) Investigations of North American cycadeoids: pollination mechanisms in *Cycadeoidea*. *American Journal of Botany*, **59**, 1048–1056.

Crepet, W.L. (1974) Investigations of North American cycadeoids: the reproductive biology of *Cycadeoidea*. *Palaeontographica B*, **148**, 144–169.

Crepet, W.L. (1996) Timing in the evolution of derived floral characters—Upper Cretaceous (Turonian) taxa with tricolpate and tricolpate-derived pollen. *Review of Palaeobotany and Palynology*, **90**, 339–359.

Crepet, W.L. and Taylor, D.W. (1985) Diversification of the Leguminosae: first fossil evidence of the Mimosoideae and Papilionoideae. *Science*, **228**, 1087–1089.

Crowson, R.W. (1981) *The Biology of the Coleoptera*. New York: Academic Press.

Delevoryas, T. (1968) Investigations of North American cycadeoids: structure, ontogeny and phylogenetic considerations of cones of *Cycadeoidea*. *Palaeontographica B*, **121**, 122–133.

Dilcher, D.L. (1979) Early angiosperm reproduction: an introductory report. *Review of Palaeobotany and Palynology*, **27**, 291–328.

Dugdale, J.S. (1996) *Chrysorthenches* new genus, conifer-associated plutellid moths (Yponomeutoidea, Lepidoptera) in New Zealand and Australia. *New Zealand Journal of Zoology*, **23**, 33–59.

Edwards, D. (1996) New insights into early land ecosystems: a glimpse of a Lilliputian world. *Review of Palaeobotany and Palynology*, **90**, 159–174.

Edwards, D., Selden, P.A., Richardson, J.B. and Axe, L. (1995) Coprolites as evidence for plant–animal interaction in Siluro-Devonian terrestrial ecosystems. *Nature*, **377**, 329–331.

Elliott, D.K. and Nations, J.D. (1998) Bee burrows in the Late Cretaceous (Late Cenomanian) Dakota Formation, northeastern Arizona. *Ichnos*, **5**, 243–253.

Farrell, B.D. (1998) 'Inordinate fondness' explained: Why are there so many beetles? *Science*, **281**, 555–559.

Farrell, B.D. and Mitter, C. (1990) Phylogenesis of insect/plant interactions: have *Phyllobrotica* leaf beetles (Chrysomelidae) and the Lamiales diversified in parallel? *Evolution*, **44**, 1389–1403.

Farrell, B.D. and Mitter, C. (1998) The timing of insect/plant diversification: might *Tetraopes* (Coleoptera: Cerambycidae) and *Asclepias* (Asclepiadaceae) have co-evolved? *Biological Journal of the Linnean Society*, **63**, 553–577.

Feeny, P. (1976) Plant apparency and chemical defense. *Recent Advances in Phytochemistry*, **10**, 1–40.

Foldi, I. (1997) Defense strategies in scale insects: phylogenetic inference and evolutionary scenarios (Hemiptera, Coccoidea). In *The Origin of Biodiversity in Insects: Phylogenetic Tests of Evolutionary Scenarios*, ed. P. Grandcolas. *Mémoires du Muséum National d'Histoire Naturelle*, **173**, 203–230.

Friis, E.M. (1985) Structure and function in Late Cretaceous angiosperm flowers. *Biologiske Skrifter*, **25**, 1–37.

Funk, D.J., Futuyma, D.J., Orti, G. and Meyer, A. (1995) A history of host associations and evolutionary diversification for *Ophraella* (Coleoptera: Chrysomelidae): new evidence from mitochondrial DNA. *Evolution*, **49**, 1008–1017.

Futuyma, D.J. and McCafferty, S.S. (1990) Phylogeny and the evolution of host plant associations in the leaf beetle genus *Ophraella* (Coleoptera, Chrysomelidae). *Evolution*, **44**, 1885–1913.

Genise, J.F. (1995) Upper Cretaceous trace fossils in permineralized plant remains from Patagonia, Argentina. *Ichnos*, **3**, 287–299.

Grauvogel-Stamm, L. and Kelber, K.-P. (1996) Plant–insect interactions and coevolution during the Triassic in western Europe. *Paleontologia Lombarda*, NS **5**, 5–23.

Greenfest, E.F. and Labandeira, C.C. (1998) Insect folivory on a Lower Permian (Sakmarian) riparian flora from north-central Texas. *Geological Society of America Abstracts with Programs*, **29**: 262, abstract.

Grimaldi, D.A., Bonwich, E., Dellanoy, M. and Doberstein, W. (1994) Electron microscopic studies of mummified tissues in amber fossils. *American Museum Novitates*, **3097**, 1–31.

Guo, S.-X. (1991) A Miocene trace fossil of an insect from Shanwang Formation in Linqu, Shandong. *Acta Palaeontologica Sinica*, **30**, 739–742.

Harland, W.B., Armstrong, R.L., Cox, A.V., Craig, L.E., Smith, A.G. and Smith, D.G. (1990) *A Geologic Time Scale—1989*. Cambridge: Cambridge University Press.

Harris, T.M. (1942) *Wonnacottia*, a new bennettitalean microsporophyll. *Annals of Botany*, NS **6**, 577–592.

Hellmund, M. and Hellmund, W. (1996) Zur endophytischen Eiablage fossiler Kleinlibellen (Insecta, Odonata, Zygoptera), mit Beschreibung eines neuen Gelegetyps. *Mitteilungen der Bayerischen Staatssammlung für Paläontologie und Historische Geologie*, **36**, 107–115.

Hickey, L.J. and Hodges, R.W. (1975) Lepidopteran leaf mine from the Early Eocene Wind River Formation of northeastern Wyoming. *Science*, **189**, 718–720.

Hill, C.R. (1976) Coprolites of *Ptilophyllum* cuticles from the Middle Jurassic of North Yorkshire. *Bulletin of the British Museum of Natural History (Geology)*, **27**, 289–294.

Hollis, D. (1987) A review of the Malvales-feeding psyllid family Carsidaridae (Homoptera). *Bulletin of the British Museum (Natural History), Entomology Series*, **56**, 87–127.

Hotton, C.L., Hueber, F.M. and Labandeira, C.C. (1996) Plant–arthropod interactions from early terrestrial ecosystems: two Devonian examples. In *Paleontological Society Special Publication*, ed. J.E. Repetski, **8**, 181, Lawrence, KS: Paleontological Society, abstract.

Hotton, N. III, Olson, E.C. and Beerbower, R. (1997) Amniote origins and the discovery of herbivory. In *Amniote Origins*, ed. S.S. Sumida and K.L.M. Martin. New York: Academic Press, pp. 207–264.

James, H.T. and Burney, D.A. (1997) The diet and ecology of Hawaii's extinct flightless waterfowl: evidence from coprolites. *Biological Journal of the Linnean Society*, **62**, 279–297.

Janz, N. and Nylin, S. (1998) Butterflies and plants: a phylogenetic study. *Evolution*, **52**, 486–502.

Jarzembowski, E.A. (1990) A boring beetle from the Wealden of the Weald. In *Evolutionary Paleobiology of Behavior and Coevolution*, ed. A.J. Boucot. Amsterdam: Elsevier, pp. 373–376.

Kelber, K.-P. (1988) Was ist *Equisetites foveolatus*? In *Neue Forschungen zur Erdgeschichte von Crailsheim*, ed. H. Hagdorn. Stuttgart: Goldschneck-Verlag Werner Weidert, pp. 166–184.

Kelber, K.-P. and Geyer, G. (1989) Lebensspuren von Insekten an Pflanzen des unteren Keupers. *Courier Forschungsinstitut Senckenberg*, **109**, 165–174.

Kerp, J.H.F. (1988) Aspects of Permian palaeobotany and palynology. X. The west- and central European species of the genus *Autunia* Krasser emend. Kerp (Peltaspermaceae) and the form-genus *Rhachiphyllum* Kerp (callipterid foliage). *Review of Palaeobotany and Palynology*, **54**, 249–360.

Kevan, P.G., Chaloner, W.G. and Savile, D.B.O. (1975) Interrelationships of early terrestrial arthropods and plants. *Palaeontology*, **18**, 391–417.

Kidston, R. and Lang, W.H. (1921) On Old Red Sandstone plants showing structure, from the Rhynie Chert Bed, Aberdeenshire. Part IV. Restorations of the vascular cryptogams, and discussion of their bearing on the general morphology of the Pteridophyta and the origin of the organisation of land-plants. *Transactions of the Royal Society of Edinburgh*, **52**, 831–854, pls. 1–5.

Koponen, M. and Nuorteva, M. (1973) Über subfossile Waldinsekten aus dem Moor Piilonsuo in Südfinnland. *Acta Entomologica Fennica*, **29**, 1–84.

Krassilov, V.A. and Rasnitsyn, A.P. (1983) A unique find: pollen in the intestine of Early Cretaceous sawflies. *Paleontological Journal*, **1982/4**, 80–95.

Krassilov, V.A. and Rasnitsyn, A.P. (1997) Pollen in the guts of Permian insects: first evidence of pollinivory and its evolutionary significance. *Lethaia*, **29**, 369–372.

Krassilov, V.A., Rasnitsyn, A.P. and Alfonin, S.A. (1999) Pollen morphotypes from the intestine of a Permian booklouse. *Review of Palaeobotany and Palynology*, **106**, 89–96.

Krassilov, V.A. and Sukacheva, I.D. (1979) Caddisfly cases from *Karkenia* seeds (Gingkophyta) in Lower Cretaceous deposits of Mongolia. *Transactions of the Soil Biology Institute*, **53**, 119–121 [in Russian].

Krassilov, V.A., Zherikhin, V.V. and Rasnitsyn, A.P. (1997) *Classopollis* in the guts of Jurassic insects. *Palaeontology*, **40**, 1095–1101.

Kristensen, N.P. (1995) Forty years' insect phylogenetic systematics: Hennig's 'Kritische Bemerkungen . . .' and subsequent developments. *Zoologische Beiträge*, NS **36**, 83–124.

Kukalová-Peck, J. (1991) Fossil history and the evolution of hexapod structures. In *The Insects of Australia*, ed. I.D. Naumann, P.B. Carne, J.F. Lawrence, E.S. Nielson, J.P. Spradbery et al., **1**, 141–179. Ithaca, NY: Cornell University Press.

Labandeira, C.C. (1990) *Use of a Phenetic Analysis of Recent Hexapod Mouthparts for the Distribution of Hexapod Food Resource Guilds in the Fossil Record*. PhD thesis, University of Chicago.

Labandeira, C.C. (1997a) Insect mouthparts: ascertaining the paleobiology of insect feeding strategies. *Annual Review of Ecology and Systematics*, **28**, 153–193.

Labandeira, C.C. (1997b) Permian pollen eating. *Science*, **277**, 1422–1423.

Labandeira, C.C. (1998a) Early history of arthropod and vascular plant associations. *Annual Review of Earth and Planetary Sciences*, **26**, 329–377.

Labandeira, C.C. (1998b) How old is the flower and the fly? *Science*, **280**, 57–59.

Labandeira, C.C. (1998c) Plant–insect associations from the fossil record. *Geotimes*, **43**, 18–24.

Labandeira, C.C. (2002) Insect diversity in deep time: implications for the Modern Era. In *Insect Diversity: Ecological, Evolutionary and Practical Considerations*, ed. W.W.M. Steiner (accepted). Washington, DC: Biological Resources Division, US Geological Survey.

Labandeira, C.C. and Beall, B.S. (1990) Arthropod terrestriality. In *Arthropods: Notes for a Short Course*, ed. D. Mikulic. Knoxville, TN: Paleontological Society, pp. 214–256.

Labandeira, C.C., Beall, B.S. and Hueber, F.M. (1988) Early insect diversification: evidence from a Lower Devonian bristletail from Québec. *Science*, **242**, 913–916.

Labandeira, C.C., Dilcher, D.L., Davis, D.R. and Wagner, D.L. (1994) 97 million years of angiosperm–insect association: paleobiological insights into the meaning of coevolution. *Proceedings of the National Academy of Sciences, USA*, **91**, 12278–12282.

Labandeira, C.C., Johnson, K.R. and Lang, P.J. (2002) Insect herbivory across the Cretaceous/Tertiary boundary: major extinction and minimum rebound. In *The Hell Creek Formation and the Cretaceous/Tertiary Boundary in the Northern Great Plains—An Integrated Continental Record at the End of the Cretaceous*, ed. J.H. Hartman, K. Johnson and D.J. Nichols (in press). Geological Society of America Special Publication.

Labandeira, C.C., Nufio, C., Wing, S. and Davis, D. (1995) Insect feeding strategies from the Late Cretaceous Big Cedar Ridge flora: comparing the diversity and intensity of Mesozoic herbivory with the present. *Geological Society of America Abstracts with Programs*, **27**, 447, abstract.

Labandeira, C.C. and Phillips, T.L. (1996a) Insect fluid-feeding on Upper Pennsylvanian tree ferns (Palaeodictyoptera, Marattiales) and the early history of the piercing-and-sucking functional feeding group. *Annals of the Entomological Society of America*, **89**, 157–183.

Labandeira, C.C. and Phillips, T.L. (1996b) A Late Carboniferous petiole gall and the origin of holometabolous insects. *Proceedings of the National Academy of Sciences, USA*, **93**, 8470–8474.

Labandeira, C.C., Phillips, T.L. and Norton, R.A. (1997) Oribatid mites and decomposition of plant tissues in Paleozoic coal-swamp forests. *Palaios*, **12**, 317–351.

Lewis, S.E. (1972) Fossil caddisfly (Trichoptera) cases from the Ruby River Basin (Oligocene) of southwestern Montana. *Annals of the Entomological Society of America*, **65**, 518–519.

Lewis, S.E. (1976) Lepidopterous feeding damage of live oak leaf (*Quercus convexa* Lesquereaux) from the Ruby River Basin (Oligocene) of southwestern Montana. *Journal of Paleontology*, **50**, 345–346.

Lewis, S.E. (1985) Miocene insects from the Clarkia deposits of northern Idaho. In *Late Cenozoic History of the Pacific Northwest*, ed. C.J. Smiley. Washington, DC: American Association for the Advancement of Science, pp. 245–264.

Lewis, S.E. (1994) Evidence of leaf-cutting bee damage from the Republic sites (Middle Eocene) of Washington. *Journal of Paleontology*, **68**, 172–173.

Lewis, S.E. and Carroll, M.A. (1991) Coleopterous egg deposition on alder leaves from the Klondike Mountain Formation (Middle Eocene), northeastern Washington. *Journal of Paleontology*, **65**, 334–335.

Linck, O. (1949) Fossile Bohrgänge (*Anobichnium simile* n.g. n.sp.) an einem Keuperholz. *Neues Jahrbuch für Mineralogie, Geologie und Paläontologie Monatschefte*, **1949B**, 180–185.

Linstow, O.V. (1906) Über Bohrgänge von Käferlarven in Braunkohlenholz. *Jahrbuch Koniglich Preußischen Geologischen Landesanstalt und Bergakademie*, **26**, 467–470.

Mädler, A.K. (1936) Eine Blattgalle an einem vorweltlichen Pappel-Blatt. *Natur und Volk*, **66**, 271–274.

Matile, L. (1997) Phylogeny and evolution of the larval diet in the Sciaroidea (Diptera, Bibionomorpha) since the Mesozoic. In *The Origin of Biodiversity in Insects: Phylogenetic Tests of Evolutionary Scenarios*, ed. P. Grandcolas. *Mémoires du Muséum National d'Histoire Naturelle*, **173**, 273–303.

Matley, C.A. (1941) The coprolites of Pijdura, central provinces. *Records of the Geological Survey of India*, **74**, 535–547, pls. 34–36.

Menken, S.B.J. (1996) Pattern and process in the evolution of insect–plant associations: *Yponomeuta* as an example. *Entomologia Experimentalis et Applicata*, **80**, 297–305.

Mikulás, R., Dvorak, Z. and Pek, I. (1998) *Lamniporichnus vulgaris* igen. et isp. nov.: traces of insect larvae in stone fruits of hackberry (*Celtis*) form the Miocene and Pleistocene of the Czech Republic. *Journal of the Czech Geologic Society*, **43**, 277–280.

Millay, M.A. and Taylor, T.N. (1979) Paleozoic seed fern pollen organs. *Botanical Review*, **45**, 301–375.

Mitter, C., Poole, R.W. and Matthews, M. (1993) Biosystematics of the Heliothinae (Lepidoptera: Noctuidae). *Annual Review of Entomology*, **38**, 207–225.

Mostovski, M.B. (1998) A revision of the nemestrinid flies (Diptera, Nemestrinidae) described by Rohdendorf, and a description of new taxa of the Nemestrinidae from the Upper Jurassic of Kazakhstan. *Paleontological Journal*, **32**, 369–375.

Müller, A.H. (1978) Zur Entomofauna des Permokarbon über die Morphologie, Taxonomie und Ökologie von *Eugereon boeckingi* (Palaeodictyoptera). *Freiberger Forschungsheft*, **334C**, 7–20.

Müller, A.H. (1982) Über Hyponome fossiler und rezenter Insekten, erster Beitrag. *Freiberger Forschungsheft*, **366C**, 7–27.

Munk, W. and Sues, H.-D. (1993) Gut contents of *Parasaurus* (Pareiasauria) and *Protorosaurus* (Archosauromorpha) from the Kupferschiefer (Upper Permian) of Hessen, Germany. *Paläontologische Zeitschrift*, **67**, 169–176.

Nambudiri, E.M.V. and Binda, P.L. (1989) Dicotyledonous fruits associated with coprolites from the Upper Cretaceous (Maastrichtian) Whitemud Formation, southern Saskatchewan, Canada. *Review of Palaeobotany and Palynology*, **59**, 57–66.

Naugolnykh, S.V. (1998) The fossil plants of Chekarda. In *Chekarda — The Locality of Permian Fossil Plants and Insects*, ed. G.Y. Ponomaryova, V.G. Novokshonov and S.V. Naugolnykh. Perm: Permian University Press, pp. 55–91 [in Russian].

Novokshonov, V.G. (1997) Some Mesozoic scorpionflies (Insecta: Panorpida = Mecoptera) of the families Mesopsychidae,

Pseudopolycentropidae, Bittacidae, and Permochoristidae. *Paleontologicheskii Zhurnal*, **1997/1**, 65–71 [in Russian].

Novokshonov, V.G. (1998a) New fossil insects (Insecta: Grylloblattida, Caloneurida, Hypoperlida?, ordinis incertis) from the Kungurian beds of the middle Urals. *Paleontological Journal*, **32**, 362–368.

Novokshonov, V.G. (1998b) The fossil insects of Chekarda. In *Chekarda — The Locality of Permian Fossil Plants and Insects*, ed. G.Y. Ponomaryova, V.G. Novokshonov and S.V. Naugolnykh. Perm: Permian University Press, pp. 25–54.

Palmer, M., Labandeira, C.C., Johnson, K. and Wehr, W. (1998) Diversity and intensity of insect herbivory on the Middle Eocene Republic Flora: comparing the fossil record with the recent. *Geological Society of America Abstracts with Programs*, **30**, 37, abstract.

Parsons, M.J. (1996) A phylogenetic reappraisal of the birdwing genus *Ornithoptera* (Lepidoptera: Papilionidae: Troidini) and a new theory of its evolution in relation to Gondwanan vicariance biogeography. *Journal of Natural History*, **30**, 1707–1736.

Pemberton, R.W. (1992) Fossil extrafloral nectaries, evidence for the ant-guard antiherbivore defense in an Oligocene *Populus*. *American Journal of Botany*, **79**, 1242–1246.

Peyerimhoff, P. (1909) Le *Cupes* de l'ambre de la Baltique (Col.). *Bulletin de la Société Entomologique de France*, **1909**, 57–60.

Plumstead, E.P. (1963) The influence of plants and environment on the developing animal life of Karoo times. *South African Journal of Science*, **59**, 147–152.

Poinar G.O., Jr. (1992) Fossil evidence of resin utilization by insects. *Biotropica*, **24**, 466–468.

Rasnitsyn, A.P. (1977) New Paleozoic and Mesozoic insects. *Paleontological Journal*, **11**, 60–72.

Rasnitsyn, A.P. and Krassilov, V.A. (1996a) First find of pollen grains in the gut of Permian insects. *Paleontological Journal*, **30**, 484–490.

Rasnitsyn, A.P. and Krassilov, V.A. (1996b) Pollen in the gut contents of fossil insects as evidence of coevolution. *Paleontological Journal*, **30**, 716–722.

Rasnitsyn, A.P. and Novokshonov, V.G. (1997) On the morphology of *Uralia maculata* (Insecta: Diaphanopterida) from the Early Permian (Kungurian) of Ural (Russia). *Entomologica Scandinavica*, **28**, 27–38.

Rayner, R.H. and Waters, S.B. (1991) Floral sex and the fossil insect. *Naturwissenschaften*, **78**, 280–282.

Ren, D. (1998) Flower-associated Brachycera flies as fossil evidence for Jurassic angiosperm origins. *Science*, **280**, 85–88.

Retallack, G.J. and Dilcher, D.L. (1988) Reconstructions of selected seed ferns. *Annals of the Missouri Botanical Garden*, **75**, 1010–1057.

Richardson, E.S. Jr. (1980) Life at Mazon Creek. In *Middle and Late Pennsylvanian Strata of [the] Margin of [the] Illinois Basin*, ed. R.L. Langenheim Jr. and C.J. Mann. Urbana: University of Illinois, pp. 217–224.

Roderick, G.K. (1997) Herbivorous insects and the Hawaiian silversword alliance: coevolution or cospeciation? *Pacific Science*, **51**, 440–449.

Roderick, G.K. and Metz, E.C. (1997) Biodiversity of planthoppers (Hemiptera: Delphacidae) on the Hawaiian silversword alliance: effects of host plant phylogeny and hybridisation. *Memoirs of the Museum of Victoria*, **56**, 393–399.

Rohdendorf, B.B. (1968) New Mesozoic nemestrinids (Diptera, Nemestrinidae). In *Jurassic Insects of Karatau*, ed. B.B. Rohdendorf. Moscow: Academy of Sciences, pp. 180–189 [in Russian].

Ronquist, F. (1995) Phylogeny and early evolution of the Cynipoidea (Hymenoptera). *Systematic Entomology*, **20**, 309–335.

Rothwell, G.W. and Scott, A.C. (1988) *Heterotheca* Benson; lyginopterid pollen organs or coprolites? *Bulletin of the British Museum of Natural History (Geology)*, **44**, 41–43.

Roubik, D.W., Segura, J.A.L. and Camargo, J.M.F. (1997) New stingless bee genus endemic to Central American cloudforests: phylogenetic and biogeographic implications (Hymenoptera: Apidae: Melponini). *Systematic Entomology*, **22**, 67–80.

Rozefelds, A.C. (1988a) Lepidoptera mines in *Pachypteris* leaves (Corystospermaceae: Pteridospermophyta) from the Upper Jurassic/Lower Cretaceous Battle Camp Formation, North Queensland. *Proceedings of the Royal Society of Queensland*, **99**, 77–81.

Rozefelds, A.C. (1988b) Insect leaf mines from the Eocene Anglesea locality, Victoria, Australia. *Acheringa*, **12**, 1–6.

Schaarschmidt, F. (1992) The vegetation: fossil plants as witnesses of a warm climate. In *Messel: An Insight into the History of Life and of the Earth*, ed. S. Schaal and W. Ziegler. Oxford: Oxford University Press, pp. 27–52.

Schedl, K.E. (1947) Die Borkenkäfer des baltischen Bernsteins. *Zentralblatt für das Gesamtgebiet der Entomologie*, **2**, 12–45.

Schoener, T.W. (1974) Resource partitioning in ecological communities. *Science*, **185**, 27–39.

Scholtz, C.H. and Chown, S.L. (1995) The evolution of habitat use and diet in the Scarabaeoidea: a phylogenetic approach. In *Biology, Phylogeny, and Classification of Coleoptera: Papers Celebrating the 80th Birthday of Roy A. Crowson*, ed. J. Pakaluk and S.A. Slipinski. Warsaw: Muzeum i Instiytut Zoologii, pp. 355–374.

Schopf, J.M. (1948) Pteridosperm male fructifications: American species of *Dolerotheca*, with notes regarding certain allied forms. *Journal of Paleontology*, **22**, 681–724.

Scott, A.C. and Paterson, S. (1984) Techniques for the study of plant/arthropod interactions in the fossil record. *Geobios Mémoire Spécial*, **8**, 449–455.

Scott, A.C., Stephenson, J. and Chaloner, W.G. (1992) Interaction and coevolution of plants and arthropods during the Palaeozoic and Mesozoic. *Philosophical Transactions of the Royal Society of London B*, **335**, 129–165.

Scott, A.C. and Taylor, T.N. (1983) Plant/animal interactions during the Upper Carboniferous. *Botanical Review*, **49**, 259–307.

Seward, A.C. (1923) The use of the microscope in palaeobotanical research. *Journal of the Royal Microscopal Society*, **1923**, 299–302.

Sharov, A.G. (1973) Morphological features and mode of life of the Palaeodictyoptera. In *Readings in the Memory of Nikolaj Aleksandrovich Kholodkovskij*, ed. G.Y. Bei-Benko. Leningrad: Science Publishers, pp. 49–63 [in Russian].

Shear, W.A. and Kukalová-Peck, J. (1990) The ecology of Paleozoic terrestrial arthropods: the fossil evidence. *Canadian Journal of Zoology*, **68**, 1807–1834.

Shields, O. and Reveal, J.L. (1988) Sequential evolution of *Euphilotes* (Lycaenidae: Scolitantidini) on their plant host *Eriogonum* (Polygonaceae: Eriogonoideae). *Biological Journal of the Linnean Society*, **33**, 51–93.

Sorensen, J.T., Campbell, B.C, Gill, R.J. and Steffen-Campbell, J.D. (1995) Non-monophyly of Auchenorrhyncha (Homoptera), based upon 18S rDNA phylogeny: eco-evolutionary and cladistic implications within pre-Heteropteroidea Hemiptera (s. l.) and a proposal for new monophyletic suborders. *Pan-Pacific Entomologist*, **71**, 31–60.

Srivastava, A.K. (1987) Lower Barakar flora of Raniganj coalfield and insect/plant relationship. *Palaeobotanist*, **36**, 138–142.

Stephenson, J. and Scott, A.C. (1992) The geological history of insect-related plant damage. *Terra Nova*, **4**, 542–552.

Straus, A. (1967) Zur Paläontologie des Pliozäns von Willershausen. *Beiheft zum Bericht der Naturhistorischen Gesellschaft Hannover*, **111**, 15–24.

Straus, A. (1977) Gallen, Minen und andere Fraßspuren im Pliokän von Willershausen am Harz. *Verhandlungen des Botanischen Vereins der Provinz Brandenberg*, **113**, 43–80.

Strong, D.R., Lawton, J.H. and Southwood, T.R.E. (1984) *Insects on Plants: Community Patterns and Mechanisms*. Cambridge, MA: Harvard University Press.

Süss, H. and Müller-Stoll, W.R. (1980) Das fossile Holz *Pruninium gummosum* Platen emend. Süss aus dem Yellowstone Nationalpark und sein Parasit *Palaeophytobia prunorum* sp. nov. nebst Bemerkungen über Markflecke. In *100 Jahre Aboretum Berlin (1879–1979), Jubiläumsschrift*, ed. W. Vent. Berlin: Humboldt-Universität, pp. 343–364.

Thomas, B.A. (1969) A new British Carboniferous calamite cone, *Paracalamostachys spadiciformis*. *Palaeontology*, **12**, 253–261.

Thulborn, R.A. (1991) Morphology, preservation and palaeobiological significance of dinosaur coprolites. *Palaeogeography, Palaeoclimatology, Palaeoecology*, **83**, 341–366.

Tidwell, W.D. and Ash, S.R. (1990) On the Upper Jurassic stem *Hermanophyton* and its species from Colorado and Utah, USA. *Palaeontographica*, **218B**, 77–92, pls. 1–6.

Tiffney, B.H. (1980) Fruits and seeds of the Brandon Lignite. V. Rutaceae. *Journal of the Arnold Arboretum*, **61**, 1–36.

Waggoner, B.M. and Poteet, M.F. (1996) Unusual oak leaf galls from the Middle Miocene of northwestern Nevada. *Journal of Paleontology*, **70**, 1080–1084.

Walker, M.V. (1938) Evidence of Triassic insects in the Petrified Forest National Monument, Arizona. *Proceedings of the United States National Museum*, **85**, 137–141.

Watson, J. (1977) Some Lower Cretaceous conifers of the Cheirolepidaceae from the U.S.A. and England. *Palaeontology*, **20**, 715–749.

Weaver, L., McLoughlin, S. and Drinnan, A.N. (1997) Fossil woods from the Upper Permian Bainmedart Coal Measures, northern Prince Charles Mountains, East Antarctica. *Journal of Australian Geology and Geophysics*, **16**, 655–676.

Wehr, W. (1998) Middle Eocene insects and plants of the Okanogan Highlands. In *Contributions to the Paleontology and Geology of* the West Coast in Honor of V. Standish Mallory. Thomas Burke Memorial Washington State Museum Research Report, **6**, 99–109.

Weiss, C.E. (1876) Steinkohlen-Calamarien, mit besonderer Berücksichtigung ihrer Fructificationen. Berlin: Nemann'schen Kartenhandlung.

Wilf, P. and Labandeira, C.C. (1999) Plant–insect associations respond to Paleocene–Eocene warming. Science, **284**, 2153–2156.

Wilf, P., Labandeira, C.C., Johnson, K.R., Coley, P.D. and Cutter, A.D. (2001) Insect herbivory, plant defense, and early cenozoic climate change. Proceedings of the National Academy of Sciences, USA, **98**, 6221–6226.

Willemstein, S.C. (1980) Pollen in Tertiary insects. Acta Botanica Neerlandica, **29**, 57–58.

Wilson, S.W., Mitter, C., Denno, R.F. and Wilson, M.R. (1994) Evolutionary patterns of host plant use by delphacid planthoppers and their relatives. In Planthoppers: Their Ecology and Management, ed. R.F. Denno and T.J. Perfect. London: Chapman & Hall, pp. 7–113.

Yeates, D.K. and Irwin, M.E. (1996) Apioceridae (Insecta: Diptera): cladistic reappraisal and biogeography. Zoological Journal of the Linnean Society, **116**, 247–301.

Zhou, Z.-Y. and Zhang, B. (1989) A sideritic Protocupressinoxylon with insect borings and frass from the Middle Jurassic, Henan, China. Review of Palaeobotany and Palynology, **59**, 133–143.

Zwölfer, H. and Herbst, J. (1988) Präadaptation, Wirtskreiserweiterung und Parallel-Cladogenese in der Evolution von phytophagen Insekten. Zeitschrift für Zoologische Systematik und Evolutionsforschung, **26**, 320–340.

References

Abrahamson, W.G. and Weis, A.E. (1997) *Evolutionary Ecology across Three Trophic Levels: Goldenrods, Gallmakers, and Natural Enemies*. Princeton, NJ: Princeton University Press.

Abrams, P.A. (2000) The evolution of predator–prey interactions: theory and evidence. *Annual Review of Ecology and Systematics*, **31**, 79–105.

Addicott, J.F. (1998) Regulation of mutualism between yuccas and yucca moths: population level processes. *Oikos*, **81**, 119–129.

Addicott, J.F. and Bao, T. (1999) Limiting the costs of mutualism: multiple modes of interaction between yuccas and yucca moths. *Proceedings of the Royal Society of London B*, **266**, 197–202.

Adler, L.S., Schmitt, J. and Bowers, M.D. (1995) Genetic variation in defensive chemistry in *Plantago lanceolata* (Plantaginaceae) and its effect on the specialist herbivore *Junonia coenia* (Nymphalidae). *Oecologia*, **101**, 75–85.

Afik, D. and Karasov, W.H. (1995) The trade-offs between digestion rate and efficiency in warblers and their ecological implications. *Ecology*, **76**, 1147–1157.

Ager, D. (1993) *The New Catastrophism: The Importance of the Rare Event in Geological History*. Cambridge: Cambridge University Press.

Agrawal, A.A. (1998a) Induced responses to herbivory and increased plant performance. *Science*, **279**, 1201–1202.

Agrawal, A.A. (1998b) Leaf damage and associated cues induce aggressive ant recruitment in a neotropical ant-plant. *Ecology*, **79**, 2100–2112.

Agrawal, A.A., Strauss, S.Y. and Stout, M. (1999) Costs of induced defenses and tolerance in male and female fitness components of wild radish. *Evolution*, **53**, 1093–1104.

Ågren, J., Danell, K., Elmqvist, T., Ericson, L. and Hjältén, J. (1999) Sexual dimorphism and biotic interactions. In *Gender and Sexual Dimorphism in Flowering Plants*, ed. M.A. Geber, T.E. Dawson and L.F. Delph. Berlin: Springer-Verlag, pp. 217–246.

Alcover, J.A., Perez-Obiol, R., Errikarta-Imanol, Y. and Bover, P. (1999) The diet of *Myotragus balearicus* Bate 1909 (Artiodactyla: Caprinae), an extinct bovid from the Balearic Islands: evidence from coprolites. *Biological Journal of the Linnean Society*, **66**, 57–74.

Al-Dabbagh, K.Y., Jiad, J.H. and Waheed, I.N. (1987) The influence of diet on the intestine length of the white-cheeked bulbul. *Ornis Scandinavica*, **18**, 150–152.

Alexander, H.M., Thrall, P.H., Antonovics, J., Jarosz, A.M. and Oudemans, P.V. (1996) Population dynamics and genetics of plant disease: a case study of anther-smut disease. *Ecology*, **77**, 990–996.

Alroy, J. (1999) The fossil record of North American mammals: evidence for a Paleocene evolutionary radiation. *Systematic Biology*, **48**, 107–118.

Alston, R.E. and Turner, B.L. (1963) *Biochemical Systematics*. Englewood Cliffs, NJ: Prentice-Hall.

Altbäcker, V., Hudson R. and Bilko, A. (1995) Rabbit-mothers' diet influences pups' later food choice. *Ethology*, **99**, 107–116.

Althoff, D.M. and Thompson, J.N. (1999) Comparative geographic structures of two parasitoid-host interactions. *Evolution*, **53**, 818–825.

Althoff, D.M. and Thompson, J.N. (2001) Geographic structure in the searching behaviour of a specialist parasitoid: combining molecular and behavioral approaches. *Journal of Evolutionary Biology*, **14**, 406–417.

Alvarez-Buylla, E.R. and Garay, A.A. (1994) Population genetic structure of *Cecropia obtusifolia*, a tropical pioneer tree species. *Evolution*, **48**, 437–453.

Andersen, A.N. (1988) Soil of the nest-mound of the seed dispersing ant, *Aphaenogaster longiceps*, enhances seedling growth. *Australian Journal of Ecology*, **13**, 469–471.

Andersen, A.N. (1989) How important is seed predation to recruitment in stable populations of long-lived perennials? *Oecologia*, **81**, 310–315.

Andersen, A.N. and Morrison, S.C. (1998) Myrmecochory in Australia's seasonal tropics—effects of disturbance on dispersal distance. *Australian Journal of Ecology*, **23**, 483–491.

Anderson, N.H. and Cargill, A.S. (1987) Nutritional ecology of aquatic detritivorous insects. In *Nutritional Ecology of Insects, Mites, Spiders, and Related Invertebrates*, ed. F. Slansky Jr and J.G. Rodriguez. New York: Wiley, pp. 903–925.

Andersson, S. (1996) Floral display and pollination success in *Senecio jacobaea* (Asteraceae): interactive effects of head and corymb size. *American Journal of Botany*, **83**, 71–75.

Ann, D.K., Lin, H.H. and Kousvelari, E. (1997) Regulation of salivary-gland-specific gene expression. *Critical Reviews in Oral Biology and Medicine*, **8**, 244–252.

Antonovics, J., Thrall, P.H. and Jarosz, A.M. (1998) Genetics and the spatial ecology of species interactions: the *Silene-Ustilago* system. In *Spatial Ecology: The Role of Space in Population Dynamics and Interspecific Interactions*, ed. D. Tilman and P. Kareiva. Princeton, NJ: Princeton University Press, pp. 158–180.

Archangelsky, S. and Cuneo, R. (1987) Ferugliocladaceae, a new conifer family from the Permian of Gondwana. *Review of Palaeobotany and Palynology*, **51**, 3–30.

Archibold, O.W. (1995) *Ecology of World Vegetation*. London: Chapman & Hall.

Arens, W. (1994) Striking convergence in the mouthpart evolution of stream-living algae grazers. *Journal of Zoology, Systematics and Evolutionary Research*, **32**, 319–343.

Arizmendi, M.C., Domínguez, C.A. and Dirzo, R. (1996) The role of an avian nectar robber and of hummingbird pollinators in the reproduction of two plant species. *Functional Ecology*, **10**, 119–127.

Armbruster, W.S. (1992) Phylogeny and the evolution of plant-animal interactions. *Bioscience*, **42**, 12–20.

Armbruster, W.S. (1997) Exaptations link evolution of plant–herbivore and plant–pollinator interactions: a phylogenetic inquiry. *Ecology*, **78**, 1661–1672.

Armbruster, W.S. and Baldwin, B. (1998) Switch from specialized to generalized pollination. *Nature*, **294**, 632.

Armbruster, W.S., Edwards, M.E. and Debevec, E.M. (1994) Floral character displacement generates assemblage structure of western Australian triggerplants (*Stylidium*). *Ecology*, **75**, 315–329.

Armbruster, W.S., Howard, J.J., Clausen, T.P. et al. (1997) Do biochemical exaptations link evolution of plant defense and pollination systems? Historical hypotheses and experimental tests with *Dalechampia* vines. *American Naturalist*, **149**, 461–484.

Armesto, J.J. and Rozzi, R. (1989) Seed dispersal syndromes in the rain forest of Chiloé: evidence for the importance of biotic dispersal in a temperate rain forest. *Journal of Biogeography*, **16**, 219–226.

Arms, K., Feeny, P. and Lederhouse, R.C. (1974) Sodium: stimulus for puddling behavior by tiger swallowtail butterflies, *Papilio glaucus*. *Science*, **185**, 372–374.

Artabe, A.E. and Stevenson, D.W. (1999) Fossil Cycadales of Argentina. *Botanical Review*, **65**, 219–238.

Ash, S. (1997) Evidence of arthropod–plant interactions in the Upper Triassic of the southwestern United States. *Lethaia*, **29**, 237–248.

Atsatt, P.R. and O'Dowd, D.J. (1976) Plant defense guilds. *Science*, **193**, 24–29.

Aucoin, R., Guillet, G., Murray, C., Philogene, B.J.R. and Arnason, J.T. (1995) How do insect herbivores cope with the extreme oxidative stress of phototoxic host plants? *Archives of Insect Biochemistry and Physiology*, **29**, 211–226.

Augspurger, C.K. (1984) Seedling survival of tropical tree species: interactions of dispersal distance, light-gaps, and pathogens. *Ecology*, **65**, 1705–1712.

Augustine, D.J. and McNaughton S.J. (1998) Ungulate effects on the functional species composition of plant communities: Herbivore selectivity and plant tolerance. *Journal of Wildlife Management*, **62**, 1165–1183.

Avise, J.C. (2000) *Phylogeography: The History and Formation of Species*. Cambridge, MA: Harvard University Press.

Bach, C.E. (1991) Direct and indirect interactions between ants (*Pheidole megacephala*), scales (*Coccus viridis*) and plants (*Pluchea indica*). *Oecologia*, **87**, 233–239.

Bach, C.E. (1994) Effects of a specialist herbivore (*Altica subplicata*) on *Salix cordata* and sand dune succession. *Ecological Monographs*, **64**, 423–445.

Baker, A.M., Thompson, J.D. and Barrett, S.C.H. (2000) Evolution and maintenance of stigma-height dimorphism in *Narcissus*. I. Floral variation and style-morph ratios. *Heredity*, **84**, 502–513.

Baker, H.G. and Baker, I. (1982) Chemical constitutents of nectar in relation to pollination mechanisms and phylogeny. In *Biochemical Aspects of Evolutionary Biology*, ed. M.H. Nitecki. Chicago: University of Chicago Press, pp. 131–171.

Baker, H.G., Baker, I. and Hodges, S.A. (1998) Sugar composition of nectars and fruits consumed by birds and bats in the tropics and subtropics. *Biotropica*, **30**, 559–586.

Bakker, R.T. (1978) Dinosaur feeding behaviour and the origin of plants. *Nature*, **274**, 661–663.

Bakker, R.T. (1986) *The Dinosaur Heresies*. New York: William Morrow.

Baldwin, I.T. (1998) Jasmonate-induced responses are costly but benefit plants under attack in native populations. *Proceedings of the National Academy of Sciences USA*, **95**, 8113–8118.

Balick, M.J., Furth, D.G. and Cooper-Driver, G. (1978) Biochemical and evolutionary aspects of arthropod predation on ferns. *Oecologia*, **35**, 55–89.

Banks, H.P. and Colthart, B.J. (1993) Plant-animal-fungal interactions in Early Devonian trimerophytes from Gaspé, Canada. *American Journal of Botany*, **80**, 992–1001.

Barkman, T.J., Chenery, G., McNeal, J.R. et al. (2000) Independent and combined analyses of sequences from all three genomic compartments converge on the root of flowering plant phylogeny. *Proceedings of the National Academy of Sciences USA*, **97**, 13166–13171.

Barrett, S.C.H. (1998) The evolution of mating strategies in flowering plants. *Trends in Plant Science*, **3**, 335–341.

Barrett, S.C.H. and Graham, S.W. (1997) Adaptive radiation in the aquatic plant family Pontederiaceae: insights from phylogenetic analysis. In *Molecular Evolution and Adaptive Radiation*, ed. T.J. Givnish and K.J. Systma. Cambridge: Cambridge University Press, pp. 225–258.

Barrick, R.E. (1998) Isotope paleobiology of the vertebrates: ecology, physiology, and diagenesis. *Paleontological Society Papers*, **4**, 101–137.

Barry, T.N. and McNabb, W.C. (1999) The implications of condensed tannins on the nutritive value of temperate forages fed to ruminants. *British Journal of Nutrition*, **81**, 263–272.

Barton, A.M. (1986) Spatial variation in the effect of ants on an extra-floral nectary plant. *Ecology*, **67**, 495–504.

Bate-Smith, E.C. and Metcalf, C.R. (1957) Leucoanthocyanins. 3. The nature and systematic distribution of tannins in dicotyledoneous plants. *Journal of the Linnean Society, Botany*, **55**, 669–705.

Batzli, G.O. (1985) Nutrition. In *Biology of New World Microtus*, ed R.H. Tamarin. Shippensburg, PA: American Society of Mammalogists, Special Publication, pp. 779–811.

Baumann, P., Munson, M.A., Lai, C.Y. et al. (1993) Origin and properties of bacterial endosymbionts of aphids, whiteflies, and mealybugs. *American Society for Microbiology News*, **59**, 21–24.

Beattie, A.J. (1982) Ants and gene dispersal in flowering plants. In *Pollination and Evolution*, ed. J.A. Armstrong, J.M. Powell and A.J. Richards. Sudney: Royal Botanic Gardens, pp. 1–8.

Beattie, A.J. (1985) *The Evolutionary Ecology of Ant–Plant Mutualisms*. Cambridge: Cambridge University Press.

Beattie, A.J. (1991) Problems oustanding in ant–plant interactions. In *Ant–Plant Interactions*, ed. C.R. Huxley and D.F. Cutler. Oxford: Oxford University Press, pp. 559–576.

Beattie, A.J. and Culver, D.C (1981) The guild of myrmecochores in the herbaceous flora of the West Virginia forests. *Ecology*, **62**, 107–115.

Beaver, R.A. (1979) Host specificity of temperate and tropical animals. *Nature*, **281**, 139–141.

Becerra, J.X. (1989) Extrafloral nectaries: a defense against ant-Homopteran mutualisms? *Oikos*, **55**, 276–280.

Becerra, J.X. (1994) Squirt-gun defense in *Bursera* and the chrysomelid counterploy. *Ecology*, **75**, 1991–1996.

Becerra, J.X. (1997) Insects on plants: macroevolutionary chemical trends in host use. *Science*, **276**, 253–256.

Beck, A.L. and Labandeira, C.C. (1998) Early Permian insect folivory on a gigantopterid-dominated riparian flora from north-central Texas. *Palaeogeography, Palaeoclimatology, Palaeoecology*, **142**, 139–173.

Behrensmeyer, A.K., Damuth, J.D., DiMichele, W.A., Potts, R., Sues, H.-D. and Wing, S.L. (1992) *Terrestrial Ecosystems through Time*. Chicago: University of Chicago Press.

Belovsky, G.E. and Schmitz, O.J. (1991) Mammalian herbivore optimal foraging and the role of plant defenses. In *Plant Defenses against Mammalian Herbivory*, ed. R.T. Palo and C.T. Robbins. Boca Raton, FL: CRC Press, pp. 1–28.

Belsky, A.J., Carson, W.P., Jensen, C.L. and Fox, G.A. (1993) Overcompensation by plants: herbivore optimization or red herring? *Evolutionary Ecology*, **7**, 109–121.

Bell, E.A. (1978) Toxins in seeds. In *Biochemical Aspects of Plant and Animals Coevolution*, ed. J.B. Harborne. London: Academic Press, pp. 143–161.

Bell, G. (1985) On the function of flowers. *Proceedings of the Royal Society of London B*, **224**, 223–265.

Benkman, C.W. (1987) Food profitability and the foraging ecology of crossbills. *Ecological Monographs*, **57**, 251–267.

Benkman, C.W. (1991) Predation, seed size partitioning and the evolution of body size in seed-eating finches. *Evolutionary Ecology*, **5**, 118–127.

Benkman, C.W. (1993) Adaptation to single resources and the evolution of crossbill (*Loxia*) diversity. *Ecological Monographs*, **63**, 305–325.

Benkman, C.W. (1995) Wind dispersal capacity of pine seeds and the evolution of different seed dispersal modes in pines. *Oikos*, **73**, 221–224.

Benkman, C.W. (1999) The selection mosaic and diversifying coevolution between crossbills and lodgepole pine. *American Naturalist*, **153**, S75–S91.

Benkman, C.W., Holimon, W.C. and Smith, J.W. (2001) The influence of a competitor on the geographic mosaic of coevolution between crossbills and lodgepole pine. *Evolution*, **55**, 282–294.

Benrey, B. and Denno, R.F. (1997) The slow-growth-high-mortality hypothesis: a test using the cabbage butterfly. *Ecology*, **78**, 987–999.

Benson, W.W. (1985) Amazon ant-plants. In *Key Environments of Amazonia*, ed. G.T. Prance and T.E. Lovejoy. New York: Pergamon Press, pp. 239–266.

Berenbaum, M.R. and Zangerl, A.R. (1992) Genetics of physiological and behavioral resistance to host furanocoumarins in the parsnip webworm. *Evolution*, **46**, 1373–1384.

Berenbaum, M.R. and Zangerl, A.R. (1998) Chemical phenotype matching between a plant and its insect herbivore. *Proceedings of the National Academy of Sciences USA*, **95**, 13743–13748.

Berenbaum, M. and Zangerl, M. (1999) Genetic variation in cytochrome P450-based resistance to plant allelochemicals and insecticides. In *Herbivores: Between Plants and Predators*, ed. H. Olff, V.K. Brown and R.H. Drent. Oxford: Blackwell, pp. 55–84.

Berenbaum, M.R., Nitao, J.K. and Zangerl, A.R. (1991) Adaptive significance of furanocoumarin diversity in *Pastinaca sativa* (Apiaceae). *Journal of Chemical Ecology*, **17**, 207–215.

Berenbaum, M.R., Zangerl, A.R. and Nitao, J.K. (1986) Constraints on chemical coevolution: wild parsnips and the parsnip webworm. *Evolution*, **40**, 1215–1228.

Berg, R.Y. (1975) Myrmecochorous plants in Australia and their dispersal by ants. *Australian Journal of Botany*, **23**, 475–508.

Bergelson, J. and Purrington, C.B. (1996) Surveying patterns in the cost of resistance in plants. *American Naturalist*, **148**, 536–558.

Bergquist, J. and Örlander, G. (1998) Browsing damage by roe deer on Norway spruce seedlings planted on clearcuts of different ages. 2. Effect of seedling vigour. *Forest Ecology and Management*, **105**, 295–302.

Bergström, R. and Danell, K. (1987) Effects of simulated browsing by moose on morphology and biomass of two birch species. *Journal of Ecology*, **75**, 533–544.

Berlocher, S.H. and Bush, G.L. (1982) An electrophoretic analysis of *Rhagoletis* (Diptera: Tephritidae) phylogeny. *Systematic Zoology*, **31**, 136–155.

Bernard, G.D. and Remington, C.L. (1991) Color vision in *Lycaena* butterflies: spectral tuning of receptor arrays in relation to behavioral ecology. *Proceedings of the National Academy of Sciences USA*, **88**, 2783–2787.

Bernardi, G. (2000) Barriers to gene flow in *Embiotoca jacksoni*, a marine fish lacking a pelagic larval stage. *Evolution*, **54**, 226–237.

Bernays, E.A. (1982) The insect on the plant—a closer look. In *Proceedings of the 5th International Symposium on Insect–Plant Relationships*, ed. J.H. Visser and A.K. Minks. Wageningen: Centre for Agricultural Publishing and Documentation, pp. 3–17.

Bernays, E.A. (1997) Feeding by lepidopteran larvae is dangerous. *Ecological Entomology*, **22**, 121–123.

Bernays, E.A. and Cornelius, M. (1992) Relationship between deterrence and toxicity of plant secondary compounds for the alfalfa weevil *Hypera brunneipennis*. *Entomologia Experimentalis et Applicata*, **64**, 289–292.

Bernays, E.A. and Funk, D.J. (1999) Specialists make faster decisions than generalists: experiments with aphids. *Proceedings of the Royal Society of London B*, **266**, 151–156.

Bernays, E.A. and Graham, M. (1988) On the evolution of host specificity in phytophagous insects. *Ecology*, **69**, 886–892.

Bertness, M.D. and Leonard, G.H. (1997) The role of positive interactions in communities: lessons from intertidal habitats. *Ecology*, **78**, 1976–1989.

Bi, J.L. and Felton, G.W. (1995) Foliar oxidative stress and insect herbivory—primary compounds, secondary metabolites, and reactive oxygen species as components of induced resistance. *Journal of Chemical Ecology*, **21**, 1511–1530.

Bigger, D.S. and Marvier, M.A. (1998) How different would a world without herbivory be? A search for generality in ecology. *Integrative Biology*, **1**, 60–67.

Björnhag, G. (1972) Separation and delay of contents in the rabbit colon. *Swedish Journal of Agricultural Research*, **2**, 125–136.

Björnhag, G. (1981) Separation and retrograde transport in the large intestine of herbivores. *Livestock Production Science*, **8**, 351–360.

Björnhag, G. (1994) Adaptations in the large instestine allowing small animals to eat fibrous foods. In *The Digestive System in Mammals: Food, Form and Function*, ed. D.J. Chivers and P. Langer. Cambridge: Cambridge University Press, pp. 287–309.

Blackburn, T.M. and Gaston, K.J. (1996) A sideways look at patterns in species richness, or why there are so few species outside the tropics. *Biodiversity Letters*, **3**, 44–53.

Blasdale, W.C. (1947) The secretion of farina by species of *Primula*. *Journal of the Royal Horticultural Society*, **72**, 240–245.

Bohannan, B.J.M., Travisano, M. and Lenski, R.E. (1999) Epistatic interactions can lower the cost of resistance to multiple consumers. *Evolution*, **53**, 292–295.

Bolser, R.C. and Hay, M.E. (1996) Are tropical plants better defended? Palatability and defenses of temperate vs. tropical seaweeds. *Ecology*, **77**, 2269–2286.

Bolton, B. (1994) *Identification Guide to the Ant Genera of the World*. Cambridge, MA: Harvard University Press.

Bond, W.J. and Breytenbach, G.J. (1985) Ants, rodents and seed predation in the Proteaceae. *South African Journal of Zoology*, **20**, 150–154.

Bond, W.J. and Slingsby, P. (1984) Collapse of an ant–plant mutualism: the Argentine ant (*Iridomyrmex humilis*) and myrmecochorous Proteaceae. *Ecology*, **65**, 1031–1037.

Borowicz, V.A. (1988) Fruit consumption by birds in relation to fat content of pulp. *American Midland Naturalist*, **119**, 121–127.

Bouchon, D., Rigaud, T. and Juchault, P. (1998) Evidence for widespread *Wolbachia* infection in isopod crustaceans: molecular identification and host feminization. *Proceedings of the Royal Society of London B*, **265**, 1081–1090.

Boulter, M.C., Spicer, R.A. and Thomas, B.A. (1988) Patterns of plant extinction from some palaeobotanical evidence. In *Extinction and Survival in the Fossil Record*, ed G.P. Larwood. *Systematics Association Special Volume*, **34**, 1–36.

Boullard, B. (1979) Considerations sur la symbiose fongique chez les Pteridophytes. *Syllogeus*, **19**, 1–59.

Boullard, B. (1988) Observations on the coevolution of fungi with hepatics. In *Coevolution of Fungi with Plants and Animals*, ed. K.A. Pirozynski and D.L. Hawksworth. London: Academic Press, pp. 107–124.

Bowers, M.D. and Puttick, G.M. (1988) Response of generalist and specialist insects to qualitative allelochemical variation. *Journal of Chemical Ecology*, **14**, 319–334.

Boyd, R.S. and Martens, S.N. (1994) Nickel hyperaccumulation by *Thlaspi montanum* var. *montanum* is acutely toxic to an insect herbivore. *Oikos*, **70**, 21–25.

Boyden, T.C. (1982) The pollination biology of *Calypso bulbosa* var. *americana* (Orchidaceae): initial deception of bumblebee visitors. *Oecologia*, **55**, 178–184.

Bradford, D.F. and Smith, C.C. (1977) Seed predation and seed number in *Scheelea* palm fruits. *Ecology*, **58**, 667–673.

Bradshaw, H.D., Wilbert, S.M., Otto, K.G. and Schemske, D.W. (1995) Genetic mapping of floral traits associated with reproductive isolation in monkeyflowers (*Mimulus*). *Nature*, **376**, 762–765.

Brantjes, N.B.M. (1978) Sensory responses to flowers in night-flying moths. In *The Pollination of Flowers by Insects*, ed. A.J. Richards. London: Academic Press, pp. 13–19.

Bretagnolle, F. and Thompson, J.D. (1996) An experimental study of ecological differences in winter growth between sympatric diploid and autotetraploid *Dactylis glomerata*. *Journal of Ecology*, **84**, 343–351.

Briese, D.T. (1982) The effects of ants on the soil of a semi-arid saltbush habitat. *Insectes Sociaux*, **29**, 375–382.

Briese, D.T. and Macauley, B.J. (1981) Food collection within an ant community in semi-arid Australia, with special reference to seed harvesters. *Australian Journal of Ecology*, **6**, 1–19.

Brink, D.E. (1980) Reproduction and variation in *Aconitum columbianum* (Ranunculaceae), with emphasis on California populations. *American Journal of Botany*, **67**, 263–273.

Broadway, R.M. (1996) Dietary proteinase inhibitors alter complement of midgut proteases. *Archives of Insect Biochemistry and Physiology*, **32**, 39–53.

Brody, A. (1992) Oviposition choices by a pre-dispersal seed predator (*Hylemya* sp.). I. Correspondence with hummingbird pollinators, and the role of plant size, density and floral morphology. *Oecologia*, **91**, 56–62.

Brody, A.K. and Waser, N.M. (1995) Oviposition patterns and larval success of a pre-dispersal seed predator attacking two confamilial host plants. *Oikos*, **74**, 447–452.

Bronstein, J.B. (1992) Seed predators as mutualists: ecology and evolution of the fig/pollinator interaction. In *Insect–Plant Interactions*, ed. E. Bernays, vol. 4. Boca Raton, FL: CRC Press, pp. 1–44.

Bronstein, J.L. (1998) The contribution of ant-plant protection studies to our understanding of mutualism. *Biotropica*, **30**, 150–161.

Bronstein, J.L. and Hossaert-McKey, M. (1996) Variation in reproductive success within a subtropical fig/pollinator mutualism. *Journal of Biogeography*, **23**, 433–446.

Brooks, D.R. and Mitter, C. (1984) Analytical approaches to studying coevolution. In *Fungus-Insect Relationships: Perspectives in Ecology and Evolution*, ed. Q. Wheeler and M. Blackwell. New York: Columbia University Press, pp. 42–53.

Brooks, D.R. and McLennan, D.A. (1993) *Parascript: Parasites and the Language of Evolution*. Washington, DC: Smithsonian Institution Press.

Brower, J.V.Z. (1958) Experimental studies of mimicry in some North American butterflies. Part I. The monarch, *Danaus plexippus*, and viceroy, *Limenitis archippus archippus*. *Evolution*, **12**, 32–47.

Brower, L.P., Brower, J.V.Z. and Corvino, J.M. (1967) Plant poisons in a terrestrial food chain. *Proceedings of the National Academy of Sciences USA*, **57**, 892–898.

Brown, D.T. and Doucet, G.J. (1991) Temporal changes in winter diet selection by white-tailed deer in a northern deer yard. *Journal of Wildlife Management*, **55**, 361–376.

Brown, H.H. and Davidson, D.W. (1977) Competition between seed-eating rodents and ants in desert ecosystems. *Science*, **196**, 880–882.

Brown, J.H. (1995) *Macroecology*. Chicago: University of Chicago Press.

Brown, J.H. and Heske, E.J. (1990) Control of a desert-grassland transition by a keystone rodent guild. *Science*, **250**, 1705–1707.

Brown, J.H., Davidson, D.W. and Reichman, O.J. (1979) An experimental study of competition between desert seed-eating rodents and ants. *American Zoologist*, **19**, 1129–1143.

Brown, M.J.F. and Human, K.G. (1997) Effects of harvester ants on plant species distribution and abundance in a serpentine grassland. *Oecologia*, **112**, 237–243.

Brown, V.K. and Gange, A.C. (1989) Herbivory by soil-dwelling insects depresses plant species richness. *Functional Ecology*, **3**, 667–672.

Brown, V.K. and Gange, A.C. (1992) Secondary plant succession: how is it modified by insect herbivory? *Vegetatio*, **101**, 3–13.

Brown, V.K., Jepsen, M. and Gibson, C.W.D. (1988) Insect herbivory: effects on early old field succession demonstrated by chemical exclusion methods. *Oikos*, **52**, 293–302.

Brues, C.T. (1924) The specificity of food-plants in the evolution of phytophagous insects. *American Naturalist*, **58**, 127–144.

Bruneau, A. (1997) Evolution and homology of bird pollination syndromes in *Erythrina* (Leguminosae). *American Journal of Botany*, **84**, 54–71.

Bryant, J.P. (1979) The regulation of snowshoe hare feeding behavior during winter by plant antiherbivore chemistry. In Proceedings of 1st International Lagomorph Conference, Guelph University, Canada.

Bryant, J.P. (1981) Phytochemical deterrence of snowshoe hare browsing by adventitious shoots of four Alaskan trees. *Science*, **213**, 889–890.

Bryant, J.P., Chapin, F.S. III and Klein, D.R. (1983) Carbon/nutrient balance of boreal plants in relation to vertebrate herbivory. *Oikos*, **40**, 357–368.

Bryant, J.P., Danell, K., Provenza, F., Reichardt, P., Clausen, T. and Werner, R.A. (1991) Effects of mammal browsing on the chemistry of deciduous woody plants. In *Phytochemical Induction by Herbivores*, ed. D. Tallamy and M.J. Raupp. New York: Wiley, pp. 135–154.

Buckley, R. (1983) Interaction between ants and membracid bugs decreases growth and seed set of host plant bearing extra-floral nectaries. *Oecologia*, **58**, 132–136.

Buckley, R.C. (1982) Ant–plant interactions: a world review. In *Ant–Plant Interactions in Australia*, ed. R.C. Buckley. The Hague: Junk, pp. 111–162.

Buchmann, S.L. (1983) Buzz pollination in angiosperms. In *Handbook of Experimental Pollination Biology*, ed. C.E. Jones and R.J. Little. New York: Van Nostrand Reinhold, pp. 73–113.

Buchmann, S.L. (1987) The ecology of oil flowers and their bees. *Annual Review of Ecology and Systematics*, **18**, 343–369.

Buchner, P. (1965) *Endosymbiosis of Animals with Plant Microorganisms*. New York: Wiley.

Buide, M.L., Sánchez, J.M. and Guitián, J. (1998) Ecological characteristics of the flora of the northwest Iberian Peninsula. *Plant Ecology*, **135**, 1–8.

Burdon, J.J. (1993) The structure of pathogen populations in natural plant communities. *Annual Review of Phytopathology*, **31**, 305–323.

Burdon, J.J. and Thrall, P.H. (1999) Spatial and temporal patterns in coevolving plant and pathogen associations. *American Naturalist*, **153**, S15–S33.

Buss, L.W. (1987) *The Evolution of Individuality*. Princeton, NJ: Princeton University Press.

Cambefort, Y. (1997) Food choice and environment occupancy in Afrotropical dung beetles: a phylogenetic study of two examples (Coleoptera, Scarabaeidae). In *The Origin of Biodiversity in Insects: Phylogenetic Tests of Evolutionary Scenarios*, ed. P. Grandcolas. *Mémoires du Muséum National d'Histoire Naturelle*, **173**, 125–134.

Campbell, D.R. (1996) Evolution of floral traits in a hermaphroditic plant: field measurements of heritabilities and genetic correlations. *Evolution*, **50**, 1442–1453.

Campbell, D.R., Waser, N.M. and Meléndez-Ackerman, E.J. (1997) Analyzing pollinator-mediated selection in a plant hybrid zone: hummingbird visitation patterns on three spatial scales. *American Naturalist*, **149**, 295–315.

Campbell, D.R., Waser, N.M., Price, M.V., Lynch, E.A. and Mitchell, R.J. (1991) Components of phenotypic selection: pollen export and flower corolla width in *Ipomopsis aggregata*. *Evolution*, **45**, 1458–1467.

Carpenter, F.M. (1971) Adaptations among Paleozoic insects. In *Proceedings of the [1st] North American Paleontological Convention*, **2**, 1236–1252. Lawrence, KS: Allen Press.

Carroll, S.P., Dingle, H. and Klassen, S.P. (1997) Genetic differentiation of fitness-associated traits among rapidly evolving populations of the soapberry bug. *Evolution*, **51**, 1182–1188.

Cartmill, M. (1972) Arboreal adaptations and the origin of the Order Primates. In *The Functional and Evolutionary Biology of Primates*, ed. R. Tuttle. Chicago: Aldine-Atherton, pp. 97–122.

Cartmill, M. (1974a) Pads and claws in arboreal locomotion. In *Primate Locomotion*, ed. F.A. Jenkins. New York: Academic Press, pp. 45–83.

Cartmill, M. (1974b) Rethinking primate origins. *Science*, **184**, 436–443.

Casper, B.B. (1988) Post-dispersal seed predation may select for wind dispersal but not seed number per dispersal unit in *Cryptantha flava*. *Oikos*, **52**, 27–30.

Castro, J., Gómez, J.M., García, D., Zamora, R. and Hódar, J.A. (1999) Seed predation and dispersal in relict Scots pine forests of southern Spain. *Plant Ecology*, **145**, 115–123.

Cebrián, J. and Duarte, C.M. (1994) The dependence of herbivory on growth rate in natural plant communities. *Functional Ecology*, **8**, 518–525.

Cerling, T.E., Harris, J.M., MacFadden, B.J. *et al.* (1997) Global vegetation change through the Miocene/Pliocene boundary. *Nature*, **389**, 153–158.

Chapman, L.J., Chapman, C.A. and Wrangham, R.W. (1992) *Balanites wilsoniana*: elephant dependent dispersal? *Journal of Tropical Ecology*, **8**, 275–283.

Chase, M.W. (1986) A reappraisal of the oncidioid orchids. *Systematic Botany*, **11**, 477–491.

Chase, M.W. and Palmer, J.D. (1997) Leapfrog radiation in floral and vegetative traits among twig epiphytes in the orchid subtribe Oncidiinae. In *Molecular Evolution and Adaptive Radiation*, ed. T.J. Givnish and K.J. Sytsma. Cambridge: Cambridge University Press, pp. 331–352.

Chaudonneret, J. (1990) *Les pièces buccales des insectes: thème et variations*. Dijon: Berthier.

Cheeke, P.R. (1987) *Rabbit Feeding and Nutrition*. Orlando, FL: Academic Press.

Cherrett, J.M. (1968) The foraging behaviour of *Atta cephalotes* L. (Hymenoptera, Formicidae). 1. Foraging pattern and plant species attacked in tropical rain forest. *Journal of Animal Ecology*, **37**, 387–403.

Cherrett, J.M. (1986) History of the leaf-cutting ant problem. In *Fire Ants and Leaf-Cutting Ants: Biology and Management*, ed. C.S. Lofgren and R.K. Vander Meer. Boulder, CO: Westview Press, pp. 10–17.

Chin, K. and Gill, B.D. (1996) Dinosaurs, dung beetles, and conifers: participants in a Cretaceous food web. *Palaios*, **11**, 280–285.

Chittka, L. (1996) Does bee color vision predate the evolution of flower color? *Naturwissenschaften*, **83**, 136–138.

Chittka, L., Shmida, A., Troje, N. and Menzel, R. (1994) Ultraviolet as a component of flower reflections, and the colour perception of Hymenoptera. *Vision Research*, **34**, 1489–1508.

Chittka, L., Thomson, J.D. and Waser, N.M. (1999) Flower constancy, insect psychology, and plant evolution. *Naturwissenschaften*, **86**, 361–377.

Christensen, K.M., Whitham, T.G. and Balda, R.P. (1991) Discrimination among pinyon pine trees by Clark's nutcrackers: effects of cone crop size and cone characters. *Oecologia*, **86**, 402–407.

Cintra, R. (1997) A test of the Janzen–Connell model with two common tree species in Amazonian forest. *Journal of Tropical Ecology*, **13**, 641–658.

Cipollini, M.L. and Levey, D.J. (1991) Why some fruits are green when they are ripe: carbon balance in fleshy fruits. *Oecologia*, **88**, 371–377.

Cipollini, M.L. and Levey, D.J. (1997a) Secondary metabolites of fleshy vertebrate-dispersed fruits: adaptive hypotheses and implications for seed dispersal. *American Naturalist*, **150**, 346–372.

Cipollini, M.L. and Levey, D.J. (1997b) Why are some fruits toxic? Glycoalkaloids in *Solanum* and fruit choice by vertebrates. *Ecology*, **78**, 782–798.

Cipollini, M.L. and Stiles, E.W. (1993) Fruit rot, antifungal defense, and palatability of fleshy fruits for frugivorous birds. *Ecology*, **74**, 751–762.

Clark, W.H. and Comanor, P.L. (1975) Removal of annual plants from the desert ecosystem by western harvester ants, *Pogonomyrmex occidentalis*. *Environmental Entomology*, **4**, 52–56.

Clay, K. (1988) Fungal endophytes of grasses: a defensive mutualism between plants and fungi. *Ecology*, **69**, 10–16.

Clay, K. (1989) Clavicipitaceous endophytes of grasses: their potential as biocontrol agents. *Mycological Research*, **92**, 1–12.

Clay, K. and Kover, P. (1996) Evolution and stasis in plant-pathogen associations. *Ecology*, **77**, 997–1003.

Clay, K., Marks, S. and Cheplick, G.P. (1993) Effects of insect herbivory and fungal endophyte infection on competitive interactions among grasses. *Ecology*, **74**, 1767–1777.

Clegg, M.T. and Durbin, M.L. (2000) Flower color variation: a model for the experimental sudy of evolution. *Proceedings of the National Academy of Sciences USA*, **97**, 7016–7023.

Clutton-Brock, T.H., Guinness, F.E. and Albon, S.D. (1982*) Red Deer: Behavior and Ecology of Two Sexes*. Chicago: University of Chicago Press.

Coddington, J.A. (1988) Cladistic tests of adaptational hypotheses. *Cladistics*, **4**, 1–22.

Coen, E. and Nugent, J.M. (1994) Evolution of flowers and inflorescences. *Development*, **1994** (Supplement), 107–116.

Coffey, K., Benkman, C.W. and Milligan, B.G. (1999) The adaptive significance of spines on pine cones. *Ecology*, **80**, 1221–1229.

Coley, P.D. (1988) Effects of plant growth rate and leaf lifetime on the amount and type of anti-herbivore defense. *Oecologia*, **74**, 531–536.

Coley, P.D. (1999) Hungry herbivores seek a warmer world. *Science*, **284**, 2098–2099.

Coley, P.D. and Barone, J.A. (1996) Herbivory and plant defenses in tropical forests. *Annual Review of Ecology and Systematics*, **27**, 305–335.

Coley, P.D., Bryant, J.P. and Chapin, F.S. III (1985) Resource availability and plant antiherbivore defense. *Science*, **230**, 895–899.

Collinson, M.E. and Hooker, J.J. (1991) Fossil evidence of interactions between plants and plant-eating mammals. *Philosophical Transactions of the Royal Society of London B*, **333**, 197–208.

Comba, L., Corbet, S.A., Hunt, H., Outram, S., Parker, J.S. and Glover, B.J. (2000) The role of genes influencing the corolla in pollination of *Antirrhinum majus*. *Plant, Cell and Environment*, **23**, 639–647.

Connell, J.H. (1971) On the role of natural enemies in preventing competitive exclusion in some marine animals and in rain forest trees. In *Dynamics of Populations*, ed. P.J. den Boer and G.R. Gradwell. Wageningen: Centre for Agricultural Publishing and Documentation, pp. 298–312.

Connell, J.H. (1978) Diversity in tropical rainforests and coral reefs. *Science*, **199**, 1302–1310.

Conner, J.K. and Rush, S. (1996) Effects of flower size and number on pollinator visitation to wild radish, *Raphanus raphanistrum*. *Oecologia*, **105**, 509–516.

Connor, E.F. and Taverner, M.P. (1997) The evolution and adaptive significance of the leaf-mining habit. *Oikos*, **79**, 6–25.

Cook, R.J. (1998) The molecular mechanisms responsible for resistance in plant-pathogen interactions of the gene-for-gene type function more broadly than previously imagined. *Proceedings of the National Academy of Sciences USA*, **95**, 9711–9712.

Coope, G.R. (1995) The effects of Quaternary climatic changes on insect populations: lessons from the past. In *Insects in a Changing*

Environment, ed. R. Harrington and N.E. Stork. London: Academic Press, pp. 29–48.

Cork, S.J. (1994) Digestive constraints on dietary scope in small and moderately-sized mammals: how much do we really understand? In *The Digestive System in Mammals: Food, Form and Function*, ed. D.J. Chivers and P. Langer. Cambridge: Cambridge University Press, pp. 337–369.

Cork, S.J. and Foley, W.J. (1991) Digestive and metabolic strategies of arboreal mammalian folivores in relation to chemical defences in temperate and tropical forests. In *Plant Defences against Mammalian Herbivory*, ed. R.T. Palo and C.T. Robbins. Boca Raton, FL: CRC Press, pp. 133–166.

Cornell, H.V. and Lawton, J.H. (1992) Species interactions, local and regional processes, and limits to the richness of ecological communities: a theoretical perspective. *Journal of Animal Ecology*, **61**, 1–12.

Cornet, B. (1996) A new gnetophyte from the Late Carnian (Late Triassic) of Texas and its bearing on the origin of the angiosperm carpel and stamen. In *Flowering Plant Origin, Evolution and Phylogeny*, ed. D.W. Taylor and L.J. Hickey. New York: Chapman and Hall, pp. 32–67.

Cottee, P.K., Bernays, E.A. and Mordue, A.J. (1988) Comparisons of deterrency and toxicity of selected secondary plant compounds to an oligophagous and a polyphagous acridid. *Entomologia Experimentalis et Applicata*, **46**, 241–248.

Coulson, R.N. and Witter, J.A. (1984) *Forest Entomology, Ecology and Management*. New York: Wiley Interscience.

Courts, S.E. (1998) Dietary strategies of Old World fruit bats (Megachiroptera, Pteropodidae): how do they obtain sufficient protein? *Mammal Review*, **28**, 185–194.

Crane, P.R. and Lidgard, S. (1989) Angiosperm diversification and paleolatitudinal gradients in Cretaceous floristic diversity. *Science*, **246**, 675–678.

Crawley, M.J. (1983) *Herbivory: The Dynamics of Animal-Plant Interactions*. Oxford: Blackwell Scientific Publications.

Crawley, M.J. (1985) Reduction of Oak *Quercus robur* fecundity by low-density herbivore populations. *Nature*, **314**, 163–164.

Crawley, M.J. (1989) The relative importance of vertebrate and invertebrate herbivores in plant-population dynamics. In *Insect–Plant Interactions*, ed. E.A. Bernays, vol. 1. Boca Raton, FL: CRC Press, pp. 45–71.

Crawley, M.J. (1992) Seed predators and plant population dynamics. In *Seeds: The Ecology of Regeneration in Plant Communities*, ed. M. Fenner. Wallingford: Commonwealth Agricultural Bureau International, pp. 157–191.

Crawley, M.J. (1997) Plant-herbivore dynamics. In *Plant Ecology*, ed. M.J. Crawley. Oxford: Blackwell, pp. 401–474.

Crawley, M.J. and Long, C.R. (1995) Alternate bearing, predator satiation and seedling recruitment in *Quercus robur* L. *Journal of Ecology*, **83**, 683–696.

Crepet, W.L. (1972) Investigations of North American cycadeoids: pollination mechanisms in *Cycadeoidea*. *American Journal of Botany*, **59**, 144–169.

Crepet, W.L. (1996) Timing in the evolution of derived floral characters—Upper Cretaceous (Turonian) taxa with tricolpate and tricolpate-derived pollen. *Review of Palaeobotany and Palynology*, **90**, 339–359.

Crepet, W.L. and Feldman, G.D. (1991) The earliest remains of grasses in the fossil record. *American Journal of Botany*, **78**, 1010–1014.

Crepet, W.L. and Friis, E.M. (1987) The evolution of insect pollination in angiosperms. In *The Angiosperms and their Biological Consequences*, ed E.M. Friis, W.G. Chaloner and P.R. Crane. Cambridge: Cambridge University Press, pp. 181–201.

Crichton, M.I. (1957) The structure and function of the mouthparts of adult caddis flies (Trichoptera). *Philosophical Transactions of the Royal Society of London B*, **241**, 45–91.

Crisp, M.D. (1994) Evolution of bird-pollination in some Australian legumes (Fabaceae). In *Phylogenetics and Ecology*, ed. P. Eggleton and R. Vane-Wright. London: Academic Press, pp. 281–309.

Crist, T.O. and MacMahon, J.A. (1992) Harvester ant foraging and shrub steppe seeds: interactions of seed resources and seed use. *Ecology*, **73**, 1768–1779.

Crome, F.H.J. (1975) The ecology of fruit pigeons in tropical Northern Queensland. *Australian Wildlife Research*, **2**, 155–185.

Crowson, R.W. (1981) *The Biology of the Coleoptera*. New York: Academic Press.

Cruden, R.W., Hermann, S.M. and Peterson, S. (1983) Patterns of nectar production and plant-pollinator coevolution. In *The Biology of Nectaries*, ed. B. Bentley and T. Elias. New York: Columbia University Press, pp. 80–125.

Cubas, P., Coral, V. and Coen, E. (1999) An epigenetic mutation responsible for natural variation in floral symmetry. *Nature*, **401**, 157–161.

Cummins, K.W. (1973) Trophic relations of aquatic insects. *Annual Review of Entomology*, **18**, 183–206.

Cummins, K.W. and Merritt, R.W. (1984) Ecology and distribution of aquatic insects. In *An Introduction to the Aquatic Insects of North America*, ed. R.W. Merritt and K.W. Cummins, 2nd edn. Dubuque, IA: Kendall-Hunt, pp. 59–65.

Curran, L.M. and Leighton, M. (2000) Vertebrate responses to spatiotemporal variation in seed production of mast fruiting Dipterocarpaceae. *Ecological Monographs*, **70**, 101–128.

Currie, C.R., Scott, J.A., Summerbell, R.C. and Malloch, D. (1999) Fungus-growing ants use antibiotic-producing bacteria to control garden parasites. *Nature*, **398**, 701–704.

Cushman, J.H. (1991) Host-plant mediation of insect mutualisms: variable outcomes in herbivore-ant interactions. *Oikos*, **61**, 138–144.

Cyr, H. and Pace, M.L. (1993) Magnitude and patterns of herbivory in aquatic and terrestrial ecosystems. *Nature*, **361**, 148–150.

da Silva, H., de Britto-Pereira, M.C. and Caramaschi, U. (1989) Frugivory and seed dispersal by *Hyla truncata*, a neotropical treefrog. *Copeia*, **1989**, 781–783.

Danell, K., Bergström, R. and Edenius, L. (1994) Effects of large mammalian browsers on woody plant architecture, biomass and nutrients. *Journal of Mammalogy*, **75**, 833–844.

Danell, K., Huss-Danell, K. and Bergström, R. (1985) Interactions between browsing moose and two species of birch in Sweden. *Ecology*, **66**, 1867–1878.

Darwin, C. (1862) *The Various Contrivances by which British and Foreign Orchids are Fertilised by Insects*. London: John Murray.

tions: Evolutionary Ecology in Tropical and Temperate Regions, ed. P.W. Price, T.M. Lewinsohn, G.W. Fernandes and W.W. Benson. New York: Wiley, pp. 119–144.

Fleming, T.H. (1992) How do fruit- and nectar-feeding birds and mammals track their food resources? In Effects of Resource Distribution on Animal-Plant Interactions, ed. M.D. Hunter, T. Ohgushi and P.W. Price. San Diego, CA: Academic Press, pp. 355–391.

Fleming, T.H. and Holland, J.N. (1998) The evolution of obligate pollination mutualisms: senita cactus and senita moth. Oecologia, 114, 368–375.

Fleming, T.H., Breitwisch, R. and Whitesides, G.H. (1987) Patterns of tropical vertebrate frugivore diversity. Annual Review of Ecology and Systematics, 18, 91–109.

Floate, K.D. and Whitham, T.G. (1993) The hybrid bridge hypothesis: host shifting via plant hybrid swarms. American Naturalist, 141, 651–662.

Floyd, T. (1996) Top-down impacts on creosotebush herbivores in a spatially and temporally complex environment. Ecology, 77, 1544–1555.

Folgarait, P.J. (1998) Ant biodiversity and its relationship to ecosystem functioning: a review. Biodiversity and Conservation, 7, 1221–1244.

Fonseca, C.R. and Ganade, G. (1996) Asymmetries, compartments and null interactions in an Amazonian ant-plant community. Journal of Animal Ecology, 65, 339–347.

Foote, M. (1993) Discordance and concordance between morphological and taxonomic diversity. Paleobiology, 19, 195–204.

Forget, P.M. (1992) Seed removal and seed fate in Gustavia superba (Lecythidaceae). Biotropica, 24, 408–414.

Forget, P.M. (1993) Post-dispersal predation and scatterhoarding of Dipteryx panamensis (Papilionaceae) seeds by rodents in Panama. Oecologia, 94, 255–261.

Forget, P.M., Kitajima, K. and Foster, R.B. (1999) Pre- and post-dispersal seed predation in Tachigalia versicolor (Caesalpiniaceae): effects of timing of fruiting and variation among trees. Journal of Tropical Ecology, 15, 61–81.

Fowler, H.G. and Claver, S. (1991) Leaf-cutter ant assemblies: effects of latitude, vegetation, and behaviour. In Ant–Plant Interactions, ed. C.R. Huxley and D.F. Cutler. Oxford: Oxford University Press, pp. 51–59.

Fox, D.L., Fischer, D.C. and Leighton, L.F. (1999) Reconstructing phylogeny with and without temporal data. Science, 284, 1816–1819.

Fox, L.R. and Morrow, P.A. (1992) Eucalypt responses to fertilization and reduced herbivory. Oecologia, 89, 214–222.

Fraenkel, G.S. (1959) The raison d'être of secondary plant substances. Science, 129, 1466–1470.

Frank, D.A. and Groffman, P.M. (1998) Ungulate vs. landscape control of soil C and N processes in grasslands of Yellowstone National Park. Ecology, 79, 2229–2241.

Freed, L.A., Smith, T.B., Carothers, J.H. and Lepson, J.K. (1996) Shrinkage is not the most likely cause of bill change in Iiwi: a rejoinder to Winker. Conservation Biology, 10, 659–660.

Freeland, W.J. (1974) Vole cycles: another hypothesis. American Naturalist, 108, 238–245.

Freeland, W.J. (1991) Plant secondary metabolites: biochemical co-evolution with herbivores. In Plant Defenses against Mammalian Herbivory, ed. R.T. Palo and C.T. Robbins. Boca Raton, FL: CRC Press, pp. 61–81.

Freeman, P.W. (1988) Frugivorous and animalivorous bats (Microchiroptera): dental and cranial adaptations. Biological Journal of the Linnean Society, 33, 249–272.

French, K. (1991) Characteristics and abundance of vertebrate-dispersed fruits in temperate wet sclerophyll forest in southeastern Australia. Australian Journal of Ecology, 16, 1–11.

French, K. (1992) Phenology of fleshy fruits in a wet sclerophyll forest in southeastern Australia: are birds an important influence? Oecologia, 90, 366–373.

Frick, C. and Wink, M. (1995) Uptake and sequestration of ouabain and other cardiac glycosides in Danaus plexippus (Lepidoptera. Danaidae): evidence for a carrier-mediated process. Journal of Chemical Ecology, 21, 557–575.

Friese, C.F. and Allen, M.F. (1993) The interaction of harvester ants and vesicular-arbuscular mycorrhizal fungi in patchy semi-arid environments: the effects of mound structure on fungal dispersion and establishment. Functional Ecology, 7, 13–20.

Fritz, R.S. (1983) Ant protection of a host plant's defoliator: consequence of an ant-membracid mutualism. Ecology, 64, 789–797.

Fritz, R.S. (1999) Resistance of hybrid plants to herbivores: genes, environment, or both? Ecology, 80, 382–391.

Frost, S.W. (1924) A study of the leaf-mining Diptera of North America. Cornell University Agricultural Experiment Station Memoirs, 78, 1–228.

Frost, S.W. (1925) The leaf-mining habit in the Hymenoptera. Annals of the Entomological Society of America, 18, 399–414, pls. 29–30.

Fry, J.D. (1990) Trade-offs in fitness on different hosts: evidence from a selection experiment with a phytophagous mite. American Naturalist, 136, 569–580.

Fry, J.D. and Rausher, M.D. (1997) Selection on a floral color polymorphism in the tall morning glory (Ipomoea purpurea): transmission success of the alleles through pollen. Evolution, 51, 66–78.

Fuentes, M. (1992) Latitudinal and elevational variation in fruiting phenology among western European bird-dispersed plants. Ecography, 15, 177–183.

Fuentes, M. & Schupp, E.W. (1998) Empty seeds reduce seed predation by birds in Juniperus osteosperma. Evolutionary Ecology, 12, 823–827.

Funk, D.J., Futuyma, D.J., Orti, G. and Meyer, A. (1995) A history of host associations and evolutionary diversification for Ophraella (Coleoptera: Chrysomelidae): new evidence from mitochondrial DNA. Evolution, 49, 1008–1017.

Funk, V.A. and Brooks, D.R. (1990) Phylogenetic systematics as the basis of comparative biology. Smithsonian Contributions to Botany, 73, 1–45.

Furniss, M.M., Harvey, A.E. and Solheim, H. (1995) Transmission of Ophiostoma ips (Ophiostomatales: Ophiostomataceae) by Ips pini (Coleoptera: Scolytidae) to ponderosa pine in Idaho. Annals of the Entomological Society of America, 88, 653–660.

Futuyma, D.J. (1998) Evolutionary Biology, 3rd edn. Sunderland: Sinauer.

Futuyma, D.J. and Mitter, C. (1996) Insect-plant interactions: the evolution of component communities. *Philosophical Transactions of the Royal Society of London B*, **351**, 1361–1366.

Futuyma, D.J., Keese, M.C. and Funk, D.J. (1995) Genetic constraints on macroevolution: the evolution of host affiliation in the leaf beetle genus *Ophraella*. *Evolution*, **49**, 797–809.

Galen, C. (1989) Measuring pollinator-mediated selection on morphometric floral traits: bumblebees and the alpine sky pilot, *Polemonium viscosum*. *Evolution*, **43**, 882–890.

Galen, C. (1999a) Flowers and enemies: predation by nectar-thieving ants in relation to variation in floral form of an alpine wildflower, *Polemonium viscosum*. *Oikos*, **85**, 426–434.

Galen, C. (1999b) Why do flowers vary? The functional ecology of variation in flower size and form within natural plant populations. *Bioscience*, **49**, 631–640.

Gandon, S., Ebert, D., Olivieri, I. and Michalakis, Y. (1998) Differential adaptation in spatially heterogeneous environments and host-parasite coevolution. In *Genetic Structure and Local Adaptation in Natural Insect Populations: Effects of Ecology, Life History, and Behavior*, ed. S. Mopper and S.Y. Strauss. New York: Chapman & Hall, pp. 325–342.

Gardner, G. (1977) The reproductive capacity of *Fraxinus excelsior* on the Derbyshire limestone. *Journal of Ecology*, **65**, 107–118.

Garin, C.F., Juan, C. and Petitpierre, E. (1999) Mitochondrial DNA phylogeny and the evolution of host-plant use in palearctic *Chrysolina* (Coleoptera, Chrysomelidae) leaf beetles. *Journal of Molecular Evolution*, **48**, 435–444.

Gaume, L., McKey, D. and Anstett, M.C. (1997) Benefits conferred by 'timid' ants: active anti-herbivore protection of the rainforest tree *Leonardoxa africana* by the minute ant *Petalomyrmex phylax*. *Oecologia*, **112**, 209–216.

Gehring, C.A., Cobb, N.S. and Whitham, T.G. (1997) Three-way interactions among ectomycorrhizal mutualists, scale insects, and resistant and susceptible pinyon pines. *American Naturalist*, **149**, 824–841.

Gervais, J.A., Noon, B.R. and Willson, M.F. (1999) Avian selection of the color-dimorphic fruits of salmonberry, *Rubus spectabilis*: a field experiment. *Oikos*, **84**, 77–86.

Gibson, D., Bazely, D.R. and Shore, J.S. (1987) Responses of brambles, *Rubus vestitus*, to herbivory. *Oecologia*, **95**, 454–457.

Gilardi, J.D., Duffey, S.S., Munn, C.A. and Tell, L.A. (1999) Biochemical functions of geophagy in parrots: detoxification of dietary toxins and cytoprotective effects. *Journal of Chemical Ecology*, **25**, 897–922.

Gilbert, F. and Jervis, M. (1998) Functional, evolutionary and ecological aspects of feeding-related mouthpart specializations in parasitoid flies. *Biological Journal of the Linnean Society*, **63**, 495–535.

Gilbert, F., Rotheray, G., Emerson, P. and Zafar, R. (1994) The evolution of feeding strategies. In *Phylogenetics and Ecology*, ed. P. Eggleton and R. Vane-Wright. London: Academic Press, pp. 323–343.

Gilbert, L.E. (1972) Pollen feeding and reproductive biology of *Heliconius* butterflies. *Proceedings of the National Academy of Science USA*, **69**, 1403–1407.

Gilbert, L.E. (1980) Food web organization and the conservation of

neotropical diversity. In *Conservation Biology: An Evolutionary-Ecological Perspective*, ed. M.E. Soulé and B.A. Wilcox. Sunderland: Sinauer, pp. 11–33.

Givnish, T.J. (1997) Adaptive radiation and molecular systematics: issues and approaches. In *Molecular Evolution and Adaptive Radiation*, ed. T.J. Givnish and K.J. Sytsma. Cambridge: Cambridge University Press, pp. 1–54.

Goldblatt, P. and Manning, J.C. (2000) The long-proboscid fly pollination system in southern Africa. *Annals of the Missouri Botanical Garden*, **87**, 146–170.

Gómez, C. and Espadaler, X. (1998) Myrmecochorous dispersal distances: a world survey. *Journal of Biogeography*, **25**, 573–580.

Gómez, J.M. and García, D. (1997) Interactions between a high-mountain shrub, *Genista versicolor* (Fabaceae), and its seed predators. *Ecoscience*, **4**, 48–56.

Gómez, J.M. and Zamora, R. (1992) Pollination by ants: consequences of the quantitative effects on a mutualistic system. *Oecologia*, **91**, 410–418.

Gómez, J.M. and Zamora, R. (1994) Top-down effects in a tritrophic system: parasitoids enhance plant fitness. *Ecology*, **75**, 1023–1030.

Gómez, J.M., Zamora, R., Hódar, J.A. and García, D. (1996) Experimental study of pollination by ants in Mediterranean high mountain and arid habitats. *Oecologia*, **105**, 236–242.

Gomulkiewicz, R., Thompson, J.N., Holt, R.D., Nuismer, S.L. and Hochberg, M.E. (2000) Hot spots, cold spots, and the geographic mosaic theory of coevolution. *American Naturalist*, **156**, 156–174.

Gordon, I.J. and Illius, A.W. (1994) The functional significance of the browser-grazer dichotomy in African ruminants. *Oecologia*, **98**, 167–175.

Gordon, I.J. and Lindsay, W.K. (1990) Could mammalian herbivores 'manage' their resources? *Oikos*, **59**, 270–280.

Gottlieb, L.D. (1984) Genetics and morphological evolution in plants. *American Naturalist*, **123**, 681–709.

Gottsberger, G. and Silberbauer-Gottsberger, I. (1983) Dispersal and distribution in the cerrado vegetation of Brazil. *Sonderbände des Naturwissenschaftlichen Vereins in Hamburg*, 7, 315–352.

Gould, S.J. (1965) Is uniformitarianism necessary? *American Journal of Science*, **263**, 223–228.

Gould, S.J. (1977) *Ontogeny and Phylogeny*. Cambridge, MA: Harvard University Press.

Grace, J.R. (1986) The influence of gypsy moth on the composition and nutrient content of litter fall in a Pennsylvania oak forest. *Forest Science*, **32**, 855–870.

Graham, R. (1986) Plant–animal interactions and Pleistocene extinctions. In *Dynamics of Extinctions*, ed. D.K. Elliott. New York: Wiley, pp. 131–154.

Grandcolas, P. (ed.) (1997) *The Origin of Biodiversity in Insects: Phylogenetic Tests of Evolutionary Scenarios*. *Mémoires du Muséum National d'Histoire Naturelle*, **173**, 1–356.

Grant, P.R. (1986) *Ecology and Evolution of Darwin's Finches*. Princeton, NJ: Princeton University Press.

Grant, P.R. and Grant, B.R. (1995) Predicting microevolutionary responses to directional selection on heritable variation. *Evolution*, **49**, 241–251.

Grant, V. (1983) The systematic and geographical distribution of hawkmoth flowers in the temperate North American flora. *Botanical Gazette*, **144**, 439–449.

Grauvogel-Stamm, L. and Kelber, K.-P. (1996) Plant–insect interactions and coevolution during the Triassic in western Europe. *Paleontologia Lombarda*, NS **5**, 5–23.

Gray, J. and Boucot, A.J. (1994) Early Silurian nonmarine animal remains and the nature of the early continental ecosystem. *Acta Palaeontologica Polonica*, **38**, 303–328.

Green, T.R. and Ryan, C.A. (1972) Wound-induced proteinase inhibitors in plant leaves: a possible defence mechanism against insects. *Science*, **175**, 776–777.

Greenwood, J.J.D. (1985) Frequency dependent selection by seed predators. *Oikos*, **44**, 195–210.

Grier, C.C. and Vogt, D.J. (1990) Effects of aphid honeydew on soil nitrogen availability and net primary production in an *Alnus rubra* plantation in western Washington. *Oikos*, **57**, 114–118.

Grimaldi, D.A., Bonwich, E., Dellanoy, M. and Doberstein, W. (1994) Electron microscopic studies of mummified tissues in amber fossils. *American Museum Novitates*, **3097**, 1–31.

Griswold, M.J. and Trumble, J.T. (1985) Consumption and utilization of celery, *Apium graveolens*, by the beet armyworm *Spodoptera exigua*. *Entomologia Experimentalis et Applicata*, **38**, 73–80.

Gronemeyer, P.A., Dilger, B.J., Bouzat, J.L. and Paige, K.N. (1997) The effects of herbivory on paternal fitness in scarlet gilia: better moms also make better pops. *American Naturalist*, **150**, 592–602.

Groombridge, B. (ed.) (1992) *Global Biodiversity: Status of the Earth's Living Resources*. London: Chapman & Hall.

Grubb, P.J., Metcalfe, D.J., Grubb, E.A.A. and Jones, G.D. (1998) Nitrogen-richness and protection of seeds in Australian tropical rainforest: a test of plant defence theory. *Oikos*, **82**, 467–482.

Guitián, J. and Sánchez, J.M. (1992) Seed dispersal spectra of plant communities in the Iberian Peninsula. *Vegetatio*, **98**, 157–164.

Guitián, J., Munilla, I., Guitián, P. and López, B. (1994) Frugivory and seed dispersal by redwings *Turdus iliacus* in southwest Iceland. *Ecography*, **17**, 314–320.

Guitián, P. (1998) Latitudinal variation in the fruiting phenology of a bird-dispersed plant (*Crataegus monogyna*) in western Europe. *Plant Ecology*, **137**, 139–142.

Gullan, P.J. and Cranston, P.S. (1994) *The Insects: An Outline of Entomology*. London: Chapman & Hall.

Gumbert, A. (2000) Color choices by bumble bees (*Bombus terrestris*): innate preferences and generalization after learning. *Behavioral Ecology and Sociobiology*, **48**, 36–43.

Gurnell, J. (1993) Tree seed production and food conditions for rodents in an oak wood in southern England. *Forestry*, **66**, 291–315.

Guthrie, R.D. (1971) Factors regulating the evolution of microtine tooth complexity. *Zeitschrift für Säugetierkunde*, **36**, 37–54.

Hairston, N.G., Smith, F.E. and Slobodkin, L.B. (1960) Community structure, population control and competition. *American Naturalist*, **94**, 421–425.

Hamilton, W.D. (1978) Evolution and diversity under bark. In *Diversity of Insect Faunas*, ed. L.A. Mound and N. Waloff. Oxford: Blackwell Scientific Publications, pp. 154–175.

Hammond, D.S. and Brown, V.K. (1998) Disturbance, phenology and life-history characteristics: factors influencing distance/ density-dependent attack on tropical seeds and seedlings. In *Dynamics of Tropical Communities*, ed. D.M. Newbery, H.H.T. Prins and N.D. Brown. Oxford: Blackwell, pp. 401–474.

Hanley, T.A. (1997) A nutritional view of understanding and complexity in the problem of diet selection by deer (Cervidae). *Oikos*, **79**, 209–218.

Hanley, T.A., Robbins, C.T. and Spalinger, D.E. (1985) *Forest Habitats and the Nutritional Ecology of Sitka Black-tailed Deer: A Research Synthesis with Implications for Forest Management*. General Technical Report PNW-GTR-230. Portland, OR: United States Department of Agriculture, Forest Service, Pacific Northwest Research Station.

Hansson, L. (1985) Damage by wildlife, especially small rodents, to North American *Pinus contorta* provenances introduced into Sweden. *Canadian Journal of Forest Research*, **15**, 1167–1171.

Hanzawa, F.M., Beattie, A.J. and Culver, D.C. (1988) Directed dispersal: demographic analysis of an ant–seed mutualism. *American Naturalist*, **131**, 1–13.

Harborne, J.B. (1991) The chemical basis of plant defense. In *Plant Defences against Mammalian Herbivory*, ed. R.T. Palo and C.T. Robbins. Boca Raton, FL: CRC Press, pp. 45–59.

Harborne, J.B. (1993) *Introduction to Ecological Biochemistry*, 4th edn. London: Academic Press.

Harder, L.D. and Wilson, W.G. (1994) Floral evolution and male reproductive success—optimal dispensing schedules for pollen dispersal by animal-pollinated plants. *Evolutionary Ecology*, **8**, 542–559.

Harmon, G.D. and Stamp, N.E. (1992) Effects of postdispersal seed predation on spatial inequalities and size variability in an annual plant, *Erodium cicutarium* (Geraniaceae). *American Journal of Botany*, **79**, 300–305.

Hartvigsen, G. and McNaughton, S.J. (1995) Tradeoff between height and relative growth rate in a dominant grass from the Serengeti ecosystem. *Oecologia*, **102**, 273–276.

Hartvigsen, G., Wait, D.A. and Coleman, J.S. (1995) Tri-trophic interactions influenced by resource availability: predator effects on plant performance depend on plant resources. *Oikos*, **74**, 463–468.

Harvey, P.H. and Nee, S. (1997) The phylogenetic foundations of behavioral ecology. In *Behavioral Ecology: An Evolutionary Approach*, ed. J.R. Krebs and N.B. Davies, 4th edn. Oxford: Blackwell Science, pp. 334–339.

Haukioja, E. (1990) Induction of defenses in trees. *Annual Review of Entomology*, **35**, 25–42.

Haukioja, E. and Neuvonen, S. (1985) Induced long-term resistance of birch foliage against defoliators: defensive or incidental? *Ecology*, **66**, 1303–1308.

Hawksworth, D.L. (1988) Coevolution of fungi with algae and cyanobacteria in lichen symbioses. In *Coevolution of Fungi with Plants and Animals*, ed. K.A. Pirozynski and D.L. Hawksworth. London: Academic Press, pp. 125–148.

Hawksworth, D.L. and Kalin-Arroyo, M.T. (1995) Magnitude and distribution of biodiversity. In *Global Diversity Assessment*, ed. V.H. Heywood and R.T. Watson. Cambridge: Cambridge University Press, pp. 107–191.

Haynes, K.F., Zhao, J.Z. and Latif, A. (1991) Identification of floral compounds from *Abelia grandiflora* that stimulate upwind flight in cabbage looper moths. *Journal of Chemical Ecology*, **17**, 637–646.

Heddi, A., Grenier, A.-M., Khatchadourian, C., Charles, H. and Nardon, P. (1999) Four intracellular genomes direct weevil biology: nuclear, mitochrondrial, principal endosymbiont, and *Wolbachia. Proceedings of the National Academy of Sciences USA*, **96**, 6814–6819.

Heil, M., Fiala, B., Linsenmair, K.E., Zotz, G., Menke, P. and Maschwitz, U. (1997) Food body production in *Macaranga triloba* (Euphorbiaceae): a plant investment in anti-herbivore defence via symbiotic ant partners. *Journal of Ecology*, **85**, 847–861.

Hering, E.M. (1951) *Biology of Leaf Miners*. The Hague: Junk.

Herms, D.A. and Mattson, W.J. (1992) The dilemma of plants: to grow or defend. *Quarterly Review of Biology*, **67**, 283–335.

Hernández, J.V. and Jaffé, K. (1995) Daño económico causado por populações de formigas *Atta laevigata* (F. Smith) em plantaçoes de *Pinus caribaea* Mor e elementos para o manejo da praga. *Anais da Sociedade de Entomologia do Brasil*, **24**, 287–298.

Herre, E.A. (1996) An overview of studies on a community of Panamanian figs. *Journal of Biogeography*, **23**, 593–607.

Herre, E.A. (1999) Laws governing species interactions? Encouragement and caution from figs and their associates. In *Levels of Selection in Evolution*, ed. L. Keller. Princeton, NJ: Princeton University Press, pp. 209–237.

Herre, E.A., Knowlton, N., Mueller, U.G. and Rehner, S.A. (1999) The evolution of mutualisms: exploring the paths between conflict and cooperation. *Trends in Ecology and Evolution*, **14**, 49–53.

Herre, E.A. and West, S.A. (1997) Conflict of interest in a mutualism: documenting the elusive fig wasp-seed trade-off. *Proceedings of the Royal Society of London B*, **264**, 1501–1507.

Herrera, C.M. (1981) Are tropical fruits more rewarding to dispersers than temperate ones? *American Naturalist*, **118**, 896–907.

Herrera, C.M. (1982a) Grasses, grazers, mutualism, and coevolution: a comment. *Oikos*, **38**, 254–258.

Herrera, C.M. (1982b) Defense of ripe fruits from pests: its significance in relation to plant-disperser interactions. *American Naturalist*, **120**, 218–241.

Herrera, C.M. (1984a) A study of avian frugivores, bird-dispersed plants, and their interaction in Mediterranean scrublands. *Ecological Monographs*, **54**, 1–23.

Herrera, C.M. (1984b) Adaptation to frugivory of Mediterranean avian seed dispersers. *Ecology*, **65**, 609–617.

Herrera, C.M. (1985a) Determinants of plant-animal coevolution: the case of mutualistic dispersal of seeds by vertebrates. *Oikos*, **44**, 132–141.

Herrera, C.M. (1985b) Habitat-consumer interaction in frugivorous birds. In *Habitat Selection in Birds*, ed. M.L. Cody. New York: Academic Press, pp. 341–365.

Herrera, C.M. (1985c) Aposematic insects as six-legged fruits: incidental short-circuiting of their defense by frugivorous birds. *American Naturalist*, **126**, 286–293.

Herrera, C.M. (1986) Vertebrate-dispersed plants: Why they don't behave the way they should? In *Frugivores and Seed Dispersal*, ed. A. Estrada and T.H. Fleming. Dordrecht: Junk, pp. 5–18.

Herrera, C.M. (1987) Vertebrate-dispersed plants of the Iberian Peninsula: a study of fruit characteristics. *Ecological Monographs*, **57**, 305–331.

Herrera, C.M. (1988a) Variation in mutualisms: the spatiotemporal mosaic of a pollinator asssemblage. *Biological Journal of the Linnaean Society*, **35**, 95–125.

Herrera, C.M. (1988b) The fruiting ecology of *Osyris quadripartita*: individual variation and evolutionary potential. *Ecology*, **69**, 233–249.

Herrera, C.M. (1989a) Seed dispersal by animals: a role in angiosperm diversification? *American Naturalist*, **133**, 309–322.

Herrera, C.M. (1989b) Frugivory and seed dispersal by carnivorous mammals, and associated fruit characteristics, in undisturbed Mediterranean habitats. *Oikos*, **55**, 250–262.

Herrera, C.M. (1993) Selection on floral morphology and environmental determinants of fecundity in a hawk moth-pollinated violet. *Ecological Monographs*, **63**, 251–275.

Herrera, C.M. (1995) Plant–vertebrate seed dispersal systems in the Mediterranean: ecological, evolutionary and historical determinants. *Annual Review of Ecology and Systematics*, **26**, 705–727.

Herrera, C.M. (1998) Long-term dynamics of Mediterranean frugivorous birds and fleshy fruits: a 12-yr study. *Ecological Monographs*, **68**, 511–538.

Herrera, C.M. (2000) Measuring the effects of pollinators and herbivores: evidence for non-additivity in a perennial herb. *Ecology*, **81**, 2170–2176.

Herrera, C.M., Jordano, P., Guitián, J. and Traveset, A. (1998) Annual variability in seed production by woody plants and the masting concept: reassessment of principles and relationship to pollination and seed dispersal. *American Naturalist*, **152**, 576–594.

Hespenheide, H.A. (1991) Bionomics of leaf-mining insects. *Annual Review of Entomology*, **36**, 535–560.

Hickman, J.C. (1974) Pollination by ants. *Science*, **184**, 1290–1292.

Higley, L.D. and Pedigo, L.P. (1993) Economic injury level concepts and their use in sustaining environmental quality. *Agriculture, Ecosystems and Environment*, **46**, 233–243.

Hillis, D.M., Moritz, C. and Mable, B.K. (1996) *Molecular Systematics*, 2nd edn. Sunderland: Sinauer.

Hjältén, J., Danell, K. and Lundberg, P. (1993) Herbivore avoidance by association: vole and hare utilization of woody plants. *Oikos*, **68**, 125–131.

Hobbs, N.T. (1996) Modification of ecosystems by ungulates. *Journal of Wildlife Management*, **60**, 695–713.

Hobbs, N.T., Schimel, D.S., Owensby, C.E. and Ojima, D.J. (1991) Fire and grazing in the tallgrass prairie: contingent effects on nitrogen budgets. *Ecology*, **72**, 1374–1382.

Hochberg, M.E. and van Baalen, M. (1998) Antagonistic coevolution over productivity gradients. *American Naturalist*, **152**, 620–634.

Hodges, S.A. (1997a) Rapid radiation due to a key innovation in columbines (Ranunculaceae: *Aquilegia*). In *Molecular Evolution and Adaptive Radiation*, ed. T.J. Givnish and K.J. Sytsma. Cambridge: Cambridge University Press, pp. 391–405.

Hodges, S.A. (1997b) Floral nectar spurs and diversification. *International Journal of Plant Sciences*, **158**, S81–S88.

Hodgson, J.G., Grime, J.P., Hunt, R. and Thompson, K. (1994) *The Electronic Comparative Plant Ecology*. London: Chapman & Hall.

Hofmann, R.R. (1973) *The Ruminant Stomach: Stomach Structure and Feeding Habits of East African Game Ruminants*. Nairobi: East African Literature Bureau.

Hofmann, R.R. (1989) Evolutionary steps of ecophysiological adaptation and diversification of ruminants: a comparative view of their digestive system. *Oecologia*, **78**, 443–457.

Hoisington, D., Khairallah, M., Reeves, T. et al. (1999) Plant genetic resources: what can they contribute toward increased crop productivity? *Proceedings of the National Academy of Sciences USA*, **96**, 5937–5943.

Holt, R.D. (1984) Spatial heterogeneity, indirect interactions, and the coexistence of prey species. *American Naturalist*, **124**, 377–406.

Holzinger, F. and Wink, M. (1996) Mediation of cardiac glycoside insensitivity in the monarch butterfly (*Danaus plexippus*): role of an amino acid substitution in the ouabain binding site of Na^+, K^+-ATPase. *Journal of Chemical Ecology*, **22**, 1921–1937.

Holland, E.A. and Detling, J.K. (1990) Plant responses to herbivory and belowground nitrogen cycling. *Ecology*, **71**, 1040–1049.

Holland, J.N. and Fleming, T.H. (1999) Mutualistic interactions between *Upiga virescens* (Pyralidae), a pollinating seed-consumer, and *Lophocereus schottii* (Cactaceae). *Ecology*, **80**, 2074–2084.

Hölldobler, B. and Wilson, E.O. (1990) *The Ants*. Cambridge, MA: Belknap Press.

Hollinger, D.Y. (1986) Herbivory and the cycling of nitrogen and phosphorus in isolated California oak trees. *Oecologia*, **70**, 291–297.

Homma, K., Akashi, N., Abe, T. et al. (1999) Geographical variation in the early regeneration process of Siebold's beech (*Fagus crenata* Blume) in Japan. *Plant Ecology*, **140**, 129–138.

Hörnicke, H. and Björnhag, G. (1980) Coprophagy and related strategies for digesta utilization. In *Digestive Physiology and Metabolism of Ruminants*, ed. Y. Ruckebush and P. Thivend. Lancaster: MTP Press, pp. 707–730.

Horovitz, I. and Meyer, A. (1997) Evolutionary trends in the ecology of New World monkeys inferred from a combined phylogenetic analysis of nuclear, mitochondrial, and morphological data. In *Molecular Evolution and Adaptive Radiation*, ed. T.J. Givnish and K.J. Sytsma. Cambridge: Cambridge University Press, pp. 189–221.

Horsfield, D. (1977) Relationships between feeding of *Philaenus spumarius* (L.) and the amino acid concentration in the xylem sap. *Ecological Entomology*, **2**, 259–266.

Horvitz, C.C. and Schemske, D.W. (1984) Effects of ants and an ant-tended herbivore on seed production of a neotropical herb. *Ecology*, **65**, 1369–1378.

Horvitz, C.C. and Schemske, D.W. (1994) Effects of dispersers, gaps, and predators on dormancy and seedling emergence in a tropical herb. *Ecology*, **75**, 1949–1958.

Hotton, C.L., Hueber, F.M. and Labandeira, C.C. (1996) Plant–arthropod interactions from early terrestrial ecosystems: two Devonian examples. In *Paleontological Society Special Publication*, ed. J.E. Repetski, **8**, 181, abstract. Lawrence, KS: Paleontological Society.

Hotton, N., Olson, E.C. and Beerbower, R. (1997) Amniote origins and the discovery of herbivory. In *Amniote Origins*, ed. S.S. Sumida and K.L.M. Martin. New York: Academic Press, pp. 207–264.

Houle, G. (1995) Seed dispersal and seedling recruitment: the missing link(s). *Ecoscience*, **2**, 238–244.

Houle, G. and Simard, G. (1996) Additive effects of genotype, nutrient availability and type of tissue damage on the compensatory response of *Salix planifolia* ssp. *planifolia* to simulated herbivory. *Oecologia*, **107**, 373–378.

Hoveland, C.S. (1993) Importance and economic significance of the *Acremonium* endophytes to performance of animals and grass plant. *Agriculture, Ecosystems and Environment*, **44**, 3–12.

Howard, D.J. and Berlocher, S.H. (eds) (1998) *Endless Forms: Species and Speciation*. Oxford: Oxford University Press.

Howe, H.F. and Smallwood, J. (1982) Ecology of seed dispersal. *Annual Review of Ecology and Systematics*, **13**, 201–228.

Howe, H.F., Schupp, E.W. and Westley, L.C. (1985) Early consequences of seed dispersal for a Neotropical tree (*Virola surinamensis*). *Ecology*, **66**, 781–791.

Hubbard, J.A. and McPherson, G.R. (1997) Acorn selection by Mexican jays: a test of a tri-trophic symbiotic relationship hypothesis. *Oecologia*, **110**, 143–146.

Hubbell, S.P. (1980) Seed predation and the coexistence of tree species in tropical forests. *Oikos*, **35**, 214–229.

Hudson, R.J. and White, R.G. (eds) (1985) *Bioenergetics of Wild Herbivores*. Boca Raton, FL: CRC Press.

Huffaker, C.B. and Kennett, C.E. (1959) A ten-year study of vegetational changes associated with biological control of Klamath weed. *Journal of Range Management*, **12**, 69–82.

Hughes, L. (1991) The relocation of ant nest entrances: potential consequences for ant-dispersed seeds. *Australian Journal of Ecology*, **16**, 207–214.

Hughes, L., Dunlop, M., French, K. et al. (1994a) Predicting dispersal spectra: a minimal set of hypotheses based on plant attributes. *Journal of Ecology*, **82**, 933–950.

Hughes, L., Westoby, M. and Jurado, E. (1994b) Convergence of elaiosomes and insect prey: evidence from ant foraging behaviour and fatty acid composition. *Functional Ecology*, **8**, 358–365.

Huignard, J., Dupont, P. and Tran, B. (1990) Coevolutionary relations between bruchids and their host plants: the influence on the physiology of the insects. In *Bruchids and Legumes: Economics, Ecology and Coevolution*, ed. K. Fujii, A.M.R. Gatehouse, C.D. Johnson, R. Mitchell and T. Yoshida. Dordrecht: Kluwer, pp. 171–179.

Hulme, P.E. (1993) Post-dispersal seed predation by small mammals. *Symposium of the Zoological Society of London*, **65**, 269–287.

Hulme, P.E. (1994a) Rodent post-dispersal seed predation in grassland: magnitude and sources of variation. *Journal of Ecology*, **82**, 645–652.

Hulme, P.E. (1994b) Seedling herbivory in grasslands — relative impact of vertebrate and invertebrate herbivores. *Journal of Ecology*, **82**, 873–880.

Hulme, P.E. (1996a) Natural regeneration of yew (*Taxus baccata* L): microsite, seed or herbivore limitation? *Journal of Ecology*, **84**, 853–861.

Hulme, P.E. (1996b) Herbivory, plant regeneration and species coexistence. *Journal of Ecology*, **84**, 609–616.

Hulme, P.E. (1996c) Herbivores and the performance of grassland plants: a comparison of arthropod, mollusc and rodent herbivory. *Journal of Ecology*, **84**, 43–51.

Hulme, P.E. (1997) Post-dispersal seed predation and the establishment of vertebrate-dispersed plants in Mediterranean scrublands. *Oecologia*, **111**, 91–98.

Hulme, P.E. (1998a) Post-dispersal seed predation: consequences for plant demography and evolution. *Perspectives in Plant Ecology, Evolution and Systematics*, **1**, 32–46.

Hulme, P.E. (1998b) Post-dispersal seed predation and seed bank persistence. *Seed Science Research*, **8**, 513–519.

Hulme, P.E. and Borelli, T. (1999) Variability in post-dispersal seed predation in deciduous woodland: relative importance of location, seed species, burial and density. *Plant Ecology*, **145**, 149–156.

Hulme, P.E. and Hunt, M.K. (1999) Rodent post-dispersal seed predation in deciduous woodland: predator response to absolute and relative abundance of prey. *Journal of Animal Ecology*, **68**, 417–428.

Hume, I.D. (1989) Optimal digestive strategies in mammalian herbivores. *Physiological Zoology*, **62**, 1145–1163.

Hume, I.D., Morgan, K.R. and Kenagy, G.J. (1993) Digesta retention and digestive performance in sciurid and microtine rodents—effects of hindgut morphology and body size. *Physiological Zoology*, **66**, 396–411.

Hunt, J.H., Brown, P.A., Sago, K.M. and Kerker, J.A. (1991) Vespid wasps eat pollen (Hymenoptera: Vespidae). *Journal of the Kansas Entomological Society*, **64**, 127–130.

Hunter, M.D., Ohgushi, T. and Price, P.W. (eds) (1992) *Effects of Resource Distribution on Animal-Plant Interactions*. San Diego, CA: Academic Press.

Hunter, M.D., Varley, G.C. and Gradwell, G.R. (1997) Estimating the relative roles of top-down and bottom-up forces on insect herbivore populations: a classic study revisited. *Proceedings of the National Academy of Sciences USA*, **94**, 9176–9181.

Huntley, B. and Webb, T. (1989) Migration: species' response to climatic variations caused by changes in the earth's orbit. *Journal of Biogeography*, **16**, 5–19.

Huntly, N. (1991) Herbivores and the dynamics of communities and ecosystems *Annual Review of Ecology and Systematics*, **22**, 477–503.

Huxley, C.R. (1998) The tuberous epiphytic Rubiaceae—the Hydnophytinae. In *The Biological Monograph*, ed. H.C.F. Hopkins, C.R. Huxley, C.M. Pannell, G.T. Prance and F. White. London: Royal Botanic Gardens, Kew, pp. 81–92.

Huxley, C.R. and Cutler, D.F. (1991) *Ant–Plant Interactions*. Oxford: Oxford University Press.

Hwang, S.Y. and Lindroth, R.L. (1997) Clonal variation in foliar chemistry of aspen: effects on gypsy moths and forest tent caterpillars. *Oecologia*, **111**, 99–108.

Iason, G.R. and Van Wieren, S.E. (1999) Digestive and ingestive adaptations of mammalian herbivores to low quality forage. In *Herbivores: Between Plants and Predators*, ed. H. Olff, V.K. Brown and R.H. Drent. Oxford: Blackwell, pp. 337–369.

Illius, A.W. and Gordon, I.J. (1999) Scaling up from functional response to numerical response in vertebrate herbivores. In *Herbivores: Between Plants and Predators*, ed. H. Olff, V.K. Brown and R.H. Drent. Oxford: Blackwell, pp. 397–425.

Irwin, R.E. and Brody, A.K. (2000) Consequences of nectar robbing for realized male function in a hummingbird-pollinated plant. *Ecology*, **81**, 2637–2643.

Itoh, A., Yamakura, T., Ogino, K. and Lee, H.S. (1995) Survivorship and growth of seedlings of four dipterocarp species in a tropical rain-forest of Sarawak, East Malaysia. *Ecological Research*, **10**, 327–338.

Iyengar, V.K. and Eisner, T. (1999) Female choice increases offspring fitness in an arctiid moth (*Utetheisa ornatrix*). *Proceedings of the National Academy of Sciences USA*, **96**, 15013–15016.

Izhaki, I. (1998) Essential amino acid composition of fleshy fruits versus maintenance requirements of passerine birds. *Journal of Chemical Ecology*, **24**, 1333–1345.

Izhaki, I. and Safriel, U.N. (1989) Why are there so few exclusively frugivorous birds? Experiments on fruit digestibility. *Oikos*, **54**, 23–32.

Jackson, S.T. and Lyford, M.E. (1999) Pollen dispersal models in Quaternary plant ecology: assumptions, parameters, and prescriptions. *Botanical Review*, **65**, 39–75.

James, H.T. and Burney, D.A. (1997) The diet and ecology of Hawaii's extinct flightless waterfowl: evidence from coprolites. *Biological Journal of the Linnean Society*, **62**, 279–297.

Janis, C.M. and Erhardt, D. (1988) Correlation of relative muzzle width and relative incisor width with dietary preference in ungulates. *Zoological Journal of the Linnean Society*, **92**, 267–284.

Janis, C.M. and Fortelius, M. (1988) On the means whereby mammals achieve increased functional durability of their dentitions, with special reference to limiting factors. *Biological Review*, **63**, 197–230.

Janson, C.H. (1983) Adaptation of fruit morphology to dispersal agents in a Neotropical forest. *Science*, **219**, 187–189.

Janz, N. and Nylin, S. (1997) The role of female search behaviour in determining host plant range in plant feeding insects: a test of the information processing hypothesis. *Proceedings of the Royal Society of London B*, **264**, 701–707.

Janz, N. and Nylin, S. (1998) Butterflies and plants: a phylogenetic study. *Evolution*, **52**, 486–502.

Janzen, D.H. (1966) Coevolution of mutualism between ants and acacias in Central America. *Evolution*, **20**, 249–275.

Janzen, D.H. (1969) Seed-eaters versus seed size, number, toxicity and dispersal. *Evolution*, **23**, 1–27.

Janzen, D.H. (1970) Herbivores and the number of tree species in tropical forests. *American Naturalist*, **102**, 592–595.

Janzen, D.H. (1971) Seed predation by animals. *Annual Review of Ecology and Systematics*, **2**, 465–492.

Janzen, D.H. (1975) Interactions of seeds and their insect predators/parasitoids in a tropical deciduous forest. In *Evolutionary Strategies of Parasitic Insects and Mites*, ed. P.W. Price. New York: Plenum Press, pp. 154–186.

Janzen, D.H. (1978) The ecology and evolutionary biology of seed chemistry as relates to seed predation. In *Biochemical Aspects of Plant and Animal Coevolution*, ed. J.B. Harborne. London: Academic Press, pp. 163–296.

Janzen, D.H. (1979) New horizons in the biology of plant defenses. In *Herbivores: Their Interactions with Secondary Plant Metabolites*, ed. G.A. Rosenthal and D.H. Janzen. New York: Academic Press, pp. 331–350.

Janzen, D.H. (1980a) When is it coevolution? *Evolution*, **34**, 611–612.

Janzen, D.H. (1980b) Specificity of seed attacking beetles in a Costa Rican deciduous forest. *Journal of Ecology*, **68**, 929–952.

Janzen, D.H. (1983a) Dispersal of seeds by vertebrate guts. In *Coevolution*, ed. D.J. Futuyma and M. Slatkin. Sunderland: Sinauer, pp. 232–262.

Janzen, D.H. (ed.) (1983b) *Costa Rican Natural History*. Chicago: University of Chicago Press.

Janzen, D.H. (1985) On ecological fitting. *Oikos*, **45**, 308–310.

Janzen, D.H. (1988) Ecological characterization of a Costa Rican dry forest caterpillar fauna. *Biotropica*, **20**, 120–135.

Janzen, D.H. (1999) Gardenification of tropical conserved wildlands: multitasking, multicropping, and multiusers. *Proceedings of the National Academy of Sciences USA*, **96**, 5987–5994.

Jaramillo, V.J. and Detling, J.K. (1988) Grazing history, defoliation, and competition: effects on shortgrass production and nitrogen accumulation. *Ecology*, **69**, 1599–1608.

Jefferies, R.L. (1999) Herbivores, nutrients and trophic cascades in terrestrial environments. In *Herbivores: Between Plants and Predators*, ed. H. Olff, V.K. Brown and R.H. Drent. Oxford: Blackwell, pp. 301–330.

Jeffries, M.J. and Lawton, J.H. (1984) Enemy-free space and the structure of ecological communities. *Biological Journal of the Linnean Society*, **23**, 269–286.

Jensen, T.S. (1982) Seed production and outbreaks of non-cyclic rodent populations in deciduous forests. *Oecologia*, **54**, 184–192.

Jermy, T. (1984) Evolution of insect/host plant relationships. *American Naturalist*, **124**, 609–630.

Jermy, T. (1993) Evolution of insect–plant relationships—a devil's advocate approach. *Entomologia Experimentalis et Applicata*, **66**, 3–12.

Jervis, M.A. and Vilhelmsen, L. (2000) Mouthpart evolution in adults of the basal, 'symphytan', hymenopteran lineages. *Biological Journal of the Linnean Society*, **70**, 121–146.

Johnson, C.D. (1990) Coevolution of Bruchidae and their hosts: evidence, conjecture, and conclusions. In *Bruchids and Legumes: Economics, Ecology and Coevolution*, ed. K. Fujii, A.M.R. Gatehouse, C.D. Johnson, R. Mitchell and T. Yoshida. Dordrecht: Kluwer, pp. 181–188.

Johnson, D.M. and Stiling, P.D. (1996) Host specificity of *Cactoblastis cactorum* (Lepidoptera: Pyralidae), an exotic *Opuntia*-feeding moth, in Florida. *Environmental Entomology*, **25**, 743–748.

Johnson, D.M. and Stiling, P.D. (1998) Distribution and dispersal of *Cactoblastis cactorum* (Lepidoptera: Pyralidae), an exotic *Opuntia*-feeding moth, in Florida. *Florida Entomologist*, **81**, 12–22.

Johnson, K.R., Nichols, D.J., Attrep, M. and Orth, C.J. (1989) High-resolution leaf-fossil record spanning the Cretaceous/Tertiary boundary. *Nature*, **340**, 708–711.

Johnson, M.L. and Gaines, M.S. (1990) Evolution of dispersal: theoretical models and empirical tests using birds and mammals. *Annual Review of Ecology and Systematics*, **21**, 449–480.

Johnson, W.C. and Webb, T. (1989) The role of blue jays (*Cyanocitta cristata* L.) in the postglacial dispersal of fagaceous trees in eastern North America. *Journal of Biogeography*, **16**, 561–571.

Johnson, W.T. and Lyon, H.H. (1991) *Insects that Feed on Trees and Shrubs*, 2nd edn. Ithaca, NY: Cornell University Press.

Jones, C.G., Lawton, J.H. and Scachak, M. (1997) Positive and negative effects of organisms as physical and ecosystem engineers. *Ecology*, **78**, 1946–1957.

Jong, P.W. and Nielsen, J.K. (1999) Polymorphism in a flea beetle for the ability to use an atypical host plant. *Proceedings of the Royal Society of London B*, **266**, 103–111.

Jordano, P. (1987a) Frugivory, external morphology and digestive system in mediterranean sylviid warblers *Sylvia* spp. *Ibis*, **129**, 175–189.

Jordano, P. (1987b) Patterns of mutualistic interactions in pollination and seed dispersal: connectance, dependence asymmetries, and coevolution. *American Naturalist*, **129**, 657–677.

Jordano, P. (1988) Diet, fruit choice and variation in body condition of frugivorous warblers in Mediterranean scrubland. *Ardea*, **76**, 193–209.

Jordano, P. (1989) Pre-dispersal biology of *Pistacia lentiscus* (Anacardiaceae): cumulative effects on seed removal by birds. *Oikos*, **55**, 375–386.

Jordano, P. (1992) Fruits and frugivory. In *Seeds: The Ecology of Regeneration in Plant Communities*, ed. M. Fenner. Wallingford: Commonwealth Agricultural Bureau International, pp. 105–156.

Jordano, P. (1993) Geographical ecology and variation of plant-seed disperser interactions: southern Spanish junipers and frugivorous thrushes. *Vegetatio*, **107/108**, 85–104.

Jordano, P. (1995) Angiosperm fleshy fruits and seed dispersers: a comparative analysis of adaptation and constraints in plant–animal interactions. *American Naturalist*, **145**, 163–191.

Juenger, T. and Bergelson, J. (1997) Pollen and resource limitation of compensation to herbivory in scarlet gilia, *Ipomopsis aggregata*. *Ecology*, 78, 1684–1695.

Juenger, T. and Bergelson, J. (1998) Pairwise versus diffuse natural selection and the multiple herbivores of scarlet gilia, *Ipomopsis aggregata*. *Evolution*, **52**, 1583–1592.

Kaitaniemi, P., Ruohomaki, K., Tammaru, T. and Haukioja, E. (1999) Induced resistance of host tree foliage during and after a natural insect outbreak. *Journal of Animal Ecology*, **68**, 382–389.

Kaltz, O. and Shykoff, J.A. (1998) Local adaptation in host-parasite systems. *Heredity*, **81**, 361–370.

Karanam, B.V., Vincent, S.H. and Chiu, S.H.L. (1994) FK 506 metabolism in human liver microsomes—investigation of the involvement of cytochrome P450 isozymes other than CYP3A4. *Drug Metabolism and Disposition*, **22**, 811–814.

Karban, R. and Baldwin, I.T. (1997) *Induced Responses to Herbivory*. Chicago: University of Chicago Press.

Karban, R. and Thaler, J.S. (1999) Plant phase change and resistance to herbivory. *Ecology*, **80**, 510–517.

Karban, R., English-Loeb, G. and Hougen-Eitzman, D. (1997) Mite vaccinations for sustainable management of spider mites in vineyards. *Ecological Applications*, 7, 183–193.

Kareiva, P.M. (1982) Experimental and mathematical analyses of herbivore movement: quantifying the influence of plant spacing and quality on foraging discrimination. *Ecological Monographs*, **52**, 262–282.

Karhu, K.J. and Neuvonen, S. (1998) Wood ants and a geometrid defoliator of birch: predation outweighs beneficial effects through the host plant. *Oecologia*, **113**, 509–516.

Karowe, D.N. and Martin, M.M. (1989) The effects of quantity and quality of diet nitrogen on the growth, efficiency of food utilization, nitrogen budget, and metabolic rate of fifth-instar *Spodoptera eridania* larvae (Lepidoptera: Noctuidae). *Journal of Insect Physiology*, **35**, 699–708.

Kaspari, M. (1993) Removal of seeds from Neotropical frugivore droppings: ant responses to seed number. *Oecologia*, **95**, 81–88.

Kato, M. and Inoue, T. (1994) Origin of insect pollination. *Nature*, **368**, 195.

Kay, C.E. (1997) Is aspen doomed? *Journal of Forestry*, **5**, 4–11.

Kearns, C.A. and Inouye, D.W. (1993) *Techniques for Pollination Biologists*. Niwot: University Press of Colorado.

Kearns, C.A., Inouye, D.W. and Waser, N.M. (1998) Endangered mutualisms: the conservation of plant-pollinator interactions. *Annual Review of Ecology and Systematics*, **29**, 83–112.

Keith, L.B. (1983) Role of food in hare population cycles. *Oikos*, **40**, 385–395.

Kelley, S.T. and Farrell, B.D. (1998) Is specialization a dead end? The phylogeny of host use in *Dendroctonus* bark beetles (Scolytidae). *Evolution*, **52**, 1731–1743.

Kelly, D. (1994) The evolutionary ecology of mast seeding. *Trends in Ecology and Evolution*, **9**, 465–470.

Kelly, V.R. and Parker, T. (1990) Seed bank survival and dynamics in sprouting and non-sprouting *Arctostaphyllos* species. *American Midland Naturalist*, **124**, 114–123.

Kevan, P.G., Chaloner, W.G. and Savile, D.B.O. (1975) Interrelationships of early terrestrial arthropods and plants. *Palaeontology*, **18**, 391–417.

Kinnaird, M.F., O'Brien, T.G. and Suryadi, S. (1996) Population fluctuation in Sulawesi red-knobbed hornbills: tracking figs in space and time. *Auk*, **113**, 431–440.

Kirk, R.S. and Sawyer, R. (1991) *Pearson's Composition and Analysis of Foods*. London: Longman.

Kirk, W.D.J. (1984) Pollen-feeding in thrips (Insecta: Thysanoptera). *Journal of Zoology*, **204**, 107–117.

Kleiber, M. (1932) Body size and metabolism. *Hilgardia*, **6**, 315–333.

Klinkhammer, P.G.L., de Jong, T.J. and van der Meijden, E. (1988) Production, dispersal and predation of seeds in the biennial *Cirsium vulgare*. *Journal of Ecology*, **76**, 403–414.

Knudsen, J.T., Tollsten, L. and Bergström, L.G. (1993) Floral scents—a checklist of volatile compounds isolated by head-space techniques. *Phytochemistry*, **33**, 253–280.

Koenig, W.D. and Knops, J.M.H. (1998) Scale of mast-seeding and tree-ring growth. *Nature*, **396**, 225–226.

Kollmann, J. (1995) Regeneration window for fleshy fruited plants during scrub development on abandoned grassland. *Ecoscience*, **2**, 213–222.

Kollmann, J., Coomes, D.A. and White, S.M. (1998) Consistencies in post-dispersal seed predation of temperate fleshy-fruited species among seasons, years and sites. *Functional Ecology*, **12**, 683–690.

Konno, K., Hirayama, C. and Shinbo, H. (1996) Unusually high concentration of free glycine in the midgut content of the silkworm, *Bombyx mori*, and other lepidopteran larvae. *Comparative Biochemistry and Physiology*, **115**, 229–235.

Konno, K., Yasui, H., Hirayama, C. and Shinbo, H. (1998) Glycine protects against strong protein-denaturation activity of oleuropein—a phenolic compound in privet leaves. *Journal of Chemical Ecology*, **24**, 735–751.

Koomen, P., van Nieukerken, E.J. and Krikken, J. (1995) Zoologische diversiteit in Nederland. In *Biodiversiteit in Nederland*, ed. E.J. van Nieukerken and A.J. van Loon. Leiden: Nationaal Natuurhistorisch Museum, pp. 49–136.

Koptur, S. (1992) Extra-floral nectary-mediated interactions between insects and plants. In: *Insect-Plant Interactions*, vol. 4, ed. E. Bernays. Boca Raton, FL: CRC Press, pp. 81–129.

Kost, G. (1988) Interactions between basidiomycetes and Bryophyta. *Endocytobiosis and Cell Research*, **5**, 287–308.

Krassilov, V.A. and Rasnitsyn, A.P. (1999) Plant remains from the guts of fossil insects: evolutionary and palaeoecological inferences. In *Proceedings of the First International Palaeoentomological Conference, Moscow*. Bratislava: AMBA Projects International, and Moscow: Paleontological Institute, pp. 65–72.

Krebs, C.J., Sinclair, A.R.E., Boonstra, R., Boutin, S., Martin, K. and Smith, J.N.M. (1999) Community dynamics of vertebrate herbivores: how to untangle the web? In: *Herbivores: Between Plants and Predators*, ed. H. Olff, V.K. Brown and R.H. Drent. Oxford: Blackwell, pp. 447–473.

Krebs, C.J., Boutin, S., Boonstra, R. et al. (1995) Impact of food and predation on the snowshoe hare cycle. *Science*, **269**, 1112–1115.

Krijger, C.L., Opdam, M., Théry, M. and Bongers, F. (1997) Courtship behaviour of manakins and seed bank composition in a French Guianan rain forest. *Journal of Tropical Ecology*, **13**, 631–636.

Kristensen, N.P. (1997) Early evolution of the Lepidoptera + Trichoptera lineage: phylogeny and the ecological scenario. In *The Origin of Biodiversity in Insects: Phylogenetic Tests of Evolutionary Scenarios*, ed P. Grandcolas. *Mémoires du Muséum National d'Histoire Naturelle*, **173**, 253–271.

Krupnick, G.A. and Weis, A.E. (1999) The effect of floral herbivory on male and female reproductive success in *Isomeris arborea*. *Ecology*, **80**, 135–149.

Krzeminski, W. (1992) Triassic and lower Jurassic stage of Diptera evolution. *Mitteilungen der Schweizerischen Entomologischen Gesellschaft*, **65**, 39–59.

Kubitzki, K. and Kurz, H. (1984) Synchronized dichogamy and dioecy in neotropical Lauraceae. *Plant Systematics and Evolution*, **147**, 253–266.

Kukalová-Peck, J. (1991) Fossil history and the evolution of hexapod structures. In *The Insects of Australia*, ed. I.D. Naumann, P.B. Carne, J.F. Lawrence et al., vol. 1. Ithaca, NY: Cornell University Press, pp. 141–179.

Kunin, W.E. (1994) Density-dependent foraging in the harvester ant *Messor ebeninus*: two experiments. *Oecologia*, **98**, 328–335.

Kunz, T.H. and Ingalls, K.A. (1994) Folivory in bats: an adaptation derived from frugivory. *Functional Ecology*, **8**, 665–668.

Kvacek, J. (1997) *Microzamia gibba* (Reuss) Corda: a cycad ovulate cone from the Bohemian Cretaceous Basin, Czech Republic—micromorphology and a reinterpretation of its affinities. *Review of Palaeobotany and Palynology*, **96**, 81–97.

Kyto, M., Niemela, P. and Larsson, S. (1996) Insects on trees: population and individual response to fertilization. *Oikos*, **75**, 148–159.

Laakso, J. and Setälä, H. (1998) Composition and trophic structure of detrital food web in ant nest mounds of *Formica aquilonia* and in the surrounding forest soil. *Oikos*, **81**, 266–278.

Labandeira, C.C. (1990) *Use of a Phenetic Analysis of Recent Hexapod Mouthparts for the Distribution of Hexapod Food-resource Guilds in the Fossil Record.* Ph.D. thesis, University of Chicago.

Labandeira, C.C. (1997) Insect mouthparts: ascertaining the paleobiology of insect feeding strategies. *Annual Review of Ecology and Systematics*, **28**, 153–193.

Labandeira, C.C. (1998a) Early history of arthropod and vascular plant associations. *Annual Review of Earth and Planetary Sciences*, **26**, 329–377.

Labandeira, C.C. (1998b) How old is the flower and the fly? *Science*, **280**, 57–59.

Labandeira, C.C. (1998c) Plant-insect associations from the fossil record. *Geotimes*, **43**, 18–24.

Labandeira, C.C. (1999) Insects and other hexapods. In *Encyclopedia of Paleontology*, ed. R. Singer and M.K. Diamond. Chicago: Fitzroy Dearborn, pp. 603–624.

Labandeira, C.C. and Phillips, T.L. (1996) Insect fluid-feeding on Upper Pennsylvanian tree ferns (Palaeodictyoptera, Marattiales) and the early history of the piercing-and-sucking functional feeding group. *Annals of the Entomological Society of America*, **89**, 157–183.

Labandeira, C.C. and Sepkoski, J.J. Jr (1993) Insect diversity in the fossil record. *Science*, **261**, 310–315.

Labandeira, C.C., Dilcher, D.L., Davis, D.R. and Wagner, D.L. (1994) 97 million years of angiosperm-insect association: paleobiological insights into the meaning of coevolution. *Proceedings of the National Academy of Sciences USA*, **91**, 12278–12282.

Labandeira, C.C., Johnson, K. and Lang, P. (2002) Insect herbivory across the Cretaceous/Tertiary boundary: major extinction and minimum rebound. In *The Hell Creek Formation and the Cretaceous/Tertiary Boundary in the Northern Great Plains: An Integrated Continental Record at the End of the Cretaceous*, ed. J.H. Hartman, K. Johnson and D.J. Nichols. Denver, CO: Geological Society of America, (in press).

Labandeira, C.C., Phillips, T.L. and Norton, R.A. (1997) Oribatid mites and decomposition of plant tissues in Paleozoic coal-swamp forests. *Palaios*, **12**, 317–351.

Lalonde, R.G. and Roitberg, B.D. (1992) On the evolution of masting behavior in trees: predation or weather? *American Naturalist*, **139**, 1293–1304.

Laman, T.G. (1995) *Ficus stupenda* germination and seedling establishment in the Bornean rain forest canopy. *Ecology*, **76**, 2617–2626.

Lammi, A. and Kuitunen, M. (1995) Deceptive pollination of *Dactylorhiza incarnata*: an experimental test of the magnet species hypothesis. *Oecologia*, **101**, 500–503.

Lamont, B.B., Le Maitre, D.C., Cowling, R.M. and Enright, N.J. (1991) Canopy seed storage in woody plants. *Botanical Review*, **57**, 277–317.

Landolt, R. (1987) Vergleichend funktionelle Morphologie des Verdauungstraktes der Tauben (Columbidae) mit besonderer Berücksichtigung der adaptiven Radiation der Fruchttauben (Treroninae). Teil 2. *Zoologische Jahrbücher Anatomie*, **116**, 285–316.

Langer, P. and Chivers, D.J. (1994) Classification of foods for comparative analysis of the gastro-intestinal tract. In *The Digestive System in Mammals: Food, Form and Function*, ed. D.J. Chivers and P. Langer. Cambridge: Cambridge University Press, pp. 74–86.

Langer, P. (1988) *The Mammalian Herbivore Stomach: Comparative Anatomy Function and Evolution.* Stuttgart: Gustav Fischer Verlag.

Larew, H.G. (1992) Fossil galls. In *Biology of Insect-Induced Galls*, ed. J.D. Shorthouse and O. Rohfritsch. New York: Oxford University Press, pp. 50–59.

Lau, D.H., Xue, L., Young, L.J., Burke, P.A. and Cheung, A.T. (1999) Paclitaxel (Taxol): an inhibitor of angiogenesis in a highly vascularized transgenic breast cancer. *Cancer Biotherapy & Radiopharmaceuticals*, **14**, 31–36.

Laurin, M. and Reisz, R.R. (1995) A reevaluation of early amniote phylogeny. *Zoological Journal of the Linnean Society*, **113**, 165–223.

Lawton, J.H., Lewinsohn, T.M. and Compton, S.G. (1993) Patterns of diversity for the insect herbivores on bracken. In *Species Diversity in Ecological Communities*, ed. R.E. Ricklefs and D. Schluter. Chicago: University of Chicago Press, pp. 178–184.

Lawton-Rauh, A.L., Alvarez-Buylla, E.R. and Purugganan, M.D. (2000) Molecular evolution of floral development. *Trends in Ecology and Evolution*, **15**, 144–149.

Leal, I.R. and Oliveira, P.S. (1998) Interactions between fungus-growing ants (Attini), fruits and seeds in cerrado vegetation in southeast Brazil. *Biotropica*, **30**, 170–178.

Lehtilä, K. (1996) Optimal distribution of herbivory and localized compensatory responses within a plant. *Vegetatio*, **127**, 99–109.

Lehtilä, K. and Strauss, S.Y. (1997) Leaf damage by herbivores affects attractiveness to pollinators in wild radish, *Raphanus raphanistrum. Oecologia*, **111**, 396–403.

Leishman, M.R., Westoby, M. and Jurado, E. (1995) Correlates of seed size variation: a comparison among five temperate floras. *Journal of Ecology*, **83**, 517–529.

Lemon, W.C. (1991) Fitness consequences of foraging behaviour in the zebra finch. *Nature*, **352**, 153–155.

Lennartsson, T., Nilsson, P. and Tuomi, J. (1998) Induction of overcompensation in the field gentian, *Gentianella campestris. Ecology*, **79**, 1061–1072.

Lennartsson, T., Tuomi, J. and Nilsson, P. (1997) Evidence for an evolutionary history of overcompensation in the grassland biennial *Gentianella campestris* (Gentianaceae). *American Naturalist*, **149**, 1147–1155.

Leston, D. (1973) The ant mosaic-tropical tree crops and the limiting of pests and diseases. *Proceedings of the National Academy of Sciences USA*, **19**, 311–341.

Letourneau, D.K. (1998) Ants, stem-borers, and fungal pathogens: experimental tests of a fitness advantage in *Piper* ant-plants. *Ecology*, **79**, 593–603.

Levin, D.A. (1976) Alkaloid-bearing plants: an ecogeographic perspective. *American Naturalist*, **110**, 261–284.

Levey, D.J. (1988) Spatial and temporal variation in Costa Rican fruit and fruit-eating bird abundance. *Ecological Monographs*, **58**, 251–269.

Levey, D.J. and Byrne, M.M. (1993) Complex ant–plant interactions: rain forest ants as secondary dispersers and post-dispersal seed predators. *Ecology*, **74**, 1802–1812.

Levey, D.J. and Cipollini, M.L. (1998) A glycoalkaloid in ripe fruit deters consumption by cedar waxwings. *Auk*, **115**, 359–367.

Levey, D.J., Place, A.R., Rey, P.J. and Martínez del Río, C. (1999) An experimental test of dietary enzyme modulation in pine warblers *Dendroica pinus*. *Physiological and Biochemical Zoology*, **72**, 576–587.

Levin, D.A. and Wilson, A.C. (1976) Rates of evolution in seed plants: net increase in diversity of chromosome numbers and species numbers through time. *Proceedings of the National Academy of Sciences USA*, **73**, 2086–2090.

Lewis, A.C. (1986) Memory constraints and flower choice in *Pieris rapae*. *Science*, **232**, 863–865.

Lewis, S.E. and Heikes, P.M. (1990) Fossil caddisfly cases (Trichoptera), Miocene of northern Idaho, USA. *Ichnos*, **1**, 143–146.

Liener, I.E. (1991) Lectins. In *Herbivores: Their Interactions with Secondary Plant Metabolites*, ed. G.A. Rosenthal and M.R. Berenbaum, vol. 1. New York: Academic Press, pp. 327–353.

Lima, S.L. and Dill, L.M. (1990) Behavioral decisions made under the risk of predation: a review and prospectus. *Canadian Journal of Zoology*, **68**, 619–640.

Lindberg, R.L.P. and Negishi, M. (1989) Alteration of mouse cytochrome P450 substrate specificity by mutation of a single amino-acid residue. *Nature*, **339**, 632–634.

Lindroth, R.L. (1989) Biochemical detoxication: mechanism of differential tiger swallowtail tolerance to phenolic glycosides. *Oecologia*, **81**, 219–224.

Lindroth, R.L. and Weisbrod, A.V. (1991) Genetic variation in response of the gypsy moth to aspen phenolic glycosides. *Biochemical Systematics and Ecology*, **19**, 97–103.

Linhart, Y.B. and Grant, M.C. (1996) Evolutionary significance of local genetic differentiation in plants. *Annual Review of Ecology and Systematics*, **27**, 237–277.

Linhart, Y.B. and Thompson, J.D. (1999) Thyme is of the essence: biochemical polymoprhism and multi-species deterrence. *Evolutionary Ecological Research*, **1**, 151–171.

Linhart, Y.B., Snyder, M.A. and Gibson, J.P. (1994) Differential host utilization by two parasites in a population of ponderosa pine. *Oecologia*, **98**, 117–120.

Lively, C.M. (1999) Migration, virulence, and the geographic mosaic of adaptation by parasites. *American Naturalist*, **153**, S34–S47.

Loconte, H. and Stevenson, D.W. (1990) Cladistics of the Spermatophyta. *Brittonia*, **42**, 197–211.

Loiselle, B.A. (1990) Seeds in droppings of tropical fruit-eating birds: importance of considering seed composition. *Oecologia*, **82**, 494–500.

Loiselle, B.A. and Blake, J.G. (1994) Annual variation in birds and plants of a tropical second-growth woodland. *Condor*, **96**, 368–380.

Lopez de Castanave, J., Cueto, V.R. and Marone, L. (1998) Granivory in the Monte desert, Argentina: is it less intense than in other arid zones of the world? *Global Biogeography and Ecology Letters*, **7**, 197–204.

Losos, J.B. (1996) Phylogenetic perspectives on community ecology. *Ecology*, **77**, 1344–1354.

Louda, S.M. (1982) Distributional ecology: variation in plant recruitment over a gradient in relation to insect seed predation. *Ecological Monographs*, **52**, 25–41.

Louda, S.M. and Rodman, J.E. (1996) Insect herbivory as a major factor in the shade distribution of a native crucifer (*Cardamine cordifolia* A. Gray, bittercress). *Journal of Ecology*, **84**, 229–237.

Louda, S.M., Kendall, D., Connor, J. and Simberloff, D. (1997) Ecological effects of an insect introduced for the biological control of weeds. *Science*, **277**, 1088–1090.

Louda, S.M., Potvin, M.A. and Collinge, S.K. (1990) Predispersal seed predation, postdispersal seed predation and competition in the recruitment of seedlings of a native thistle in sandhills prairie. *American Midland Naturalist*, **124**, 105–113.

Lumaret, R., Guillerm, J.-L., Maillet, J. and Verlaque, R. (1997) Plant species diversity and polyploidy in islands of natural vegetation isolated in extensive cultivated lands. *Biodiversity and Conservation*, **6**, 591–613.

Lunau, K. (1996) Signalling functions of floral colour patterns for insect flower visitors. *Zoologischer Anzeiger*, **235**, 11–30.

Lutz, H. (1993) The Middle Eocene 'Fossillagerstätte Eckfelder Maar' (Eifel, Germany). *Kaupia*, **2**, 21–25.

Lloyd, D.G. and Webb, C.J. (1986) The avoidance of interference between the presentation of pollen and stigmas in angiosperms I. Dichogamy. *New Zealand Journal of Botany*, **24**, 135–162.

Ma, R., Cohen, M.B., Berenbaum, M.R. and Schuler, M.A. (1994) Black swallowtail (*Papilio polyxenes*) alleles encode cytochrome P450s that selectively metabolize linear furanocoumarins. *Archives of Biochemistry and Biophysics*, **310**, 332–340.

MacArthur, R.H. (1972) *Geographical Ecology: Patterns in the Distribution of Species*. New York: Harper & Row.

MacArthur, R.H. and Wilson, E.O. (1967) *The Theory of Island Biogeography*. Princeton, NJ: Princeton University Press.

Mack, A.L. (1993) The sizes of vertebrate-dispersed fruits: a Neotropical–Paleotropical comparison. *American Naturalist*, **142**, 840–856.

Mack, R.N. and Thompson, J.N. (1982) Evolution in steppe with few large, hooved mammals. *American Naturalist*, **119**, 757–773.

Mackenzie, A. and Dixon, A.F.G. (1991) An ecological perspective of host alternation in aphids (Homoptera: Aphidinea: Aphididae). *Entomologia Generalis*, **16**, 265–284.

Maddock, L. (1979) The 'migration' and grazing succession. In *Serengeti. Dynamics of an Ecosystem*, ed. A.R.E. Sinclair and M. Norton-Griffiths. Chicago: University of Chicago Press, pp. 104–129.

Madej, C.W. and Clay, K. (1991) Avian seed preference and weight loss experiments: the effect of fungal endophyte-infected tall fescue seeds. *Oecologia*, **88**, 296–302.

Maddox, G.D. and Root, R.B. (1990) Structure of the encounter between goldenrod (*Solidago altissima*) and its diverse insect fauna. *Ecology*, **71**, 2115–2124.

Maddrell, S.H.P. and Gardiner, B.O.C. (1976) Excretion of alkaloids by malpighian tubules of insects. *Journal of Experimental Biology*, **64**, 267–281.

Majer, J.D. (1990) Rehabilitation of disturbed land: long-term prospects for the recolonization of fauna. *Proceedings of the Ecological Society of Australia*, **16**, 509–519.

Maloof, J.E. and Inouye, D.W. (2000) Are nectar robbers cheaters or mutualists? *Ecology*, **81**, 2651–2661.

Mardulyn, P., Milinkovitch, M.C. and Pasteels, J.M. (1997) Phylogenetic analyses of DNA and allozyme data suggest that *Gonioctena* leaf beetles (Coleoptera: Chrysomelidae) experienced convergent evolution in their history of host-plant family shifts. *Systematic Biology*, **46**, 722–747.

Margulis, L. and Fester, R. (eds) (1991) *Symbiosis as a Source of Evolutionary Innovation: Speciation and Morphogenesis*. Cambridge, MA: MIT Press.

Maron, J.L. (1998) Insect herbivory above and below ground: individual and joint effects on plant fitness. *Ecology*, **79**, 1281–1293.

Marquis, R.J. (1984) Leaf herbivores decrease fitness of a tropical plant. *Science*, **226**, 537–539.

Marshall, J.E.A., Astin, T.R. and Clack, J.A. (1999) East Greenland tetrapods are Devonian in age. *Geology*, **27**, 637–640.

Marty, M.A. and Krieger, R.I. (1984) Metabolism of uscharidin, a milkweed cardenolide, by tissue homogenates of monarch butterfly larvae, *Danaus plexippus* L. *Journal of Chemical Ecology*, **10**, 945–956.

Maschinski, J. and Whitham, T.G. (1989) The continuum of plant responses to herbivory: the influence of plant association, nutrient availability, and timing. *American Naturalist*, **134**, 1–19.

Mathews, S. and Donoghue, M.J. (1999) The root of angiosperm phylogeny inferred from duplicate phytochrome genes. *Science*, **286**, 947–950.

Matsuki, M. and MacLean, S.F. (1994) Effects of different leaf traits on growth rates of insect herbivores on willows. *Oecologia*, **100**, 141–152.

Mattson, W.J. (1980) Herbivory in relation to plant nitrogen content. *Annual Review of Ecology and Systematics*, **11**, 119–161.

Mattson, W.J. (1986) Competition for food between two principal cone insects of red pine, *Pinus resinosa*. *Environmental Entomology*, **15**, 88–92.

Maynard Smith, J. and Szathmáry, E. (1995) *The Major Transitions in Evolution*. San Francisco: Freeman.

McAdoo, J.K., Evans, C.C., Roundy, B.A., Young, J.A. and Evans, R.A. (1983) Influence of heteromyid rodents on *Oryzopsis hymenoides* germination. *Journal of Range Management*, **36**, 61–64.

McArthur, C., Hagerman, A.E. and Robbins, C.T. (1991) Physiological strategies of mammalian herbivores against plant defences. In *Plant Defenses against Mammalian Herbivory*, ed. R.T. Palo and C.T. Robbins. Boca Raton, FL: CRC Press, pp. 103–114.

McCoy, E.D. (1990) The distribution of insects along elevational gradients. *Oikos*, **58**, 313–322.

McGinley, M.A., Dhillion, S.S. and Neumann, J.C. (1994) Environmental heterogeneity and seedling establishment: ant–plant–microbe interactions. *Functional Ecology*, **8**, 607–615.

McKey, D. (1974) Adaptive patterns in alkaloid physiology. *American Naturalist*, **108**, 305–320.

McKey, D. (1975) The ecology of coevolved seed dispersal systems. In *Coevolution of Animals and Plants*, ed. L.E. Gilbert and P.H. Raven. Austin: University of Texas Press, pp. 159–191.

McMillan, W.O., Weigt, L.A. and Palumbi, S.R. (1999) Color pattern evolution, assortative mating, and genetic differentiation in brightly colored butterflyfishes (Chaetondontidae). *Evolution*, **53**, 247–260.

McNaughton, S.J. (1978) Serengeti ungulates: feeding selectivity influences the effectiveness of plant defense guilds. *Science*, **199**, 806–807.

McNaughton, S.J. (1979) Grazing as an optimization process: grass–ungulate relationships in the Serengeti. *American Naturalist*, **113**, 691–703.

McNaughton, S.J. (1984) Grazing lawns: animals in herds, plant form and coevolution. *American Naturalist*, **124**, 863–886.

McNeill, S. and Lawton, J.H. (1970) Annual production and respiration in animal populations. *Nature*, **225**, 472–474.

McPeek, M.A., Schrot, A.K. and Brown, J.M. (1996) Adaptation to predators in a new community: swimming performance and predator avoidance in damselflies. *Ecology*, **77**, 617–629.

McPheron, B.A. and Han, H.Y. (1997) Phylogenetic analysis of North American *Rhagoletis* (Diptera: Tephritidae) and related genera using mitochondrial DNA sequence data. *Molecular Phylogenetics and Evolution*, **7**, 1–16.

Menzel, R. and Backhaus, W. (1991) Colour vision in insects. In *Vision and Visual Dysfunction: The Perception of Color*, ed. P. Gouras. London: Macmillan, pp. 262–288.

Menzel, R. and Müller, U. (1996) Learning and memory in honeybees: from behavior to neural substrates. *Annual Review of Neurobiology*, **19**, 379–404.

Messina, F.J. (1981) Plant protection as a consequence of an ant–membracid mutualism: interactions on Goldenrod (*Solidago* sp.). *Ecology*, **62**, 1433–1440.

Metcalf, R.L., Metcalf, R.A. and Rhodes, A.M. (1980) Cucurbitacins as kairomones for diabroticite beetles. *Proceedings of the National Academy of Sciences USA*, **77**, 3769–3772.

Mitchell-Olds, T. and Bradley, D. (1996) Genetics of *Brassica rapa*. 3. Costs of disease resistance to three fungal pathogens. *Evolution*, **50**, 1859–1865.

Meyer, G.A. (1993) A comparison of the impacts of leaf and sap-feeding insects on growth and allocation of goldenrod. *Ecology*, **74**, 1101–1116.

Meyer, G.A. and Root, R.B. (1993) Effects of herbivorous insects and soil fertility on reproduction of goldenrod. *Ecology*, **74**, 1117–1128.

Michener, C.D. (1979) The biogeography of the bees. *Annals of the Missouri Botanical Garden*, **66**, 277–347.

Milchunas, D.G. and Lauenroth, W.K. (1993) Quantitative effects of grazing on vegetation and soils over a global range of environments. *Ecological Monographs*, **63**, 327–366.

Milewski, A.V. and Bond, W.J. (1982) Convergence of myrmecochory in mediterranean Australia and South Africa. In *Ant–Plant Interactions in Australia*, ed. R.C. Buckley. The Hague: Junk, pp. 89–98.

Milton, K., Windsor, D.M., Morrison, D.W. and Estribi, M.A. (1982) Fruiting phenologies of two neotropical *Ficus* species. *Ecology*, **63**, 752–762.

Milton, S.J. (1992) Plants eaten and dispersed by adult leopard tortoises *Geochelone pardalis* (Reptilia: Chelonii) in the southern Karoo. *South African Journal of Zoology*, **27**, 45–49.

Milton, S.J. (1995) Effects of rain, sheep and tephritid flies on seed production of two arid Karoo shrubs in South Africa. *Journal of Applied Ecology*, **32**, 137–144.

Miller, R.B. (1981) Hawkmoths and the geographic patterns of floral variation in *Aquilegia caerulea*. *Evolution*, **35**, 763–774.

Mitchell, R. (1977) Bruchid beetles and seed packaging by palo verde. *Ecology*, **58**, 644–651.

Mitchell, R.J., Shaw, R.G. and Waser, N.M. (1998) Pollinator selection, quantitative genetics, and predicted evolutionary responses of floral traits in *Penstemon centranthifolius* (Scrophulariaceae). *International Journal of Plant Sciences*, **159**, 331–337.

Mitter, C. and Brooks, D.R. (1983) Phylogenetic aspects of coevolution. In *Coevolution*, ed. D.J. Futuyma and M. Slatkin. Sunderland: Sinauer, pp. 65–98.

Mitter, C. and Farrell, B. (1991) Macroevolutionary aspects of insect–plant relationships. In *Insect–Plant Interactions*, ed. E. Bernays, vol. 3. Boca Raton, FL: CRC Press, pp. 35–78.

Mitter, C., Farrell, B. and Futuyma, D.J. (1991) Phylogenetic studies of insect–plant interactions: insights into the genesis of diversity. *Trends in Ecology and Evolution*, **6**, 290–293.

Mitter, C., Farrell, B. and Wiegmann, B. (1988) The phylogenetic study of adaptive zones: has phytophagy promoted insect diversification? *American Naturalist*, **132**, 107–128.

Mode, C.J. (1958) A mathematical model for the co-evolution of obligate parasites and their hosts. *Evolution*, **12**, 158–165.

Moegenburg, S.M. (1996) *Sabal palmetto* seed size: causes of variation, choices of predators, and consequences for seedlings. *Oecologia*, **106**, 539–543.

Møller, A.P. (1995) Bumblebee preference for symmetrical flowers. *Proceedings of the National Academy of Sciences USA*, **92**, 2288–2292.

Møller, A.P. (2000) Developmental stability and pollination. *Oecologia*, **123**, 149–157.

Montandon, R., Stipanovic, R.D., Williams, H.J., Sterling, W.L. and Vinson, S.B. (1987) Nutritional indices and excretion of gossypol by *Alabama argillacea* (Hubner) and *Heliothis virescens* (F.) (Lepidoptera: Noctuidae) fed glanded and glandless cotyledonary cotton leaves. *Journal of Economic Entomology*, **80**, 32–36.

Mopper, S. (1996) Adaptive genetic structure in phytophagous insect populations. *Trends in Ecology and Evolution*, **11**, 235–238.

Mopper, S. and Strauss, S.Y. (eds) (1997) *Genetic Structure and Local Adaptation in Natural Insect Populations*. London: Chapman & Hall.

Morales, M.A. and Heithaus, E.R. (1998) Food gained from seed dispersal mutualism shifts sex ratios in colonies of the ant *Aphaenogaster rudis*. *Ecology*, **79**, 734–739.

Moran, N.A. (1988) The evolution of host-plant alternation in aphids: evidence for specialization as a dead end. *American Naturalist*, **132**, 681–706.

Moran, N.A. (1989) A 48-million-year-old aphid–host plant association and complex life cycle: biogeographic evidence. *Science*, **245**, 173–175.

Moran, N.A., Munson, M.A., Baumann, P. and Ishikawa, H. (1993) A molecular clock in endosymbiotic bacteria is calibrated using the insect hosts. *Proceedings of the Royal Society of London B*, **253**, 167–171.

Moran, N.A. and Telang, A. (1998) Bacteriacyte-associated symbionts of insects. *Bioscience*, **48**, 295–304.

Moreno, J.M. and Oechel, W.C. (1990) Fire intensity and herbivory effects on post-fire resprouting of *Adenostoma fasciculatum* in southern California chaparral. *Oecologia*, **85**, 429–433.

Morris, M.G. (1974) Oak as a habitat for insect life. In *The British Oak*, ed. M.G. Morris and F.H. Perring. Faringdon: E.W. Classey, pp. 274–294.

Morrow, P.A. and LaMarche, J.V.C. (1978) Tree ring evidence for chronic insect suppression of productivity in subalpine *Eucalyptus*. *Science*, **201**, 1244–1246.

Mostovski, M.B. (1998) A revision of the nemestrinid flies (Diptera, Nemestrinidae) described by Rohdendorf, and a description of new taxa of the Nemestrinidae from the Upper Jurassic of Kazakhstan. *Paleontological Journal*, **32**, 369–375.

Mothershead, K. and Marquis, R.J. (2000) Fitness impacts of herbivory through indirect effects on plant–pollinator interactions in *Oenothera macrocarpa*. *Ecology*, **81**, 30–40.

Mueller, U.G., Rehner, S.A. and Schultz, T.R. (1998) The evolution of agriculture in ants. *Science*, **281**, 2034–2038.

Mulder, C.P.H. (1999) Vertebrate herbivores and plants in the Arctic and subarctic: effects on individuals, populations and ecosystems. *Perspectives in Plant Ecology, Evolution and Systematics*, **2**, 29–55.

Mumford, J.D. and Knight, J.D. (1997) Injury, damage and threshold concepts. In *Methods in Ecological and Agricultural Entomology*, ed. D.R. Dent and M.P. Walton. Wallingford: Commonwealth Agricultural Bureau International, pp. 203–220.

Munk, W. and Sues, H.-D. (1993) Gut contents of *Parasaurus* (Pareiasauria) and *Protorosaurus* (Archosauromorpha) from the Kupferschiefer (Upper Permian) of Hessen, Germany. *Paläontologische Zeitschrift*, **67**, 169–176.

Mutikainen, P. and L.F. Delph (1997) Effects of herbivory on male reproductive success in plants. *Oikos*, **75**, 353–358.

Myster, R.W. (1997) Seed predation, disease and germination on landslides in neotropical lower montane wet forest. *Journal of Vegetation Science*, **8**, 55–64.

Myster, R.W. and Pickett, S.T.A. (1993) Effect of litter, distance, density and vegetation patch type on postdispersal tree seed predation in old fields. *Oikos*, **66**, 381–388.

Nambudiri, E.M.V. and Binda, P.L. (1989) Dicotyledonous fruits associated with coprolites from the Upper Cretaceous (Maastrichtian) Whitemud Formation, southern Saskatchewan, Canada. *Review of Palaeobotany and Palynology*, **59**, 57–66.

Narayana, Y.D. and Muniyappa, V. 1996. Evaluation of techniques for the efficient transmission of sorghum stripe virus by vector (*Peregrinus maidis*) and screening for disease resistance. *Tropical Agriculture*, **73**, 119–123.

Nason, J.D., Herre, E.A. and Hamrick, J.L. (1998) The breeding structure of a tropical keystone plant resource. *Nature*, **391**, 685–687.

Nault, L.R. (1997) Arthropod transmission of plant viruses: a new synthesis. *Annals of the Entomological Society of America*, **90**, 521–541.

Neal, J.J. (1989) Methylenedioxyphenyl-containing alkaloids and autosynergism. *Phytochemistry*, **28**, 451–453.

Neal, J.J. and Wu, D. (1994) Inhibition of insect cytochromes P450 by furanocoumarins. *Pesticide Biochemistry and Physiology*, **50**, 43–50.

Ngisong, A.J., Overholt, W.A., Njagi, P.G.N., Dicke, M., Ayertey, J.N. and Lwande, W. (1996) Volatile infochemicals used in host and host habitat location by *Cotesia flavipes* Cameron and *Cotesia sesamiae* (Cameron) (Hymenoptera, Braconidae), larval parasitoids of stemborers on Gramineae. *Journal of Chemical Ecology*, **22**, 307–323.

Nilsson, L.A. (1980) The pollination ecology of *Dactylorhiza sambucina* (Orchidaceae). *Botaniska Notiser*, **133**, 367–385.

Nilsson, L.A. (1988) The evolution of flowers with deep corolla tubes. *Nature*, **334**, 147–149.

Nilsson, L.A., Jonsson, L., Ralison, L. and Randrianjohany, E. (1987) Angraecoid orchids and hawkmoths in central Madagascar: specialized pollination systems and generalist foragers. *Biotropica*, **19**, 310–318.

Nilsson, S.G. and Wastljung, U. (1987) Seed predation and cross pollination in mast-seeding beech (*Fagus sylvatica*) patches. *Ecology*, **68**, 260–265.

Noma, N. and Yumoto, T. (1997) Fruiting phenology of animal-dispersed plants in response to winter migration of frugivores in a warm temperate forest in Yakushima Island, Japan. *Ecological Research*, **12**, 119–129.

Norman, D. (1985) *An Illustrated Encyclopedia of Dinosaurs*. London: Salamander Books.

Norment, C.J. and Fuller, M.E. (1997) Breeding-season frugivory by Harris' sparrows (*Zonotrichia querula*) and white-crowned sparrows (*Zonotrichia leucophrys*) in a low-arctic ecosystem. *Canadian Journal of Zoology*, **75**, 670–679.

Norstog, K.J. and Nicholls, T.J. (1997) *The Biology of the Cycads*. Ithaca, NY: Cornell University Press.

North, R.D., Jackson C.W. and Howse, P.E. (1997) Evolutionary aspects of ant–fungus interactions in leaf-cutting ants. *Trends in Ecology and Evolution*, **12**, 386–389.

Norton, D.A. and Kelly, D. (1988) Mast seeding over 33 years by *Dacrydium cupressinum* Lamb. (rimu) (Podocarpaceae) in New Zealand: the importance of economies of scale. *Functional Ecology*, **2**, 399–408.

Novak, S.J., Soltis, D.E. and Soltis, P.S. (1991) Ownbey's tragopogons: 40 years later. *American Journal of Botany*, **78**, 1586–1600.

Novokshonov, V.G. (1997) Some Mesozoic scorpionflies (Insecta: Panorpida = Mecoptera) of the families Mesopsychidae, Pseudopolycentropidae, Bittacidae, and Permochoristidae. *Paleontologicheskii Zhurnal*, **1997/1**, 65–71 [in Russian].

Novokshonov, V.G. (1998) New fossil insects (Insecta: Grylloblattidae, Caloneurida, Hypoperlida?, ordinis incertis) from the Kungurian beds of the middle Urals. *Paleontological Journal*, **32**, 362–368.

Nowak, R.M. (1999) *Walker's Mammals of the World*, 6th edn. Baltimore, MD: John Hopkins University Press.

Nowak, R.S., Nowak, C.L., DeRocher, T., Cole, N. and Jones, M.A. (1990) Prevalence of *Oryzopsis hymenoides* near harvester ant mounds: indirect facilitation by ants. *Oikos*, **58**, 190–198.

Nuismer, S.L., Thompson, J.N. and Gomulkiewicz, R. (1999) Gene flow and geographically structured coevolution. *Proceedings of the Royal Society of London B*, **266**, 605–609.

Nystrand, O. and Granström, A. (1997) Post-dispersal predation on *Pinus sylvestris* seeds by *Fringilla* spp: ground substrate affects selection for seed color. *Oecologia*, **110**, 353–359.

O'Dowd, D.J. and Gill, A.M. (1984) Predator satiation and site alteration: mass reproduction of alpine ash (*Eucalyptus delegatensis*) following fire in southeastern Australia. *Ecology*, **65**, 1052–1066.

O'Dowd, D.J., Brew, C.R., Christophel, D.C. and Norton, R.A. (1991) Mite–plant associations from the Eocene of southern Australia. *Science*, **252**, 99–101.

Oksanen, L. and Oksanen, T. (2000) The logic and realism of the hypothesis of exploitation ecosystems. *American Naturalist*, **155**, 703–723.

Oksanen, L., Fretwell, S.D., Arruda, J. and Niemelä, P. (1981) Exploitation ecosystems in gradients of primary productivity. *American Naturalist*, **115**, 240–261.

Olff, H. and Ritchie, M.E. (1998) Effects of herbivores on grassland plant diversity. *Trends in Ecology and Evolution*, **13**, 261–265.

Oliveira, P.S., Galetti, M., Pedroni, F. and Morellato, L.P.C. (1995) Seed cleaning by *Mycocepurus goeldii* ants (Attini) facilitates germination in *Hymenaea courbaril* (Caesalpiniaceae). *Biotropica*, **27**, 518–522.

Omura, H., Honda K. and N. Hayashi, N. (2000) Floral scent of *Osmanthus fragrans* discourages foraging behavior of cabbage butterfly, *Pieris rapae*. *Journal of Chemical Ecology*, **26**, 655–666.

Opler, P.A. (1973) Fossil lepidopterous leaf mines demonstrate the age of some insect–plant relationships. *Science*, **179**, 1321–1323.

Ortega, A. (1997) *Niche Reconstruction of an Extinct South American Mammal using Stable Isotopes and Microwear*. B.Sc. thesis, Princeton University.

Osmond, C.H. and Monro, J. (1981) Prickly pear. In *Plants and Man in Australia*, ed. D.J. Carr and S.G. Carr. New York: Academic Press, pp. 194–222.

Overpeck, J.T., Webb, R.S. and Webb, T. III (1992) Mapping eastern North American vegetation change of the past 18 ka: no analogs and the future. *Geology*, **20**, 1071–1074.

Owen, D.F. and Wiegert, R.G. (1976) Do consumers maximize plant fitness? *Oikos*, **27**, 488–492.

Pagel, M.D. (1999) Inferring the historical patterns of biological evolution. *Nature*, **401**, 877–884.

Paige, K.N. (1992) Overcompensation in response to mammalian herbivory: from mutualistic to antagonistic interactions. *Ecology*, **73**, 2076–2085.

Painter, R.H. (1958) Resistance of plants to insects. *Annual Review of Entomology*, **3**, 267–290.

Palumbi, S.R. (1985) Spatial variation in an algal–sponge commensalism and the evolution of ecological interactions. *American Naturalist*, **126**, 267–274.

Palumbi, S.R. (1994) Genetic divergence, reproductive isolation, and marine speciation. *Annual Review of Ecology and Systematics*, **25**, 547–572.

Parker, I.M. (1997) Pollinator limitation of *Cytisus scoparius* (Scotch broom), an invasive exotic shrub. *Ecology*, **78**, 1457–1470.

Parker, M.A. (1993) Constraints on the evolution of disease resistance in an annual legume. *Heredity*, **71**, 290–294.

Parker, M.A. (1999) Mutualism in metapopulations of legumes and rhizobia. *American Naturalist*, **153**, S48–S60.

Parra, R. (1978) Comparison of foregut and hindgut fermentation in herbivores. In *The Ecology of Arboreal Folivores*, ed. G.G. Montgomery. Washington, DC: Smithsonian Institution Press, pp. 205–229.

Pastor, J., Dewey, B., Moen, R., Mladenoff, D.J., White, M. and Cohen, Y. (1998) Spatial patterns in the moose-forest-soil ecosystem on Isle Royale, Michigan, USA. *Ecological Applications*, **8**, 411–424.

Pastor, J., Dewey, B., Naiman, R.J., McInnes, P.F. and Cohen, Y. (1993) Moose browsing and soil fertility in the boreal forests of Isle Royale National Park. *Ecology*, **74**, 467–480.

Pastor, J., Moen, R. and Cohen, Y. (1997) Spatial heterogeneities, carrying capacity, and feedbacks in animal-landscape interactions. *Journal of Mammalogy*, **78**, 1040–1052.

Paton, T.R., Humphreys, G.S. and Mitchell, P.B. (1995) *Soils: A New Global View*. New Haven: Yale University Press.

Peakall, R. (1989) The unique pollination of *Leporella fimbriata* (Orchidaceae): pollination by pseudocopulating male ants (*Myrmecia urens*, Formicidae). *Plant Systematics and Evolution*, **167**, 137–148.

Peakall, R. and Beattie, A.J. (1989) Pollination of the orchid *Microtis parviflora* R. Br. by flightless worker ants. *Functional Ecology*, **3**, 515–522.

Peakall, R., Handel, S.N. and Beattie, A.J. (1991) The evidence for, and importance of, ant pollination. In *Ant–Plant Interactions*, ed. C.R. Huxley and D.F. Cutler. Oxford: Oxford University Press, pp. 423–429.

Peigler, R.S. (1986) Worldwide predilection of resiniferous host-plants by three unrelated groups of moths in the genera *Actias*, *Citheronia* (Saturniidae) and the subfamily Euteliinae (Noctuidae). *Tyô to Ga*, **37**, 45–50.

Pellmyr, O. (1986) Three pollination morphs in *Cimicifuga simplex*: incipient speciation due to inferiority in competition. *Oecologia*, **78**, 304–307.

Pellmyr, O. (1988) Bumblebees (Hymenoptera: Apidae) assess pollen availability in *Anemonopsis macrophylla* (Ranunculaceae) through floral shape. *Annals of the Entomological Society of America*, **81**, 792–797.

Pellmyr, O. (1989) The cost of mutualism: interactions between *Trollius europaeus* and its pollinating parasites. *Oecologia*, **78**, 53–59.

Pellmyr, O. (1992) The phylogeny of a mutualism: evolution and coadaptation between *Trollius* and its seed-parasitic pollinators. *Biological Journal of the Linnean Society*, **47**, 337–365.

Pellmyr, O. (1999) Systematic revision of the yucca moths in the *Tegeticula yucasella* complex (Lepidoptera: Prodoxidae) north of Mexico. *Systematic Entomology*, **24**, 243–270.

Pellmyr, O. and Huth, C.J. (1994) Evolutionary stability of mutualism between yuccas and yucca moths. *Nature*, **372**, 257–260.

Pellmyr, O. and Leebens-Mack, J. (2000) Reversal of mutualism as a mechanism for adaptive radiation in yucca moths. *American Naturalist*, **156**, S62–S76.

Pellmyr, O. and Thien, L.B. (1986) Insect reproduction and floral fragrances: keys to the evolution of the angiosperms? *Taxon*, **35**, 76–85.

Pellmyr, O., Leebens-Mack, J. and Huth, C.J. (1996a) Non-mutualistic yucca moths and their evolutionary consequences. *Nature*, **380**, 155–156.

Pellmyr, O., Thompson, J.N., Brown, J. and Harrison, R.G. (1996b) Evolution of pollination and mutualism in the yucca moth lineage. *American Naturalist*, **148**, 827–847.

Pemberton, R.W. (1992) Fossil extrafloral nectaries, evidence for the ant-guard antiherbivore defense in an Oligocene *Populus*. *American Journal of Botany*, **79**, 1242–1246.

Pemberton, R.W. (1998) The occurrence and abundance of plants with extrafloral nectaries, the basis for antiherbivore defensive mutualisms, along a latitudinal gradient in east Asia. *Journal of Biogeography*, **25**, 661–668.

Persson, I.L., Danell, K. and Bergström, R. (2000) Disturbance by large herbivores in boreal forests with special reference to moose. *Annales Zoologici Fennici*, **37**, 251–263.

Peters, R.H. (1983) *The Ecological Implications of Body Size*. Cambridge: Cambridge University Press.

Pettersson, M.W. (1991) Pollination by a guild of fluctuating moth populations: option for unspecialization in *Silene vulgaris*. *Journal of Ecology*, **79**, 591–604.

Picker, M.D. and Midgley, J.J. (1996) Pollination by monkey beetles (Coleoptera: Scarabaeidae: Hopliini): flower and colour preferences. *African Entomology*, **4**, 7–14.

Picó, F.X. and Retana, J. (2000) Temporal variation in the female components of reproductive success over the extended flowering season of a Mediterranean perennial herb. *Oikos*, **89**, 485–492.

Pico, B., Diez, M.J. and Nuez, F. (1998) Evaluation of whitefly-mediated inoculation techniques to screen *Lycopersicon esculentum* and wild relatives for resistance to tomato yellow leaf curl virus. *Euphytica*, **101**, 259–271.

Pilson, D. (1992) Aphid distribution and the evolution of goldenrod resistance. *Evolution*, **46**, 1358–1372.

Pilson, D. (1999) Plant hybrid zones and insect host range expansion. *Ecology*, **80**, 407–415.

Pitelka, F.A. (1964) The nutrient-recovery hypothesis for arctic microtine cycles. 1. Introduction. In *Grazing in Terrestrial and Marine Environments*, ed. D. Crisp. Oxford: Blackwell, pp. 55–56.

Pizo, M.A. and Oliveira, P.S. (1998) Interaction between ants and seeds of a nonmyrmecochorous neotropical tree, *Cabralea canjerana* (Meliaceae), in the Atlantic forest of southeast Brazil. *American Journal of Botany*, **85**, 669–674.

Poinar, G.O., Jr (1992) *Life in Amber*. Stanford, CA: Stanford University Press.

Polis, G.A. and Strong, D.R. (1996) Food web complexity and community dynamics. *American Naturalist*, **147**, 813–846.

Ponomarenko, A.G. (1996) Evolution of continental aquatic ecosystems. *Paleontological Journal*, **30**, 705–709.

Portnoy, S. and Willson, M.F. (1993) Seed dispersal curves: behavior of the tail of the distribution. *Evolutionary Ecology*, **7**, 25–44.

Poulin, B., Wright, S.J., Lefebvre, G. and Calderón, O. (1999) Interspecific synchrony and asynchrony in the fruiting phenologies of congeneric bird-dispersed plants in Panama. *Journal of Tropical Ecology*, **15**, 213–227.

Powell, J.A., Mitter, C. and Farrell, B. (1999) Evolution of larval food preferences in Lepidoptera. In *Lepidoptera, Moths and Butterflies*, vol. 1: *Evolution, Systematics, and Biogeography*, ed. N.P. Kristensen. *Handbuch der Zoologie*, vol. 4. Berlin: Walter de Gruyter, pp. 403–422.

Price, P.W. (1980) *Evolutionary Biology of Parasites*, Princeton, NJ: Princeton University Press.

Price, P.W. (1981) Semiochemicals in evolutionary time. In *Semiochemicals: Their Role in Pest Control*, ed. D.A. Nordlund, R.L. Jones and W.J. Lewis. New York: Wiley, pp. 251–279.

Price, P.W. (1991a) The web of life: development over 3.8 billion years of trophic relationships. In *Symbiosis as a Source of Evolutionary Innovation: Speciation and Morphogenesis*, ed. L. Margulis and R. Fester. Cambridge, MA: MIT Press, pp. 262–272.

Price, P.W. (1991b) Plant vigor hypothesis and herbivore attack. *Oikos*, **62**, 244–251.

Price, P.W. (1994) Phylogenetic constraints, adaptive syndromes, and emergent properties: from individuals to population dynamics. *Researches in Population Ecology*, **36**, 3–14.

Price, P.W. (1996) *Biological Evolution*. Philadelphia: Saunders.

Price, P.W. (1997) *Insect Ecology*, 3rd edn. New York: Wiley.

Price, P.W., Cobb, N., Craig, T.P. et al. (1990) Insect herbivore population dynamics on trees and shrubs: new approaches relevant to latent and eruptive species and life table development. In *Insect–Plant Interactions*, ed. E.A. Bernays, vol. 2. Boca Raton, FL: CRC Press, pp. 1–38.

Price, P.W., Craig, T.P. and Roininen, H. (1995) Working toward theory on galling sawfly population dynamics. In *Population Dynamics: New Approaches and Synthesis*, ed. N. Cappuccino and P.W. Price. San Diego, CA: Academic Press, pp. 321–338.

Primack, R.B. (1987) Relationships among flowers, fruits, and seeds. *Annual Review of Ecology and Systematics*, **18**, 409–430.

Proctor, M., Yeo, P. and Lack, A. (1996) *The Natural History of Pollination*. Portland, OR: Timber Press.

Provenza, F.D. and Cincotta, R.P. (1993) Foraging as a self-organizational learning process: accepting adaptability at the expense of predictability. In *Diet Selection: An Interdisciplinary Approach to Foraging Behaviour*, ed. R.N. Hughes. Oxford: Blackwell, pp. 78–101.

Provenza, F.D., Pfister, J.A. and Cheney, C.D. (1992) Mechanisms of learning in diet selection with reference to phytotoxicosis in herbivores. *Journal of Range Management*, **45**, 36–45.

Pulliainen, E., Helle, P. and Tunkkari, P. (1981) Adaptive radiation of the digestive system, heart and wings of *Turdus pilaris, Bombycilla garrulus, Sturnus vulgaris, Pyrrhula pyrrhula, Pinicola enucleator* and *Loxia pytyopsittacus. Ornis Fennica*, **58**, 21–28.

Pulliam, H.R. and Dunning, J.B. (1987) The influence of food supply on local density and diversity of sparrows. *Ecology*, **68**, 1009–1014.

Puterbaugh, M.N. (1998) The roles of ants as flower visitors: experimental analysis in three alpine plant species. *Oikos*, **83**, 36–46.

Pyke, G.H. (1982) Local geographic distributions of bumblebees near Crested Butte, Colorado: competition and community structure. *Ecology*, **63**, 555–573.

Qiu, Y.-L., Lee, J., Bernasconi-Quadroni, F. et al. (1999) The earliest angiosperms: evidence from mitochondrial, plastid and nuclear genomes. *Nature*, **402**, 404–407.

Quinlan, R.J. and Cherrett, J.M. (1979) The role of fungus in the diet of the leaf-cutting ant *Atta cephalotes* (L.). *Ecological Entomology*, **4**, 151–160.

Raguso, R.A. (2001) Floral scent, olfaction, and scent-driven foraging behavior. In *Cognitive Ecology of Pollination: Animal Behaviour and Floral Evolution*, ed. J.D. Thomson and L. Chittka. Cambridge: Cambridge University Press, 83–105.

Raguso, R.A. and Pichersky, E. (1999) A day in the life of a linalool molecule: chemical communication in a plant-pollinator system. Part 1: Linalool biosynthesis in flowering plants. *Plant Species Biology*, **14**, 95–120.

Ramsay, J. and Schemske, D.W. (1998) Pathways, mechanisms, and rates of polyploid formation in flowering plants. *Annual Review of Ecology and Systematics*, **29**, 467–501.

Ramsey, M. (1995) Ant pollination of the perennial herb *Blandfordia grandiflora. Oikos*, **74**, 265–272.

Raup, D.M. (1972) Taxonomic diversity during the Phanerozoic. *Science*, **177**, 1065–1071.

Raup, D.M. and Sepkoski, J.J., Jr (1982) Mass extinctions in the marine fossil record. *Science*, **215**, 1501–1503.

Reader, R.J. (1993) Control of seedling emergence by ground cover and seed predation in relation to seed size for some old-field species. *Journal of Ecology*, **81**, 169–175.

Reddy, N.R., Pierson, M.D., Sathe, S.K. and Salunkhe, D.K. (1989) *Phytates in Cereals and Legumes*. Boca Raton, FL: CRC Press.

Ree, R.H. and Donoghue, M.J. (1999) Inferring rates of change in flower symmetry in asterid angiosperms. *Systematic Biology*, **48**, 633–641.

Regal, P.J. (1977) Ecology and evolution of flowering plant dominance. *Science*, **196**, 622–629.

Reinthal, P.N. and Meyer, A. (1997) Molecular phylogenetic tests of speciation models in Lake Malawi cichlid fishes. In *Molecular Evolution and Adaptive Radiation*, ed. T.J. Givnish and K.J. Sytsma. Cambridge: Cambridge University Press, pp. 375–390.

Remsen, J.V., Hyde, M.A. and Chapman, A. (1993) The diets of Neotropical trogons, motmots, barbets and toucans. *Condor*, **95**, 178–192.

Rensberger, J.M. (1986) Early chewing mechanisms in mammalian herbivores. *Paleobiology*, **12**, 474–494.

Rey, P.J. (1995) Spatio-temporal variation in fruit and frugivorous bird abundance in olive orchards. *Ecology*, **76**, 1625–1635.

Rey, P.J., Gutiérrez, J.E., Alcántara, J. and Valera, F. (1997) Fruit size in wild olives: implications for avian seed dispersal. *Functional Ecology*, **11**, 611–618.

Ricklefs, R.E. and Schluter, D. (eds) (1993) *Species Diversity in Ecological Communities*. Chicago: University of Chicago Press.

Rickson, F.R. (1971) Glycogen plastids in Mullerian body cells of *Cecropia peltata*, a higher green plant. *Science*, **173**, 344–347.

Rickson, F.R. (1979) Absorption of animal tissue breakdown products into a plant stem — the feeding of plants by ants. *American Journal of Botany*, **66**, 87–90.

Rico-Gray, V., Barber, J.T., Thien, L.B., Ellgaard, E.G. and Toney, J.J. (1989) An unusual animal–plant interaction: feeding of *Schomburgkia tibicinis* (Orchidaceae) by ants. *American Journal of Botany*, **76**, 603–608.

Richardson, K.C. and Wooller, R.D. (1988) The alimentary tract of a specialist frugivore, the mistletoebird, *Dicaeum hirundinaceum*, in relation to its diet. *Australian Journal of Zoology*, **36**, 373–382.

Ridley, H.N. (1930) *The Dispersal of Plants throughout the World*. Ashford: Reeve.

Ridley, M. (1996) *Evolution*, 2nd edn. Cambridge, MA: Blackwell.

Ridley, P.S., Howse, P.E. and Jackson, C.W. (1996) Control of the behaviour of leaf-cutting ants by their 'symbiotic' fungus. *Experientia*, **52**, 631–635.

Rigaud, T. (1999) Further *Wolbachia* endosymbiont diversity: a tree hiding in the forest? *Trends in Ecology and Evolution*, **14**, 212–213.

Risch, S.J. (1981) Insect herbivore abundance in tropical monocultures and polycultures: an experimental test of two hypotheses. *Ecology*, **62**, 1325–1340.

Risch, S.J. and Carroll, C.R. (1986) Effects of seed predation by a tropical ant on competition among weeds. *Ecology*, **67**, 1319–1327.

Risch, S.J. and Rickson, F.R. (1981) Mutualism in which ants must be present before plants produce food bodies. *Nature*, **291**, 149–150.

Risley, L.S. and Crossley, V. (1993) Contribution of herbivore-caused greenfall to litterfall nitrogen flux in several southern Appalachian forested watersheds. *American Midland Naturalist*, **129**, 67–74.

Rissing, S.W. (1981) Foraging specializations of individual seed-harvester ants. *Behavioral Ecology and Sociobiology*, **9**, 149–152.

Ritchie, M.E. (1990) Optimal foraging and fitness in Columbian ground squirrels. *Oecologia*, **82**, 56–67.

Ritchie, M.E. and Olff, H. (1999) Herbivore diversity and plant dynamics: compensatory and additive effects. In *Herbivores: Between Plants and Predators*, ed. H. Olff, V.K. Brown and R.H. Drent. Oxford: Blackwell, pp. 175–204.

Ritchie, M.E., Tilman, D. and Knops, J.M.H. (1998) Herbivore effects on plant and nitrogen dynamics in oak savanna. *Ecology*, **79**, 165–177.

Robbins, C.T. (1993) *Wildlife Feeding and Nutrition*, 2nd edn. San Diego, CA: Academic Press.

Robbins, C.T., Spalinger, D.E. and van Hooven, W. (1995) Adaptation of ruminants to browse and grass diets: are anatomical-based browser-grazer interpretations valid? *Oecologia*, **103**, 208–213.

Roderick, G.K. (1997) Herbivorous insects and the Hawaiian silversword alliance: coevolution or cospeciation? *Pacific Science*, **51**, 440–449.

Roderick, G.K. and Metz, E.C. 1997. Biodiversity of planthoppers (Hemiptera: Delphacidae) on the Hawaiian silversword alliance: effects of host plant phylogeny and hybridisation. *Memoirs of the Museum of Victoria*, **56**, 393–399.

Rohdendorf, B.B. and Rasnitsyn, A.P. (1980) Historical development of the class Insecta. *Transactions of the Paleontological Institute*, **85**, 1–258 [in Russian].

Roininen, H., Price, P.W. and Tahvanainen, J. (1996) Bottom-up and top-down influences in the trophic system of a willow, a galling sawfly, parasitoids and inquilines. *Oikos*, **77**, 44–50.

Room, P.M., Harley, K.L.S., Forno, I.W. and Sands, P.P.D. (1981) Successsful biological control of the floating weed, *Salvinia*. *Nature*, **294**, 78–80.

Root, R.B. (1973) Organization of a plant arthropod association in simple and diverse habitats: the fauna of collards (*Brassica oleracea*). *Ecological Monographs*, **43**, 95–124.

Rose, H.A. (1985) The relationship between feeding specialization and host plants to aldrin epoxidase activities of midgut homogenates in larval Lepidoptera. *Ecological Entomology*, **10**, 455–467.

Rosenthal, G.A. & Bell, E.A. (1979) Naturally occurring, toxic nonprotein amino acids. In *Herbivores: Their Interactions with Secondary Plant Metabolites*, ed. G.A. Rosenthal and D.H. Janzen. Orlando, FL: Academic Press, pp. 353–385.

Rosenthal, G.A. and Berenbaum, M.R. (eds) (1991) *Herbivores: Their Interactions with Secondary Plant Metabolites*. 2nd edn. New York: Academic Press.

Rossi, M., Goggin, F.L., Miligan, S.B., Kaloshian, I., Ullman, D.E. and Williamson, V.M. (1998) The nematode resistance gene *Mi* of tomato confers resistance against the potato aphid. *Proceedings of the National Academy of Sciences USA*, **95**, 9750–9754.

Rossow, L.J., Bryant, J.P. and Kielland, K. (1997) Effects of aboveground browsing by mammals on mycorrhizal infection in an early successional taiga ecosystem. *Oecologia*, **110**, 94–98.

Rowell-Rahier, M. and Pasteels, J.M. (1992) Third trophic level influences of plant allelochemicals. In: *Herbivores: Their Interactions with Secondary Plant Metabolites*, ed. G.A. Rosenthal and M.R. Berenbaum. New York: Academic Press, pp. 243–277.

Roy, B.A. (1994) The effects of pathogen-induced pseudoflowers and buttercups on each others insect visitation. *Ecology*, **75**, 352–358.

Roy, B.A. and Bierzychudek, P. (1993) The potential for rust infection to cause natural selection in apomictic *Arabis holboellii* (Brassicaceae). *Oecologia*, **95**, 533–541.

Rozefelds, A.C. (1988) Lepidoptera mines in *Pachypteris* leaves (Corystospermaceae: Pteridospermatophyta) from the Upper Jurassic/Lower Cretaceous Battle Camp formation, North Queensland. *Proceedings of the Royal Society of Queensland*, **99**, 77–81.

Ruhren, S. and Dudash, M.R. (1996) Consequences of the timing of seed release of *Erythronium americanum* (Liliaceae), a deciduous forest myrmecochore. *American Journal of Botany*, **83**, 633–640.

Sadras, V.O. (1996) Cotton compensatory growth after loss of reproductive organs as affected by availability of resources and duration of recovery period. *Oecologia*, **106**, 432–439.

Saether, B.-E. (1985) Annual variation in carcass weight of Norwegian moose in relation to climate along a latitudinal gradient. *Journal of Wildlife Management*, **49**, 977–983.

Sallabanks, R. (1993) Hierarchical mechanisms of fruit selection by an avian frugivore. *Ecology*, **74**, 1326–1336.

Sallabanks, R. and Courtney, S.P. (1992) Frugivory, seed predation, and insect–vertebrate interactions. *Annual Review of Ecology and Systematics*, **37**, 377–400.

Samson, D.A., Philippi, T.E. and Davidson, D.W. (1992) Granivory and competition as determinants of annual plant diversity in the Chihuahuan desert. *Oikos*, **65**, 61–80.

Sanderson, M.J. and Donoghue, M.J. (1994) Shifts in diversification rate with the origin of angiosperms. *Science*, **264**, 1590–1593.

Sargent, S. (1995) Seed fate in a tropical mistletoe: the importance of host twig size. *Functional Ecology*, **9**, 197–204.

Sarukhan, J. (1986) Studies on the demography of tropical trees. In *Tropical Trees as Living Systems*, ed. P.B. Tomlinson and M.H. Zimmerman. Cambridge: Cambridge University Press, pp. 163–184.

Sazima, I., Buzato, S. and Sazima, M. (1995) The saw-billed hermit *Ramphodon naevius* and its flowers in southeastern Brazil. *Journal für Ornithologie*, **136**, 195–206.

Scarpati, M.L., Scalzo, R.L. and Vita, G. (1993) *Olea europaea* volatiles attractive and repellent to the olive fruit fly (*Dacus oleae*, Gmelin). *Journal of Chemical Ecology*, **19**, 881–891.

Schaal, S. and Ziegler, W. (eds) (1992) *Messel: An Insight into the History of Life and of the Earth*. Oxford: Oxford University Press.

Schemske, D.W. and Bradshaw, H.D. (1999) Pollinator preference and the evolution of floral traits in monkeyflowers. *Proceedings of the National Academy of Sciences USA*, **96**, 11910–11915.

Schiestl, F.P., Ayasse, M., Paulus, H.F. et al. (1999) Orchid pollination by sexual swindle. *Nature*, **399**, 421–422.

Schluter, D. and Repasky, R.R. (1991) Worldwide limitation of finch densities by food and other factors. *Ecology*, **72**, 1763–1774.

Schmidt-Adam, G., Young, A.G. and Murray, B.G. (2000) Low outcrossing rates and shift in pollinators in New Zealand pohutukawa (*Metrosideros excelsa*: Myrtaceae). *American Journal of Botany*, **87**, 1265–1271.

Schmitz, O.J. (1993) Trophic exploitation in grassland food chains: simple models and a field experiment. *Oecologia*, **93**, 327–335.

Schoen, D.J., Johnston, M.O., L'Heureux, A. and Marsolais, J.V. (1997) Evolutionary history of the mating system in *Amsinckia* (Boraginaceae). *Evolution*, **51**, 1090–1099.

Schoener, T.W. (1986) Mechanistic approaches to community ecology: a new reductionism? *American Zoologist*, **26**, 81–106.

Scholes, R.J. and Walker, B.H. (1993) *An African Savanna: Synthesis of the Nylsvley Study*. Cambridge: Cambridge University Press.

Scholtz, C.H. and Chown, S.L. (1995) The evolution of habitat use and diet in the Scarabaeoidea: a phylogenetic approach. In *Biology, Phylogeny, and Classification of Coleoptera: Papers Celebrating the 80th Birthday of Roy A. Crowson*, ed. J. Pakaluk and S.A. Lipiski. Warsaw: Muzeum i Instytut Zoologii PAN, pp. 355–374.

Schoonhoven, L.M., Jermy, T. and van Loon, J.J.A. (1998) *Insect–Plant Biology: From Physiology to Evolution*. London: Chapman & Hall.

Schuler, M.A. (1996) The role of cytochrome P450 monooxygenases in plant–insect interactions. *Plant Physiology*, **112**, 1411–1419.

Schuler, T.H., Martinez-Torres, D., Thompson, A.J. et al. (1998) Toxicological, electrophysiological, and molecular characterisation of knockdown resistance to pyrethroid insecticides in the diamondback moth, *Plutella xylostella* (L.). *Pesticide Biochemistry and Physiology*, **59**, 169–182.

Schultz, J.C. (1989) Tannin–insect interactions. In *Chemistry and Significance of Condensed Tannins*, ed. R.W. Hemingway and J.J. Karchesy. New York: Plenum Press, pp. 417–433.

Schupp, E.W. and Feener, D.H., Jr (1991) Phylogeny, lifeform, and habitat dependence of ant-defended plants in a Panamanian forest. In *Ant–Plant Interactions*, ed. C.R. Huxley and D.F. Cutler. Oxford: Oxford University Press, pp. 175–197.

Schupp, E.W., Howe, H.F., Augspurger, C.K. and Levey, D.J. (1989) Arrival and survival in tropical treefall gaps. *Ecology*, **70**, 562–564.

Schuster, J.C. (1974) Saltatorial Orthoptera as common visitors to tropical flowers. *Biotropica*, **6**, 138–140.

Scott, A.C. and Taylor, T.N. (1983) Plant/animal interactions during the Upper Carboniferous. *Botanical Review*, **49**, 259–307.

Scott, A.C., Stephenson, J. and Collinson, M.E. (1994) The fossil record of leaves with galls. In *Plant Galls—Organisms, Interactions, Populations*, ed. M.A.J. Williams. *Systematics Association Special Volume*, **49**, 447–470. Oxford: Clarendon Press.

Scriber, J.M. and Feeny, P. (1979) Growth of herbivorous caterpillars in relation to feeding specialization and to the growth form of the their food plants. *Ecology*, **60**, 829–850.

Segraves, K.A. and Thompson, J.N. (1999) Plant polyploidy and pollination: floral traits and insect visits to diploid and tetraploid *Heuchera grossulariifolia*. *Evolution*, **53**, 1114–1127.

Segraves, K.A., Thompson, J.N., Soltis, P.S. and Soltis, D.E. (1999) Multiple origins of polyploidy and the geographic structure of *Heuchera grossulariifolia*. *Molecular Ecology*, **8**, 253–262.

Seiber, J.N., Tuskes, P.M., Brower, L.P. and Nelson, C.J. (1980) Pharmacodynamics of some individual milkweed cardenolides fed to larvae of the monarch butterfly. *Journal of Chemical Ecology*, **6**, 321–339.

Seilacher, A. (1990) Taphonomy of fossil Lagerstätten: overview. In *Paleobiology: A Synthesis*, ed. D.E.G. Briggs and P.R. Crowther. Oxford: Blackwell, pp. 266–270.

Seldal, T., Andersen, K.-J. and Högstedt, G. (1994) Grazing-induced proteinase inhibitors: a possible cause for lemming population cycles. *Oikos*, **70**, 3–11.

Senft, R.L., Coughenour, M.B., Bailey, D.W., Rittenhouse, L.R., Sala, O.E. and Swift, D.M. (1987) Large herbivore foraging and ecological hierarchies. *Bioscience*, **37**, 789–799.

Sereno, P.C. (1999) The evolution of dinosaurs. *Science*, **284**, 2137–2147.

Service, R.F. (1997) Microbiologists explore life's rich, hidden kingdoms. *Science*, **275**, 1740–1742.

Setsuda, K. (1995) Ecological study of beetles inhabiting *Cryptoporus volvatus* (Peck) Shear (II). Relationship between development of the basidiocarps and life cycles of five major species of beetle inhabiting the fungus, with discussion of the spore dispersal. *Japanese Journal of Entomology*, **63**, 609–620.

Seuter, F. (1970) Ist eine endozoische Verbreitung der Tollkirsche durch Amsel und Star möglich? *Zoologische Jahrbücher Physiologie*, **75**, 342–359.

Shattuck, S.O. (1999) *Australian Ants: Their Biology and Identification*. Collingwood: CSIRO.

Shcherbakov, D.E., Lukashevich, E.D. and Blagoderov, V.A. (1995) Triassic Diptera and initial radiation of the order. *Dipterological Research*, **6**, 75–115.

Shear, W.A. and Kukalová-Peck, J. (1990) The ecology of Paleozoic terrestrial arthropods: the fossil evidence. *Canadian Journal of Zoology*, **68**, 1807–1834.

Sherbrooke, W.C. (1976) Differential acceptance of toxic jojoba seed (*Simonsia chinesis*) by four Sonoran Desert heteromyid rodents. *Ecology*, **57**, 596–602.

Shipley, L.A., Gross, J.E., Spalinger, D.E., Hobbs, N.T. and Wunder, B. (1994) The scaling of intake rate in mammalian herbivores. *American Naturalist*, **143**, 1055–1082.

Shorthouse, J.D. and Rohfritsch, O. (eds) (1992) *Biology of Insect-Induced Galls*. New York: Oxford University Press.

Shykoff, J.A., Bucheli, E. and Kaltz, O. (1997) Anther smut disease in *Dianthus silvester* (Caryophyllaceae): natural selection on floral traits. *Evolution*, **51**, 383–392.

Sibley, C.G. and Ahlquist, J.E. (1990) *Phylogeny and Classification of Birds*. New Haven: Yale University Press.

Silvertown, J. (1989) The paradox of seed size and adaptation. *Trends in Ecology and Evolution*, **4**, 24–26.

Silvertown, J.W. (1980) The evolutionary ecology of mast seeding in trees. *Biological Journal of the Linnean Society*, **14**, 235–250.

Simberloff, D. and Stiling, P. (1987) Larval dispersion and survivorship in a leaf-mining moth. *Ecology*, **68**, 1647–1657.

Simms, E.L. and Rausher, M.D. (1993) Patterns of selection on phytophage resistance in *Ipomoea purpurea*. *Evolution*, **47**, 970–976.

Simpson, B.B. and Neff, J.L. (1983) Evolution and diversity of floral rewards. In *Handbook of Experimental Pollination Biology*, ed. C.E. Jones and R.J. Little. New York: Van Nostrand Reinhold, pp. 142–159.

Singer, M.C. and Thomas, C.D. (1996) Evolutionary responses of a butterfly metapopulation to human- and climate-caused environmental variation. *American Naturalist*, **148**, S9–S39.

Sirur, G.M. and Barlow, C.A. (1984) Effects of pea aphids (*Acyrthosiphon pisum*; Homoptera: Aphididae) on the nitrogen-fixing activity of bacteria in the root nodules of pea plants (*Pisum sativum*). *Journal of Economic Entomology*, **77**, 606–611.

Slansky, F. and Rodriguez, J.G. (1987) *Nutritional Ecology of Insects, Mites, Spiders, and Related Invertebrates*. New York: Wiley Interscience.

Smedley, S.R. and Eisner, T. (1996) Sodium: a male moth's gift to its offspring. *Proceedings of the National Academy of Sciences USA*, **93**, 809–813.

Smith, C.C. (1970) The coevolution of pine squirrels (*Tamiasciurus*) and conifers. *Ecological Monographs*, **40**, 349–371.

Smith, J.J. and Bush, G.L. (1997) Phylogeny of the genus *Rhagoletis* (Diptera: Tephritidae) inferred from DNA sequences of mitochondrial cytochrome oxidase II. *Molecular Phylogenetics and Evolution*, **7**, 33–43.

Smith, T.B., Freed, L.A., Lepson, J.K. and Carothers, J.H. (1995) Evolutionary consequences of extinctions in populations of a Hawaiian honeycreeper. *Conservation Biology*, **9**, 107–113.

Snow, B.K. and Snow, D.W. (1988) *Birds and Berries*. Waterhouses: T. & A.D. Poyser.

Snow, D.W. (1971) Evolutionary aspects of fruit-eating by birds. *Ibis*, **113**, 194–202.

Snyder, M.J., Walding, J.K. and Feyereisen, R. (1994) Metabolic fate of the allelochemical nicotine in the tobacco hornworm *Manduca sexta*. *Insect Biochemistry and Molecular Biology*, **24**, 837–846.

Solomon, B.P. (1983) Autoallelopathy in *Solanum carolinense*: reversible delayed germination. *American Midland Naturalist*, **110**, 412–418.

Soltis, D.E. and Soltis, P.S. (1993) Molecular data and the dynamic nature of polyploidy. *Critical Reviews in Plant Sciences*, **12**, 243–273.

Soltis, D.E. and Soltis, P.S. (1999) Polyploidy: recurrent formation and genome evolution. *Trends in Ecology and Evolution*, **14**, 348–352.

Soltis, P.S., Plunkett, G.M., Novak, S.J. and Soltis, D.E. (1995) Genetic variation in *Tragopogon* species: additional origins of the allotetraploids *T. mirus* and *T. miscellus* (Compositae). *American Journal of Botany*, **82**, 1329–1341.

Soltis, P.S., Soltis, D.E. and Chase, M.W. (1999) Angiosperm phylogeny inferred from multiple genes as a tool for comparative biology. *Nature*, **402**, 402–404.

Somerville, C. and Somerville, S. (1999) Plant functional genomics. *Science*, **285**, 380–383.

Sorensen, A.E. (1986) Seed dispersal by adhesion. *Annual Review of Ecology and Systematics*, **17**, 443–463.

Southwood, T.R.E. (1973) The insect/plant relationship — an evolutionary perspective. *Symposia of the Royal Entomological Society of London*, **6**, 3–30.

Souza-Stevaux, M.C., Negrelle, R.R.B. and Citadini-Zanette, V. (1994) Seed dispersal by the fish *Pterodoras granulosus* in the Paraná River Basin, Brazil. *Journal of Tropical Ecology*, **10**, 621–626.

Spencer, K.A. (1990) *Host Specialization in the World Agromyzidae* (Diptera). Dordrecht: Kluwer.

Sperber, I., Björnhag, G. and Ridderstråle, Y. (1983) Function of proximal colon in lemming and rat. *Swedish Journal of Agricultural Research*, **13**, 243–256.

Spiller, D.A. and Schoener, T.W. (1990) A terrestrial field experiment showing the impact of eliminating top predators on foliage damage. *Nature*, **347**, 469–472.

Spiller, D.A. and Schoener, T.W. (1994) Effects of top and intermediate predators in a terrestrial food web. *Ecology*, **75**, 182–196.

Sprengel, C.K. (1995) Discovery of the secret of nature in the structure and fertilization of flowers. In *Floral Biology*, ed. D.G. Lloyd and S.C.H. Barrett. New York: Chapman & Hall, pp. 3–43. [Translation by P. Haase of the introduction by C. K. Sprengel (1793) to *Das entdeckte Geheimniss der Natur im Bau und in der Befruchtung der Blumen. I.* Berlin: Vieweg.]

Srivastava, A.K. (1987) Lower Barakar flora of Raniganj Coalfield and insect/plant relationship. *Palaeobotanist*, **36**, 138–142.

Stanton, M.L., Palmer, T.M., Young, T.P., Evans, A. and Turner, M.L. (1999) Sterilization and canopy modification of a swollen thorn acacia tree by a plant-ant. *Nature*, **401**, 578–581.

Stanton, M.L., Snow, A.A., Handel, S.N. and Bereczky, J. (1989) The impact of a flower-color polymorphism on mating patterns in experimental populations of wild radish (*Raphanus raphanistrum* L.). *Evolution*, **43**, 335–346.

Steadman, D.W. (1995) Prehistoric extinctions of Pacific island birds: biodiversity meets zooarchaeology. *Science*, **267**, 1123–1131.

Steele, M.A., Knowles, T., Bridle, K. and Simms, E.L. (1993) Tannins and partial consumption of acorns — implications for dispersal of oaks by seed predators. *American Midland Naturalist*, **130**, 229–238.

Steinberg, P.D., Estes, J.A. and Winter, F.C. (1995) Evolutionary consequences of food chain length in kelp forest communities. *Proceedings of the National Academy of Sciences USA*, **92**, 8145–8148.

Steiner, K.E. and Whitehead, V.B. (1991) Oil flowers and oil bees: further evidence for pollinator adaptation. *Evolution*, **45**, 1493–1501.

Stelzl, M. (1991) Untersuchungen zu Nahrungsspektren mitteleuropäischer Neuropteren-Imagines (Neuropteroidea, Insecta). *Journal of Applied Entomology*, **111**, 469–477.

Stephens, D.W. and Krebs, J.R. (1986) *Foraging Theory*. Princeton, NJ: Princeton University Press.

Stephenson, J. (1992) *Evidence of Plant/Insect Interactions in the Late Cretaceous and Early Tertiary*. Ph.D. thesis, University of London.

Stevens, C.E. (1988) *Comparative Physiology of the Vertebrate Digestive System*. Cambridge: Cambridge University Press.

Stevens, G.C. (1989) The latitudinal gradient in geographical range: how so many species coexist in the tropics. *American Naturalist*, **133**, 240–256.

Stiles, E.W. (1982) Fruit flags: two hypotheses. *American Naturalist*, **120**, 500–509.

Stiles, E.W. (1989) Fruits, seed and dispersal agents. In *Plant–Animal Interactions*, ed. W.G. Abrahamson. New York: McGraw Hill, pp. 87–122.

Stiling, P. and Rossi, A.M. (1997) Experimental manipulations of top-down and bottom-up factors in a tri-trophic system. *Ecology*, **78**, 1602–1606.

Stout, M.J., Workman, K.V. and Duffey, S.S. (1996) Identity, spatial distribution, and variability of induced chemical responses in tomato plants. *Entomologia Experimentalis et Applicata*, **79**, 255–271.

Stowe, K.A. (1998) Experimental evolution of resistance in *Brassica rapa*: correlated response of tolerance in lines selected for glucosinolate content. *Evolution*, **52**, 703–712.

Stowe, M.K. (1988) Chemical mimicry. In *Chemical Mediation of Coevolution*, ed. K. Spencer. San Diego, CA: Academic Press, pp. 513–580.

Strauss, S.Y. (1987) Direct and indirect effects of host-plant fertilization on an insect community. *Ecology*, **68**, 1670–1678.

Strauss, S.Y. (1991) Direct, indirect, and cumulative effects of three native herbivores on a shared host plant. *Ecology*, **72**, 543–558.

Strauss, S.Y. (1994) Levels of herbivory and parasitism in host hybrid zones. *Trends in Ecology and Evolution*, **9**, 209–214.

Strauss, S.Y. (1997) Floral characters link herbivores, pollinators, and plant fitness. *Ecology*, **78**, 1640–1645.

Strauss, S.Y. and Agrawal, A.A. (1999) The ecology and evolution of plant tolerance to herbivory. *Trends in Ecology and Evolution*, **14**, 179–185.

Strauss, S.Y., Conner, J.K. and Rush, S.L. (1996) Foliar hebivory affects floral characters and plant attractiveness to pollinators: implications for male and female plant fitness. *American Naturalist*, **147**, 1098–1107.

Strong, D.R. (1992) Are trophic cascades all wet? Differentiation and donor control in species ecosystems. *Ecology*, **73**, 747–754.

Strong, D.R. (1999) Predator control in terrestrial ecosystems: the underground food chain of bush lupine. In *Herbivores: Between Plants and Predators*, ed. H. Olff, V.K. Brown and R.H. Drent. Oxford: Blackwell, pp. 577–602.

Strong, D.R., Lawton, J.H. and Southwood, T.R.E. (1984) *Insects on Plants: Community Patterns and Mechanisms*. Cambridge, MA: Harvard University Press.

Stuessy, T.F., Spooner, D.M. and Evans, K.A. (1986) Adaptive significance of ray corollas in *Helianthus grosseserratus* (Compositae). *American Midland Naturalist*, **115**, 191–197.

Sues, H.-D. and Reisz, R.R. (1998) Origins and early evolution of herbivory in tetrapods. *Trends in Ecology and Evolution*, **13**, 141–145.

Sukacheva, I.D. (1999) The Lower Cretaceous caddisfly (Trichoptera) case assemblages. *Proceedings of the First International Palaeoentomological Conference, Moscow* [1968]. Bratislava: AMBA Projects International, and Moscow: Paleontological Institute, pp. 163–165.

Summerhayes, V.S. (1941) The effect of voles (*Microtus agrestis*) on vegetation. *Journal of Ecology*, **29**, 14–48.

Summers, C.B. and Felton, G.W. (1996) Peritrophic envelope as a functional antioxidant. *Archives of Insect Biochemistry and Physiology*, **32**, 131–142.

Sun, C. and Moermond, T.C. (1997) Foraging ecology of three sympatric turacos in a montane forest in Rwanda. *Auk*, **114**, 396–404.

Suominen, O., Danell, K. and Bergström, R. (1999) Moose, trees and ground-living invertebrates: indirect interactions in Swedish pine forests. *Oikos*, **84**, 215–226.

Szentesi, A. and Jermy, T. (1995) Predispersal seed predation in leguminous species: seed morphology and bruchid distribution. *Oikos*, **73**, 23–32.

Taylor, T.N. and Millay, M.A. (1979) Pollination biology and reproduction in early seed plants. *Review of Palaeobotany and Palynology*, **27**, 329–355.

Temeles, E.J., Pan, I.L., Brennan, J.L. and Horwitt, J.N. (2000) Evidence for ecological causation of sexual dimorphism in a hummingbird. *Science*, **289**, 441–443.

Terborgh, J., Robinson, S.K., Parker, T.A., Munn, C.A. and Pierpont, N. (1990) Structure and organization of an Amazonian forest bird community. *Ecological Monographs*, **60**, 213–238.

Thaler, J.S. (1999) Jasmonate-inducible plant defences cause increased parasitism of herbivores. *Nature*, **399**, 686–688.

Thompson, J.N. (1982) *Interaction and Coevolution*. New York: Wiley.

Thompson, J.N. (1989) Concepts of coevolution. *Trends in Ecology and Evolution*, **4**, 179–183.

Thompson, J.N. (1994) *The Coevolutionary Process*. Chicago: University of Chicago Press.

Thompson, J.N. (1996) Conserving interaction biodiversity. In *The Ecological Basis of Conservation: Heterogeneity, Ecosystems, and Biodiversity*, ed. S.T.A. Pickett, R.S. Ostfeld, M. Shachak and G.E. Likens. New York: Chapman & Hall, pp. 285–293.

Thompson, J.N. (1998a) Coping with multiple enemies: 10 years of attack on *Lomatium dissectum* plants. *Ecology*, **79**, 2550–2554.

Thompson, J.N. (1998b) The evolution of diet breadth: monophagy and polyphagy in swallowtail butterflies. *Journal of Evolutionary Biology*, **11**, 563–578.

Thompson, J.N. (1998c) Rapid evolution as an ecological process. *Trends in Ecology and Evolution*, **13**, 329–332.

Thompson, J.N. (1999a) Specific hypotheses on the geographic mosaic of coevolution. *American Naturalist*, **153**, S1–S14.

Thompson, J.N. (1999b) The raw material for coevolution. *Oikos*, **84**, 5–16.

Thompson, J.N. (1999c) What we know and do not know about coevolution: insect herbivores and plants as a test case. In *Herbivores: Between Plants and Predators*, ed. H. Olff, V.K. Brown and R.H. Drent. Oxford: Blackwell, pp. 7–30.

Thompson, J.N. (1999d) The evolution of species interactions. *Science*, **284**, 2116–2118.

Thompson, J.N., Cunningham, B.M., Segraves, K.A., Althoff, D.M. and Wagner, D. (1997) Plant polyploidy and insect/plant interactions. *American Naturalist*, **150**, 730–743.

Thompson, R.S., Van Devender, T.R., Martin, P.S., Foppe, T. and Long, A. (1980) Shasta ground sloth (*Nothrotheriops shastense* Hoffstetter) at Shelter Cave, New Mexico: environment, diet and extinction. *Quaternary Research*, **14**, 360–376.

Thorp, R.W. (1979) Structural, behavioral, and physiological adaptations of bees (Apoidea) for collecting pollen. *Annals of the Missouri Botanical Garden*, **66**, 788–812.

Thulborn, R.A. (1991) Morphology, preservation and palaeobiological significance of dinosaur coprolites. *Palaeogeography, Palaeoclimatology, Palaeoecology*, **83**, 341–366.

Tibbets, T.M. and Faeth, S.H. (1999) *Neotyphodium* endophytes in grasses: deterrents or promoters of herbivory by leaf-cutting ants. *Oecologia*, **118**, 297–305.

Tiffney, B.H. (1986) Evolution of seed dispersal syndromes according to the fossil record. In *Seed Dispersal*, ed. D.R. Murray. North Ryde, New South Wales: Academic Press, pp. 273–305.

Tofts, R. and Silvertown, J. (2000) A phylogenetic approach to community assembly from a local species pool. *Proceedings of the Royal Society of London B*, **267**, 363–369.

Tolvanen, A., Laine, K., Pakonen, T., Saari, E. and Havas, P. (1993) Above-ground growth responses of the bilberry (*Vaccinium myrtillus* L.) to simulated herbivory. *Flora (Jena)*, **188**, 197–202.

Tollrian, R. and Harvell, C.D. (1999) *The Ecology and Evolution of Inducible Defenses*. Princeton, NJ: Princeton University Press.

Tomlinson, P.B., Braggins, J.E. and Rattenbury, J.A. (1991) Pollination drop in relation to cone morphology in Podocarpaceae: a novel reproductive mechanism. *American Journal of Botany*, **78**, 1289–1303.

Tovée, M.J. (1995) Ultra-violet photoreceptors in the animal kingdom: their distribution and function. *Trends in Ecology and Evolution*, **10**, 455–460.

Traveset, A. (1990) Postdispersal predation of *Acacia farnesiana* seeds by *Stator vachelliae* (Bruchidae) in Central America. *Oecologia*, **84**, 506–512.

Traveset, A. (1991) Pre-dispersal seed predation in Central American *Acacia farnesiana*: factors affecting the abundance of co-occurring bruchid beetles. *Oecologia*, **87**, 570–576.

Traveset, A. (1994) Cumulative effects on the reproductive output of *Pistacia terebinthus* (Anacardiaceae). *Oikos*, **71**, 152–162.

Traveset, A. (1998) Effects of seed passage through vertebrate frugivores' guts on germination: a review. *Perspectives in Plant Ecology, Evolution and Systematics*, **1**, 151–190.

Turchin, P., Taylor, A.D. and Reeve, J.D. (1999) Dynamical role of predators in population cycles of a forest insect: an experimental test. *Science*, **285**, 1068–1071.

Turgeon, J.J., Roques, A. and de Groot, P. (1994) Insect fauna of coniferous seed cones: diversity, host plant interactions, and management. *Annual Review of Ecology & Systematics*, **39**, 179–212.

Turlings, T.C.J. and Benrey, B. (1998) Effects of plant metabolites on the behavior and development of parasitic wasps. *Ecoscience*, **5**, 321–333.

Turlings, T.C.H., Tumlinson, J.H., Heath, RR., Proveaux, A.T. and Doolittle, R.E. (1991) Isolation and identification of allelochemicals that attract the larval parasitoid, *Cotesia marginiventris* (Cresson), to the microhabitat of one of its hosts. *Journal of Chemical Ecology*, **17**, 2235–2251.

Turnbull, L.A., Crawley, M.J. and Rees, M. (2000) Are plant populations seed limited? A review of seed sowing experiments. *Oikos*, **88**, 225–238.

Tutin, C.E.G., Ham, R.M., White, L.J.T. and Harrison, M.J.S. (1997) The primate community of the Lopé Reserve, Gabon: diets, responses to fruit scarcity, and effects on biomass. *American Journal of Primatology*, **42**, 1–24.

Valido, A. and Nogales, M. (1994) Frugivory and seed dispersal by

the lizard *Gallotia galloti* (Lacertidae) in a xeric habitat of the Canary Islands. *Oikos*, **70**, 403–411.

van Baalen, M. (1998) Coevolution of recovery ability and virulence. *Proceedings of the Royal Society of London B*, **265**, 317–325.

Van Dam, N.M. and Hare, J.D. (1998) Differences in distribution and performance of two sap-sucking herbivores on glandular and non-glandular *Datura wrightii*. *Ecological Entomology*, **23**, 22–32.

van der Meijden, E., Wijn, M. and Verkaar, H.J. (1988) Defense and regrowth, alternative plant strategies in the struggle against herbivores. *Oikos*, **51**, 355–363.

van der Pijl, L. (1969) *Principles of Dispersal in Higher Plants*. Berlin: Springer-Verlag.

van Loon, J.J.A., de Boer, J.G. and Dicke, M. (2000) Parasitoid–plant mutualism: parasitoid attack of herbivore increases plant reproduction. *Entomologia Experimentalis et Applicata*, **97**, 219–227.

van Schaik, C.P., Terborgh, J.W. and Wright, S.J. (1993) The phenology of tropical forests: adaptive significance and consequences for primary consumers. *Annual Review of Ecology and Systematics*, **24**, 353–377.

Van Soest, P.J. (1994) *Nutritional Ecology of the Ruminant*, 2nd edn. Ithaca, NY: Cornell University Press.

Van Soest, P.J. (1996) Allometry and ecology of feeding behavior and digestive capacity in herbivores: a review. *Zoo Biology*, **15**, 455–479.

Van Soest, P.J. and Jones, L.H.P. (1968) Effect of silica in forages upon digestibility. *Journal of Dairy Science*, **51**, 1644–1648.

Van Valen, L. (1973) A new evolutionary law. *Evolutionary Theory*, **1**, 1–30.

Van Zandt, P.A. and Mopper, S. (1998) A meta-analysis of adaptive deme formation in phytophagous insect populations. *American Naturalist*, **152**, 595–604.

Vander Wall, S.B. (1990) *Food Hoarding in Animals*. Chicago: University of Chicago Press.

Vander Wall, S.B. (1992) The role of animals in dispersing a 'wind-dispersed' pine. *Ecology*, **73**, 614–621.

Vander Wall, S.B. (1994) Seed fate pathways of antelope bitterbrush: dispersal by seed-caching yellow pine chipmunks. *Ecology*, **75**, 1911–1926.

Väre, H., Ohtonen, R. and Mikkola, K. (1996) The effect and extent of heavy grazing by reindeer in oligotrophic pine heaths in eastern Fennoscandia. *Ecography*, **19**, 245–253.

Vasconcelos, H.L. (1991) Mutualism between *Maieta guianensis* Aubl., a myrmecophytic melastome, and one of its ant inhabitants: ant protection against insect herbivores. *Oecologia*, **87**, 295–298.

Vasconcelos, H.L. and Cherrett, J.M. (1997) Leaf-cutting ants and early forest regeneration in central Amazonia: effects of herbivory on tree seedling establishment. *Journal of Tropical Ecology*, **13**, 357–370.

Vega, F.E., Dowd, P.F. and Bartelt, R.J. (1995) Dissemination of microbial agents using an autoinoculating device and several insect species as vectors. *Biological Control*, **5**, 545–552.

Verhoef, H.A., Prast, J.E. and Verweij, R.A. (1988) Relative importance of fungi and algae in the diet and nitrogen nutrition of *Orchesella cincta* (L.) and *Tomocerus minor* (Lubbock) (Collembola). *Functional Ecology*, **2**, 195–201.

Vermeij, G.J. (1987) *Evolution and Escalation: An Ecological History of Life*. Princeton, NJ: Princeton University Press.

Vilhelmsen, L. (1996) The preoral cavity of lower Hymenoptera (Insecta): comparative morphology and phylogenetic significance. *Zoologica Scripta*, **25**, 143–170.

von Helversen, D. and von Helversen, O. (1999) Acoustic guide in bat-pollinated flower. *Nature*, **398**, 759–760.

Vrieling, K., Saumitou-Laprade, P., Cuguen, J., van Dijk, H., de Jong, T.J. and Klinkhamer, P.G.L. (1999) Direct and indirect estimates of selfing rate in small and large individuals of the bumble-bee pollinated *Cynoglossum officinale* L (Boraginaceae). *Ecology Letters*, **2**, 331–337.

Waggoner, B.M. and Poteet, M.F. (1996) Unusual oak leaf galls from the Middle Miocene of northwestern Nevada. *Journal of Paleontology*, **70**, 1080–1084.

Wagner, D.W., Brown, M.J.F. and Gordon, D.M. (1997) Harvester ant nests, soil biota and soil chemistry. *Oecologia*, **112**, 232–236.

Wagner, L. and Funk, V.A. (eds) (1995) *Hawaiian Biogeography: Evolution on a Hot Spot Archipelago*. Washington, DC: Smithsonian Institution Press.

Waldorf, E. (1981) The utilization of pollen by a natural population of *Entomobrya socia*. *Revue d'Écologie et de Biologie du Sol*, **18**, 397–402.

Waser, N.M. and Price, M.V. (1983) Pollinator behaviour and natural selection for flower colour in *Delphinium nelsonii*. *Nature*, **302**, 422–424.

Watkinson, E.R., Lonsdale, W.M. and Andrew, M.H. (1989) Modelling the population dynamics of an annual plant *Sorghum intrans* in the wet-dry tropics. *Journal of Ecology*, **77**, 162–181.

Wcislo, W.T. and Cane, J.H. (1996) Floral resource utilization by solitary bees (Hymenoptera: Apoidea) and exploitation of their stored foods by natural enemies. *Annual Review of Entomology*, **41**, 257–280.

Weaver, L., McLoughlin, S. and Drinnan, A.N. (1997) Fossil woods from the Upper Permian Bainmedart Coal Measures, northern Prince Charles Mountains, East Antarctica. *Journal of Australian Geology and Geophysics*, **16**, 655–676.

Weckerly, F.W. and Kennedy, M.L. (1992) Examining hypotheses about feeding strategies of white-tailed deer. *Canadian Journal of Zoology*, **70**, 432–439.

Weintraub, J.D., Lawton, J.H. and Scoble, M.J. (1995) Lithinine moths on ferns: a phylogenetic study of insect-plant interactions. *Biological Journal of the Linnean Society*, **55**, 239–250.

Weishampel, D.B. and Norman, D.B. (1989) Vertebrate herbivory in the Mesozoic: jaws, plants, and evolutionary metrics. *Geological Society of America Special Paper*, **238**, 87–100.

Weiss, M.R. (1995) Floral color change: a widespread functional convergence. *American Journal of Botany*, **82**, 167–185.

Wellnhofer, P. (1991) *The Illustrated Encyclopedia of Pterosaurs*. London: Salamander Books.

Wenny, D.G. and Levey, D.J. (1998) Directed seed dispersal by bellbirds in a tropical cloud forest. *Proceedings of the National Academy of Sciences USA*, **95**, 6204–6207.

Westoby, M., Leishman, M. and Lord, J. (1996) Comparative ecology of seed size and dispersal. *Philosophical Transactions of the Royal Society of London B*, **351**, 1309–1318.

Westoby, M., Rice, B. and Howell, J. (1990) Seed size and plant growth form as factors in dispersal spectra. *Ecology*, **71**, 1307–1315.

Wetschnig, W. and Depisch, B. (1999) Pollination biology of *Welwitschia mirabilis* Hook. F. (Welwitschiaceae, Gnetopsida). *Phyton*, **39**, 167–183.

Wheelwright, N.T. (1983) Fruits and the ecology of Resplendent Quetzals. *Auk*, **100**, 286–301.

Wheelwright, N.T. (1985a) Competition for dispersers, and the timing of flowering and fruiting in a guild of tropical trees. *Oikos*, **44**, 465–477.

Wheelwright, N.T. (1985b) Fruit size, gape width, and the diets of fruit-eating birds. *Ecology*, **66**, 808–818.

Wheelwright, N.T. (1993) Fruit size in a tropical tree species: variation, preference by birds, and heritability. *Vegetatio*, **107/108**, 163–174.

Wheelwright, N.T. and Janson, C.H. (1985) Colors of fruit displays of bird-dispersed plants in two tropical forests. *American Naturalist*, **126**, 777–799.

Wheelwright, N.T. and Orians, G.H. (1982) Seed dispersal by animals: contrasts with pollen dispersal, problems of terminology, and constraints on coevolution. *American Naturalist*, **119**, 402–413.

Wheelwright, N.T., Haber, W.A., Murray, K.G. and Guindon, C. (1984) Tropical fruit-eating birds and their food plants: a survey of a Costa Rican lower montane forest. *Biotropica*, **16**, 173–192.

Whelan, C.J. and Willson, M.F. (1994) Fruit choice in migrating North American birds: field and aviary experiments. *Oikos*, **71**, 137–151.

Whelan, C.J., Schmidt, K.A., Steele, B.B., Quinn, W.J. and Dilger, S. (1998) Are bird-consumed fruits complementary resources? *Oikos*, **83**, 195–205.

Whitford, W.G., Schaeffer, D. and Wisdom, W. (1986) Soil movement by desert ants. *Southwestern Naturalist*, **31**, 273–274.

Whitford, W.G., Van Zee, J., Nash, M.S., Smith, W.E. and Herrick, J.E. (1999) Ants as indicators of exposure to environmental stressors in North American desert grasslands. *Environmental Monitoring and Assessment*, **54**, 143–171.

Whitham, T.G., Morrow, P.A. and Potts, B.M. (1991) Conservation of hybrid plants. *Science*, **254**, 779–780.

Whitham, T.G., Morrow, P.A. and Potts, B.M. (1994) Plant hybrid zones as centers of biodiversity: the herbivore community of two endemic Tasmanian eucalypts. *Oecologia*, **97**, 481–490.

Whitham, T.G., Martinsen, G.D., Floate, K.D., Dungey, H.S., Potts, B.M. and Keim, P. (1999) Plant hybrid zones affect biodiversity: tools for a genetic-based understanding of community stucture. *Ecology*, **80**, 416–428.

Whittaker, R.J. and Jones, S.H. (1994) The role of frugivorous bats and birds in the rebuilding of a tropical forest ecosystem, Krakatau, Indonesia. *Journal of Biogeography*, **21**, 245–258.

Wiegmann, B.M., Mitter, C. and Farrell, B. (1993) Diversification of carnivorous parasitic insects: extraordinary radiation or specialized dead end? *American Naturalist*, **142**, 737–754.

Wilcox, J.A. (1979) *Leaf Beetle Host Plants in Northeastern North America*. Marlton, NJ: World Natural History Publications.

Wilding, N., Collins, N.M., Hammond, P.M. and Webber, J.F. (eds) (1989) *Insect–Fungus Interactions*. London: Academic Press.

Wilf, P. and Labandeira, C.C. (1999) Response of plant-insect associations to Paleocene-Eocene warming. *Science*, **284**, 2153–2156.

Wilf, P., Labandeira, C.C., Kress, W.J. et al. (2000) Timing the radiations of leaf-beetles: hispines on gingers from latest Cretaceous to Recent. *Science*, **289**, 291–294.

Williams, R.T. (1974) Inter-species variations in the metabolism of xenobiotics. *Biochemical Society Transactions*, **2**, 359–377.

Williams, A.G. and Whitham, T.G. (1986) Premature leaf abscission: an induced plant defense against gall aphids. *Ecology*, **67**, 1619–1627.

Willmer, P.G. and Stone, G.N. (1997) How aggressive ant-guards assist seed-set in *Acacia* flowers. *Nature*, **388**, 165–167.

Wills, C., Condit, R., Foster, R.B. and Hubbell, S.P. (1997) Strong density- and diversity-related effects help to maintain tree species diversity in a neotropical forest. *Proceedings of the National Academy of Sciences USA*, **94**, 1252–1257.

Willson, M.F. (1983) *Plant Reproductive Ecology*. New York: Wiley.

Willson, M.F. (1986) Avian frugivory and seed dispersal in eastern North America. *Current Ornithology*, **3**, 223–279.

Willson, M.F. (1991) Dispersal of seeds by frugivorous animals in temperate forests. *Revista Chilena de Historia Natural*, **64**, 537–554.

Willson, M.F. (1993) Mammals as seed-dispersal mutualists in North America. *Oikos*, **67**, 159–176.

Willson, M.F. and O'Dowd, D.J. (1989) Fruit color polymorphism in a bird-dispersed shrub (*Rhagodia parabolica*) in Australia. *Evolutionary Ecology*, **3**, 40–50.

Willson, M.F. and Thompson, J.N. (1982) Phenology and ecology of color in bird-dispersed fruits, or why some fruits are red when they are 'green'. *Canadian Journal of Botany*, **60**, 701–713.

Willson, M.F. and Whelan, C.J. (1989) Ultraviolet reflectance of fruits of vertebrate-dispersed plants. *Oikos*, **55**, 341–348.

Willson, M.F. and Whelan, C.J. (1990) The evolution of fruit color in fleshy-fruited plants. *American Naturalist*, **136**, 790–809.

Willson, M.F., Rice, B.L. and Westoby, M. (1990) Seed dispersal spectra: a comparison of temperate plant communities. *Journal of Vegetation Science*, **1**, 547–562.

Wilson, E.O. (1991) *The Diversity of Life*. New York: Norton.

Wilson, P. and Thomson, J.D. (1991) Heterogeneity among floral visitors leads to discordance between removal and deposition of pollen. *Ecology*, **72**, 1503–1507.

Windsor, D., Ness, J., Gómez, L.D. and Jolivet, P.H. (1999) Species of *Aulacoscelis* Duponchel and Chevrolat (Chrysomelidae) and *Nomotus* Gorham (Languriidae) feed on fronds of Central American cycads. *Coleopterists Bulletin*, **53**, 217–231.

Wing, S.L. and Tiffney, B.H. (1987) Interactions of angiosperms and herbivorous tetrapods through time. In *The Origins of Angiosperms and their Biological Consequences*, ed. E.M. Friis, W.G. Chaloner and P.R. Crane. Cambridge: Cambridge University Press, pp. 203–224.

Wisner, T., Smedley, S.R., Young, D.K., Eisner, M., Roach, B. and Meinwald, J. (1996) Chemical basis of courtship in a beetle (*Neopyrochroa flabellata*): cantharidin as 'nuptial gift'. *Proceedings of the National Academy of Sciences USA*, **93**, 6499–6503.

Witmer, M.C. (1996a) Annual diet of cedar waxwings based on U.S.

Biological Survey records (1885–1950) compared to diet of American robins: contrasts in dietary patterns and natural history. *Auk*, **113**, 414–430.

Witmer, M.C. (1996b) Consequences of an alien shrub on the plumage coloration and ecology of cedar waxwings. *Auk*, **113**, 735–743.

Witmer, M.C. (1998) Ecological and evolutionary implications of energy and protein requirements of avian frugivores eating sugary diets. *Physiological Zoology*, **71**, 599–610.

Witmer, M.C. and Cheke, A.S. (1991) The dodo and the tambalacoque tree: an obligate mutualism reconsidered. *Oikos*, **61**, 133–137.

Wolf, P.G., Soltis, D.E. and Soltis, P.S. (1990) Chloroplast-DNA and allozymic variation in diploid and autotetraploid *Heuchera grossularifolia* (Saxifragaceae). *American Journal of Botany*, 77, 232–244.

Wootton, R.J. (1988) The historical ecology of aquatic insects: an overview. *Palaeogeography, Palaeoclimatology, Palaeoecology*, **62**, 477–492.

Worthington, A.H. (1989) Adaptations for avian frugivory: assimilation efficiency and gut transit time of *Manacus vitellinus* and *Pipra mentalis*. *Oecologia*, **80**, 381–389.

Yates, C.J., Taplin, R., Hobbs, R.J. and Bell, R.W. (1995) Factors limiting the recruitment of *Eucalyptus salmonophloia* in remnant woodlands. II. Post-dispersal seed predation and soil seed reserves. *Australian Journal of Botany*, **43**, 145–155.

Young, H.J., Stanton, M.L., Ellstrand, N.C. and Clegg, J.M. (1994) Temporal and spatial variation in heritability and genetic correlations among floral traits in *Raphanus sativus*, wild radish. *Heredity*, **73**, 298–308.

Yu, D.W. and Pierce, N.E. (1998) A castration parasite of an ant-plant mutualism. *Proceedings of the Royal Society of London B*, **265**, 375–382.

Yumoto, T. (1987) Pollination systems in a warm temperate evergreen broad-leaved forest on Yaku Island (Japan). *Ecological Research*, **2**, 133–146.

Zagt, R.J. and Werger, M.J.A. (1997) Spatial components of dispersal and survival for seeds and seedlings of two codominant tree species in the tropical rain forest of Guyana. *Tropical Ecology*, **38**, 343–355.

Zangerl, A.R. and Berenbaum, M.R. (1992) Plant chemistry, insect adaptations to plant chemistry, and host plant utilization patterns. *Ecology*, **74**, 47–54.

Zangerl, A.R. and Berenbaum, M.R. (1997) Cost of chemically defending seeds: furanocoumarins and *Pastinaca sativa*. *American Naturalist*, **150**, 491–504.

Zangerl, A.R., Berenbaum, M.R. and Nitao, J.K. (1991) Parthenocarpic fruits in wild parsnip: decoy defense against a specialist herbivore. *Evolutionary Ecology*, **5**, 136–145.

Zhou, W., Rousset, F. and O'Neil, S. (1998) Phylogeny and PCR-based classification of *Wolbachia* strains using wsp gene sequences. *Proceedings of the Royal Society of London B*, **265**, 509–515.

Zinkler, D., Götze, M. and Fabian, K. (1986) Cellulose digestion in 'primitive insects' (Apterygota) and oribatid mites. *Zoologische Beiträge*, NS **30**, 17–28.

Zizka, G. (1990) *Pflanzen und Ameisen*. Frankfurt an Main: Stadt.

Index

Page numbers in *italics* refer to figures and boxes, those in **bold** refer to tables. Plates are indicated by their numbers.

abomasum 115
abscission 87
Acacia 87–8
  ant-guard systems 233
  elephant impact 121, *Plate 4.1*
  feeding loops 124
  leaf availability 108
  spinescence 123
Acacia–Pseudomyrmex system 233
Aconitum 171
Aconitum columbianum (monkshood) 172, *174*
Aconitum senanense Plate 6.2
Acromyrmex 212, 215, *216*
actinomorphy 163–4
adaptation, reciprocal 98
adaptive radiation 19–21
  adaptive shift analysis 22
  angiosperms 4
  herbivory 20
  insects 3
  plants 15
'advance payment' 206
Aepyceros melampus (impala) 115
agouti 138
Ailuravus (rodent fossil) 42
air, conquest of 3
  see also flight
Alces alces (moose) 109, *Plate 4.1*
aldehyde reductase 97
alder 106
  red 105
algae
  earliest eukaryotes 33
  macroalgal communities 244
algal accumulations, prokaryotic 33
alkaloids 8, *86*, **87**, 111, 197–8, 203
  metabolism by P450s 90
  seeds 139
allelochemicals 85, 92
  concentration 93
  delivery 85, 87

enzymatic conversion 90
excretion 92
food plants 111
insect adaptation 97
intolerance 92
nutrient cycling through ecosystems 105–6
  see also plant defences; secondary compounds; toxins
allelopathy 186
Allomerus demerarae 233
allopolyploidy 242, 243
Alnus rubra (red alder) 105
Alnus tenuifolia (alder) 106
alternation, coevolutionary 238
amber deposits 33–5, 41–2, 72, 85
Amborella flowers 160
amino acids 167
  essential 82
  non-protein 83, *86*, **87**, 136, 148
amphibians 62, 192
Amphicarpaea bracteata 238
Amsinckia furcata *183*
amylase inhibitors 136
Anagrapha falcifera (noctuid moth) *90*
Anagrus (egg parasite) 104
ancestral habit 23
ancient relationship survival 26
Anemonopsis macrophylla 164
angiosperms
  ancestral floral structure 160
  animal dispersal of seeds 3, 185, *186*
  bud-feeding 45
  dinosaur coprolites 67
  divergence of lineages 4
  diversity 36, 161
  dominance 3
  early radiation 39
  ecological relationships 4
  fleshy fruit production 190
  fossil galls 50
  fossil pollinators 54
  hermaphroditism 162
  leaf-mining lineages 50
  mite associations *30*
  nectar 161

origins 181
phenotypic diversity 177
phytophagous beetle associations *66*
pollination 36, 158
regenerative strategies 142
self-incompatibility 162
selfing rate 162
species numbers 4, *6*
survival strategies 64, 67
  see also floral entries; flowers
animal-seed predators 187
anomodonts 64
ant(s)
  acacia 87–8
  arboreal aggregations 218
  biology 211–12
  carpenter 45
  colonialism/colonies 211, 231–2, 234–5
  desert biomass 227
  ecological indicators 232
  elaiosome-bearing seed removal 226
  extra-floral nectary associations 88
  fire 227
  foot morphology 234
  fossil plant damage 39
  granivory 138
  greenhead 224
  habitats 211
  harvester *140*, 226–8
  morphological innovations 211–12
  mosaics 235
  mutualism 61, 211, 232
  nests
    building 218
    faunal composition 229
    refuse piles 230
    specialists 229
  omnivory 232
  parasitism 233
  plant dispersal 187
  plant mediation of behaviour by chemical signals 234
  pollination 221–3, 233, 234, 235
  predation on tree pests 230
  protective functions to plants 219

seedling establishment 225
seeds 223–8
  dispersal 223, 226, 228
  germination facilitation 226
  predation 223
self-pollination of flowers 222
social organization 211
soil effects 228–31
soldier caste 232
specialization 235
territoriality 232
weaponry 232
weaver 217–18
wood 229, 230
workers 215, 232
see also leaf-cutter ants
ant–acacia mutualism 87–8
ant–caterpillar interactions 220–1
ant gardens 211, 215, 218
ant-guard systems 216–21
  Acacia–Pseudomyrmex system 233
  chemical specialization 233
  density 217, *218*
  domatia 217–20
  efficiency 234
  food bodies 216, 217
  fungal activity suppression 219
  nest sites 217
  opportunist nest-building 217–18
  protective functions to plants 219
ant–homopteran–plant interactions 221
ant–lepidopteran–plant interactions 221
ant mounds 228
ant–plant interactions 211–35, *Plate 8.1*
  antagonism 232
  asymmetry 233–4
  competition 233–4
  convergent evolution 233–4
  habitats 235
  incidence 232
  mutualism 61, 232
  outcome 232–3
ant–plant–soil relationship evolution
    230–1
antagonism
  ant–plant interactions 232
  pollination 157
antagonistic associations 27, 237
anteaters 7
antelopes 10
anthers 163
  poricidal 167
anthocyanins 165
anthoxanthins 165
anthropogenic overkill in Quaternary
    extinctions 68
antibiotics, attine 216
anti-feedants 83, 136
antimicrobial secretions, ant-borne
    222–3
Antron (fossil gall wasp) 50
Aphaenogaster rudis 225, 235
aphid/sumach/moss life cycle *30*
aphids 220, 243

diversity 30
feeding 78
gall-making 78
goldenrod feeding 93
honeydew 105
host switching 80, 82–3
legume attacks 95
premature leaf abscission 87
seasonal movements 82–3
specialist 80
visual mimicry of clusters 168
Apis mellifera (honeybee) *Plate 6.2*
Apocynceae 8
  butterfly feeding 17
apomorphic groups 23
apparency 95
aquatic ecosystems 58, 60
Aquilegia (columbine) *182*
Aquilegia caerulea 172
archipelagos, mainland 19
arctic environment 127, *128*
  frugivorous birds 192
arginyl-tRNA-synthetase 83
arid ecosystems 138
aromatic compounds 167–8
arthropods 33
  mouthparts 58
  prepollen transport 37
  see also insects; mites; spiders
artiodactyls, rumen 10, 12
Asclepias (milkweed) 17, 180
Asclepias syriaca (milkweed) 89, *90*, 98–9
ascorbic acid 91
asexual reproduction 158
ash *140*, 143
aspen regeneration 121
asphalt pools 35
assemblage *73*
associational resistance 95, 104–5, 117
associations 26, 27
  origin 27–8, *29–30*, 30
  study of past 35–73
aster floral structure 171–2
  sterile ray flowers 164
Asteraceae 190
Astrocaryum murumuru (palm) 104
ATPase 92
Atropa belladonna (deadly nightshade)
    198, 203
atropine 111, 203
Atta 212, 215
Atta sexdens 228
Attini 212, 214
  fruit pulp removal 226
  see also leaf-cutter ants
attractants, floral 163–6, 181–2
attraction mechanisms 39
aucubin 18–19, 93
autecology 69
autoallelopathy 186
autopolyploidy 242, 243
autotoxicity 148
autotrophy 3, 5
Autunia (seed-fern) 48

bacteria
  antibiotic-producing 215
  plant carbohydrate digestion enzymes
      113
  symbiotic *14*, 82
Baeolophus griseus (juniper titmouse) 147
bagworms 61
Balanites wilsoniana 204
bark-beetles 45, 81
batrachosaurs 64
bats
  echolocation 166
  fruit-eating 159, 200, 201, 202
    reproduction time shift 203, *204*
  plant dispersal 3
  pollination 158, 159
  seed dispersal 192
beaver 110, 123
beech 136
beeflies 158
bees
  anthophorid 39
  apid 169
  columbine adaptive radiation 19–20
  diversity 30
  euglossine 20, 168
  eusocial 158
  halictid 20
  pollination 58, 171, 177
    Dalechampia vines 71, *72*
    of legumes 39
    syndromes 169, **170**
  repulsion by flower pigments 180
  resin antimicrobial properties 95
  resin-collecting 236, 237
  social 171
  solitary 167
    plant species visited 169, 171
    pollen removal/deposition *173*
  see also bumblebees; honeybees
beetles
  adaptive radiation 3
  ambrosia 12
  bark 45, 81
  boring 39, 45, 243
  bruchid 138, 139, 148–9
  chewing 77
  chrysomelid 77, 243
    Bursera feeding 85, 87
    defensive compounds 80
    milkweed 89
    plant tissue preference 83
  cucurbit 97
  elm-bark 94
  fossil
    aquatic feeders 56
    external foliage-feeders 42–3
    plant damage 39
  fungus-carrying 12
  galls 50
  herbivory on willow 102
  leaf 19
    plant–host associations *66*, 71
    sunflower family hosts 71, *72*

beetles (*cont'd*)
  leaf-mining 48
  longhorn 158
    milkweed host 71, *72*
    plant–host associations *66*, 71
  mouthparts for pollen consumption 56
  nitidulid 172
  Phytophaga plant-host associations *66*
  plant-piercing ovipositors 61
  plant virus vectors 94
  pollenivore 56
  pollinators 54, 158, 161, 171
  pre-dispersal granivory 137
  scarab 70
  scarabaeoid 15, *16*
  seed eating 132
  siphoning 52
  soldier 158
  wood-boring 243
  see also Blepharida (flea beetles); weevils
bellbird 187
Beltian bodies 88
bennettitaleans
  dinosaur coprolites 67
  fossil plant damage 39
  galls 50
  pollination 54
bergapten 148
betaxanthines 165
Betula (birch) 121, *122*
Betula pubescens (mountain/silver birch)
  117, *118*, 134, 230
Betulaceae 248
bilateral symmetry 163, 177
bilberry 120
biodiversity 4–5, 7–10, *11*, 12–13, *14*
  coevolutionary process 238–9
  creation 13, 15–21
  estimates 22
  evolutionary history 26
  interaction 236, 247
  organization 237
  plant–animal mutualism 244–5
  species interactions 240
biological control 24, 102, 247
biological perspectives of associations 26,
    69–71
  complementarity with palaeobiological
    approach 71–3
  temporal dimensions *38*, 248
biomass
  availability 107–8
  pyramid 7
  temporal variation 109–10
biomechanics *73*
biota composition 4–5, *6*
birch 117, *118*
  browsing 121, *122*
  herbivory 248
  mountain 117, *118*, 230
  silver 134
birds
  bill morphology 137, 150, *Plate 5.1*
    sexual dimorphism 178

specialization for nectar feeding 159,
    179
  bill size 147, 241
  community structure 178
  fossilized gut contents 42
  frugivorous
    Arctic species 192
    bill morphology 200
    detoxification ability 203
    diet composition 207
    digestive system 200–1
    fruit size 195
    gape width 195, *196*, 200
    migration 203
  granivory 137, 149
  migration 244
  nectar feeding 159
  plant dispersal 3
  pollination 158, 159
    syndromes 70
  seeds
    dispersal 192, 194–5
    eating 132
    fluctuation tracking 151
    hoarding 144
  tongue morphology 159
  toxic seed avoidance 149
bison 10
blackbird, European 203
blackcap see *Sylvia atricapilla*
Blandfordia grandiflora 222
Blepharida (flea beetles) 17, 85, 87, 89,
    99
body-fossil *73*
  morphology 70
Bombycilla cedrorum (cedar waxwing)
    196, 202
Bombycilla garrulus (waxwing) 203
boreal ecosystems 127
boring
  beetles 39, 45, 243
  fossil insects 45, *47*, 252
  shelter 61
bracken 27
Bradyrhizobium 238
bramble 123
Brassica rapa tolerance and resistance traits
    89
Brassicaceae 8
  butterfly coevolution 17
brood site mimicry 168
browsing
  birch 121, *122*
  compensatory growth 127
  high 64
  by moose 122
  trees 120
  vegetation composition change 127
Bucerotidae 203
Buchnera 243
bud-feeding fossil insects 42, *45*
buffalo
  African 115
  grazing 104

bugs
  lygaeid 132
  soapberry 150
bulbuls 192, 201
bullfinch 84
bumblebees 165, 168
  community structure 178
  nectar robbing 180
  pollen collection 164
  pollen removal/deposition *173*
  pollination 171, 172, 176
  pollinator guilds 172
bunchgrass 228
Bursera 17, 99
Bursera schlechtendalii 85, 87, 89
butterflies
  adaptive radiation 3
  birdwing 249
  buckeye 93
  cabbage white 16–17, 85, 171
  coevolution with plants 16–17
  lesser cabbage white 171
  lycaenid 165
  male requirement for sodium 83
  monarch 9, 92, 97
  nymphalid 17, 166
  oviposition preferences 89
  pollination 166
  swallowtail 236
[13]C isotope of carbon 68
C3 and C4 pathways 68
caching 123, 137, 144, 151, 189, 227
cacti, senita 181
Cactoblastis cactorum (moth) 102
caddisfly
  case construction 50, 61
  sponging 52
caecum 113, 114
Caenozoic *29*
  galls 50
  Late 35
  mammalian herbivory 67–8
  plant damage 39
  seed predation 50
Caenozoic (Early) Thermal Maximum
    30
caffeine 84
Calathea 221
Calhoun Coal Flora (Illinois, USA) 38, 58
Callicarpa saccata Plate 8.1
Calochortus (lily) 172
Calvaria major 204
Calypso bulbosa (orchid) 168
cambium-borers 45
camphene 17
Camponotus 211
Camponotus schmitzi Plate 8.1
L-canavanine 83
canids 208
Capparidaceae 17
Capreolus capreolus (roe deer) 109
captorhinids 64
carbohydrates, plant structural 113

carbon allocation to roots/above-ground
        parts 127
carbon/nutrient hypotheses 117
carbon–nitrogen ratio, faeces 127
Cardamine cordifolia (mustard) 102
cardenolides 84, 97, 99
cardiac glycosides 8, 9
    butterfly feeding 17
    sequestering by monarch butterflies 92
Carduelis chloris (greenfinch) 203
Carduus nutans (musk thistle) 102
carib, purplethroated 178
carnivores/carnivory
    adaptive radiation 15
    geometrid moths 237
    herbivorous insect utilization 3
    plant interactions 240
    trophic systems 3
carotenoid pigment 165, 196
    bee repulsion 180
carpenter worms, larval borers 45
Carya ovata (hickory nut) 134
Caryedes brasiliensis (beetle) 83
Casearia corymbosa 7
caseids 64
Cassia fasciculata 217
Castor (beaver) 110
catalpol 18–19
catastrophism 31–2, *73*
cattle, feeding 115
Cecropia 219
    food bodies 217, 234
Cecropia–Azteca ant system 233
Cecropia obtusifolia 205
cellulase 10, 12, 113
cellulose 82, 110
Ceratocystis ulmi (Dutch elm disease) 94
Ceratosolen (fig wasp) *Plate 6.2*
Cercidium pods 141
Cervus elephus (red deer) 107, **108**
character-state evolution 70
charcoal 37
chats 159
'cheater' taxa 245
cheating by flower-visitors 180
Chelaner 226
chemical reactor theory 115
Chihuahuan Desert (Mexico) 147
chimpanzee 204
chipmunk 144
chromosomal evolution rates 4, *7*
Chrysanthemoides 190
chrysomelid beetles 243
    cospeciation 18
chymotrypsin inhibitors 123
cicadas 78, 82
Cicadoidea 78
cichlid fish, adaptive radiation 19
Cimicifuga simplex 166
Cirsium (thistle) 102
civets 13
clades 35–6
cladogenesis, parallel 18, 35, 71, *72*, 98,
        99

cladogram 35, *73*
    character-state evolution 70
    parallel 98, *99*
    synapomorphy 69
Cladomyrma petalae *Plate 8.1*
Clarkia 177
Clethrionomys glareolus (bank vole) 184
Clibadium 190
climate
    plant migration response to
            modifications 187
    trends 30
    see also weather conditions
Clusia fruit *Plate 7.1*
Clusia resins 167
coal-ball permineralizations 39, 50, *73*
coalescence *29*
coccids 220
Coccoloba uvifera (sea grape) *8*
coconut, double 134
coevolution 16–17, 124, 238–9
    community constituents *29*
    diffuse 71, *72*
    diversifying **18**
    gene-for-gene **18**, 238
    granivory 149
    insect herbivores 98
    interacting populations 98
    plant and herbivore at population level
            100–1
    plants and pollinators 178–80
    types **18**
    see also escape and radiate coevolution;
            geographic mosaic theory of
            coevolution
cold spots, coevolutionary 238
Coleoptera *see* beetles
collards crop 104, 105
Collembola, pollenivore 56
Colobus satanas (black colobus) 204
colon, sacculated 114
colonialism in ants 211, 231–2, 234–5
colonization
    granivory effects 132
    herbivore effects on plant community
            125
    of land 15
    seed at distance from parent population
            144–5
colour pigments 165, 180
colour vision 165–6
columbines, terrestrial 19
commensalism 27, 157
community *73*
    antagonistic/mutualistic interactions
            237
    component 7, *73*
    compound 7
    decomposer 229, 243
    ecological 101–6
    ecology 241
    evolution *29*
    interaction biodiversity 247
    past biological *29–30*

structure 236
    see also plant communities
competition
    granivores 139
    herbivory effect on plant species
            richness 125
    plants 121–2, 243
competitive exclusion 125
compression *73*
compression deposits 72
    fossil pollinators 54
    insect mouthparts 42
    piercing-and-sucking insect feeding 45,
        46
concentrate selectors 115
conifers
    dinosaur coprolites 67
    fossil plant damage 39
    gnetalean 158, 161
    phytophagous beetle associations *66,*
        71
    seed crops 139
conjugation 91–2
Connochaetes (gnu) 115
Connochaetes taurinus (wildebeest) 110
conservation initiatives 24
conspicuousness 95
continental plates *35*
convergence *73*
    ecological 26, 30
    evolution of floral traits 169, **170,**
        181–2
    evolutionary of fruits *Plate 7.1*
copepods 241
co-pollinators 181
coprolites, dinosaur 67
coprolites, dispersed 37, *38,* 41
    arthropod assemblages 41
    fossil pollen consumers 54
    pollen-containing 41
    prepollen 37
    quantitative analysis 249
    spheroidal 37
    spore-packed 57
    vertebrate 62
coprophagy 41, 58
Cordia nodosa 233
Corydalis aurea 187
cospeciation 17–19
Costa Rica neotropical forests 7, *10*
Cotesia (wasp) 88
cottonwood, hybrid zone 100
coumarins 85, *86,* **87**
counter-defence 98–101
Crataegus monogyna (hawthorn) 134,
        193
Crematogaster 211
Crematogaster nigriceps 233
Cretaceous, Early 48
Cretaceous–Tertiary (K/T) boundary 30
    extinction 33
        galls 50
        leaf-miners 50
        plant damage 39

crop plants
  economic injury level (EIL) 89
  pollination 24
cross-pollination, animal-mediated 158
crossbill 137, 141
  bill *152*, 154
  red 137, *152*, 153–4, 238, *Plate 5.1*
  seed fluctuation tracking 151
Cucurbitaceae 97
cucurbitacins 85, 97
cutin 110
cutworms 45
cyanogenic compounds 85, *86*, **87**, 136, 197
Cycadales 190
cycadophytes 38
cycads
  dioecious 161
  phytophagour beetle associations *66*, 71
  pollination 54, 158
  species-richness 161
Cynomys (prairie dog) 110
Cyphomyrmex 212
Cyrtobagus salvinae (Brazilian weevil) 102
cytochrome P450 90
  inhibition 85

Dactylorhiza incarnata 168
Dalechampia vines 71, *72*, 94–5, 236, 237, *Plate 3.1*
  resins 167
  specialization in pollination 182–3
Danaus plexippus (monarch butterfly) 97
Darwin, Charles 157
Datura wrightii 84
death assemblages, entombed 33, *34*
debarking 121
deception
  apid bees 169
  'cheater' taxa 245
  cheating by flower-visitors 180
  food-based 169
  pollination 157, 161, 168–9
  see also mimicry
decomposer community 229, 243
decoys, plant defences 83
deer
  red 107, **108**, 115
  roe 109, 115
  rumen 10
  white-tailed 118
defoliation, partial 120
dehiscence 134
Delphinium nelsonii 165
Dendroctonus (bark beetle) 81
Dendroica coronata (yellow-rumped warbler) 202
Dendroica pinus (pine warbler) 201, 202
dentition
  frugivory 200
  see also molars; teeth
Depressaria pastinacella (parsnip webworm) 100, 148

derived groups 23
deterrents *see* allelochemicals; plant defences; toxins
detoxification 90–1
  benefits 96
  capability 147
detritivory
  aquatic ecosystems 56
  dispersed coprolites 41
  generalized 43
Devonian
  biota 33
  Early 57
  Late
    megaspores 37
    vertebrate emergence on land 62
    piercing-and-sucking insect feeding 45, 46
diadectomorphs 64
diapause timing 241
  extension 151
diapsids, herbivory 64
Diaptomus sanguineus 241
diaspore, ants carrying 223
Dibolia (leaf beetle) 19
Diceros bicornis (black rhino) 108
Dieffenbachia seguine Plate 7.1
diet
  abrasive 111
  breadth limitation in insect herbivores 80
  Caenozoic herbivorous mammals 68
  food types 57
  high-fibre of vertebrate herbivores 64
  Pennsylvanian arthropod 41
  shift in Lepidoptera 21
dietary convergence 56
dietary guilds 28, 37
  insect 56–8
digestion theory 115
digestive system
  frugivorous birds 200–1
  insects 82
  intestine length seasonal fluctuation 201
  processing rate/efficiency 202
  seed passage through 204
dinocephalians 64
dinosaurs
  adaptive radiation 5
  coprolites 41, 67
  diversification into feeding niches 64, 67
  feeding on insects 5
  fossil chewing mechanisms 42
  fossilized gut contents 42
  gizzard stones 67
  gut contents 67
  herbivory 64, *65*, 67
  phylogeny *65*
  teeth modifications 67
Dioclea megacarpa 83
dioecy 161
Diptera
  aquatic feeding 56

diapause extension 151
fossil borers 45
galls 50
leaf-mining 48
mouthpart classes 60
pollinators 54, 158, 161
sponging labellum 58
Dipteryx micrantha (legume) 104
Disa ferruginea (orchid) 168
disease vectors 93–4
dispersal
  directed 187
  offspring of plants 185
  see also seed dispersal
diversification
  feeding niches 64, 67
  floral spurs 161–2, *182*
  morphological in fruit 207, *Plate 7.1*
  plant communities 177
  plants 176, 181
  pollination by animals 160–2, 181
  reciprocal 161–2
  species 242
diversity
  of interactions 5, 7, *8*
    biotic 12, *13*
  latitudinal trends 30
dodo 204
dog, prairie 110
domatia *73*, 217–20, 232, 234
  ant communities 235
  ant competition for 220
  definition 214
  exploitation by ants 234
  Homoptera 221
  lepidopterans 221
  mites 61
  obligate ant species 234
  occupation by ants 220
  orchid 233
Drosophila adaptive radiation 19
Dutch elm disease 94

Early Caenozoic Thermal Interval 248
Early Caenozoic Thermal Maximum 30
earthworms, seed eating 132
earworm, corn 85
ecological approaches to interactions 245, 246
ecological associations 26, 30
ecological attributes
  mapping onto clades 35, *36*
  non-phylogenetic 70
  phylogeny association 70
ecological communities 101–6
ecological convergence detection 70
ecological dynamics, rapid 240–1
ecological engineering 119
ecological indicators 232
ecological saturation hypothesis 28
ecological transfer 69
ecology
  chemical 22
  community 241

evolutionary 241
  of interactions 236–42
  physiological 241
  transfer 69–70
ecomorphological feature *73*
economic injury level (EIL) 89
ecosystem
  mammalian herbivory effects 126–7,
    128
  population cycles 128, *129*, 130
  ungulate impact 127
ectomycorrhiza 122
edaphosaurs 64
Edaphosaurus (fossil sail-backed reptile)
    64
Ediacarian fauna 33
elaiosome 234
  ant colony sex ratio 235
  ant communities 235
  ant gardens 218
  composition 225
  definition 214
  herbaceous species 224
  morphological origins 225
  myrmecochory 223
elephants
  debarking of trees 121, *Plate 4.1*
  seed dispersal 192, 204
elm trees 94, *140*
empirical approaches to interactions 245,
    246
endopeptidase inhibitors 149
enemies
  hypothesis 104
  multiple 239–40
energy
  digestible 113
  herbivorous insects 82–3
  loss with microbial fermentation
    113
  seeds 134
entombment 33, *34*
entomophily *73*
environment, changing 27, 127
environmental degradation 68
environmental factors, vegetation
    succession 126
Eocene, middle *30*
Ephemeroptera, fossil aquatic feeders
    56
Epilobium angustifolium (fireweed) 165
epiphytes
  ant-associated 218
  orchids 20
epistatic interactions 239
Equus burchellii (zebra) 110
ergot alkaloids 88
Erithacus rubecula (European robin) 204,
    207
erosion *35*
Erythrina (legume) 182
Erythroneura (leaf hopper) 15
escalation hypothesis 27
escape and radiate coevolution **18**, 71, *72*,
    238, 240

granivory 149
Escherichia coli 239
Escovopsis 215–16
eucalyptus
  insect herbivory 93
  inter-fire establishment 228
  unburnt woodland 227
Eucalyptus baxteri 227
eukaryotes 240
  earliest 33
eukaryotic cell 12
Eulampis jugularis (purplethroated carib)
    178
Euonymus europaeus (spindle tree) 134
Euonymus sachalinensis Plate 7.1
Eurybia 221
Euura (willow gallwasp) 83
evening primrose 177
evolution
  convergent of floral traits 169, **170**
  of interactions 236–42
  parallel of plant species 179
  sequential 71, *72*
evolutionary biology 236
evolutionary change
  angiosperms 4
  reciprocally induced 16
evolutionary clock, post-event 28
evolutionary dynamics, rapid 240–1
evolutionary ecology 241
evolutionary rate variation 241–2
evolutionary stable strategy (ESS) 151
excretion, allelochemicals 92
expanding resources hypothesis 28
exploitative mechanisms **9**
extinctions
  end-Cretaceous 33
  herbivore effects on plant community
    125
  major floral 30
  megafaunal of Quaternary 68–9
  risks of specialization 80
  role in plant–animal associations 28,
    29, 30
  taxonomic 26

faecal pellets 114
faeces
  carbon–nitrogen ratio 127
  deposition 127
  seed removal 226
Fagus sylvatica (beech) 136
fatty acid derivatives 166
feeding
  aquatic 56
  endophytic 61
  fossil record of mechanisms *38*,
    42
  functional groups 27, 28, 37, 42–5
    fossil record 42–5, *46–7*, 45–56, *55*,
      56
  granivore habits 132
  habits of herbivorous mammals 115
  heritability of traits affecting abilities
    147

insect guilds 77
insect herbivores 88
  knowledge acquisition 119
  learning across generations 119
  loops 124, 127
  opportunist mixed 115
  patches 118, 124, 127
  permineralized deposit sources
    72
  phylogenetic analysis of clades 70
  plant-related attributes 249
  processes 27
  rate 128
  relationship diversity 7
  sampling 119
  specialized habit evolution 81
  stimulants 97
  see also food
Fergliocladus (fossil conifer) 39
fermentation
  chambers 12
  foregut 12–13, 113, *114*–15
  hindgut 12–13, 113
  microbial in herbivorous mammals
    113–14
  sauropod digestion 67
  vertebrate herbivore digestion 64
  see also gut microbiota
ferns
  bracken 27
  dinosaur coprolites 67
  floating 102
  fossil plant damage 39
  marattialean 27, 30, 50
  spores 3
fertility loss 15
Festuca arundinacea (tall fescue) 88
fibre, digestible 113
Ficus 227
Ficus congesta Plate 6.2
field studies 22–3
figs 207
  fig wasps 71, 181, 238, 245
filtering 56
finches
  Darwin's 19, 144, 150
  ground-foraging 139, 144
fir, Douglas 141
fire
  exposure 120–1
  grazing relationship 127
  landscape heterogeneity 127
  serotiny 151
  vegetation succession 126
fireweed 165
fish, seed dispersal 192
fissure fills 35
fitness, optimal foraging 119
flavones 165
flavonoids **87**, 111
flavonols 165
flax 16
  Australian wild 241
flea beetles *see Blepharida* and
    Chrysomelidae

flies
　adaptive radiation 3
　anthomyiid 180, 181
　carrion 169, **170**
　fossil plant damage 39
　gall-making 78
　Hessian 243
　leaf-miner 48
　nemestrinid 180
　pollination 171
　　syndromes 169, **170**
　seed parasites 180
　sponging 52
　syrphid 172
　tabanid 180
　tachinid 9
　tongue length 180
　see also sawflies
flight 3
　pollinator groups 158
floral attractants 163–6
floral constancy 171
floral display 94
floral parasites 238
floral resources, interspecific competition
　　178
floral spurs
　coevolution with insect tongues 178,
　　179, 180
　diversification *182*
　elongation 172
　probing pollinators 181
floral symmetry 163, 164–5, 177
floral traits 94, 158
　animal pollination 162–3
　control by single genes 177
　convergent evolution 169, **170**, 181–2
　covariance 175
　evolution 181–3
　phenotypic selection 175
　pollination efficacy 172
　pollinators 69
　visitor efficacy increase 172
florivores 180
flowerpeckers 159
flower-piercers 180
flower-visitors
　lifetime fitness 177
　parasitic/predatory 180
flowers
　asymmetry 164–5
　auditory cues 166
　colour 165–6
　　phenotypes 177
　　polymorphism 168
　damage by visitors 180
　hermaphroditic 160, 161, 162
　mimicry 168, 169, 176–7
　olfactory cues 166
　pendant 164
　pigments 165
　predators 180
　reward 23
　　accessibility 172

　concealment 163–4
　robbing 180
　scents 166, 176, 177
　sexual phase synchrony 162
　shape 163–4
　size 164
　spatial separation of parts 162–3
　specialized tissues 168
　staggered opening 162
　symmetry 163, 164–5
　temporal offset 162
　visual cues 163–6
　see also angiosperms; pollination
fluid-feeding, surface
　fossil record 52–3, 254
　mouthparts 58
foliage
　consumption rate 82
　external feeding fossil record 42–3, *45*,
　　251
　nitrogen content 82
　see also leaves
foliar flag 194
food
　abundance 117–18
　choice by herbivorous mammals 115
　compensatory consumption 82
　harvesting methods 77–8
　predator interactions 130
　quality 118–19, 128
　　enhancement 123–4
　quantity 118–19
　resource management by herbivores
　　123–4
　retention time 114
　seasonal variations 118
　types 57, 58
　webs 237
　see also feeding
food bodies
　ant-guard systems 216, 217, *218*
　definition 214
　exploitation by ants 234
　Homoptera 221
　lepidopterans 221
　specialized 216
food chain length 126
food plants 107–11
　abrasive 111
　abundance 108, 109
　availability 108, 110
　chemical composition *109*, 110, 111
　components 110, 111
　digestibility 111
　plant defences 111
　production 108
　quality 108–9, 111
　quantity 108–9
　resource estimation 107
　seasonal changes 110
　spatial variation 108–10
　standing crop 108
　temporal variation 108–10
　weather conditions 109–10

foraging
　fitness maximization 138–9
　frequency-dependent 141
　hierarchical 116
　models 118–19
　random 127
　time spent 139
forest gaps 147
forest types 191
Formica aquilonia (wood ant) 229, 230
Formica fusca 221
Formica neorufibarbus 222
Formicinae 226
fossils/fossil record
　animal-mediated damage to plants 36
　aquatic feeding 56
　degradation *35*
　deposits 33, *34*
　external foliage feeding 42–3, *45*, 251–2
　feeding mechanisms *38*, 42, 251–6
　functional feeding groups 42–5, *45–6*,
　　45–56, *55*, 56
　galls 50, *51*, 253
　gut contents *38*, 41–2, 249
　impression *73*
　insect dietary guilds 56–8
　insect herbivores 69
　leaf-mining 48–50, 252–3
　long-term prediction testing 30
　mouthparts *38*, 42
　occurrences *38*
　oviposition 61, *62*, 255
　piercing-and-sucking insect feeding 45,
　　46, 251–2
　plant damage 39, 41
　plant–arthropod associations *29–30*, *40*
　plant–insect associations 39
　pollination 54, *55*, 56, 254–5
　preservation modes *34–5*
　quality 36, *72*
　seed predation 50, *52*, 253–4
　surface fluid-feeding 52–3, 254
　three-dimensional impregnation 33,
　　37
　trace *74*
　vertebrates 61–2, *63*, 64, *65*, 67–9,
　　255–6
foxes 192
frass trails 48
Fraxinus excelsior (ash) *140*, 143
free-feeding, fossil insects 42–3, *45*
fructose 167
frugivores/frugivory
　animal food ingestion 202
　assemblage restriction by plants 207
　bats 200
　behaviour 203, *204*
　Caenozoic mammals 67–8
　dentition 200
　dietary variation 202–3, 207
　digestive physiology 201–3, *204*
　ecological distribution 192–3
　extensive 203
　external morphology 200–1

food-retention time 201
fruit size selection 196
gene flow 205
habitat shifts 203
heavy 199–200
internal morphology 200–1
lek-mating birds 189–90
occasional 199
plant dependence on array 204
primate 203
protein requirement 202
range of plants eaten 204
reciprocal specialization 205–7
reproduction time shift 203, *204*
seasonal migration 203
seed dispersal 145
specialization-limiting factors 207
taxonomic distribution 192–3
trophic systems 3
tropical distribution 192
vertebrates 189, 191
frugivores/granivores 112
frugivores/herbivores 112
frugivores/omnivores 112
fruit(s) 188–90
  acceptability to dispersers 198
  advertisement 194–5
  ant seed dispersal 226
  availability peaks 193
  chemical signals 194
  colouration 194–5
  conspicuousness 194
  defensive chemical compounds 198
  diet composition 207
  evolutionary convergence *Plate 7.1*
  functional convergence 189
  lipid content 196, 197
  morphological diversification 207
  nitrogen content 202
  nutrients 196–7
  nutritional imbalance 202–3
  palatability 198
  phenotypic changes 189
  photosynthetic ability 195
  pigments 195
  plant phylogeny 197, *198*, 199
  poisonous 203
  protective structures *135*
  pulp composition 196–8
  ripening season 197
  seasonal variation 197
  secondary metabolites 197–8, 202–3
    toxic 203
  seed defence 134, *135*
  size 195–6
  structural features 150
  syndromes 198–9
  traits 193, 199
  visual signals 194–5
  volatiles 194
  see also frugivores/frugivory
fruit flies 17, 18, *19*
  cytochrome P450 genes 91
  leaf-mining 48

fruiting
  abiotic factors 193–4
  phenology 193–4
  seasons 193, 207
  synchronous 193
fundatrix, aphid 80
fungi
  blue-stain 12, 94
  cellulose 82
  control over ant foraging 215, 234
  cultivation by leaf-cutter ants 214
  deliquescing fruiting bodies 52
  endophytic 88, 243
    grasses 111, 139
  fungal infections 215
  growth promotion 215
  lignin 82
  pathogens
    fruit rot 198
    seed survival and distance from
      parent plant 187
  plant carbohydrate digestion enzymes
    113
  spores 3
  symbiotic *14*, 243
  woody tissues 57, 58
  see also mycorrhizal associations;
    vesicular arbuscular (VA)
    mycorrhizal fungi
furanocoumarins 85, 148
  detoxification 100
  metabolism by P450s 90
  wild parsnip 100–1

Galapagos Islands 19
Galium 190
gallflies, fungal inoculation 12
galls/galling *29*, 78, *79*
  cottonwood 100
  fossil record 45, 50, *51*, 253
  insect diversity 30
  life habit origins 50
gardenification of conserved wildlands
  247
gardening, ants 211, 215, 218
Gazella thomsonii (Thomson's gazelle)
  104, 110, 115
geitonogamy 162, 164
gene-for-gene coevolution **18**, 238
gene-for-gene interactions 243
generalists
  extreme 236–7
  herbivore fitness reduction 92–3
genes
  extra-nuclear 240
  flow in frugivory 205
  nuclear 240
  resistance 241
  self-replicating 240
  virulence 241
genetic resources 247
genome
  interspecific interactions 240
  species interactions 246

geochemistry, isotope 68, *74*
geochronology *31*, 33, 248
  feeding-related characters 70
geographic mosaic theory of coevolution
  208, 238, 242
geographical structure of interactions 242
geological time-scale 33
geometrid moths 237
geophagy 149
Geospiza (Darwin's finches) 150
geotectonic activity *35*, *73*
germination inhibition 186
gingko foliage, dinosaur coprolites 67
gingkophyte seeds 50
Giraffa camelopardalis (giraffe) 108, 115
gizzard stones, dinosaurs 67
gizzards 132, 134
glandular trichomes 84
glider, greater 113
glucose 167
glucosinolates 8, 17, 85, *86*, **87**
glycine 91–2
glycogen 217, 234
β-glycosidases 91
gnat, fungus 45, 168
gnu 115
goats
  feeding 115
  food preference 123
  rumen 10
golden rod 93, 221
Gomesa (orchid) 23
gongylidia, definition 214
Goodyera repens (orchid) 134
gorillas, seed dispersal 192
granivores/granivory 132–54, *Plate 5.1*
  alternative food source availability 141
  characteristics 137–9, *140*, 141–2
  coevolution 149, 151, *152*, 153–4
  colonization at distance from parent
    population 144–5
  costs 139
  counter-adaptations to seed size
    149–50
  density-responsive feeding 147
  distance-responsive feeding 147
  eradication of individuals 136
  evolutionary implications 147–51,
    152, 153–4
  food-limitation 144
  frequency-dependence 146
  guilds 137–8, *Plate 5.1*
  harvester ants 227
  natural selection 147–8
  parasitoids 144
  percentage of crop eaten 141
  phytophagous 141
  plant demography effects 144–6
  plant species diversity 146–7
  pollination 168
  population dynamics/size 142–4
  post-dispersal 132, **133**, 137–8, 141,
    143
  pre-dispersal 132, **133**, 137, 145, 153

granivores/granivory (*cont'd*)
    predator-limitation 144
    resource storage 151
    satiation 143, 148, 151
    seeds
        crop variability 150–1
        dispersal 145
        movement 144
        production depression 147
    spatial heterogeneity 141–2, 146–7
    spatial patterns 145–6
    temporal heterogeneity 141–2
    tradeoffs
        with plant competitive ability 146
        with seed dispersal/germination
            147–8
    trophic systems 3
grapevines, spider mites 93
grass/roughage eaters 115
grasses
    coevolved mutualism with grazers 120
    ecological spread 68
    fungal endophytes 111, 139
    grazing areas 122
    prostrate growth form 122
    savanna 120
    seeds 139
    Soay sheep 128
    tillers 120
grasshoppers 80
grasslands 127, 244
    mammalian herbivores 68
    temperate 147
grazed patches 118, 124, 127
grazing 237
    coevolution mutualism with grasses
        120
    cyclic 124
    fire relationship 127
    large mammals 127, 244
    lawns 124
    optimization concept 123–4
    optimization hypothesis 121
    plant–soil system effects 127
    pressure 125–6
    trypsin inhibitors 123
Greater River Basin (Wyoming, USA) 30
greenfall 105
greenfinch 203
Grevillea 226
Greya moth 181, 239
growth
    clonal 185
    compensatory 120, 121, 127
    limitations of herbivorous insects 81–3
    relationship to defence investment 96
Grylloblattida 41
Guatapo National Park (Venezuela) 7, *11*
guild coevolution **18**
gumivores 112
gut contents 37
    dinosaurs 67
    fossil records *38*, 41–2, 249
    mammalian 67

gut microbiota
    endosymbionts 82
    fossil arthropod 45
    symbionts 113
    see also fermentation
guttation 52
Guttiferae 190
gymnosperms
    animal dispersal of seeds 185, *186*
    diversity 161

habitat
    ant pollination 223
    islands 19
    shifts 203
Hadena bicruris (noctuid moth) 166
hands, prehensile 13, 15
Haplopappus squarrosus (maritime shrub)
    101–2
hare 123
    Arctic 108
    mountain 117, *118*
    population cycles 128, *129*, 130
    snowshoe 130
harvesting, stored plants 123
hawkmoths 177
    columbine adaptive radiation 19–20
    long-tongued 180
    pollination 172, 175
hawthorn 134, 193
'haypiles' 123
Heliconia 7, 178
Heliothis zea (corn earworm) 85
hemicellulase 113
hemicellulose 110
Hemiptera 243
    fluid siphoning 78
    fossil feeding mechanisms 45, 56
    fossil pollinators 54
    galls 50
    mutualistic gut endosymbionts 82
    plant-piercing ovipositors 61
    plant virus vectors 94
    pre-dispersal granivory 137
herbivores/browsers 112
herbivores/grazers 112
herbivores/herbivory
    above-ground production 122
    adaptive radiation 20
    ant protection of plants 216
    aquatic ecosystems 56
    attraction by plant volatile compounds
        9
    below-ground biomass 122
    birch 248
    bottom-up regulation 126
    clades 35
    control 244
    counteradaptations of attack 26
    dinosaurs 64, *65*, 67
    dispersed coprolites 41
    divergence of radiations in clades 35
    evolutionary responses of plants 119
    fitness 110

geometrid moths 237
host-plant damage patterns 61
impact
    measures 36
    on slow-growing plants 96
    initial on land 26
    insects 243
    invertebrate 107, **108**
    long-term patterns 26
    mammals 107
    multiple 93, *94*
    nitrogen 105
    nutrient cycling effects 105–6
    origins 61
    phytochemicals in defensive strategies
        19
    plant community evolution 61
    plant damage 69
        quantitative analysis 60–1
    plant-host's response 37
    plant module size 120
    plant responses 120–1
    plant shape/size effects 121–2
    pollination relationship 175
    roots 122
    seed-head 101–2
    seed/seedling 77
    seedling mortality 120
    selective 212
    species numbers *6*
    specificity measures 36
    tolerance 244
    tree seed location 103
    trophic systems 3
    vertebrate 107, **108**
        Late Palaeozoic *63*, 64
    see also insect herbivores; mammals,
        herbivorous
herbs 4
herkogamy 162–3
hermaphroditism 160, 161, 162
hermit birds 159
Heterobathmiidae, leaf-mining 48
Heuchera grossulariifolia 242, 243
hexapod mouthparts 58
hickory nuts 134
hole-feeding, fossil insects 42, *45*
holistic approach 24
holly, European 123
holometabolous larvae 48, 60, *73*
home range 110
Homoptera 82
    ant-guard systems 220
    domatia 221
    food bodies 221
    viral disease transmission to plants 220
honeybees 171
    flower damage 180
    pollen removal/deposition *173*
honeycreepers 19, 159, 179
honeydew 78, 220
    ant communities 235
    definition 214
    nitrogen cycling 105

honeyeaters 159
Hormathophylla spinosa (crucifer) 144
hornbills 192, 203
horntails, larval borers 45
hornworm, tobacco 92
horse, caecal fermentation 114
host plant
    identification 97
    range 19
    specialization on 236
hot spots
    coevolutionary 238
    grazing 124
hoverflies 48, 158
humans
    population 246–7
    traits 13
hummingbirds 159
    beak sexual dimorphism 178
    columbine adaptive radiation 20
    nectar source exploitation 171
    neotropical forests 7
    pollination 171, 174–5, 176
        syndromes 169, **170**
hybrid bridge hypothesis 99–100
hybridization, interspecific 242
Hydnophytum 219–20
Hydroponera 211
Hyles lineata (hawkmoth) 172
Hymenaea courbaril 226
Hymenoptera 21
    chewing 77
    colour vision evolution 70
    diapause extension 151
    fossil external foliage-feeders 43
    galls 50
    larval borers 45
    leaf-mining 48, *49*
    maxillolabiate apparatus 58
    mouthparts for pollen consumption 56
    plant-piercing ovipositors 61
    pollinators 54, 158
    pre-dispersal granivory 137
    surface fluid-feeding 52
    see also ant(s); bees; wasps
Hypericum 190
Hypericum perforatum (European Klamath
        weed) 102
Hypoperlida
    fossil external foliage-feeders 43
    ovipositors 61
hypsodonty 115

idiographic approach 23–4
i'iwi 179
Ilex aquifolium (European holly) 123
impala 115
Impatiens pallida 173, Plate 6.2
indirect interactions 239–40
insect(s)
    adaptive radiation 3
    beak length 147
    carnivorous
        diversity of parasitic 36

species numbers 4, *6*
case-making 61
chewing 77–8
diapause 151, 241
diversity 26, 98–101
eggs 61
fossil external foliage feeders 42–3, *45*,
        251–2
fossil gut contents 41–2
geochronological changes in family-
        level diversity *31*
granivores 137, 138, 139, 141, 148–9
    diapause extension 151
insecticide resistance 241–2
local specialization 238
methods of consuming land plants 27
mouthparts 37, 172, 178, 179, 180
    fossil 58, *59*, 60, 249
parasitic 5
phylogeny 69
phytophagous 241, 243
plant damage 248
plant dispersal 3
pollination 158–9, 243
predatory 5, 239
pre-dispersal granivory 137
primitive 20–1
sap-sucking 52
saprophagy 4, *6*
seed-parasitic 181
sex pheromones 176, 183–4
shelter 61
species numbers *6*
tongue coevolution with floral spurs
        178, 179, 180
toxin sequestration 97
wood-boring 30
see also oviposition
insect herbivores 3, 243
    adaptations for acquiring resources from
        plants 77–81
    behaviour 89–90
    bottom-up forces 81
    bracken 27
    chemical composition 81
    coevolution 98
    detoxification 90–1
    diet limitation 80
    dispersal rates 105
    eating 88
    energy 82–3
    external feeding 77
    fossil record 69
    generalists 79–81
    growth limitations 81–3
    harvesting methods 77–8
    host identification 97
    internal feeding 77
    macronutrients 82–3
    marattialean ferns 27, 30
    micronutrients 83
    mutualistic symbionts 13
    nutrient cycling through ecosystems
        105–6

nutritional requirements 82–3
phylogenetic congruence with plant
        hosts 70–1
plant community
    diversity effects 104–5
    effects 102–4
    evolution 61
plant defences 83–5, *86*, 87–9
    adaptations 89–92
plant distribution/abundance 101–2
plant fitness reduction 93–5
    disease vectors 93–4
    impeding relationships with
        mutualists 94
plant resistance 35
plant tissue preference 83
plant toxin use 97
population size limitations 81–3
protein restriction 82
rate of attack 105
reproduction limitations 81–3
resistance evolution 96–8
seed-heads 101–2
specialization 78–81
species numbers 4, *6*
top-down forces 81
insectary plants 104
interacting species 7, **9**
intercropped plants 104
intermediate disturbance hypothesis 125
interspecific interactions 237–8
    genome 240
intertidal communities 245
intestine *see* digestive system
intrinsic trend of diversification hypothesis
        28
introduced plants 102, 247
invertebrates
    fruit consumption 194
    plant defence 198
    seed dispersal 187
Ipomoea 177
Ipomopsis aggregata 120, 174–5, 180,
        Plate 6.2
Ipomopsis arizonica 121
Ips pini 94
iridoid glycosides *86*, **87**, 93
Iridomyrmex Plate 8.1
Iridomyrmex colonies 219–20
ironstone nodules 42
island biogeography theory 24
island habitat 19
isoprenoids 111

Jadera haematoloma (soapberry bug) 150
Janzen–Connell hypothesis 103–4, 145,
        147
jay 144, 145
jojoba shrub 136
Juniperus osteosperma (juniper) 147
Junonia coenia (buckeye butterfly) 93

katydids, pollenivore 56
kdr (knock-down resistance) genes 92

kelp forests, marine 244
ketone reductase 97
keystone mutualists 7, *10*
Klamath weed, European 102
koalas 13, 113
kudu 115

Labidomera clivicollis (chrysomelid beetle) 89
Lactuca serriola (prickly lettuce) *90*
Lagerstätten 33, *34–5*, 37–9
    insect mouthparts 58
    mammalian gut contents 67
    plant damage evidence 39
    plant–insect associations 249
lagomorphs 12, 114
lake deposits 33
land colonization 15
land surface of earth 26
lapping 52
larder-hoarding 144
Lasiacis 190
latex-canal defences 36
Lathyrus venosus 105
Laurales 190
Lavandula latifolia 173, Plate 6.2
leaf litter 105
leaf-cutter ants 212, 213–16
    antibiotic-producing bacterium 215
    distribution *213*
    foraging 212, 214
    fungus control over foraging 215, 234
    fungus gardens 215
    mutualism between fungus-growing
        ants and fungi 212, 214–15
    mycophagy 214
    phylogenetic analyses 215
    selective herbivory 212
    soil deposition 228
    survivorship of tree seedlings 212,
        213
leafhoppers 13, 15, 220
    grape 104
leaf-miners/-mining *29*
    blotch 48
    diversity 30
    fossil record 43–5, 48–50, 252–3
    larval borers 45
    modern 48
    premature leaf abscission 87
    serpentine 48
    shelter 61
leapfrog radiation 20
leaves
    damage and ant recruitment 219
    glandular hairs 38
    toughness 84
lectins **87**
legumes
    brush blossom pollination 39
    endopeptidase inhibitors 149
    floral feature analysis 70
    Rhizobium 95
lek mating systems 189–90

lemming
    population cycles 128, *129*, 130
    wood-lemming 108
lemurs 159, 192
Leonardoxa–Aphomomyrmex system 233
Lepidoptera
    adaptive radiation 20–1
    chewing 77
    diapause extension 151
    diet shift 21
    ditrysian 158
    domatia 221
    extra-floral nectaries 221
    food bodies 221
    fossil aquatic feeders 56
    fossil external foliage-feeders 43
    galls 50
    generalists 80
    glycine concentration in gut 91–2
    larval ant-guard systems 220
    larval borers 45
    leaf-mining 48
    mouthparts for pollen consumption 56
    plant-piercing ovipositors 61
    pollinators 54, 158–9, 161
    pre-dispersal granivory 137
    range of plant species for feeding 77
    seed eating 132
    siphons 52, 58
    specialization 80
Leporella fimbriata 222–3, 233
Lepus americanus (snowshoe hare) 130
Lepus arcticus (Arctic hare) 108
Lepus timidus (mountain hare) 117, *118*
Leucoagaricus gongylophorus 214
Liebig's law of the minimum 82
lignin **87**, 110
    digestion 48, 82
    leaf 84
Ligustrum obtusifolium 92
limonene 17
lineages, adaptive shift analysis 22
Linepithema humile 231–2
linolenic acid 82
Linum usitatissimum (flax) 16
lipids
    elaiosomes 225
    food bodies 217
    toxic 148
litter detritus 57
lizards
    anolis 7, *8*
    seed dispersal 192
lobeliads 179
Loidocea maldivica (double coconut) 134
Lomatium dissectum 239
Lonicera splendida Plate 7.1
Loxia curvirostra (crossbill) 137, *152*,
        153–4, Plate 5.1
Loxodonta africana (elephant) 121
Lupinus arboreus (bush lupin) 93
Lycaenidae 221
Lymantria dispar (gypsy moth) 93

Macaranga–Crematogaster system 233
Macaranga triloba 217, *218*
macroalgal communities 244
macroecology 24
macroevolution *74*
Macroglossum stellatarum (hawkmoth)
        175
Macroneuropteris (fossil leaf miner) 48
macronutrients, herbivorous insects 82–3
MADS developmental genes 160
magmatism *35*
Magnoliales 190
Malpighian tubules 92
mammalian herbivore–plant systems
        128
mammals
    arboreal 113
        folivores 12–13, 113
    body mass 112–13
    carnivorous 7
        seed-dispersal 208
    coprolites 41–2
    dietary specialization 112
    dispersal syndromes 68
    feeding relationships 7, *11*
    fossilized gut contents 41–2
    frugivorous 200
    granivory 137, 138, 141
        toxic seed avoidance 149
    grazing 127, 244
    metabolic requirements 112–13
    pollination 159, 168
    seed dispersal 192
        fruit coloration 194
    seed hoarding 144
    tannin conjugation 91
    tropical forest 7, *11*
mammals, herbivorous 7, 106, 107,
        Plate 4.1
    abundance 128, 130
    Caenozoic 67–8
    classification 111–12
    coevolution 124
    constraints 112–13
    digestive apparatus 113–15
    digestive system morphology 115
    distribution 111–12
    ecological impact 119–20
    exploratory movements 119
    feedback of plant responses 128, *129*,
        130
    feeding 111–19
        rate 128
        types 115
    feeding patch 118, 124, 127
    fermentation chambers 113
    folivores 113
    food
        abundance 117–18
        choice 115
        passage rate 115
        resource management 123–4
        retention time 114
        selection 115–19

food plant genotypic variation detection 109
foraging models 118–19
genetic variation in plant characteristics 117
genetically determined nutritional wisdom 115
hierarchical foraging 116
indirect effects on other organisms 126
intensity of herbivory 121
microbical fermentation 113–14
migration 110
mobility 110
plant characteristics 116–18
plant chemistry changes 122–3
plant mortality 120
plant response 121, 123
plant species diversity 125
pre-/post-gastric fermentation 113
roots 122
ruminants 114
sampling 119
seasonal preference 118
selection 116–18
size 113
spinescence response of plants 123
substrate specialization 113
timing of herbivory 121
vegetation succession 126
manakins 201
    protein requirement 202
    seed dispersal 192
mandibles, toothed 77–8
mandibular gland antimicrobial substance 215
Manduca sexta (tobacco hornworm) 92
margin-feeding, fossil insects 42–3, *45*
marginal value theorem 118
marine species, geographic structure 245
marsupials, forest-dwelling 113
mastication 113, 114
masting 143
mate-mimicking flowers 169, 176–7
mathematical approaches to interactions 245, 246
mechanistic explanation 24
megafauna *74*
Megaloptera, fossil aquatic feeders 56
megaspores 37
Melampsora lini (rust fungus) 16
Meranoplus (ant) *140*
Mertensia paniculata 180
Mesozoic, leaf-mining origination 48
Messel Oil Shale (Germany) 67
Messor 227
metapleural gland 211–12
    antimicrobial substance 215, 219
    definition 214
    secretion actions on pollen grains 222
methane 113
Metrosideros nectar 179
mevalonic acid pathway 85
Miconia 193
Miconia affinis 226

Miconia nervosa 230
microarthropods
    ant nests 229
    Devonian 45
microbes
    diversity of associations 9–10, 12–13
    fruit destruction 194
    microbiota of fossil arthropod gut 45
    mutualism 9–10, 12–13
    pathogens 198
    symbiotic associations 9–10
    see also fermentation
microcommunity of associated species 245
microfossils, earliest 33
microhabitats, granivory variation 141
micronutrients, insect herbivores 83
micropyle-pollination drop mechanism 39
microspores, *Xyela* feeding 21
Microtis parviflora 222, 223
midge, gall 50
migration
    avian 244
    mammalian herbivores 110
    seasonal 203
milkweed/milkweed beetle system 71, *72*, 89, *90*, 98–9
mimicry
    flowers 168, 176–7
    see also deception
Mimulus 176, 177
Mimulus cardinalis 180, *Plate 6.1*
Mimulus lewisii Plate 6.1
Mississippian era
    batrachosaurs 64
    megaspores 38
    radiations of amphibians and reptiles 62
mistletoe seedling establishment 187
mites
    angiosperm hosts *30*
    coprophagy 41, 58
    leaf domatia 61
    oribatid 41, 45, 58
    plant associations 37
    plant virus vectors 94
    spider 80, 93
mitochondrial DNA analysis 22
mobility of plants 185
moisture availability, vegetation structure 192
molars
    complexity 115
    seed eating 132
molecular approaches to interactions 245, 246
molecular clock hypothesis *74*
monkshood 172, *174*, *Plate 6.2*
monoecy evolution 163
Monomorium 226
monoterpenes 166
moose 109, 115
    browsing 122

ecological engineering 119
    rebrowsing 124
morphological data 22
mosaics
    ants 235
    selection 238
    see also geographic mosaic theory of coevolution
moss, forest floor 108
moths
    adaptive radiation 3
    arctiid 89, 97
    biological control 102
    California oak 105
    geometrid 230
    granivory 138
    gypsy 93, 105
    hadenine 181
    larval borers 45
    leaf-mining 48
    noctuid 166
    pollination 179
        passive 181
        syndromes 169, **170**
    seed predation *143*
    senita 181
    tongue length 179, 180
    yucca 168
    see also hawkmoths
mouse 145
    pocket 136
    pollination 159
mouthparts
    classes 28, 37
        geochronological history 58
        time of establishment 60
    fossil record *38*, 42, 58, *59*, 60
        stylet *45*, *46*
    haustellate structures 58
    insects 37, 172, 178, 179, 180
        fossil 58, *59*, 60, 249
    mandibulate 45
    pollen consumption 56
    sap feeding 60
movement, exploratory 119
Mucuna holtonii (liana) 166
multitrophic level interactions 7–9
mustard oils 8, 17
mustelids 208
mutualism 7, *10*, 239–40
    ant–plant interactions 61, 211, 232
    beneficial arrangement 27
    biotic diversity 12, *13*
    coevolved 120
    convergent 245
    dissolution 184
    insect herbivore interactions 94–5
    interactions 237, 247
    keystone 7, *10*
    microbial 9–10, 12–13
    non-symbiotic 233, 237
    obligate 183–4
    phylogenetic structure 245
    plant–animal 244–5

mutualism (*cont'd*)
  pollination 53, 54, 56, 157–8, 161–2,
    180–1
    fossil evidence 249
  seed dispersal 148
    in animal interiors 188–90
  strong 232, 233
  symbiotic 13, 217, 233, 237
  weak 232
  see also yucca/yucca moth mutualism
mycetocytes 243
mycophagy 214
mycorrhizal associations 95, 229, 243
Myopus schisticolor (wood-lemming) 108
Myotragus balearicus (fossil goat) 69
myriapod mouthparts 58
myristicin 85
Myrmecia urens 223, 233
myrmecochory 223–4
  characteristics **224**
  definition 214
  effects on plants 225–6
  seed release timing 225
  vertebrate predation 226
Myrmecodia 219–20, *Plate 8.1*
myrmecophytes 219
  definition 214
  diversity 220
  feeding by ants 219–20
Myrmicaria Plate 8.1
Myrmicinae 226
Myrmicocrypta 212

natural selection 239
  fluctuating 239, 242
  gene combinations 247
  granivory 147–8
nectar 158, 166–7
  ant communities 235
  extra-floral 216–17, 218
    ant communities 235
  feeding 159
  floral 161
  robbing 180
nectar guides 165
nectaries 53
  ant-tended 61
  floral 52
  spur length 172
nectaries, extra-floral 52, 61, 221, 234
  ant associations 88
  ant effects on plants **217**
  ant-guard systems 220
  definition 214
  evolution 221
  occurrence 216
nectarivores 58, 112, 249
nematodes, phytophagous 243
neotropical forests
  Costa Rica 7, *10*
  gaps 147
Nepenthes bicalcarata Plate 8.1
Nesosydne planthoppers 71, 100
nest sites, ant 216

nettle, hedge 18
neurotoxins 83
nickel 85
Nicotiana attenuata (wild tobacco) 96
nicotine 84, 92, 111
nightshade, deadly 198, 203
nitrogen
  allocation to roots/above-ground parts
    127
  availability 82
  cycling 105, 106
  fixation 95, 238
  limitation 82
  loss with microbical fermentation 113
  pollen 167
  seed content 148
  soil 127
nitrogen-containing compounds 111
nomadic behaviour 203
nomothetic approach 24
nuclear techniques 22
nuptial gifts 83
nut hoarding 67
nutcracker 144
nutrients
  cycling through ecosystem 105–6,
    127
  herbivore feeding 117, 127
  recovery hypothesis 130
  seasonal distribution in plants 83
  seeds 134
nutrition, genetically determined wisdom
  115
nuts 134
Nylsvley savanna (southern Africa) 107,
  108
Nymphalidae butterflies 17, 166

oak
  food sources 5
  gypsy moth defoliation 105
  habitats 5
  insect herbivory 93
  pedunculate 134, 136
oak–birch woodland 134
obligate interactions 243
oceanic archipelagos 19
oceanic islands 35
Ochotona princeps (pika) 123
Ocotea endresiana 187
Odocoileus virginianus (white-tailed deer)
  118
Odonata, fossil aquatic feeders 56
Odontomachus 226
Oecophylla (weaver ant) 217–18
oil glands, floral 53
oils 158
  consumption 58
  fatty 167
oleic acid 225
Olmeca 190
omasum 114–15
omnivores, mammalian 111–12
Oncidium (orchid) 23

ontogenic partitioning of interactions 239
Ophiostoma ips 94
Ophraella (leaf beetles) 71, *72*, 80
Ophrys (fly orchid) 168, 176
  mutualism 183–4
opossums 13
opportunists
  ecological 18
  mixed feeders 115
optimal defence theory 95–6
optimal foraging theory 117–18, 119
Opuntia stricta (cactus) 102
orchids
  Australian 222
  caladeniine 168, 176–7
  deception 168
  epiphytes 23
  floral spur length 179, 180
  fly 168, 176
  growth habits *21*
  non-rewarding 168
  oncidiine 20, *21*
  pollination 20, 222–3, 233
  pseudobulb domatia 220
  seed size 134
origination, relatively recent *29*
Orthoptera
  chewing 77
  fossil external foliage-feeders 43
  fossil pollinators 54
  plant-piercing ovipositors 61
Oryzopsis hymenoides (bunchgrass)
  228
Osyris alba Plate 7.1
Osyris quadripartita 207
otter, sea 244
ouabain 92
outgroups 23
overcompensation 89
oviposition 61, *62*
  cues 7, **9**
  fossil record 61, *62*, 255
  insect granivores 149
  plant substrates 61, *62*
  preferences 89
  site choice 138–9
  site mimicry 168
  stimulants 97
oxidative burst 91

pacas 138
Pachipteris (seed-fern) 48
Pachycondyla 226
Paeonia broteroi 175
palaeobiology 26, 27, 36
  complementarity with biological
    approach 71–3
  temporal dimensions of associations
    38, 248
palaeodictyopteroids *46*, 45
  ovipositors 61
  seed predation 50
Palaeozoic 28
  Late 35, *63*, 64

palatability
  herbivory effect on plant species
      richness 125
  plant gender 117
  pruning of juvenile-form plants 123
palatability–defence tradeoff hypothesis
      198
palm nut 141
  Tagua 136
Pan troglodytes (chimpanzee) 204
panmixis breakdown 238
Parareptilia 64
parasites/parasitism 3, 237
  antagonistic interaction 27
  ants 233
  floral 238
  phytochemical sequestering 9
  prey presence cues 88
  rate of evolution 241–2
parasitoids 239
  food web 7, 8
  insect granivores 144
Paratrichoptera 43
pareiasaurs 42, 64
Paridae 189
parrots 149
  brushtongued 159
Pastinaca sativa (wild parsnip) 100–1,
      148, 153, 239
pathogens
  dispersal by pollinators 175
  fungal 187, 198
  see also plant pathogens
pattern identification 24
pearl bodies 217
peccary 138
pectinase 113
Pennsylvanian era
  coal-balls 39
  coal-swamp forests 48, 50
  diadectomorphs 64
  piercing-and-sucking insect feeding 45,
      46
  radiations of amphibians and reptiles 62
  seed predation 50
  vertebrate herbivores 64
perissodactyls, caecum 12
Permian
  aquatic feeding 56
  Early
    Chekarda deposits (Urals, Russia) 41
    seed-ferns 38–9
  Late 64
  oribatid mite damage 48
  terminal extinction 45
permineralized deposits 33, 37
  feeding patterns 72
  insect mouthparts 42
Perognathus (pocket mouse) 136
persistence 29
Petauroides volans (greater glider) 113
Pharomacrus moccino (resplendent quetzal)
      203
Phascolarctos cinereus (koala) 113

Phasmatodea 43
Pheidole 211, 226
  arboreal species 227
  nest moving 228
  refuse pile soils 230
Pheidole bicornis 217
phellandrene 17
phenolic compounds 111, 166, 197
  antifungal 195
phenolic glucosides 8
phenology, shifted 16
phenotypic plasticity, adaptive 96
phenylalanine 83
phenylpropanoid pathway 85
phlorotannin 244
phlox 174
phosphorus cycling 105
photosynthesis ability of large fruits 195
Phryganidia californica (California oak
      moth) 105
Phyllobrotica (chrysomelid beetles) 18, 19
Phyllotreta (beetle) 105
phylogenesis, parallel 101
phylogenetic analysis 23
phylogenetic congruence 23, 70–1
phylogenetic constraint 19
phylogenetic tracking 18, 71
phylogeny 69, 237
  generalist ancestors of specialists 80–1
  hierarchical structure of interactions
      246
  past associations 35, 36
Phytelephas macrocarpa (Tagua palm nut)
      136
phytochemicals 8–9
  adaptive radiation of plant-related taxa
      15
  diversity 17
  four-trophic-level system of plants 12
  sequestering by parasites 9
Phytophaga 66, 71
phytophagy 243, 249
piercing-and-sucking insect feeding, fossil
      record 45, 46, 45, 251–2
Pierinae butterflies 16, 17
Pieris (cabbage butterfly) 85, 171
pigeon 203
pika 123
pine
  cone spines 150
  lodgepole 109, 141, 152, 153–4, 238
  pinyon 95
  Scots 120
pinenes 17
Pinus contorta (lodgepole pine) 109, 141,
      152, 153–4
Pinus sylvestris (Scots pine) 120
Piper 219
Piper arieianum (pepper plant) 93
Piper cenocladum 217
piperonyl butoxide 85
pith-borers 45
plant(s)
  adaptive radiation 15

animals associated with species 3
  associations 117
  breeding system evolution 185
  cell walls 110
  chemical defence 8
  chemistry changes with herbivory
      122–3, 126
  competitive ability 146
  cuticle 176, 183–4
  demography and seed predation 144
  disruptive selection 175–7
  divergence of radiations in clades 35
  diversification 176, 181
  diversity 98–101
    granivory 146–7
  evolution under herbivore pressure 119
  factors limiting specialization 206
  flight 3
  green 4, 6
    autotrophic 5
  herbivory responses 120–1
  introduced 102, 247
  modules 120
  non-rewarding 168–9
  palatability and gender 117
  parallel evolution of functional
      structures 179
  phylogenetic congruence with insect
      herbivores 70–1
  phylogeny 69
    fruit size correlation 195–6
  pruning of juvenile form 123
  refuge 117
  regeneration 121, 132, 145
  repellant 104
  reproductive biology 37, 38, 39, 185
  spatial distribution 132
  specialization/specificity for pollination
      172–3
  stored 123
  structure changes 126
  succession 126
  unfavourable growth periods 146
  see also food plants; host plant
plant–arthropod associations 29, 37
  evidence 37–42
  fossil record 40
plant–carnivore interactions 240
plant communities
  composition 102–3
  diversification 177
  evolution 61
  insect herbivore effects 61, 102–4
  mammalian herbivore effects 124–6
  population cycles 128, 129, 130
  species diversity 102–3
  spur development variation 179
  tubular corolla development variation
      179
plant damage 37, 39, 41
  earliest occurrences 39, 41
  fossil record 39, 41
  herbivores 69, 130
  insect 248

plant damage (*cont'd*)
  oxidative burst 91
  prey presence cues 88
  quantitative analysis 60–1
  trypsin inhibitors 123
  wound responses 57
plant defences 243–4
  abscission 87
  adaptive landscapes 243
  anti-feedants 83
  associational 117
  biogeographical patterns 244
  chemical 8
  compensatory growth 120
  compounds
    chrysomelid beetles 80
    herbivore attack 122–3, 130
  constitutive 84, 96
  cycles of adaptations 26
  deactivation 89
  decoys 83
  deterrents 84–5, 111, 147
  diversity of plant and animal species
      98–101
  evolution 95–6
  food plants 111
  granivory 138
  guilds 117
  herbivore fitness reduction 92–3
  induced 96, 122–3, 130
  insect herbivores 83–5, *86*, 87–9
  investment relationship to growth rate
      96
  mechanisms 239
  physical barriers 83, 84
  regrowth 120
  seed toxins 136
  squirt gun defence 85, 87
  third-party 87–8
  tolerance 88–9, 120
  toxins 83, 84–5
  see also allelochemicals; secondary
      compounds; toxins
plant fitness
  herbivore feeding 93–5
  herbivore reduction by impeding
      relationships with mutualists
      94–5
plant–insect associations 77
  categories of evidence 250
  ecological communities 101–6
  evolutionary trajectory 98
  fossil records 39, 249, 250
  response to Early Caenozoic Thermal
      Interval 248
  see also ant–plant interactions
plant–mammalian herbivory interactions
    124
plant–pathogen interactions 243
plant pathogens 243–4
  floral biology 243
plant–pollinator systems 157, 204–5
  multispecies 184
  population fragmentation 184

preadaptations 184
  rewards 95
plant populations
  population cycles *129*, 130
  seed influences 136
  vertebrate herbivore impact 107
plant protection 24
plant–soil system 127
plant tissue value 95–6
plant–vertebrate associations, fossil history
    255–6
plant vigour index 117
plant virus vectors 94
Plantago insularis 228
Plantago lanceolata (long-leaf plantain) 93
planthoppers 71, 100
Platanthera (orchid) 179
Plecoptera, aquatic feeding 56
Pleistocene interglacials 68–9
  forest plant recolonization 187
plesiomorphy 23
pluralistic theory 24
Poaceae 190
Pogonomyrmex 227, 228
  refuse piles 230
  soil effects 229
  vegetation clearance 227–8
Pogonomyrmex occidentalis 227
Polemonium viscosum 222
pollen 158
  animal vectors 185
  carriage 158
  complete removal 162
  consumption 54, *55*, 56
    fossil record 254–5
  coprolites 41
  dinosaur coprolites 67
  dispersal 185, 206
  flight 3
  gene flow mediation 176
  grains and ant metapleural gland
      secretions 222
  heterospecific transfer 171
  nitrogen 167
  pickup control 162
  prestigmatic mechanisms 162
  reward 167
pollination
  water/wind 158
  see also self-pollination
pollination by animals/pollinators
    157–69, **170**, 171–84
  antagonism 157, 161
  ants 221–3, 233, 234, 235
  assortative behaviour 176
  attractants 163–6, 181–2
  auditory cues 166
  availability 172
  bats 158, 159
  behaviour manipulation 206
  benefits 158
  birds 158, 159
  colour perception 165–6
  colour preferences 165

commensalism 157
community structure 178
convergent evolution 181–2
costs 158
crop plants 24
deception 157, 161, 168–9
determination of flower-visitor role
    160
disruptive selection within plant species
    175–7
diversification 181
  early 160–1
drop mechanism 39, *74*
early history of modern types 39
floral traits 162–3
flying 3
food reward for animal 157
foraging behaviour 94
foraging pattern 171
fossil record 54, *55*, 56, 254–5
granivory 168
guilds 167, 172
handling time 171
herbivory relationship 175
indirect actions 94
insects 158–9, 243
interspecific competition 178
lifetime fitness 177–8
mammals 159, 168
mouthparts 172
mutual adaptations with animal-
    pollinated plants 206
mutualism 53, 54, 56, 157–8, 161–2,
    180–1
  fossil evidence 249
olfactory cues 166
orchids 20, 222–3, 233
origins 158–62, 181
passive 181
pathogen dispersal 175
phylogenetic methods 181–4
plant adaptations 181, 185, 206–7
pollen placement 177
process 157–8
reciprocal diversification 161–2
relative cost *173*
resin-collecting bees on *Dalechampia*
    236, 237
rewards 95, 157, 158, 161, 166–8,
    181–2
  concealment 163–4
  harvesting rate 171
seed dispersal mutualism 244
selection on traits 177–8
selective attraction 172
self-pollination *183*, 184
sex-specific territories 178
short-term selectivity 171
spatiotemporal variation in visitor
    abundance 172
specialization 169, 171–2
  irreversibility 182–3
specialized floral characters 69
specificity among visitors 169, 171–2

styles 36
syndromes 169, **170**
  bird 70
  tradeoffs 158
  trophic systems 3
  visitor behaviour modification 166
pollinia 158, 180
polyphenolic compounds 110
polyploidy, plant 242–3
Polyrachis 220
Pometia pinnata Plate 8.1
Ponerinae 226
poplar, balsam 106
population dynamics, granivory 142–4
populations
  cycles 128, *129*, 130
  geographic patterns of differentiation
    237–8
  human 246–7
  interacting 98
  limitations of herbivorous insects
    81–3
  spatial distribution of genetic variability
    205–6
Populus (cottonwood) 100
Populus balsamifera (balsam poplar) 106
possums 13, 159
post-ingestive feedback 119
potato plants, protease inhibitors 96
prairie, North American tall-grass 124
precipitation trends 30
predators/predation 3, 237
  antagonistic interaction 27
  flowers 180
  food interactions 130
  food web 7, *8*
  mammalian 111–12
  overt 27
  prey presence cues 88
  seed/seedling herbivory 77
  sunfish 241
  see also seed predation
prepollen 37, *74*
primary production, net
  removal by herbivores 107, *108*
  savanna grasses 120
primates
  frugivorous 67, 203, 204
  origin 13, 15
  pollination 159, 168
Procnias tricarunculata (bellbird) 187
productivity of area 126
prokaryotic algal accumulations 33
Propalaeotherium (horse fossil) 42
prosauropods, high-browsing 64
protease inhibitors 92, 96, 136
protein 82
  dietary 113
α-proteobacteria 240
Protorosaurus (fossil reptile) *63*, 64
Protorthoptera 43
Prototaxites (fossil basidiomycete) 48
protozoa
  ant nests 229

plant carbohydrate digestion enzymes
    113
  symbiotic *14*
Pseudomyrmex ferruginea (acacia ant)
    87–8
Pseudotsuga menziesii (Douglas fir) 141
Psocoptera 41
Psychotria 193
Psychotria nervosa 205
pteridosperms 190
Pterocallis alni (aphid) 105
pterodactyls 5
Pteropodiae 201
pterosaurs 3
Publilia concava 221
'pull of the Recent' 36, *74*
punctuated equilibrium theory 242
Pycnonotus leucogenys (bulbul) 201
Pyrameis cardui (butterfly) 171
pyrethrins 92
pyrethroid resistance 92
Pyrrhula pyrrhula (bullfinch) 184
pyrrolizidine alkaloids 97

quantitative trait locus (QTL) mapping
    176, 184
Quaternary extinctions 68–9
Quercus (oak) 105
Quercus robur (pedunculate oak) 134, 136
quetzals 202
  resplendent 203

raccoons 13
radial symmetry 163, 177
radiation *74*
radish, wild 96, 174–5
rain forest, closed-canopy 158
range contraction, Quaternary extinctions
    68
Rangifer tarandus (reindeer) 127
Raphanus raphanistrum (wild radish) 89,
    96, 174–5
Raphus cucullatus (dodo) 204
Red Queen hypothesis 27
reductionist approach 23–4
re-evolution *29*
refoliation 120
regional patterns of interactions 244
reindeer 127
repellant plants 104
reproduction
  limitations of herbivorous insects 81–3
  supra-annual synchrony 150–1
  time shift in frugivores 203, *204*
reproductive isolation 15
  evolution 15–16
reptiles
  early radiations 62
  fossil chewing mechanisms 42
  seed dispersal 192
Reptilia 64
resin-secreting glands 53
resins 85
  antimicrobial properties 95

Dalechampia vines 94–5, 167
diterpenoid 167
resistance
  associational 95, 104–5, 117
  evolution in insects 96–8
  genes 241
  mechanism benefit exceeding costs 96
  traits 89
resource availability hypothesis 96
resource concentration hypothesis 104
resources
  competitive interaction for 177–8
  floral 178
  food plants 107
  management by herbivores 123–4
  storage in granivory 151
resprouting 120–1
reticulum 114
Rhagodia parabolica 195
Rhagoletis (fruit flies) 17, 18
  phylogeny *19*
rhinoceros
  black 108
  seed dispersal 192
Rhinocyllus conicus (weevil) 102
Rhiodinidae 221
Rhizobium 95
Rhizophora propagules 138
Rhynie Chert (Scotland) 39, *40*
Rhytidoponera Plate 8.1
Rhytidoponera metallica (greenhead ant)
    224
rift-valley lakes (East Africa) 19
river floodplains, primary succession 106
robin, European 204, 207
rodents
  bilberry branch-cutting 120
  caecal fermentation 114
  caviomorph 144
  frugivorous 67
  granivory 138, *140*, 141, 147
    toxic seeds 149
  microtine 109, 123
  seed burial 146
  seed density 141
  seed eating 132, 139
Rodriguezia (orchid) 23
roller insects 61
roots, resprouting 120
Rosa sicula Plate 7.1
Rubiaceae 190
Rubus spectabilis 194
Rubus vestitus (bramble) 123
rumen/ruminants 114
  artiodactyls 10, 12
rust fungus 16
  flax 241

Salicaceae 8
saliva 113
  rumination 114
salivary glands 115
Salix planifolia (willow) 121
saltational events 12

Salvinia molesta (floating fern) 102
Sanguinaria canadensis 225
sap feeding 57, 58
   aphids 78
   mouthparts 60
   sucking insects 52
sap flows 52
saponins 197
   autotoxicity 148
Saraca thaipingensis   Plate 8.1
sauropods 5
   fossil chewing mechanisms 41
   gut fermentation 67
savanna 68, 120
   grazing systems 128
sawflies
   adaptive radiation 3
   fossilized gut contents 41–2
   larval borers 45
   leaf-mining 48
saxifrage 239, 242
scale insects 220
scatter-hoarding 144
scent
   plume 166
   see also flowers, scent
Scheelea palm 141
Schismus arabicus 228
scientific theory 24
Sciurus (squirrel) 108, 144
Sclerolaena diacantha   Plate 8.1
scraping 56
Scutellaria (skullcap) 18–19
sea grape *see Coccoloba uvifera*
seasonality trends 30
secondary compounds 84, 85, *86*, **87**
   insect granivores 148–9
   nutrient cycling effects 106
   pollinator foraging behaviour 94
   synergism 85
   see also allelochemicals; toxins, plant
      defences
seed(s) 132–54
   abundance 136, 139
   animal-dispersed 148
   animal vectors 185
   ants 223–8
   burial 144, 146
   colonization at distance from parent
      population 144–5
   composition *135*
   conifer crops 139
   crop sizes 141, 148
   decoys 147
   defence 134, 148
   density 139, *140*, 141
      distance from parent plant 186
      harvester ant effects 227
   deposition by harvester ants 227
   detectability 139
   deterrents 147
   detoxification costs 139
   digestion costs 139
   distribution 139

empty 147
endosperm 134, 136
energy content 134
flight 3
germination
   ant nest refuse piles 230
   facilitation by ants 226
   inhibition 198
   passage through digestive tract 204
   requirements 187
   tradeoff with granivory 147–8
granivores
   life cycle synchrony with availability
      137
   movement by 144, 145
handling 139
harvesting
   by ants 226–8
   deliberate 188
   imperfect 188, 189
herbivory 77
incidental picking up 188
limitation in communities of perennials
      143
microsites 145, 146
   availability 143
nitrogen content 148
nutrients 134
nutritional quality 139
parasites 180, 181
passage rate through gut 198, 201–2
percentage of crop eaten 141
plant population influence 136
   impact by granivores 132
predation risk reduction 226
production
   depression by granivores 147
   strategies 132
recovery after storage 144
recruitment distance from plant 103–4
regeneration 145
release timing in myrmecochory 225
removal from faeces 226
retention in canopy 151
seasonal phenology of production 151
shadows 103–4
shape 134
size 134, 139, 148, 190, 192
   counter-adaptations of granivores
      149–50
   oviposition 149
structural features 150
supra-annual reproductive synchrony
      150–1
survival at distance from parent plant
      145, 147, 187
synchronous crop production 143,
      150–1
toxins 136, 148–9
traits 132
   physical 149–50
   physiological 148–9
transporting 139
tree species 103–4

see also caching
seed banks, soil 142, 146
seed-caching *see* caching
seed dispersal
   on animal exteriors 189
   ants 223, 224, 225–6
      community composition 231–2
      disturbance 231–2
      efficiency 234
      secondary 226
   avian 192
   binary 188
   chemical influence of parent plant 186
   depature-related advantages 186, 206,
      208
   determination by maternal parent traits
      185
   external adhesion 188
   granivory 144, 145
   harvest-based 188
   imperfect harvesting 188, 189
   invertebrates 187
   long-distance 3–4
   mutualism with pollination 244
   mutualists 148
   passive 185
   plant benefits 186
      arrival 186, 187, 206, 208
      departure 186–7
   plant–animal relationships 187
   primary 188, 189
   secondary 189, 226
   selection 150–1
   syndromes 132, 198–9
   trade-off with granivory 147–8
   wind 188
seed dispersal by vertebrates 187–208
   animal adaptations 199–203, *204*
   animal behaviour 203, *204*
   animal interiors 188–93
   carnivorous mammals 208
   crop sizes 207
   digestive physiology 201–3
   disperser abundance 193
   fleshy-fruited plants 189
   food-retention time 201
   frugivores 192–3
   fruit(s) *Plate 7.1*
      acceptability 198
      advertisement 194–5
      availability peaks 193
      coloration 194–5
      nutritional characteristics 197
      pigments 195
      pulp composition 196–8
      size 195–6
      traits 193
   fruit-related traits 193
   functional categories 187–8
   gene flow 205
   genetic consequences 208
   genetic homogeneity of populations
      205
   geographical location 190, 191–2

gut retention time of seeds 198, 201–2
habitat type 190, 191
mammals 192
mutual dependence patterns 204–5
mutual relationship spatial/temporal
    variation 208
nutritional imbalance 202–3
plant adaptations 193–9
plant growth form association 190–1
reciprocal specialization 205–7
relationship between plant and disperser
    189
seed-plant lineages 190
seed size 190, 192
seed-ferns 190
dinosaur coprolites 67
fossil plant damage 39
leaf-miners 48–50
medullosan 37
seed predation 50
seed-head herbivory 101–2
seed plants 26
mouthpart classes 60
seed predation 187
ants 223, 228
density-dependent 103, 104
determinants 138–9
fossil record 50, 52, 253–4
harvester ants 228
seed-predators see granivores/granivory
seedlings
density 143
    distance from parent plant 186
    harvester ant effects 227
density-dependent predation 103, 104
establishment 187
herbivory 77
    mortality from 120
recruitment in ant nest refuse piles 230
regeneration 145
survival
    after seed caching 144–5
    distance from parent plant 187
survivorship with ant dispersal 225
selection mosaics see geographic mosaic
    theory of coevolution
self-fertilization 147, 158
self-incompatibility, angiosperms 162
selfing rate 162
self-learning 119
self-pollination 183, 184
ant visitors to flowers 222
semi-arid ecosystems 138
semiochemicals 215, 234
senita cacti 181
sequential evolution 71, 72
Serengeti grasslands 124
Serengeti–Mara system 130
serotiny 151, 153
sesquiterpenes 166
sexual reproduction, plants 185
sexual selection in arctiid moths 97
sheep 115
    grass grazing 128

rumen 10
shrub grazing 120
shelter, plant substrates 61
shield-bearer insects 61
shredding 56
Sierra Nevada (California, USA) 176
Sierra Nevada (Spain) 144
Silene latifolia (white campion) 166
silica 111
Silurian, Late
    earliest biota 33
    plant feeding development 56–7
silversword, Hawaiian 71, 100
Simmondsia chinensis (jojoba shrub) 136
simmondsin 136
siphoning 52, 58, 78
sister-taxon comparison 36
Sitophilus oryzae (rice weevil) 240
Sitophilus oryzae principal endosymbiont
    (SOPE) 240
skullcap 18–19
sloths, tree 13
Soay sheep 128
sodium 83
sodium channel receptor proteins 92
soil 127
    ants 228–31
    decomposer community 229
    fertility and vegetation structure 192
    seed bank 142, 146
Solanaceae
    alkaloids 8
    butterfly feeding 17
Solenopsis geminata (fire ant) 227
Solidago (golden rod) 221
Solidago altissima 239
Sorghum intrans (sorghum) 140
sound reflection 166
special-case approach 23–4
specialists
    extreme 236–7
    generalist ancestors 80–1
    herbivore fitness reduction 92–3
    insect herbivores 78–81
specialization 236–7
    extinction risk 80
    feeding habit 81
    non-adaptiveness 80–1
    patterns 237
speciation 15–16
    allopatric 15
    angiosperms 4, 7
    geographic 15
    instantaneous 242
    sympatric 15
species
    closely related 236
    diversification 242
    geographic structure 237–8
    interactions
        in diversity of life 240
        genome 246
        hierarchical structure 245–6
    microcommunity of associated 245

numbers 4–5, 6
specialization 236–7, 239
specific coevolution 18
specificity, mutual 204–5
Spermophilus parryii (Arctic ground
    squirrel) 130
Sphecomyrma freyi 211
sphenopsids, calamitalean 50, 51
sphondin 148
spiders
    food web 7, 8
    nectar use 217
spindle tree 134
spinescence 123
spittlebugs 78
sponging 52
    labellum 58
spores
    consumption 54, 55, 56
    flight 3
sporivory 132, 134, 249
    fossil record 254–5
Sporobolus kentrophyllus (grass) 122
squirrel 13, 108, 145
    Arctic ground 130
    coevolution with seeds 152, 153–4
    pine 141, 152, 153–4
    red 238
    tree 144
squirt gun defence 85, 87, 89
St Kilda (Scotland) 128
Stachys (hedge nettle) 18
stamen 163
staphylae 214, 215
starling 203
Stemmadenia Plate 7.1
sterols 82
stigma–anther distance 163
Streptomyces bacteria 215, 216
study methods 22–3
Sturnus vulgaris (starling) 203
stylet tracks, fossil 45, 46
Stylidium 177
subarctic environment 127, 128
subdermal tissue, exposed 52
subsidence, eustatic 35
substrate specialization 113
succession 126
sucrose 167
sugarbirds 159
sulfides 166
sumach 30
    multiple herbivores 93, 94
sunbirds 159
sundew 52
sunfish predation 241
Sylvia atricapilla (blackcap warbler) 203,
    207
symbionts 243
    eukaryote uses 240
    gut organisms 113
    single-celled 240
symbiosis
    microbial associations 9–10

symbiosis (*cont'd*)
  mutualistic 13, 217, 233, 237
symmetry, floral 163, 164–5
  control of switching 177
synapomorphy 36
  cladogram 69
  feeding characters 70
Synapsida 64
Synceros caffer (African buffalo) 115
synergism, secondary compounds 85
syrphids, larval diet 70
systematic studies 22

Tamiasciurus (pine squirrel) 141
tanagers, fringillid 159
tannins 8, **87**, 111
  autotoxicity 148
  conjugation 91
  defence investment 96
  poplar 106
taphonomic filters 33
taphonomic loss *35*
tapir 138
target site insensitivity 92
Taxus baccata (yew) Plate 7.1
Taxus brevifolia (Pacific yew) 84
Technomyrmex Plate 8.1
tectonism *35*
teeth
  dinosaurs 67
  vertebrate fossils *63*, 64
  see also dentition; molars
Tegeticula yuccasella (yucca moth) 245,
    Plate 6.2
temperature trends, environmental 30
Tenthredinoidea 48
tent-making insects 61
termites
  fossil borers 45
  fossil plant damage 39
  paunch 12
terpenes 17, 84
  insect adaptation 99
  pressurized canals 85, 89
terpenoids 85, *86*, **87**, 111, 167–8, 215
  metabolism by P450s 90
terrestrial ecosystems 77–106
  counter-defence 98–101
  defence 98–101
  ecological communities 101–6
  insect herbivores
    adaptations 77–81
    adaptations to plant defences 89–92
    fitness reduction by plant defences
        92–3
    growth limitations 81–3
    plant defences 83–9
    plant fitness reduction 93–5
    population size limitations 81–3
    reproduction limitations 81–3
    resistance evolution 96–8
  plant defence evolution 95–6
terrestrial life 26, 33
  earliest evidence 33

Tetraopes (longhorn beetle) 71, *72*, *90*,
    98–9
Thambetochin chauliodous (fossil
    herbivorous duck) 69
Thecodiplosis (fossil gall midge) 50
Themada triandra (savannah grass) 104
third-party defences 87–8
thistle 102
Thlaspi montanum var *montanum* 85
thorns 84
thrips
  gall-making 78
  plant virus vectors 94
  pollenivore 56
thrushes 192
Thysanoptera
  fossil feeding mechanisms 45
  galls 50
  pollinators 54
tier insects 61
time-averaging *74*
titmice
  fruit predation 189
  juniper 147
  seed dispersal 189
tobacco, wild 96
tolerance, plant defence 88–9, 120
tomato plants, protease inhibitors 96
tortoises, seed dispersal 192
toucans 192, 202
touracos, frugivory 192
toxins
  animal learning process 111
  binding by soil minerals 149
  delivery 85, 87
  detoxification 136, 148–9
  employment by insects 97
  food plants 111
  herbivore attack 130
  nutrient cycling through ecosystems
      105–6
  plant defences 83, 84–5
  seeds 136, 139, 148–9
  sequestration by insects 97, 148–9
  storage products 148
  synergism 85
  target site insensitivity 92
  see also allelochemicals; plant defences;
      secondary compounds
Tragelaphus (kudu) 115
Tragopogon 243
trait remixing 238
transfer ecology 69–70, *74*
transferases 91
transposons 240
trees
  browsing 120
  mass-flowering 168
  seed size 134
  shoots 120
  stump resprouting 120
Triassic, galls 50, *51*
trichomes 84
Trichoptera 56, 60

Trirhabda virgata 221
tritrophic interactions 239–40
trogons 192
tropanes 197–8
trophic cascades 237
trophic levels 126
trophic systems 3
trophic webs 7, *8*
trophollaxis 232
tropics
  diversity of organisms 30
  granivory 137–8
  see also neotropical forests
trypsin inhibitors, grazing 123
Tupiocoris notatus 84
Turdus merula (European blackbird) 203
turkey 134
turnover, coevolutionary 238

Ulmus glabra (elm) *140*
ultraviolet light, reflectance by ripe fruits
    194
ungulates
  ecosystem impact 127
  mobility 110
  seed eating 132
  Serengeti–Mara system 130
uniformitarianism 31, *74*
urine deposition 127
Utetheisa ornatrix (arctiid moth) 97

Vaccinium corymbosum 198
Vaccinium macrocarpon 198
Vaccinium myrtillus (bilberry) 120
vascular plants, taxonomic turnover 30
vertebrates
  chewing apparatus 62, 64
  fossil chewing mechanisms 42
  fossil record of plant associations 61–2,
      63, 64, *65*, 67–9
  herbivores of Late Palaeozoic *63*, 64
vesicular arbuscular (VA) mycorrhizal
    fungi 229
Viburnum inflorescence 164
Viburnum opulus 194
Viburnum tinus 204
Viola 168
Viola cazorlensis (violet) 175, 184
Viola odorata (sweet violet) Plate 8.1
Virola surinamensis 187
virulence genes 241
viruses, symbiotic *14*
vision, binocular 13, 15
visual pigments 165–6, 180
visual predation hypothesis 13
vitamins 82
volatile substances, plant 104, 239
vole
  bank 184
  plant mortality 120
  population cycles 128, *129*, 130

warblers
  blackcap 203, 207

pine 201, 202
  seed dispersal 192
  yellow-rumped 202
wasps
  adaptive radiation 3
  fig 168, 181, 238, 245
    plant species visited 169, 171
  fossil plant damage 39
  gall 50, 78
  orchid pollination 20
  parasitic 60
  parasitoid 8, 217
  prey presence cues 88
  sex pheromones 176, 183–4
  thynnine 176–7
  wood 45
water 158, 167
water hyacinth 19
water-pollination 158
waxwings 192, 203

cedar 196, 202
weather conditions
  food plants 109–10
  vegetation succession 126
  see also climate
webworm, parsnip 100–1, 148, 153, 239
weevils 243
  biological control 102
  Brazilian 102
  cycad pollination 54
  granivory 144
  mouthparts 60
  plant–host associations 66, 71
  rice 240
wheat 243
whiteflies 78, 84
wildebeest 104, 110
willows 102, 122
wind-pollination 158
wing polymorphism, insect 20

Wolbachia 240, 243
wood-boring 30, 39
woodland, temperate 137
wood-lemming 108
woody browse–vertebrate systems 123

Xyela 21
xylem sap feeding 78

Yellowstone National Park (USA) 121, 127
yew, Pacific 84
Yucca intermedia Plate 6.2
yucca/yucca moth mutualism 71, 181, 237, 238, 240, 245
  obligate 183

zebra 104, 110
Zenillia adamsoni (tachinid fly) 9
zygomorphy 163, 171